Renewable-to-Vehicle Smart Charging

RENEWABLE-TO-VEHICLE SMART CHARGING

PLANNING, MODELING, AND LOAD FORECASTING FOR GRID INTEGRATION

Edited by

SALEH AGHAJAN-ESHKEVARI

Department of Electrical Engineering, Shahid Beheshti University, Tehran, Iran

SASAN AZAD

Department of Electrical Engineering, Shahid Beheshti University, Tehran, Iran

MORTEZA NAZARI-HERIS

Department of Electrical Engineering, East Carolina University, Greenville, NC, United States

MIADREZA SHAFIE-KHAH

Research and Innovation Division, Nowocert, Dublin, Ireland

HAMDI ABDI

Department of Electrical Engineering, Faculty of Engineering, Razi University, Kermanshah, Iran

ELSEVIER

Elsevier
Radarweg 29, PO Box 211, 1000 AE Amsterdam, Netherlands
125 London Wall, London EC2Y 5AS, United Kingdom
50 Hampshire Street, 5th Floor, Cambridge, MA 02139, United States

ISBN: 978-0-443-36684-0

For information on all Elsevier publications visit our website at
https://www.elsevier.com/books-and-journals

Publisher: Priscilla Braglia
Acquisitions Editor: Dan Egan
Editorial Project Manager: Avni Gupta
Production Project Manager: Bharadwaj Varatharajan
Cover Designer: Arumugam Kothandan

Typeset by TNQ Tech

Contents

Contributors

Hamdi Abdi
Department of Electrical Engineering, Faculty of Engineering, Razi University, Kermanshah, Iran

Gibran David Agundis-Tinajero
Center for Global Cooperation, Danish Energy Agency, Copenhagen, Denmark

Sina Ahmadi
Faculty of Electrical Engineering Shahid Beheshti University, Tehran, Iran

Mohammad Mehdi Amiri
Department of Electrical Engineering, Shahid Beheshti University, Tehran, Iran; Electrical Networks Research Institute, Shahid Beheshti University, Tehran, Iran

Rajamallaiah Anugula
School of Aeronautics and Astronautics, Zhejiang University, Hangzhou, China; Center for Research on Microgrids (CROM), Huanjiang Laboratory, Shaoxing, Zhejiang, China

Hamidreza Arasteh
Center for Research on Microgrids (CROM), Huanjiang Laboratory, Shaoxing, Zhejiang, China; Power Systems Operation and Planning Research Department, Niroo Research Institute (NRI), Tehran, Iran

Sercan Aygun
School of Computing and Informatics, University of Louisiana at Lafayette, Lafayette, LA, United States

Sasan Azad
Department of Electrical Engineering, Shahid Beheshti University, Tehran, Iran

Mohsen Bahmani
Department of Electrical Engineering, Shahid Beheshti University, Tehran, Iran

Ismail Baklouti
National School of Engineers of Monastir, University of Monastir, Av. Ibn El Jazzar, Monastir, Tunisia

Abbas Babaei Birag
Department of Electrical Engineering, Shahid Beheshti University, Tehran, Iran

Robbert Claeys
EnSy/Lemcko, Department of Electromechanical, Systems and Metal Engineering, Ghent University, Kortrijk, Belgium

Basir Darsareh
Department of Electrical Engineering, Shahid Beheshti University, Tehran, Iran

Jan Desmet
EnSy/Lemcko, Department of Electromechanical, Systems and Metal Engineering, Ghent University, Kortrijk, Belgium

Cesar Diaz-Londono
Center for Research on Microgrids (CROM), Huanjiang Laboratory, Shaoxing, Zhejiang, China

Touria El Mezyani
Electrical Engineering Department, University of North Florida, Jacksonville, FL, United States

Farzad Ferdowsi
Electrical and Computer Engineering Department, University of Louisiana at Lafayette, Lafayette, LA, United States

Josep M. Guerrero
School of Aeronautics and Astronautics, Zhejiang University, Hangzhou, China; Center for Research on Microgrids (CROM), Huanjiang Laboratory, Shaoxing, Zhejiang, China

Hossein Jafari
Department of Electrical, Electronic, and Information Engineering, University of Bologna, Bologna, Italy

Kamran Jalilpoor
EnSy/Lemcko, Department of Electromechanical, Systems and Metal Engineering, Ghent University, Kortrijk, Belgium

Hamed Jafari Kaleybar
Politecnico di Milano, Energy Department-En:Lab, Milano, MI, Italy

Ali Khaleghi
Department of Electrical & Software Engineering Schulich School of Engineering University of Calgary, Calgary, AB, Canada

Md Zahirul Islam Khan
Electrical and Computer Engineering Department, University of Louisiana at Lafayette, Lafayette, LA, United States

Muhammad Abdul Mujeebu
Department of Building Engineering, College of Architecture and Planning, Imam Abdulrahman Bin Faisal University, Al-Dammam, Saudi Arabia

Mehrdad Setayesh Nazar
Faculty of Electrical Engineering Shahid Beheshti University, Tehran, Iran

Morteza Nazari-Heris
Department of Electrical Engineering, East Carolina University, Greenville, NC, United States

Ali Mosalla Nejad
Department of Electrical Engineering, Shahid Beheshti University, Tehran, Iran

Ali Paeizi
Department of Electrical Engineering, Shahid Beheshti University, Tehran, Iran; Electrical Networks Research Institute, Shahid Beheshti University, Tehran, Iran

Shantharam Patil
Manipal School of Architecture and Planning, Manipal Academy of Higher Education, Manipal, Karnataka, India

Nazanin Pourmoradi
Department of Electrical Engineering, Shahid Beheshti University, Tehran, Iran

Abdultawab M. Qahtan
Architectural Engineering Department, College of Engineering, Najran University, Najran, Saudi Arabia

Anand R.
Center for Research on Microgrids (CROM), Huanjiang Laboratory, Shaoxing, Zhejiang, China

Ahmad Salemnia
Department of Electrical Engineering, Shahid Beheshti University, Tehran, Iran

Khalil Moradi Sani
Department of Electrical Engineering, Shahid Beheshti University, Tehran, Iran

Mohammad Setareh
Department of Electrical Engineering, Shahid Beheshti University, Tehran, Iran

Amirreza Shafieinejad
Department of Electrical Engineering, Shahid Beheshti University, Tehran, Iran

Alireza Sobbouhi
Department of Electrical Engineering, Shahid Beheshti University, Tehran, Iran

Daogui Tang
School of Transportation and Logistics Engineering, Wuhan University of Technology, Wuhan, China

Nishant Thakkar
School of Aeronautics and Astronautics, Zhejiang University, Hangzhou, China; Center for Research on Microgrids (CROM), Huanjiang Laboratory, Shaoxing, Zhejiang, China

Yupeng Yuan
School of Transportation and Logistics Engineering, Wuhan University of Technology, Wuhan, China

Alireza Zare
Department of Electrical Engineering, Shahid Beheshti University, Tehran, Iran

Liu Zhang
School of Aeronautics and Astronautics, Zhejiang University, Hangzhou, China; Center for Research on Microgrids (CROM), Huanjiang Laboratory, Shaoxing, Zhejiang, China

CHAPTER ONE

Concepts, technical challenges, emerging trends, and opportunities in the interconnected smart renewable-to-vehicle (R2V) ecosystem

Ali Paeizi, Nazanin Pourmoradi, Sasan Azad
Department of Electrical Engineering, Shahid Beheshti University, Tehran, Iran

1.1 Introduction

The electrical grid's integration of renewable energy sources (RESs), like solar Photovoltaic (PV) and wind energy, has significantly expanded in recent years. To effectively meet the demand for Electric vehicle (EV) charging, however, policies, programs, and technologies are needed due to their intermittent nature. Fast charging and the development of power electronic converters in recent years have increased the potential for EV charging by significantly cutting down on charging time and optimizing the charge. The utility power grid may eventually experience grid capacity collapse due to the charging behavior of various EVs (Gupta et al., 2020). By installing a planned network of RESs for future charging stations and strategically utilizing the current power plants, precise charge management can be accomplished (Barman et al., 2023).

Achieving a sustainable and efficient energy future requires the integration of RE Systems into Smart Grids, particularly when EVs are included. The carbon footprint of energy generation is decreased by RESs, which offer a sustainable substitute for fossil fuels. By generating employment in the manufacturing, installation, and maintenance sectors, the use of renewable energy technologies can stimulate economic growth, and in the long term, it also aids in lowering the operational costs of energy production. With Vehicle to Grid (V2G) technologies, EVs can serve as mobile energy storage devices in smart grids, providing special chances for grid balancing.

Renewable-to-Vehicle Smart Charging
ISBN: 978-0-443-36684-0
https://doi.org/10.1016/B978-0-443-36684-0.00014-X

This enables EVs to feed the electricity back to the grid during peak hours and store extra renewable energy when supply outpaces demand. The efficiency and affordability of integrating RESs into the grid are steadily increasing due to developments in energy storage technologies, smart grid infrastructure, and RE technologies. A more resilient, efficient, and sustainable energy future is promised by the energy sector's crucial but challenging evolution toward smart grids with integrated RESs and EVs (Mojumder et al., 2022).

An important step toward sustainable transportation is the integration of EVs into the electrical grid, which could reduce CO2 emissions and dependency on fossil fuels. However, the current power grid infrastructure faces both opportunities and challenges as a result of this integration (Manousakis et al., 2023). Global EV adoption may put strain on the grid during peak charging periods, resulting in problems like capacity limitations, higher peak loads, and the need for grid upgrades. The implementation of V2G systems and smart grid technologies, on the other hand, can mitigate some of these issues. Through load balancing, demand response, and ancillary services, V2G allows EVs to function as dynamic energy storage units, improving the stability and resilience of the grid. Additionally, by combining RESs with EV charging infrastructure, the carbon footprint of electricity generation and transportation can be further reduced, promoting a more resilient and sustainable energy landscape (Sarda et al., 2024).

The growing number of EVs directly affects peer-to-peer (P2P) energy trading, which is made possible by digital platforms, and is the direct trading of electricity between energy producers and consumers. During times of high demand or when renewable energy production is at its lowest, EVs can supply electricity to the grid via V2G technology, which has the potential to improve grid stability, energy management, and cost savings (Pothumsetty and Thomas, 2020). This capability can stabilize the system, increase the use of renewable energy sources, and give EV owners access to new sources of income. Throughout the charging process, smart charging strategies and optimization are used to maximize the use of RESs and maximize electromobility (Gopinathan and Shanmugam, 2022). EVs with extra energy could sell that energy directly to other EVs that need it. Vehicle-to-Vehicle (V2V) energy trading is another name for this direct energy exchange between EVs, which has great social and economic benefits (Wang et al., 2023). While Vehicle to Home (V2H) technology allows EVs to deliver power directly to a home, potentially acting as a backup power source during outages, and V2G technology allows EVs to return electricity to the power grid, helping to manage grid loads and support grid stability, these technologies primarily interact with stationary energy systems (Xu et al., 2024).

R2V systems bridge the gap between clean power generation and smart mobility solutions by allowing vehicles to use energy that is either directly or indirectly derived from renewable resources such as solar,and wind. In addition to lowering reliance on fossil fuels, this integration promotes the growth of sustainable smart cities and decentralized energy ecosystems. Despite major advancements in technology, it is still challenging to create a scalable and effective R2V ecosystem. Technical problems that still impede large-scale deployment include RE energy intermittency, scalability, and infrastructure upgrade. Emerging trends, including sustainable charging solutions, blockchain-based P2P trading, and artificial intelligence-driven energy management, are changing the way renewable energy and vehicle technologies work together. In order to address these intricate challenges, a thorough comprehension that integrates viewpoints from power systems, renewable integration, and vehicular engineering is needed to accelerate the shift toward an intelligent and sustainable RE-integrated ecosystem. In this regard, this chapter focuses on conceptual foundations, challenges, opportunities, and future research directions. The following are the main contributions of this chapter.

1. This chapter presents a conceptual foundation for understanding interconnected smart R2V ecosystem key terminologies and principles.
2. Main challenges include power capacity and grid stability, control and communication complexities, interoperability, cybersecurity, and grid interaction are provided. In this regard, the applicability of technological enablers such as AI-based optimization, V2G integration, and blockchain is discussed.
3. Opportunities, potential research development perspectives that can guide future academic and potential future research directions are highlighted.

The rest of the chapter is as follows: in **Section II**, conceptual foundations for understanding smart R2V and interconnected technologies are presented. In **Section III**, the main challenges and concerns in this domain are provided. In **Section IV**, research opportunities and potential research development, including environmental, economic, and industrial opportunities, are discussed. Future directions are provided in **Section V**, and lastly, **Section VI** presents the chapter's conclusion. The overview of this chapter, which includes challenges and opportunities in the smart R2V ecosystem, is shown in Fig. 1.1.

Figure 1.1 The overview of the chapter.

1.2 Conceptual foundation and enabling interrelated technologies in the smart R2V ecosystem

The potential of EVs as mobile energy assets became more significant as the global focus shifted to renewable energy and grid modernization (Kumar et al., 2025b). In this section, an overview of these conceptual foundations, such as V2G and V2V, is provided along with more sophisticated paradigms like smart charging, P2P energy trading, and blockchain technology. Additionally, the section explores how vehicle-to-everything (V2X) communication and Internet of Things (IoT) connectivity can be integrated.

1.2.1 Fundamental concepts and enabling technologies

The simplest basic type of smart charging is unidirectional vehicle-to-grid (V1G), which enables EVs to dynamically adjust charging rates and times because the EV and charging stations are connected via a data connection. This technology's benefits include lower charging costs, improved safety, easy identification of charging locations, and the ability to measure vehicle electricity consumption (Barman et al., 2023). V2G permits bidirectional power flow between EVs and the power grid; in contrast to V1G and V2G's ability to balance the grid, particularly when combined with RE, is a major benefit. It provides EV owners with an affordable service that allows them to sell the extra power generated by their vehicles to the grid at a set price. Furthermore, if utilized properly, EVs' on-board batteries can be combined with V2G technology to create an effective energy storage system for the utility grid (Barman et al., 2023).

In V2G systems, bidirectional charging enables EVs to return stored energy and receive electricity from the grid. Centralized and decentralized networks are the two primary approaches for P2P trading with EVs that have been discussed in the literature (Soto et al., 2021). Centralized P2P trading networks manage and supervise the electricity exchange through a central organization, usually a utility or a third-party middleman. This model ensures a controlled and regulated business environment by having a central authority verify and regulate transactions (Liu and Li, 2021). Distributed ledger technology, like blockchain, is the foundation of decentralized P2P trading networks, in which a network of participants validates transactions without the involvement of a central authority (Khan et al., 2024).

A more decentralized electrical network is made possible by the huge rise in energy prosumers, which is encouraged by the V2V energy trading framework's implementation, which promotes multifaceted energy transactions within a local geographic area. In the local energy market, this framework balances the relationship between energy supply and demand, promotes the use of distributed energy resources, and lessens the demand pressure on the grid during periods of peak energy consumption (Xu et al., 2024).

One strategy for improving security and decentralizing transactions is blockchain technology (Soto et al., 2021). Blockchain acts as a distributed ledger, safely capturing and preserving important data, events, contracts, and financial transactions. This technology prevents unauthorized changes and promotes a safe, decentralized transaction environment by giving multiple parties access to a shared, unchangeable record of transactions, each encrypted and connected to the one before it (Xu et al., 2024). Reference (Umoren et al., 2020) concentrated on creating an application to remotely monitor energy trading activities between trading entities and utilizing blockchain technology to create a reliable energy trading ecosystem.

A strong mathematical framework for simulating and analyzing the strategic interactions between autonomous agents with conflicting interests, like EVs, is provided by game theory (Ho et al., 2022). In the context of V2V energy trading systems, this method is crucial because it makes it easier to comprehend and forecast participant behavior in a variety of scenarios. Game theory can be used as a stand-alone mathematical basis for creating and executing V2V energy trading strategies (Yu et al., 2023) or it can be integrated into other technological solutions, such as the establishment and enforcement of trading rules through the use of smart contracts in blockchain technology (Xu et al., 2024).

Between individual EV owners, the electrical grid, and other stakeholders like utility companies, RESs, and regulators, EV aggregators are crucial middlemen. Effectively integrating EVs into the energy system is essential given their growing usage in order to preserve grid stability, maximize energy use, and promote the use of RESs. EV aggregators' primary role is to participate in demand response programs, whereby the entire EV fleet can be managed collectively to satisfy grid demands. Aggregators can temporarily reduce the amount of electricity used to charge connected EVs during periods of high demand. On the other hand, aggregators can increase billing rates to utilize the extra energy when demand declines and there is an excess of renewable energy. By taking this step, renewable energy waste is reduced and grid stability is preserved (Sarda et al., 2024). The coordination of the combined energy resources of numerous EVs is managed by the aggregator platform. It interacts with the grid operator to improve the power grid's efficiency by offering services like frequency regulation and peak energy demand reduction (Kumar et al., 2025b).

EV charging modes include dumb charging, which begins charging as soon as the car is connected to the grid, regardless of any excess or deficit in the system's electricity generation or the price signals from the corresponding electricity market, and smart charging, which allows EV charging power to be adjusted based on the cost and availability of electricity (Bogdanov and Breyer, 2024). An EV and a charging device share a common network with a charging operator in an ecosystem known as smart charging. Smart charging devices, as opposed to traditional dumb chargers, have the ability to connect to the cloud, which enables the owner of the charging station to control, monitor, and limit the use of their devices in order to maximize energy distribution (Barman et al., 2023). Table 1.1

Table 1.1 Different EV charging types with description and features.

Method	Description	Features
Network charging	Permits users to use renewable energy to charge EVs.	It encourages off-peak charging and matches charging times with RE generation.
Shift charging	Uses time-of-use (ToU) tariffs during off-peak hours or when renewable energy generation is high.	Aims to align the timing of renewable energy generation with EV charging schedules by shifting charging loads to off-peak times.
Charging with excess renewable	Synchronizes EV charging with renewable energy surpluses, when there is an excess of solar or wind power.	It encourages charging when the production of renewable energy surpasses the demand by using ToU rates.
On-site renewables	Renewable energy is generated on-site in conjunction with EV charging.	Methods such as managed charging to control load, or integrating battery storage, can be used to integrate solar panels and storage to charge EVs.
Managed charging	It uses communication between the EV, charging station, and utility to coordinate EV charging rates and times with grid demand and RE availability.	Giving utilities the ability to regulate charging schedules and even recover power from EVs when necessary.

shows different EV charging types with corresponding descriptions and features.

1.2.2 Interconnected technologies fundamental concepts

R2V systems rely on Internet of Things (IoT) technology, and large volumes of data are gathered and processed using sensors, networked devices, and communication systems installed in infrastructure and vehicles (Ahmadzadeh et al., 2021). The gathered data is analyzed in real time using edge and cloud computing, which serves as the foundation for quick decision-making in the areas of route optimization and energy management (Hu et al., 2024). Artificial intelligence (AI) algorithms, for instance, can forecast energy consumption trends and vehicle charging requirements while preserving the network's electricity flow balance (Zhang et al., 2025).

Connected EVs to the grid can serve as distributed energy sources and boost grid stability in addition to being energy consumers (Mojumder et al., 2022). The most well-known instances of this interaction are V2H and V2G technologies. These technologies make it possible to store extra energy and use it again when consumption is at its highest (Bonfiglio et al., 2025). Additionally, combining EVs with RESs like solar and wind power could lower greenhouse gas emissions and boost energy efficiency (Barman et al., 2023).

Smart grid technologies in cooperation with R2V, V2X, and IoT could form an ecosystem that lowers operating costs, improves environmental sustainability, and permits efficient energy management (Kumar et al., 2025a). R2V systems are becoming more capable of making intelligent and automated decisions at the energy grid level thanks to the development of new algorithms like deep reinforcement learning and predictive analytics (Mansour et al., 2025). Additionally, in this intricate ecosystem, the use of blockchain and security technologies can improve data exchange security (Sharma et al., 2023).

Table 1.2 compares R2V and other V2X subsets and highlights the differences and purposes.

Fig. 1.2 shows an ecosystem of smart R2V and V2G blockchain-based energy systems, showing the relationship between R2V, V2G, the Internet of Things, and electric vehicles.

The purpose of Fig. 1.2 is to clarify how these technologies interact and the data and energy flows between them. While EVs engage with the grid through bidirectional energy exchange via R2V and V2G mechanisms, renewable energy systems provide power to the grid. In addition to using

Table 1.2 Comparison of R2V and other V2X subsets.

Technology	Interaction domain	Data type	Functional purpose	Examples
V2V	Vehicle to vehicle	Energy status, charge level, scheduling data	Emergency charge, P2P energy sharing	Mobile energy sharing
V1G	Unidirectionalvehicle to grid (one-way)	Control signal of charging, energy signal	Demand peak reduction by controlling unidirectional charging	Load management, smart scheduling
V2G	Bidirectional Vehicle to Grid	Energy and frequency signals	Energy flow for grid stability enhancement and RE integration	Energy Storage During Peak Hours (Mojumder et al., 2022), ancillary service, peak shaving
R2V	RESs to the vehicle	Energy flow and its control	RE-based charging, sustainability promotion	Solar and wind powered charging station (CS) and RE charging
V2X	Vehicle to everything	Energy, scheduling, coordination, and control data	Supporting a sustainability, flexibility, and resilience system	Mobility energy and smart grid system

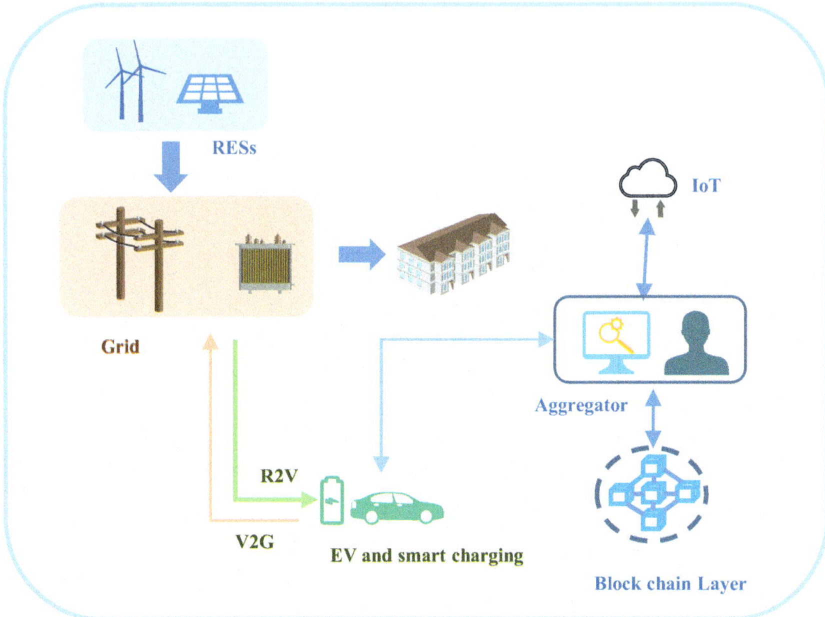

Figure 1.2 An ecosystem of smart R2V and V2G blockchain-based energy system.

energy from the distribution network, prosumers may also contribute extra power through distributed generation. In order to enable intelligent monitoring and control, the IoT layer gathers operational data in real time from physical components. In order to maximize charging, discharging, and power dispatch, the aggregator coordinates a number of distributed assets, including EVs, residences, and renewable energy units, while interacting with the market operator and the blockchain network. A transparent and safe framework for logging energy transactions, confirming bids, and carrying out smart contracts between market participants, prosumers, and aggregators is offered by the blockchain layer. Modern smart grids can operate efficiently, decentralized, thanks to this multi-layered structure.

1.2.3 Smart R2V versus traditional vehicle communication paradigms

Over the past decade, automotive communication paradigms have changed significantly. Intelligent and integrated solutions, like smart R2V systems, represent a more modern concept as an emerging paradigm in comparison to V2V and V2G frameworks. The primary focus of traditional automotive communication has been on sharing fundamental data, such as battery life, traffic patterns, or straightforward network interactions. Despite their dependability, these techniques do not fully realize their potential in smart grids and are unable to adapt and optimize in real time (Aruna and Vasan, 2019).

Innovative R2V systems, on the other hand, use cutting-edge energy management techniques and communication technologies to arrange the interactions between vehicles and RESs in a more dynamic manner. By utilizing real-time data analytics, AI, and IoT networks, these systems optimize energy flows, lessen grid pressure, and raise the percentage of renewable energy consumption (Barman et al., 2023; Kumar et al., 2025a). AI-based scheduling, for instance, can balance the grid load, forecast charging demand, and even return excess energy to the grid, all of which greatly increase system efficiency (Hu et al., 2024).

The systems' scalability and flexibility represent yet another key distinction. Traditional models primarily rely on manual control or set schedules, which limit their capacity to react swiftly to abrupt shifts in demand or the production of renewable energy. By emphasizing energy storage, dynamically modifying the charging rate, and smoothly integrating with commercial, industrial, and residential microgrids, the R2V ecosystem could surpass the limitations (Bilal et al., 2024). As the proportion of RESs rises, this capability becomes more crucial because their variable generation must be matched to patterns of consumption (Anisie et al., 2023).

The two most crucial features of these systems are security and dependability. Traditional systems are vulnerable to cyberattacks and infrastructure disruptions because their performance is primarily dependent on constrained and standardized protocols (Acharya et al., 2020). On the other hand, cutting-edge R2V systems use blockchain technology, multi-layered security, and real-time monitoring to guarantee safe and continuous energy transfer (Sharma et al., 2023). Additionally, the widespread use of EVs and the integration of RESs will be made possible by the integration of these systems with 5G and next-generation 6G networks, which could boost speed, lower latency, and improve reliability (Varga et al., 2025).

The advantages of cutting-edge R2V systems for the economy and environment are becoming more and more clear. These systems minimize grid pressure and reduce carbon emissions by controlling the charging process and adjusting it to the availability of renewable energy sources. They are also the most cost-effective option for utilities and consumers (Karmaker et al., 2023). On the other hand, considering the reactive nature of traditional models, they cannot fully utilize the advantages, although they are

Table 1.3 Comparison of traditional and Smart R2V paradigms.

Criterion/Feature	Traditional paradigms	Intelligent R2V paradigms
Type of communication	Basic V2V and V2G (Aruna and Vasan, 2019)	IoT- and AI-driven (Kumar et al., 2025a; Barman et al., 2023b)
Energy optimization	Limited, Reactive	Predictive and Adaptive (Hu et al., 2024; Zhang et al., 2025)
Flexibility and scalability	Fixed Programming	Dynamic Integration with microgrids (Bonfiglio et al., 2025; Bilal et al., 2024)
Security	Standard Protocols (Acharya et al., 2020)	Multi-layer Security and blockchain-based (Sharma et al., 2023; Rao et al., 2023)
Integration with renewable energy	Limited	Optimized (Anisie et al., 2023; Bilal et al., 2024)
Economic and environmental impact	Moderate (Aruna and Vasan, 2019)	Optimized and Low-Carbon (Bilal et al., 2024; Dahiwale et al., 2024; Subashini and Vijayan, 2023; Karmaker et al., 2023)

dependable and simple. Table 1.3 provides a comparison between traditional and intelligent R2V models to clarify their differences and advantages.

1.3 Challenges and emerging trends in the interconnected smart R2V ecosystem

To successfully integrate electric vehicles into the grid, it is essential to recognize and address any shortcomings and challenges in the power transfer process between vehicles and the power grid. The regulatory framework for V2G system integration is insufficient and inconsistent across regions. In order to address technological, economic, and legal issues, comprehensive and standardized policies must be developed. In order to optimize power flow while taking grid constraints into account, complex control systems must be developed. One major concern is the security of V2G systems, which calls for robust defenses against potential cyberattacks. Battery degradation in V2G can be accelerated by frequent cycles of charge and discharge. Modern Battery Management Systems (BMSs) use predictive algorithms to optimize charging schedules and discharge depth in order to reduce degradation. Blockchain-based protocols for cybersecurity are starting to appear as safe V2G network solutions (Kumar et al., 2025b). There are numerous obstacles to EV load coordination, including local voltage instability, power rerouting to reduce circuit distance, upstream transformer capacity enhancement, grid thermal capacity upgrades, reactive power compensation, and grid architecture changes (Barman et al., 2023). Table 1.4 shows key challenges in smart R2V with solutions (Sarda et al., 2024; Kumar et al., 2025b; Bharathidasan et al., 2022).

Table 1.4 Key challenges in smart R2V ecosystem with solutions.

Research challenges and gaps	Proposed potential solutions
Battery degradation due to frequent charge and discharge cycles	Using ML-based methods for charge optimization
Infrastructure availability with limited CS and points	Infrastructure development and investment increase
Lack of standardization and policy limitations	Development of universal standards for charging, communication, and grid services, and supportive policies and regulatory frameworks considering RE integration
Maintaining grid stability due to the high demand for EV charging	Enhancement of capacity and upgrading of infrastructure for appropriate demand response and energy management
Intermittent RE and EV integration	Developing energy storage and demand response programs
Cybersecurity concerns	Using protection policies and advanced algorithms, and technologies such as ML methods and blockchain
Uncertainty in demand, time, and capacity	Using forecasting methods, penalty function, and state of charge estimation
Fluctuations in voltage and frequency	Using mechanisms for EVs motivation

The following highlights the key challenges that need to be addressed to ensure the smart R2V ecosystem is efficient and reliable (Sadeghian et al., 2022; Barman et al., 2023).

- Battery degradation: Degradation causes the battery to exhibit increased resistance, which shortens its lifespan. Two elements that increase this resistance are low temperatures and high state-of-charge levels. Long-term battery degradation results from charging EVs with fast charging options. Two elements that may hasten battery deterioration and adversely impact its condition are discharge depth and cycling frequency.
- Power capacity and grid stability: The infrastructure for EV charging needs a huge amount of power capacity, which can put stress on the electrical grid and cause instability. V2G systems can also inject or absorb electricity to maintain grid frequency.
- Loss of power: Due to the longer distances EVs must travel to reach charging stations, they will bring additional load on the grid, so they can increase power losses. Power loss in the distribution network can be improved by integrating electricity back into the grid during periods of peak demand or grid instability.
- Electrical lines congestion and Scalability: Simultaneously, EVs charging could cause congestion, particularly during peak hours; however, this can be lessened by effectively managing EV charging in CSs. Building additional infrastructure is necessary due to the congestion at the current charging stations in city centers.
- Intermittent nature of RES and stochastic behavior of EV owners: Although EVs can store RES energy in batteries and release it when needed, the intermittent nature of RES (such as wind and solar) generation as auxiliary resources in the electrical supply can impact EV charging and discharging. Additionally, Deterministic charging planning is unrealistic due to the stochastic behavior of EV owners in terms of travel times, SOC, and arrival times. To increase vehicle owners' willingness to participate, the V2G concept's technical and cost-effective designs are essential.
- Infrastructure upgrade and grid integration: Upgrading the current infrastructure is a major challenge for cities with high EV and RE growth. Through V1G or V2G, EVs can participate in local or bulk flexibility, which could aid in boosting the uptake of RESs. To create a robust ecosystem for EV charging, significant renovations are required.

- Comprehensive standardization: EV owners encounter challenges when trying to charge their cars at various manufacturers' stations. Thus, the widespread adoption of EVs in this market is threatened by the lack of benchmarks and unified standards.
- Maintenance: The integration of multiple RE sources has significantly increased the sophistication and complexity of hardware and software design in smart grid systems. Updating and maintaining the smart grid framework has an impact on the network as a whole and the runtime services.
- Data security: Information about EVs, transactions, charging station identity, and personal information should all be protected in a V2G communication. Legal frameworks should be specifically developed for smart charging infrastructures in order to protect privacy. The framework needs to specify how private information should be gathered, safeguarded, kept, and shared.
- Charge Availability: The EV owner's access to chargers is restricted at certain times due to the stochastic nature of the RE. When EV deployment increases over time and the power grid is unable to supply the necessary load, this problem may become more severe.
- Interoperability functions: These functions facilitate smooth communication and data exchange between various EV stakeholders are necessary for effective EV-grid management.
- Gridfluctuation: Large-scale electric vehicle discharge in a short amount of time and disordered electric vehicle charging behavior could be the reason for the power grid's fluctuations and instability (Xie et al., 2025).

V2G technology's cost optimization is a major contributor to its extensive use since it improves the flexibility and financial advantages of grid dispatch while lowering energy consumption and operating expenses. However, the design and capacity constraints of charging infrastructure, the technology of electric vehicle batteries, and the ongoing modifications to grid regulations and policies are all important factors that impact cost optimization in the process of cost optimization for V2G systems (Xie et al., 2025). In this regard, key economic challenges that need to be addressed are listed below (Alaee et al., 2023).

- Considering the interactions with the electricity market, guaranteeing that charging stations can be financially successful in offering charging services is a challenge.
- Using aggregators to provide service in a scheduling system to provide service for assigned EV charging.

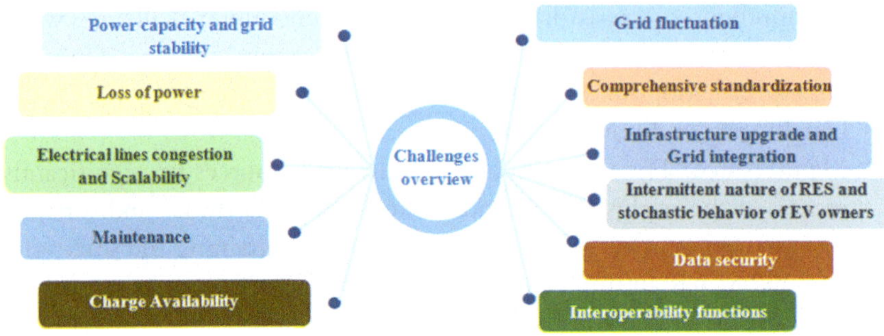

Figure 1.3 The overview of challenges in the interconnected smart R2V ecosystem.

- CS installation and maintenance have high expenses, and effective real-time pricing strategies are needed.
- Optimizing the advantages of EV welfare and charging costs in the electricity market in V2G mode.

The overview of challenges in the interconnected Smart R2V ecosystem is illustrated in Fig. 1.3.

1.3.1 Concerns in R2V and EV integrated energy trading ecosystem

Security concerns regarding cyber vulnerability in the domain of smart charging may focus more on critical cyberattacks such as tampering or falsifying charging data to prevent billing loss, stopping EV power supplies, and encrypting charging data, which could result in revealing charging account and associated location (Sadeghian et al., 2022).

The scalability of current blockchain platforms may have an impact on the flexibility of blockchain in electric vehicle energy trading. Processing of millions of EVs supported by blockchain technology in real-time is a challenging task. The data immutability feature of the ledger guarantees that no adversary can unintentionally or purposely alter or remove data. As a result, irreversibility is a key component of blockchain that ensures data security and openness. New security issues for blockchain platforms have emerged due to the exceptional computational capabilities of quantum computing. Since data and transactions are open and transparent on public blockchain platforms, protecting data privacy is one of their challenges (Al-Saif et al., 2021).

In order to promote P2P energy trading, policies should first concentrate on the current grids rather than introducing standalone systems. The P2P energy trading network should be able to accommodate

Table 1.5 V2V energy trading operationalization challenges.

Challenge	Description	Features
Matching challenge	Involves several buyers and sellers; in order to match participants and carry out transactions, a rule is needed.	Rule creation for V2V energy trading that matches buyers and sellers.
Pricing challenge	After buyers and sellers have been matched, a fair clearing price is determined by taking supply and demand, fairness, and profitability into account.	Fairness in transactions and transparent pricing based on market conditions.
Optimization challenge	Optimizing social welfare and financial gains for both parties and framework operators.	Economic gains and overall social welfare optimization.
Demand forecasting challenge	Estimating future energy consumption to enhance the V2V energy trading framework and facilitate well-informed decision-making.	AI methods for predicting energy demand and their effects on V2V trading systems.

decentralization of connections, RES integration, and EV consumer extension (Khan et al., 2024). In order to facilitate prosumers and consumers, protocols must be in place to manage P2P energy trading at the local, governmental, and private levels without the assistance of any third parties (Li et al., 2021). Table 1.5 provides a summarization of V2V energy trading operationalization challenges (Xu et al., 2024).

1.4 Research opportunities in an interconnected smart R2V ecosystem

Smart R2V systems' quick development has created a wealth of research opportunities that connect communication networks and renewable energy integration. Exploring these opportunities helps to achieve global decarbonization objectives and improves the resilience of future smart grids, in addition to facilitating the creation of more sustainable vehicle-energy ecosystems. Despite all of the progress made in EV technology, there are still some areas that need to be improved, and researchers are being asked to introduce methods for addressing these challenges. Some of the prominent issues are discussed in the following (Khalid et al., 2022).

- System integration: More studies are needed to integrate advanced and modern charging technologies (battery switching, wireless, and plug-in charging technology) with green energy sources like solar and wind. It is difficult to deploy, maintain, and oversee this sophisticated integrated system in smart cities.
- Scheduling EV charging efficiently: For efficient charge scheduling, the charging system's real-time communication with mobile vehicles must be effective. As a result, it is difficult to integrate charging management work with current mobility projects since this requires a solid design.
- Security concerns: The security issue is crucial in addition to the technical difficulties in managing EV smart charging. Providing security at the user and charging station levels is one of these challenges. When working with other aggregators, information about EVs and users must be kept secure and confidential. Additionally, since the anonymous group may send malicious information to EVs and CSs for manipulation and undesired events, communication between EVs and CSs must be secure.
- Operation of V2G: Power can be transferred both from the grid to the vehicle and from the vehicle to the grid. Depleted EV batteries are usually charged in real-time at CSs, which takes time and causes congestion because there aren't enough ports. For better management and time savings, the remaining energy in the replaced depleted battery could be sold back to the EV.
- Battery life and scheduling: Battery charging and discharging are made possible by bidirectional power flow. EV researchers and engineers are very interested in examining how discharging affects battery life and taking this effect into account when solving optimization problems.
- Heterogeneity of EVs: CS needs to have a variety of batteries that can be used to replace a wide range of EVs. An intriguing issue to be tackled is the development of an algorithm and the variety of batteries offered at CSs to support heterogeneous EVs.

- Cost optimization: In addition to the technical deployment of CSs and V2G operations, a business model that accounts for price adjustment (variable for peak and regular hours) and maximizes profit for both aggregators and end users is required.

There are also opportunities for utilizing blockchain technology in EV energy trading methods. Managing the asset record during asset processing by the participating organizations is the goal of the blockchain traceability feature. Owners of electric vehicles can verify the data provenance of RESs thanks to the comprehensive and unchangeable transaction records. Reducing energy trading prices, maximizing participant profit, and meeting network constraints are the goals of an effective system for an optimized energy trading market for electric vehicles. Because owners of electric chargers can be given incentives to bring their private chargers online for public use, blockchain technology can play a significant role in the expansion of the electric vehicle energy trading market (Li et al., 2021). An overview of key research opportunities in smart R2V is illustrated in Fig. 1.4.

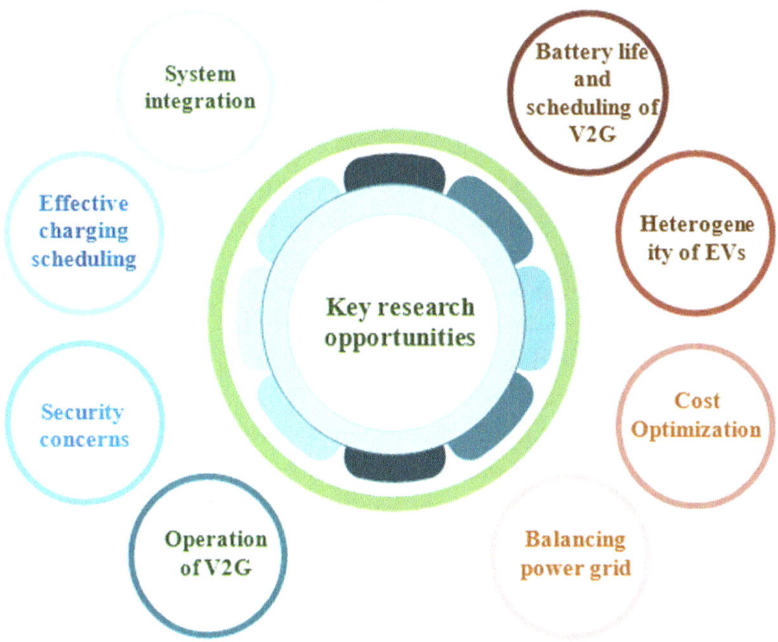

Figure 1.4 An overview of key research opportunities.

1.4.1 Environmental and sustainability opportunities

According to recent research, charging demand forecasting and energy distribution in the grid can be optimized by expanding the system and utilizing machine learning and predictive algorithms, trying to address issues such as charging station congestion and variations in renewable energy (Zhang et al., 2025).

A clean and sustainable energy cycle can be created by combining EVs with RESs like solar and wind. In addition to being energy consumers, vehicles can also act as distributed storage systems by returning their excess energy to the grid thanks to V2G and V2X technologies. This capability lessens the need for emergency power plants, improves grid load management, and increases grid flexibility and stability (Mojumder et al., 2022).

The integration of electric vehicles with smart grids presents significant economic and social opportunities. Increased public acceptance of electric vehicles, lower energy costs, and better air quality all contribute to the growth of sustainable infrastructure and social capital. Additionally, the best use of energy resources in residential and commercial settings is made possible by integrated "car-to-home" and "car-to-building" services (Bonfiglio et al., 2025). These creative approaches not only support environmental objectives but also provide users and grid managers with real financial advantages (Anisie et al., 2023).

In order to optimize vehicle energy consumption, AI and machine learning algorithms are essential. Energy waste can be decreased and the efficiency of renewable resources can be increased through the use of energy consumption prediction models, charging scheduling algorithms, and reinforcement learning-based energy management systems (Mansour et al., 2025). These technologies lower operating costs, increase grid reliability, and match energy consumption with clean energy production by optimizing the charging and discharging processes of EVs.

The integration of EV technologies with REs and smart grids presents a wide range of social, economic, and environmental benefits. In addition to improving grid flexibility and efficiency, this strategy offers a clear route toward sustainable development and fewer environmental effects (Mansour et al., 2025).

To better understand the environmental and economic impacts, the primary advantages of smart charging and V2G technologies can be summed up in Table 1.6.

Table 1.6 Key benefits of smart charging and V2G technologies.

Technology	Carbon emission reduction	Energy efficiency improvement	Energy cost reduction	Grid flexibility
Smart EV charging	✓	✓	✓	×
V2G/V2X	✓	✓	✓	✓
Integration with renewable energy	✓	✓	×	✓

The vehicle and renewable energy sources can be integrated using a multi-layered approach thanks to the Smart R2V architecture/system, which offers advantages like better energy efficiency, lower carbon emissions, and more grid flexibility (Colak et al., 2024).

1.4.2 Economic and industrial opportunities

Global economic and industrial opportunities have been greatly enhanced by the quick expansion of EV adoption and their incorporation into smart energy management systems (SEMS). EVs are crucial in lowering greenhouse gas emissions and redefining industrial value chains as the energy sector transitions to a low-carbon future (Kumar et al., 2025a). Additionally, the development of EV-related infrastructure, including battery production, charging stations, and the use of V2G technologies, is creating new markets and opportunities for manufacturers, technology companies, and the energy sector (Barman et al., 2023).

Incorporating RESs into electric vehicle charging networks improves system sustainability and energy efficiency from an industrial standpoint (Bilal et al., 2024). Innovative charging techniques, like demand-responsive charging and AI-based optimization techniques, make it possible to better align the use of electric vehicles with the production of renewable energy, which reduces pressure on the grid during peak hours (Hu et al., 2024). In addition to lowering utility operating costs, the optimization techniques open doors for the monetization of distributed energy resources and the provision of grid services via dynamic pricing.

Additionally, electric vehicles can serve as mobile energy storage units in addition to critical loads and backup services thanks to V2X technologies, such as vehicle-to-building (V2B) and V2H systems (Bonfiglio et al., 2025). These technologies' industrial applications, which range from home-level energy management to commercial microgrids, present new chances for both established businesses and start-ups to create integrated and effective solutions (Subashini and Vijayan, 2023).

More job opportunities in manufacturing, installation, and maintenance services could be the result of EV infrastructure development (Ghani et al., 2025). Increased production of batteries, fast-charging stations, and other cutting-edge network elements will create industrial clusters and promote regional investment and economic expansion. Additionally, advances in machine learning and AI in energy management will maximize battery life and lower the total cost of ownership, which will boost industrial demand and consumer adoption (Bhupathi and Chinta, 2022).

The economic and industrial opportunities related to EVs and SEMS are schematically depicted in Fig. 1.5. An integrated industrial and financial ecosystem is created by the connections between each opportunity area, as depicted in the figure.

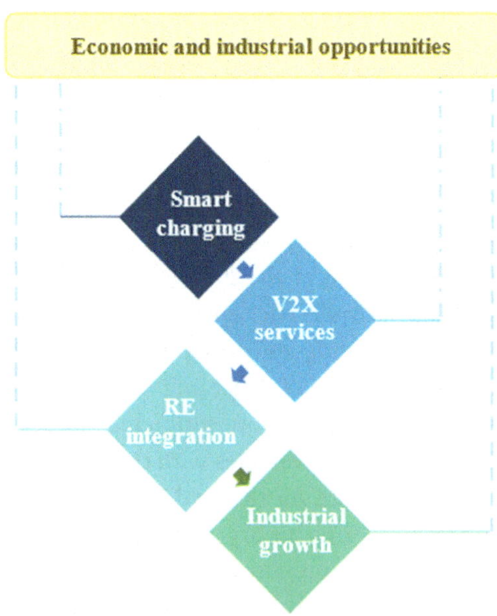

Figure 1.5 Schematic of economic and industrial opportunities related to electric vehicles and SEMS.

The quick integration of EVs into contemporary energy systems has brought attention to how crucial standardization and coherent policy frameworks are to ensuring sustainable adoption and optimal performance. In addition to speeding up the adoption of EVs, effective policies can facilitate V2G and V2X interactions, which are essential for reducing RES variability and boosting grid resilience (Kumar et al., 2025a).

Three main issues are the main focus of policy opportunities: promoting EV adoption, making it easier to integrate RESs, and assisting in the construction of smart charging infrastructure. Establishing regulatory mechanisms to support demand-responsive charging strategies is crucial, according to numerous studies. By optimizing the energy flow between electric cars and the grid, these strategies help keep the grid from excessive strain (Barman et al., 2023). In order to improve the grid's technical and economic efficiency, supportive policies can also boost investment in distributed energy storage systems, fast-charging stations, and smart grids (Bilal et al., 2024).

The standardization of electric vehicle-related home services and smart buildings is another important area. This standardization can boost the effectiveness of RESs and optimize energy management at the microgrid scale (Bonfiglio et al., 2025). The standardization of IoT protocols for smart grids opens up new possibilities for data analysis and energy optimization as new communication technologies like 5G and Beyond-5G are developed (Varga et al., 2025). Compatibility issues between energy management systems, communication protocols, and charging infrastructure may restrict these advantages in the absence of harmonized standards (Colak et al., 2024).

In order to guarantee technology harmonization and establish a single market, standardization is essential. The sustainable use of V2G and V2X capabilities will be made possible by the creation of international standards for cybersecurity, communication protocols, and data exchange between EVs and the grid (Rao et al., 2023). Furthermore, energy management algorithms and other hardware and software standards guarantee system compatibility and lower operating costs (Dahiwale et al., 2024). The key policy and standardization areas, as well as the opportunities related to them, are displayed in Table 1.7.

Taken together, the combination of creative regulations and thorough standardization will not only speed up the adoption of EVs but also allow for the optimal use of smart grids and RESs, opening the door to low-emission and sustainable cities and societies.

Table 1.7 Key policy and standardization areas and their associated opportunities.

Domain	Opportunities	References
EV adoption promotion	Financial incentives, tax reductions, and infrastructure support	(Barman et al., 2023; Ghani et al., 2025; Bilal et al., 2024)
Smart charging and V2G/V2X	Demand-responsive charging strategies, optimization algorithms	(Kumar et al., 2025a)
Hardware and software standards	Communication protocols, cybersecurity, grid interoperability	(Dahiwale et al., 2024; Rao et al., 2023)
Building and home services	Energy management in microgrids, utilization of renewable energy	Bonfiglio et al. (2025)
Emerging communication technologies	5G and Beyond-5G, Internet of Things, energy optimization	Varga et al. (2025)

1.5 Future research directions

EVs can take part in various electricity markets and ancillary services from the perspective of future market mechanisms. For instance, by offering demand response services, EVs can participate in energy markets. This lessens the need for additional power generation during peak hours and helps to balance the grid. By modifying their charging rate in reaction to variations in grid frequency, EVs can also take advantage of frequency regulation services. By using EVs to supply backup power in an emergency and to discharge during peak hours to assist with peak loads, EVs can also take part in capacity markets. Additionally, by serving as a backup power source in the event of unplanned grid outages, EVs can offer spinning reserve services. Future studies could also focus on how to improve the competitive environment by coordinating charges within each CS as well as between CSs collectively. One trend that is currently being assessed is the requirement for reservation systems to oversee EV charging. Sustainable charging solutions are necessary due to the EV market's explosive growth, the demand for charging infrastructure, and the incorporation of RES.

Future electric transportation system trends also include new technology developments like V2G and smart charging (Khan et al., 2024).

More research is required to show the applicability of P2P trading via blockchain, which can be applied to V2G in the future. Additionally, blockchain-based internet of vehicles (IoV) is a hot topic for upcoming studies (Hildebrand et al., 2023). Enhancing the sustainability and accessibility of EV charging infrastructure is the goal of future trends in charging station design. Managing CSs presents a difficult set of obstacles that call for further study and creativity in order to support the expansion of the EV market and the creation of a sustainable transportation network.

Researchers are working on EV energy trading as a result of the growing number of EVs. Blockchain technology plays a significant role in P2P energy trading frameworks because of offering prosumers and consumers trust, transparency, safe transactions, and information privacy. Real-time settlements in P2P energy trading are made possible by smart contracts. Developing smart contract platforms based on blockchain that use various algorithms for P2P trading is one of the current research focuses. In this regard, some of the potential future directions are listed below (Khan et al., 2024):

- There will be significant financial gains from creating a blockchain-supported P2P electricity trading system with an energy-efficient consensus algorithm.
- The establishment of new smart contracts for P2P energy trading that creatively handle environmental compliance requirements is possible using optimized energy.
- Demand-side management will be transformed for energy management and dynamic load forecasting in order to maximize energy use. Blockchain technology will be used to build a responsive and effective system that instantly adjusts to changing energy needs. By facilitating the smooth integration of RESs into electrical grids, this approach will not only improve energy efficiency but also advance sustainability.
- AI integration can improve V2G operations by facilitating predictive analytics and real-time decision-making. For example, machine-learning algorithms can minimize battery degradation during cycles of charging and discharging, optimize energy dispatch from EVs, and forecast patterns in grid demand. Research might concentrate on creating energy management systems with AI capabilities that are suited for extensive system deployments.

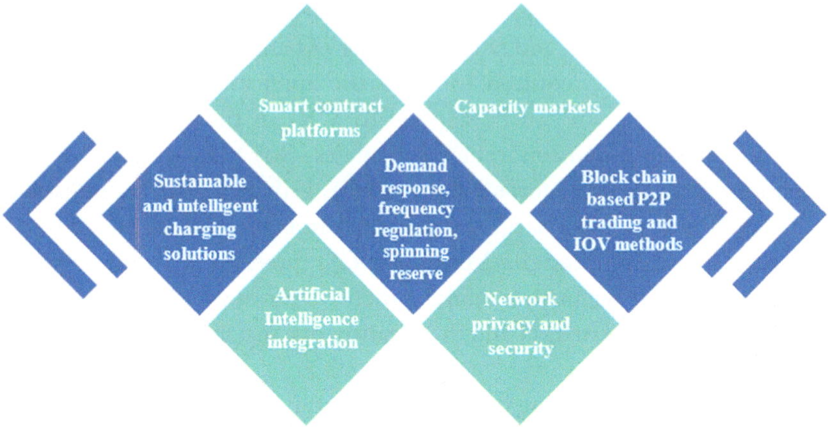

Figure 1.6 The overview of future research direction.

Data from micro-grids and V2G networks is continuously collected by the smart grid's power system, which may result in some participants' privacy being violated. A practical method to guarantee a tamper-proof, secure, and private data sharing between prosumers and consumers is through blockchain-based P2P energy trading. Blockchain-based P2P preserves network security and offers immutability, transparency, and protection from attacks. P2P energy trading allows traders to exchange electricity in any direction based on supply and demand. But it is also necessary to show how P2P energy trading benefits the distribution grid. Therefore, P2P energy trading could be another field of potential future studies v (Aggarwal et al., 2021). The overview of future research direction is depicted in Fig. 1.6.

1.6 Conclusion

The conceptual foundation, an overview of challenges, and prospects in the developing field of smart R2V smart ecosystems are provided in this chapter. RESs and vehicle networks will integrate more smoothly with the advancement of emerging technologies such as IoT and AI. The importance of combining EVs with RE technologies to create a sustainable, decentralized, and adaptable energy system is clarified in this chapter. Additionally, it highlights how crucial technological convergence and cross-disciplinary innovation are to the development of smart R2V ecosystems in terms of challenges and opportunities. There are some noteworthy points, as follows, that emerge from the chapter's insights and the state of the research.

- Further research is necessary for the advancement of these early-stage technologies. Even with impressive developments, there are still open issues such as cybersecurity, charge availability, interoperability, stability, and standardization.
- AI and blockchain-based methods are among the most promising research future directions, along with ancillary services, smart charging solutions, and smart contract platforms.

Ultimately, interdisciplinary cooperation, creative regulatory frameworks, and ongoing research to remove current obstacles are necessary to realize the full potential of the smart R2V ecosystem. By taking proactive measures to address these issues, a smart R2V ecosystem can significantly contribute to the acceleration of the world's shift toward intelligent, sustainable, and profitable energy ecosystems.

References

Acharya S, Dvorkin Y, Pandzic H, Karri R. Cybersecurity of smart electric vehicle charging: a power grid perspective. IEEE Access 2020;8:214434–53. https://doi.org/10.1109/ACCESS.2020.3041074.

Aggarwal S, Kumar N, Tanwar S, Alazab M. A survey on energy trading in the smart grid: taxonomy, research challenges and solutions. IEEE Access 2021;9:116231–53. https://doi.org/10.1109/access.2021.3104354.

Ahmadzadeh S, Parr G, Zhao W. A review on communication aspects of demand response management for future 5G IoT- based smart grids. IEEE Access 2021;9:77555–71. https://doi.org/10.1109/ACCESS.2021.3082430.

Alaee P, Bems J, Anvari-Moghaddam A. A review of the latest trends in technical and economic aspects of EV charging management. Energies 2023;16(9):3669. https://doi.org/10.3390/en16093669.

Al-Saif N, Ahmad RW, Salah K, Yaqoob I, Raja J, Omar M. Blockchain for electric vehicles energy trading: requirements, opportunities, and challenges. IEEE Access 2021;9:156947–61. https://doi.org/10.1109/access.2021.3130095.

Anisie A, Navarro JPJ, Antic T, Pasimeni F, Blanco H. Innovation landscape for smart electrification: Decarbonising end-use sectors with renewable power. 2023.

Aruna P, Vasan PV. Review on energy management system of electric vehicles. In: 2019 2nd International Conference on Power and Embedded Drive Control (ICPEDC). IEEE; 2019. p. 371–4.

Barman P, Dutta L, Bordoloi S, Kalita A, Buragohain P, Bharali S, et al. Renewable energy integration with electric vehicle technology: a review of the existing smart charging approaches. Renew Sustain Energy Rev 2023;183:113518. https://doi.org/10.1016/j.rser.2023.113518.

Bharathidasan M, Indragandhi V, Suresh V, Jasiński M, Leonowicz Z. A review on electric vehicle: technologies, energy trading, and cyber security. Energy Rep 2022;8:9662–85. https://doi.org/10.1016/j.egyr.2022.07.145.

Bhupathi HP, Chinta S. Smart charging revolution: AI and ML strategies for efficient EV battery use. ESP J Eng Technol Advance 2022;2(2):154–67.

Bilal M, Oladigbolu JO, Mujeeb A, Al-Turki YA. Cost-effective optimization of on-grid electric vehicle charging systems with integrated renewable energy and energy storage: an economic and reliability analysis. J Energy Storage 2024;100. https://doi.org/10.1016/j.est.2024.113170.

Bogdanov D, Breyer C. Role of smart charging of electric vehicles and vehicle-to-grid in integrated renewables-based energy systems on country level. Energy 2024;301:131635. https://doi.org/10.1016/j.energy.2024.131635.

Bonfiglio A, Minetti M, Loggia R, Mascioli L, Golino A, Moscatiello C, et al. Integrated vehicle-to-building and vehicle-to-home services for residential and worksite microgrids. Smart Cities 2025;8(3):101. https://doi.org/10.3390/smartcities8030101.

Colak I, Irmak E, Tepe IF. A brief overview of future mobility and energy management strategies with vehicle-to-everything (V2X) concept on smart grid. In: 13th International Conference on Renewable Energy Research and Applications, ICRERA 2024. Türkiye: Institute of Electrical and Electronics Engineers Inc; 2024. p. 1687–92. https://doi.org/10.1109/ICRERA62673.2024.10815234.

Dahiwale PV, Rather ZH, Mitra I. A comprehensive review of smart charging strategies for electric vehicles and way forward. IEEE Trans Intell Transport Syst 2024;25(9):10462–82. https://doi.org/10.1109/TITS.2024.3365581.

Ghani AA, Abdullah N, Saidur R, Pandey AK. Assessing technological-driven challenges and policies associated with electric vehicle (EV) adoption. Transp Policy 2025;169:167–77. https://doi.org/10.1016/j.tranpol.2025.04.030.

Gopinathan N, Shanmugam P. Energy anxiety in decentralized electricity markets: a critical review on EV models. Energies 2022;15(14):5230. https://doi.org/10.3390/en15145230.

Gupta V, Reddy Konda S, Kumar R, Panigrahi BK. Collaborative multi-aggregator electric vehicle charge scheduling with PV-assisted charging stations under variable solar profiles. IET Smart Grid 2020;3(3):287–99. https://doi.org/10.1049/iet-stg.2019.0088.

Hildebrand B, Mohamed B, Salman T, Tabassum S, Konatham B, Amsaad F, et al. A comprehensive review on blockchains for Internet of Vehicles: challenges and directions. Comput Sci Rev 2023;48:100547. https://doi.org/10.1016/j.cosrev.2023.100547.

Ho E, Rajagopalan A, Skvortsov A, Arulampalam S, Piraveenan M. Game theory in defence applications: a review. Sensors 2022;22(3):1032. https://doi.org/10.3390/s22031032.

Hu J, Lin Y, Li J, Hou Z, Chu L, Zhao D, et al. Performance analysis of AI-based energy management in electric vehicles: a case study on classic reinforcement learning. Energy Convers Manag 2024;300:117964. https://doi.org/10.1016/j.enconman.2023.117964.

Karmaker AK, Hossain MA, Pota HR, Onen A, Jung J. Energy management system for hybrid renewable energy-based electric vehicle charging station. IEEE Access 2023;11:27793–805. https://doi.org/10.1109/ACCESS.2023.3259232.

Khalid M, Ahmad A, Ullah Z, Ahmad N, Aljaidi M, Khan MK. Electric vehicle charging modes, technologies and applications in smart cities. SSRN Electron J 2022. https://doi.org/10.2139/ssrn.4129040.

Khan S, Amin U, Abu-Siada A. P2P energy trading of EVs using blockchain technology in centralized and decentralized networks: a review. Energies 2024;17(9):2135. https://doi.org/10.3390/en17092135.

Kumar GMS, Guo X, Zhou S, Luo H, Qi W, Liu Y, et al. State-of-the-art review of smart energy management systems for supporting zero-emission electric vehicles with X2V and V2X interactions. Renew Sustain Energy Rev 2025a;208:115050. https://doi.org/10.1016/j.rser.2024.115050.

Kumar P, Kaur Channi H, Kumar R, Rajiv A, Kumari B, Singh G, et al. A comprehensive review of vehicle-to-grid integration in electric vehicles: Powering the future. Energy Convers Manag X 2025b;25:100864. https://doi.org/10.1016/j.ecmx.2024.100864.

Li H, Xiao F, Yin L, Wu F. Application of blockchain technology in energy trading: a review. Front Energy Res 2021;9. https://doi.org/10.3389/fenrg.2021.671133.
Liu C, Li Z. Comparison of centralized and peer-to-peer decentralized market designs for community markets. IEEE Trans Ind Appl 2021;58(1):67–77.
Manousakis NM, Karagiannopoulos PS, Tsekouras GJ, Kanellos FD. Integration of renewable energy and electric vehicles in power systems: a review. Processes 2023;11(5):1544. https://doi.org/10.3390/pr11051544.
Mansour SH, Azzam SM, Hasanien HM, Tostado-Véliz M, Alkuhayli A, Jurado F. Deep reinforcement learning-based plug-in electric vehicle charging/discharging scheduling in a home energy management system. Energy 2025;316. https://doi.org/10.1016/j.energy.2025.134420.
Mojumder MdRH, Antara FA, Hasanuzzaman M, Alamri B, Alsharef M. Electric vehicle-to-grid (V2G) technologies: impact on the power grid and battery. Sustainability 2022;14(21):13856. https://doi.org/10.3390/su142113856.
Pothumsetty R, Thomas MR. Bharat stage IV to VI-challenges and strategies. Int J Recent Technol Eng 2020;8:2614–23.
Rao PM, Jangirala S, Pedada S, Das AK, Park Y. Blockchain integration for IoT-enabled V2X communications: a comprehensive survey, security issues and challenges. IEEE Access 2023;11:54476–94. https://doi.org/10.1109/ACCESS.2023.3281844.
Sadeghian O, Oshnoei A, Mohammadi-ivatloo B, Vahidinasab V, Anvari-Moghaddam A. A comprehensive review on electric vehicles smart charging: solutions, strategies, technologies, and challenges. J Energy Storage 2022;54:105241. https://doi.org/10.1016/j.est.2022.105241.
Sarda J, Patel N, Patel H, Vaghela R, Brahma B, Kumar Bhoi A, et al. A review of the electric vehicle charging technology, impact on grid integration, policy consequences, challenges and future trends. Energy Rep 2024;12:5671–92. https://doi.org/10.1016/j.egyr.2024.11.047.
Sharma G, Joshi AM, Mohanty SP. sTrade: blockchain based secure energy trading using vehicle-to-grid mutual authentication in smart transportation. Sustain Energy Technol Assessments 2023;57. https://doi.org/10.1016/j.seta.2023.103296.
Soto EA, Bosman LB, Wollega E, Leon-Salas WD. Peer-to-peer energy trading: a review of the literature. Appl Energy 2021;283:116268. https://doi.org/10.1016/j.apenergy.2020.116268.
Subashini M, Vijayan S. Smart charging for zero emission vehicles – a comprehensive review. Renew Energy Focus 2023;46:57–67. https://doi.org/10.1016/j.ref.2023.05.005.
Umoren IA, Ali Jaffary SS, Shakir MZ, Katzis K, Ahmadi H. Blockchain-based energy trading in electric-vehicle-enabled microgrids. IEEE Consumer Electr Magaz 2020;9(6):66–71. https://doi.org/10.1109/mce.2020.2988904.
Varga P, Jászberényi ÁI, Pásztor D, Nagy B, Nasar M, Raisz D. How beyond-5G and 6G makes IIoT and the smart grid green—a survey. Sensors 2025;25(13). https://doi.org/10.3390/s25134222.
Wang Y, Yuan L, Jiao W, Qiang Y, Zhao J, Yang Q, et al. A fast and secured vehicle-to-vehicle energy trading based on blockchain consensus in the internet of electric vehicles. IEEE Trans Veh Technol 2023;72(6):7827–43. https://doi.org/10.1109/tvt.2023.3239990.
Xie H, Song G, Shi Z, Zhang J, Lin Z, Yu Q, et al. Reinforcement learning for vehicle-to-grid: a review. Adv Appl Energy 2025;17:100214. https://doi.org/10.1016/j.adapen.2025.100214.
Xu Y, Peralta AA, Balta-Ozkan N. Vehicle-to-vehicle energy trading framework: a systematic literature review. Sustainability 2024;16(12):5020. https://doi.org/10.3390/su16125020.

Yu Y, Li G, Liu Y, Li Z. V2V energy trading in residential microgrids considering multiple constraints via bayesian game. IEEE Trans Intell Transport Syst 2023;24(6):5946–57. https://doi.org/10.1109/tits.2023.3250649.

Zhang T, Peng Q, Zeng S. Predicting EV charging demand in renewable-energy-powered grids using explainable machine learning. Sustainability 2025;17(9):4158. https://doi.org/10.3390/su17094158.

CHAPTER TWO

Bibliometric analysis and case study investigation of renewable-based electric vehicle charging: Trends, advances, and future directions

Nishant Thakkar[1,2], Hamidreza Arasteh[2,3], Josep M. Guerrero[1,2], Anand R.[2], Liu Zhang[1,2], Rajamallaiah Anugula[1,2]
[1]School of Aeronautics and Astronautics, Zhejiang University, Hangzhou, China
[2]Center for Research on Microgrids (CROM), Huanjiang Laboratory, Shaoxing, Zhejiang, China
[3]Power Systems Operation and Planning Research Department, Niroo Research Institute (NRI), Tehran, Iran

2.1 Introduction

The global shift towards electric vehicles (EVs) is progressively unfolding in the area of transportation (Sathyan et al., 2025). According to the Global EV Outlook 2023, there were some 14 million registered electric cars worldwide in 2023, which increased the global stock to 40 million. This growth signals a huge growth in market share; in 2023, EVs made up about 18% of all vehicle sales compared to only 14% in 2022 and 2% in 2018, a clear sign that the industry is developing strongly (Colares et al., 2023). This growing usage demands that EV charging infrastructure be deployed quickly (Gharbaoui et al., 2013). In response to sustainable and large-scale requirement of EV charging, the Renewable-to-Vehicle (R2V) system is an emerging contender (Alkawsi et al., 2021). A schematic diagram of R2V system is presented in Fig. 2.1.

A faster coalescence of renewable energy (RE) such as solar photovoltaic (PV) and wind power has occurred in power systems for the last decade. However, the intermittence of these sources demonstrates the importance of supportive policies and enable technologies in order for EV charging to be reliable (Yazdani et al., 2025; Thakkar and Paliwal, 2023a, 2023b). The development of RECS started as a result of the increasing adoption of EVs and was originally focused on wind and solar-based designs (Hou

Renewable-to-Vehicle Smart Charging
ISBN: 978-0-443-36684-0
https://doi.org/10.1016/B978-0-443-36684-0.00003-5

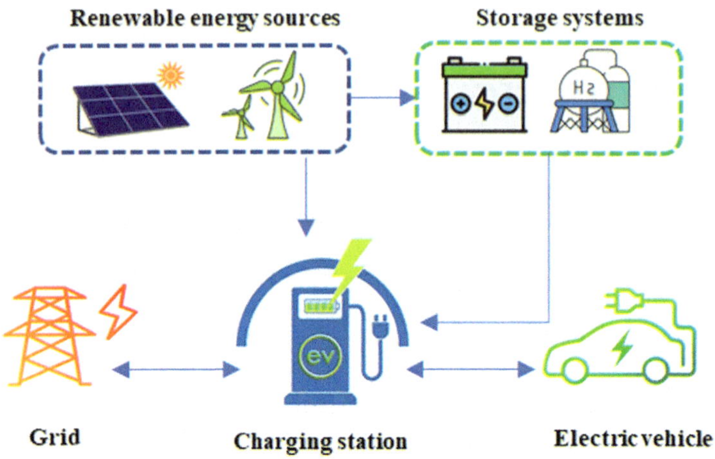

Figure 2.1 Renewable-to-vehicle (R2V) system.

et al., 2023; Sathyan et al., 2024). They have great potential for reducing carbon emissions and increasing the share of RE. In addition, RECS installations can provide savings in costs by mitigating the need for grid electric power (Darshi et al., 2023; Thakkar and Paliwal, 2024). Substantial work has been done to further the development of RECS. Khazali et al. (2024), for instance, presented an optimization approach for creating a standalone microgrid that strikes a compromise between dependability and cost in EV charging. This model determines the ideal PV and battery capacity while addressing uncertainty in EV demand and renewable generation through the application of decision theory. In the same direction, Dimitriou et al. (Dimitriou and Gkiotsalitis, 2025) put up a model for incorporating solar and wind energy into freight EV charging systems with the goals of lowering operational expenses, preserving delivery service efficiency, and reducing grid dependence. In order to determine the most economical option for EV charging in Adana, Turkey, Güven et al. (2025) used HOMER Pro to examine six renewable hybrid configurations. The best option was a mix of solar, biomass, hydrogen, and grid sources. A model with mixed-integer linear programming (MILP) originated by de Lima et al. (2021) in order to strategically distribute RECS in distribution networks. This model, which was verified on a 24-node system using AMPL and CPLEX, incorporates a forecasting method for EV charging behavior in addition to addressing demand uncertainties and environmental consequences. In addition to these analytical investigations, a number of review papers extend the field of RECS. An extensive review of the role of

RE in sustainable EV charging was given by Barman et al. (2023), who also examined smart charging techniques, grid integration, and associated difficulties. Shekari et al. (2022) offers a comprehensive investigation into EV charging behaviors and their implications for power grid stability, employing machine learning methodologies to discern and categorize prevalent charging patterns with the objective of optimizing energy management practices. Aghapour et al. (2022) investigates the integration of pure EVs in smart microgrids, emphasizing optimized charging strategies for enhanced grid support, while also addressing cyber-physical security challenges, including potential attack vulnerabilities and corresponding defense mechanisms. In other study, Aghapour et al. (2020) strategically optimizes the capacity and allocation of EV fast-charging stations by integrating traffic flow dynamics and stochastic charging demand. The objective centers on minimizing capital expenditure and energy losses through advanced probabilistic modeling, with simulation results corroborating the efficacy of the proposed framework. In their discussion of the technical challenges associated with EV integration into the power grid, Ibrahim et al. (2024) highlighted how Vehicle-to-Grid technology might improve grid stability and encourage the use of RE sources. Upon reviewing the previously listed literature, it is noticeable that no research work has carried out a bibliometric evaluation that focuses exclusively on RE to EV charging station technologies. However, some studies on relevant subjects, including EV (Singh et al., 2024a; Abdullaev et al., 2024; Yao et al., 2023; Purwanto and Irawan, 2023; Swarnkar et al., 2022; Ayodele and Mustapa, 2020), grid integration with storage (Tahir, 2024; Irham et al., 2024; Wang et al., 2024), EV charging stations (Gbey et al., 2022; Hildebrand and Kummer, 2024; Purwanto et al., 2024; Varghese et al., 2025; Barbosa et al., 2022) and V2G (Veza et al., 2024), have conducted bibliometric analyses. The present study utilizes a bibliometric approach to address this research gap by presenting a comprehensive review of various case study-based literature from the collected dataset. This bibliometric evaluation of international research publications on EV changing stations that are specifically based on RE aims to determine the regional contributions of leading authors, institutions, nations, and publishers as well as to gauge the technology's growth and subresearch areas within RECS technology.

The remaining sections are as follows: The data and methodology for extracting information from the SCOPUS dataset for bibliometric evaluation are presented in Section 2.2. The results of the bibliometric assessment are discussed and presented in Section 2.3. Analysis of various existing case studies has been discussed in Section 2.4. The investigation's summary along with future aspects and conclusions are provided in Section 2.5 and Section 2.6 respectively.

2.2 Methodology and data

In order to find pertinent works on RECS technology, this study conducted a bibliographic assessment using the SCOPUS dataset in accordance with PRISMA (Preferred Reporting Items for Systematic Reviews and Meta Analyses) criteria. PRISMA is a well-known and esteemed collection of guidelines for doing systematic reviews and meta-analyses in a variety of domains (Sarkis-Onofre et al., 2021; Page et al., 2021). PRISMA guidelines offers a methodical way to verify openness, precision, and consistency in the reporting of systematic reviews, which are in-depth examinations of earlier study findings (Rethlefsen et al., 2021).

Beyond alternative databases, the SCOPUS dataset is selected for this study because it contained a greater number of relevant papers on the issue. The use of SCOPUS dataset ensures a thorough representation of global studies from several locations by offering access to a large number of international publications on a wide range of topics (Odoi-Yorke et al., 2024). Bibliometrics evaluation is a quantitative examination of published data in specific academic subjects. This transdisciplinary discipline has advantages for practically every scientific field (van Nunen et al., 2018). To achieve the study's objectives, a complete literature search was conducted using the SCOPUS dataset, with an emphasis on studies on RECS technology and EVs published between 2015 and 2025. Fig. 2.2 depicts the four unique processes of the analytical technique.

In Stage 1, the foundational parameters for the bibliometric study are determined, such as the database selection, search query development, and inclusion and exclusion criteria definition. The SCOPUS dataset was examined with a structured combination of keywords: "Renewable" OR "Solar" OR "Wind" AND "EV" OR "Electric Vehicle" AND "Charging" AND "Station" applied particularly to article titles. This strategic search produced a starting corpus of 4199 items from SCOPUS dataset of duration 2015 to 2025. In Stage 2, an extensive screening method was carried out to determine the significance and acceptability of the retrieved documents in relation to the set eligibility standards. Following a thorough review, 1843 papers were considered relevant and saved for future investigation. On April 16, 2025, critical metadata such as journal titles, publication years, author

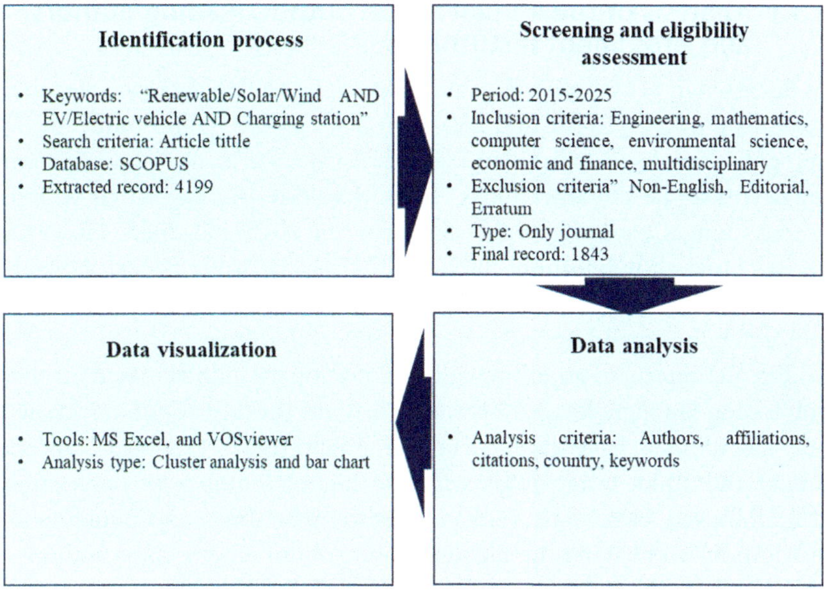

Figure 2.2 Framework of bibliometric appraisal.

affiliations, country of origin, source titles, abstracts, citation counts, keywords, and bibliographic references were systematically extracted and exported into a CSV file for use in bibliometric analysis. Stage 3 consisted of processing the information and visualization via two analytical tools: Excel from Microsoft and VOSviewer. VOSviewer, professional bibliometric mapping software, is utilized for developing visual networks exhibiting citation connections, co-authorship structures, co-citation pattern, and topic clustering in the literature. Simultaneously, Excel extensive data structuring, statistical evaluation, and visualization capabilities were used for activities such as formatting datasets, making dynamic charts, filtering entries, and building pivot tables. In the last stage, a detailed bibliometric and thematic analysis was performed to outline significant research trajectories, theme concentrations of substances, and existing gaps in the fields of RECS infrastructure.

2.3 Bibliometric analysis and discussion

The obtained bibliometric appraisal results of final relevant document of RECS via Microsoft Excel and VOSviewer are discussed in following subsections.

2.3.1 Analysis of the annual article output, leading authors, and prominent institutions

As discussed in Section 2.2, this study uses the SCOPUS data set and gathers 1843 documents on RES and EVs from 2015 to 2025. The annual development of articles is depicted in Fig. 2.3. An analysis of annual publications suggests that the literature on RECS technology is going to grow at an average annual pace of roughly 5% between 2015 and 2025. However, yearly counts of publications fluctuated during the study period. Notably, the most recent 3 years have seen the greatest increase in publications, indicating an uptrend in research on RECS technology.

Fig. 2.4 illustrates an examination of the top 10 authors based on their publication frequencies and citations. Among the 6014 authors studied, only 56 (0.93%) have more than five publications. Notably, Francisco Jurado and Bhim Singh leads with the highest number of publications (12), followed closely by Furkan Ahmad, who has contributed to 11 publications and having the highest citation count among these authors.

Drawing upon an author and co-author network analysis—restricted to scholars with a publication count exceeding five. The investigation reveals that only 16 authors exhibit discernible scholarly interlinkages, culminating in the formation of seven distinct collaborative clusters. These clusters, emblematic of intellectual synergy and academic alliance, are graphically represented in Fig. 2.5, delineating the structural affiliations and thematic orientations of the field's most prolific contributors.

Fig. 2.6 showcases the 10 leading institutions based on the number of publications linked to them between 2015 and 2025. Notably, Zhejiang University in Hangzhou, China, and Kermanshah University emerged as

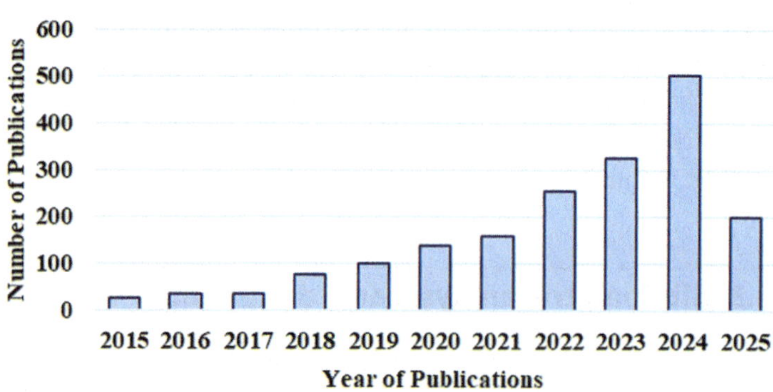

Figure 2.3 Annual publication growth.

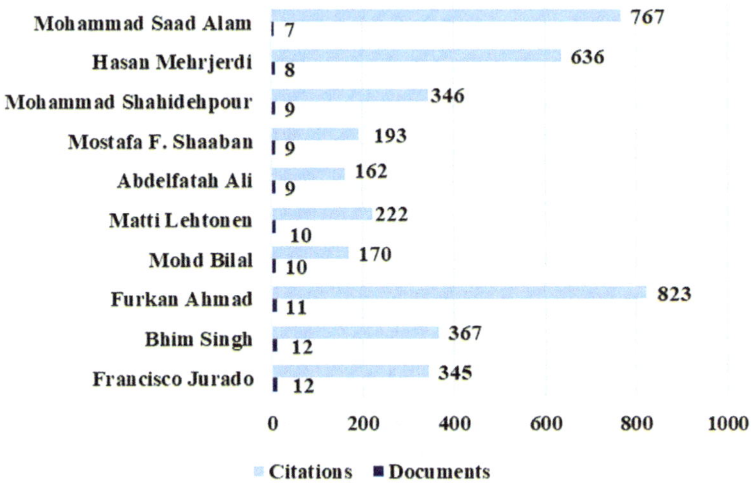

Figure 2.4 Publications and total citations of top authors.

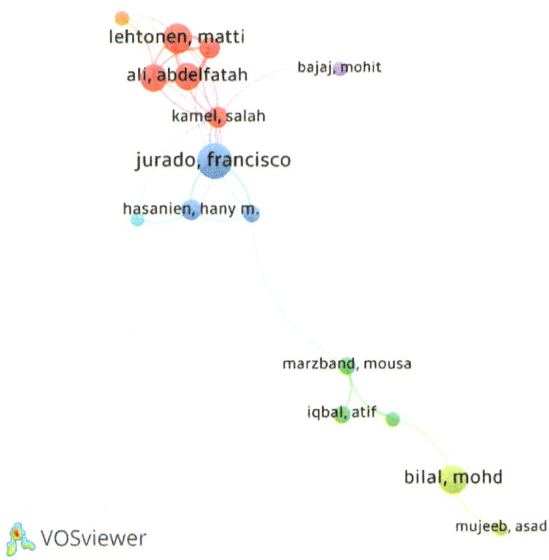

Figure 2.5 Author and co-author connections.

the most prolific in terms of research contributions. From a total of 4084 organizations, identified through author and co-author affiliations, only 10 institutions recorded five or more publications, highlighting a concentration of academic activity among a relatively small group. As depicted in Fig. 2.6, the top 10 institutions comprise three from both China and

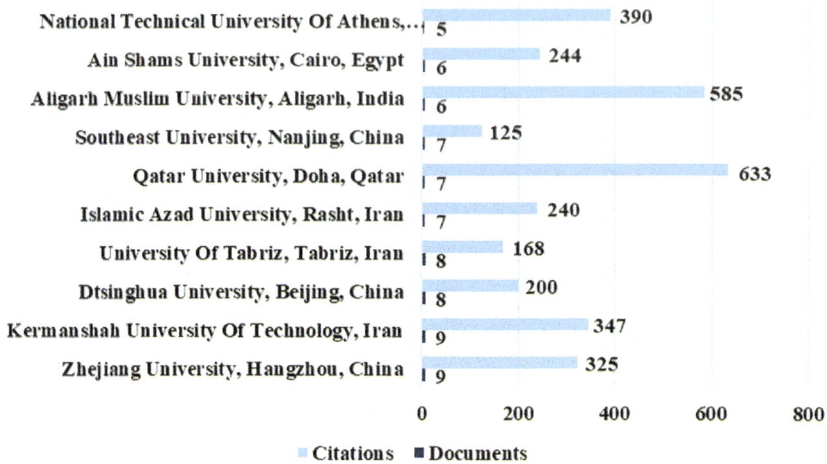

Figure 2.6 Top 10 institutes based on the number of publications.

Iran, and one each from India, Qatar, Egypt, and Greece, indicating a broad international presence yet a clustered pattern of scholarly productivity.

2.3.2 Analysis based on country origins

Fig. 2.7 depicts the country-based assessment of published works. India and China have the greatest number of publications. The authors of these two centuries published around 40% of all publications worldwide. Aside from China and India, the top five publishing countries include the United States, Saudi Arabia, Iran, and the United Kingdom. A percentage-based analysis with respect to total number of articles of dataset, reveals that India (22.47%), China (17.21%), the United States (9.42%), Saudi Arabia (5.06%), and Iran (4.83%) collectively account for more than 59% of the articles published on RECS technology between 2015 and 2025. This large number of articles highlights the growing significance of RECS technology in the nations mentioned.

The examination of data from 2015 to 2025 based on document citations showed that all 1843 documents were cited 72,083 occasions. More than 82% of all document citations come from the top 25 nations worldwide. Fig. 2.8 depicts the citations of papers from the top 20 nations from 2015 to 2025. In terms of country-based article citations, China's publications garnered the most citations, followed by the United States and India. Articles from China account for roughly 14%, articles from the United States for 12%, and articles from India for approximately 10% of total citations.

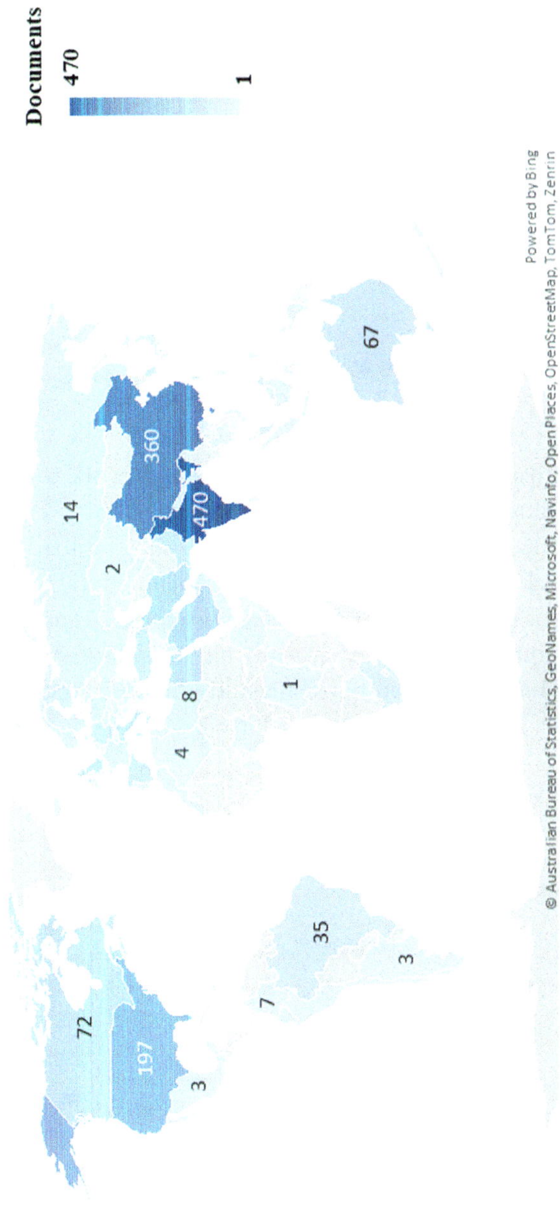

Figure 2.7 Number of articles from different countries.

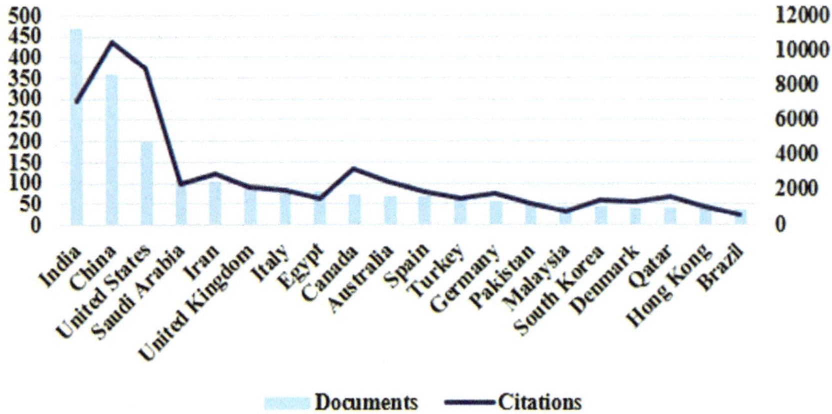

Figure 2.8 Top 25 country citations.

Aside from nation-specific publications, authors work together worldwide across multiple countries. The examination of co-authorship among authors from 106 countries from 2015 to 2025 showed that authors from 61 nations collaborated on more than five documents. Fig. 2.9 displays the clustering of papers and co-authors across these 61 nations. Fig. 2.8 shows nine unique groupings. After studying Fig. 2.8, it is clear that contributors from India, followed by China and United States, engage in the most significant amounts of co-authorship in works connected to RECS technology.

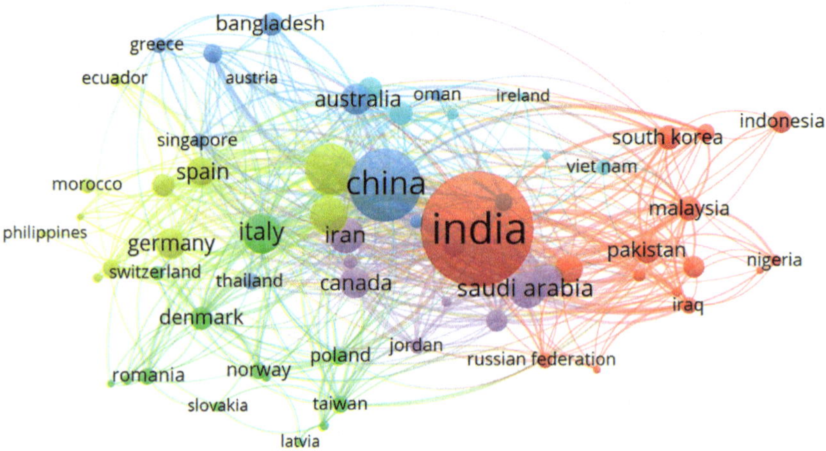

Figure 2.9 Country wise author and co-author relations network visualization.

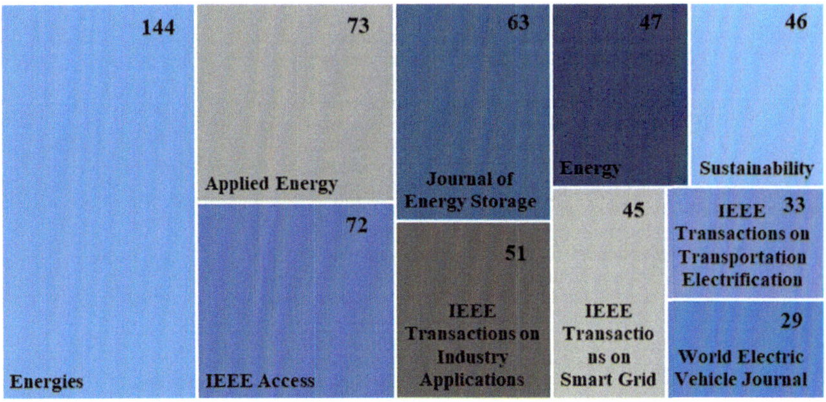

Figure 2.10 Number of articles from different journals.

2.3.3 Analysis of journal articles and their citations

This section examines the most significant journals, highly referenced journals, and clusters of journal citations. Fig. 2.10 shows the top 10 journals based on publishing activity. The analysis shows that the Energies journal has published the most articles in the RECS area. There have been 603 articles published from the top 10 sources, with more than 32% visible in leading energy-focused journals such as Energies, Applied Energy, IEEE Access, Journal of Energy Storage, and other journals such as Industry Applications, Energy, IEEE Transactions on Smart Grid, Sustainability, IEEE Transactions on Transportation Electrification, and World Electric Vehicle Journal. This demonstrates the enormous research potential in the RECS discipline.

Shows the citation numbers for the top 20 journals. According to the data, IEEE Transactions on Smart Grid has the most citations of any of the journals. Notably, IEEE-published publications account for eight of the top 25 cited sources. Further examination of journal-based citations reveals that just 59 of 391 sources have received more than 100 citations. Interestingly, more than 71% of these citations are concentrated in the top 25 sites, as seen in Fig. 2.11. This demonstrates the significant influence and scholarly impact of these top publications in the field of RECS technology.

2.3.4 Analysis of keywords

The keyword assessment tries to identify sub-areas and research trends in RECS technology by focusing on common phrases. The examination of

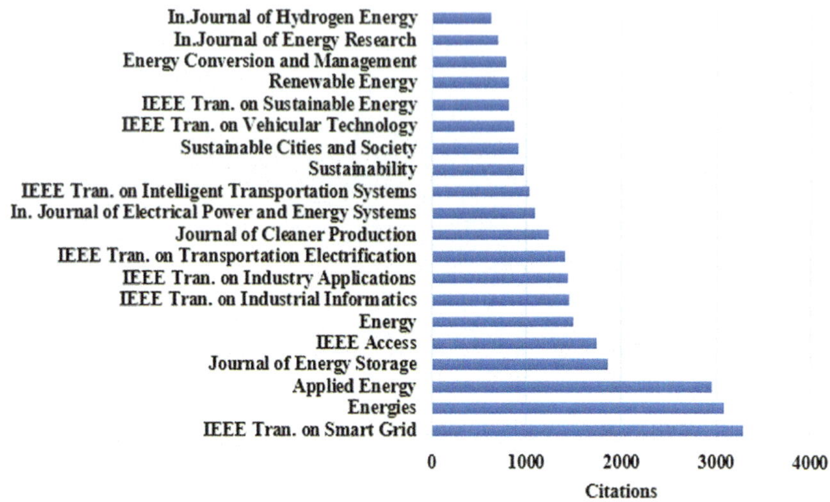

Figure 2.11 Top 25 journal citations.

authors' keywords in the retrieved documents revealed that only 102 keywords out of a total of 4468 keywords had at least 10 keyword occurrences. The major search terms "Renewable/Solar/Wind," "Electric vehicles/EV," and "Charging station" exist with varied frequencies with comparable search phrases like: "Electric vehicle/EV" happens 797 times, "Charging station" occurs 503 times, and "Renewable/solar/wind" appears 384 times. Furthermore, Fig. 2.12 shows that the most frequently used terms are

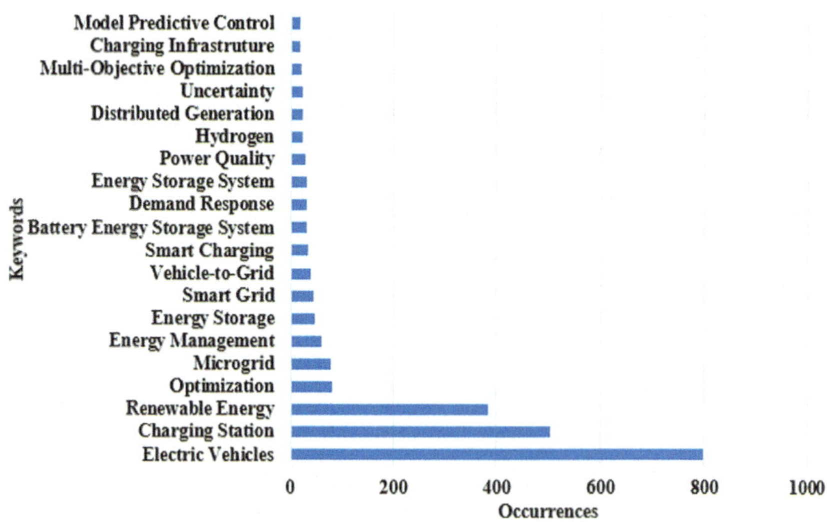

Figure 2.12 Occurrence of top 25 keyword.

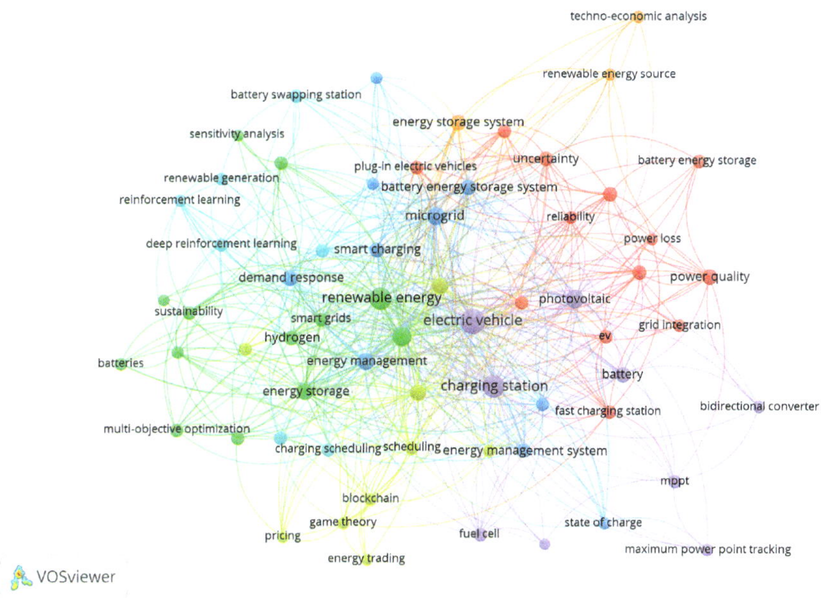

Figure 2.13 Keyword relation network visualization.

"optimization," "Microgrid," "Energy management," "Energy storage," "Smart grid," and "Vehicle to grid." These phrases refer to the major research hotspots in the RECS technology field.

Fig. 2.13 displays the clustering of keyword occurrences in associated papers, as retrieved from the SCOPUS dataset. The study finds seven unique clusters generated by the interactions between terms. As illustrated in Fig. 2.13, the word "Electric vehicle" has strong ties with "charging station," "battery," and "Photovoltaic." Furthermore, it is closely associated with keywords such as "bidirectional converter," "MPPT," and "fuel cell," suggesting the broad applicability of sub-related domains. Fig. 2.13 additionally demonstrates other essential features of RECS technology, highlighting its importance and interconnection in with other research fields.

2.4 Case studies

This section discusses and categorizes various RECS station case studies according to their analysis horizons. Based on the analysis aspect different case studies are categorized into following sections.

2.4.1 Based on the source and grid connection

Numerous case studies explore the integration of RESs and storages in EV charging infrastructure. Hossain et al. (2024) investigates the economic and environmental feasibility of a solar-powered EV charging station in Gazipur, Bangladesh, showcasing its potential to reduce emissions and meet local EV demand. Das (2024) assesses a solar-wind hybrid EV charging station for grid integration in coastal cities of India, focusing on performance in high-humidity coastal environments. Singh et al. (2024b) estimates energy generation from a solar-wind tower installed along Indian highways, emphasizing highway-based charging for long-distance transport. Salazar and Gómez (2024) details the design and implementation of a PV-based EV charging station at Universidad Tecnológica de Pereira, Colombia, contributing to campus sustainability. Peerlings et al. (2024) estimates PV potential along highways in the Netherlands, supporting highway corridor electrification. Nafeez et al. (2024) presents a solar PV parking infrastructure study at Ufa International Airport, Russia, assessing its suitability for airport EV support. Viswambaran et al. (2024) assesses residential solar-powered charging in the UAE, demonstrating applicability in hot climates. Kassem et al. (2023) studies the dual-use of solar energy for EV charging and residential building needs in Northern Cyprus. Costa and Cobas (2023) focuses on solar-PV charged EVs in Brazil, forecasting 2030 adoption scenarios. Ramanan et al. (2023) proposes solar PV and second-life battery systems for EV charging in India, promoting sustainability and recycling. Oladigbolu et al. (2023) optimizes renewable-based charging station design in Nigeria, addressing power reliability issues. Pai and Senjyu (2022) simulates park-and-ride systems using PV and battery storage in Fuxin, China, focusing on integrated transport-energy models. Podder et al. (2022) assesses a hybrid PV-biogas charging system for rickshaws and e-bikes in a developing nation, emphasizing grassroots-level applications. Wangsupphaphol and Chaitusaney (2020) explores levelized cost of electricity for PV-battery powered EV charging in Thailand, offering a financial comparison. Salameh et al. (2021) proposes a PV-based dynamic charging station for short-run vehicles in unspecified urban settings. Erdemir and Dincer (2020) assesses a flywheel-integrated EV bus fast-charging station in a renewable setup, enhancing power reliability. Zhou et al. (2020) applies GIS and MCDM to assess urban PV charging station siting in Beijing, combining location science with renewables. Calise et al. (2019) presents case studies from Naples and Salerno, Italy, integrating PVs, storage, and

EVs for urban mobility. Ahmad et al. (2019) evaluates a community microgrid-integrated EV station within the Indian power market, targeting local energy resilience.

According to the findings of this section, the vast majority of case studies focus on PV-based systems for RECS, as opposed to other RECS technologies. These studies focus on crucial issues such as local environmental circumstances, economic feasibility, and grid integration challenges, emphasizing solar PV's global applicability and scalability in a variety of geographic and climatic scenarios.

2.4.2 Based on the analysis criteria

Several case studies apply optimization, techno-economic analysis, and simulation-based assessments. Gnann et al. (2025) examines the impact of smart home charging on EV market share in Germany, integrating behavioral and economic modeling. Ref. 4 explores smart charging for a commercial EV fleet to reduce costs and maximize renewables in Germany. Tostado-Véliz et al. (2024) proposes best-case aware PV-battery planning for multimodal charging stations without a specific location. Međugorac et al. (2025) analyzes AC optimal power flow models for EV and PV hosting in Bosnia and Herzegovina, enhancing grid planning. Pastorelli et al. (2024) compares smart EV management and battery storage in Danish energy communities, emphasizing distributed control. Deshmukh et al. (2024) models EV integration and optimal charging under uncertainty in Qatar, considering planning resilience. Jirdehi and Tabar (2023) discusses long-term EV infrastructure planning under uncertainty using wind and solar in a real test case scenario, likely in Europe or China. Alsaidan et al. (2023) performs a novel evolutionary algorithm-based techno-economic analysis of EV charging in Qassim, Saudi Arabia. Alwesabi and Mohammed (2023) models self-sufficient solar EV adoption in New York State, USA, providing a decarbonization forecast. Zhao et al. (2023) uses GIS and MCDM to select EV station sites in China, merging spatial and decision-making models. Nazari-Heris et al. (2022) investigates mobile EV charging equity in Los Angeles, USA, focusing on transport justice. Syed Mohammed et al. (2022) assesses a hydrogen-based EV charging station feasibility, likely in a developing or urbanizing country. Al Wahedi and Bicer (2020) develops an optimal energy management strategy for a multi-renewable fast-charging station in Qatar, addressing power diversity. Bernal Heredia et al. (2020) evaluates V2B smart charging at the building level in a case study setup, improving demand response. Cazzato et al. (2020) introduces a new dispatching paradigm including EVs and renewables in a real power system test case. Han et al. (2018) assesses V2G-based peak shaving using idle stations on Jeju Island, South Korea, balancing local loads. Codani et al. (2015) surveys EV-RE coupling in France, synthesizing national-level insights. Menchaca Santos et al. (2025) develops business cases for V2G and degradation-aware charging for homes and fleets, location assumed to be Europe.

These case studies use thorough analysis, optimization, economic evaluation, and simulation techniques to cover multiple regions and investigate smart charging, grid unity, and energy management. Several studies take into account user behavior, location planning, and emphasizing an integrative approach to EV infrastructure development.

2.4.3 Based on design aspect analysis

This category includes physical layout, system architecture, and techno-design configurations. Zhang et al. (2024) explores synergies between variable renewables and battery swapping stations in Beijing, addressing urban electrification. Heidari et al. (2024) presents a hybrid EV station using INVELOX wind turbines in Kermanshah, Iran, focusing on unique wind capture. Fresia et al. (2024) discusses industrial microgrid optimization in a Northern Italy greentech firm, using community energy concepts. De Rosa et al. (2024) evaluates RE infrastructure in decarbonized urban districts in Spain, aligning with city design. Massana et al. (2024) introduces a multi-vector energy scheduler for Port of Borg, focusing on port-based electrification. Tikka et al. (2025) discusses public bidirectional EV smart charging (V2G) using architectural modeling. Sannigrahi (2024) analyzes a standalone hybrid system for power and water needs in an off-grid location, likely Africa or Middle East. Manaa Barhoumi (2023) studies solar-wind hybrid systems for hydrogen refueling in a rural or off-grid context. Chmielewski et al. (2023) investigates sustainable EV infrastructure development in Poland, with a national policy framework. Lv et al. (2022) proposes optimal energy control for a net-zero building in Hong Kong, integrating EV charging. Krishnan Nair et al. (2021) compares snubber circuit designs in PV-based off-grid EV stations, focusing on hardware improvement. Kumar et al. (2020) investigates solar PV tree design using multilayer cells in an urban environment, promoting space-efficient solar. Harighi et al. (2019) evaluates EV charging station siting using GIS in

Ankara, Turkey, focusing on urban spatial analysis. Agaton et al. (2019) compares diesel versus electric jeepney investments in the Philippines, using a real-options financial approach. Grzanic et al. (2019) introduces a distribution network platform via a case study, laying groundwork for broader EV-grid simulation.

These case studies highlight a variety of EV charging concepts, with a focus on system layout, hybrid renewables, and smart control. They address urban, rural, and off-grid situations, and use technologies like as V2G, battery swapping, and GIS-based siting to optimize architecture and sustainability.

2.5 Summary and future research aspects

The incorporation of RECS is a game-changing solution to decarbonizing the transportation industry while reducing reliance on fossil-fuel-based grid electricity. According to a study of existing research, solar PV is the most extensively used renewable power source for RECS owing to its capacity for expansion and low cost, which is frequently complemented with BSS to offset solar generation's intermittency. However, real-world deployment faces numerous technical, economic, and regulatory hurdles. Current research has made significant progress in developing appropriate sizing approaches for solar arrays and battery systems to successfully satisfy load needs while minimizing costs and increasing RE usage. Hybrid RECS configurations, which include biomass, wind, and hydrogen-based storage systems, have also sparked research attention, particularly in areas with a broad renewable resource availability (Yazdani et al., 2025; Güven et al., 2025; Das, 2024; Singh et al., 2024b; Podder et al., 2022; Heidari et al., 2024; Sannigrahi, 2024; Manaa Barhoumi, 2023). Energy management systems are the most studied topic in the area of RECS technology analysis; integrating real-time control, load forecasting, and prioritization algorithms are essential to ensuring operational effectiveness and charging reliability (Singh et al., 2024a; Fresia et al., 2024; Chmielewski et al., 2023). Different authors have studied optimal energy management system development. Recent material focuses on energy management (Sathyan et al., 2024; Darshi et al., 2023; Al Wahedi and Bicer, 2020; Massana et al., 2024), grid-independence analysis for off-grid RECS (Sannigrahi, 2024; Krishnan Nair et al., 2021), demand management strategies (Hou et al., 2023; Salameh et al., 2021; Jirdehi and Tabar, 2023), and lifetime techno-economic assessments (Sathyan et al., 2025; Colares et al.,

2023; Oladigbolu et al., 2023; Alsaidan et al., 2023; Syed Mohammed et al., 2022). Furthermore, the integration of RECS with smart grid frameworks and IoT-based monitoring systems is being investigated to improve dynamic control and predictive maintenance capabilities (Aghapour et al., 2020). Despite these advances, future research must fill various gaps. These include standardized RECS structures, effective regulations for grid-connected versus freestanding systems, extreme weather resilience, and expandable frameworks suited to both urban and rural settings. Additionally, techno-economic models must account for changing battery chemicals, local renewable supply, seasonal demand changes, and the possible involvement of second-life EV batteries. To promote widespread adoption of RECS, additional interdisciplinary research involving electrical infrastructure, green energy, transportation planning, and policy formulation is required to provide sustainable, user-friendly, economically feasible solutions.

2.6 Conclusions

This chapter presents an overview of Renewable Energy-based Charging Systems (RECS), encompassing both the underlying technologies and the current state of research. The sharp rise in publications over recent years reflects the growing global focus on this domain. Drawing on insights from bibliometric analysis and case studies, the following key conclusions are outlined.

- The analysis of article publications shows an average annual growth of around 5%, reflecting sustained research expansion in this domain.
- India, China, and the USA emerge as major research hubs in RECS technology, in accordance with the highest number of publications and citations.
- Among top publishers, MDPI's Energies journal stands out in terms of publication volume, while IEEE Transactions on Smart Grid receives the highest citations, highlighting the impact of engineering-focused research in this area.
- Keyword analysis reveals that "RE-based EV charging stations" is closely associated with terms like "Energy storage," "Microgrid," "Demand response," "Smart charging," and "Power quality," underscoring the interdisciplinary nature and broad applicability of RECS.

- The resource type based case studies emphasize the prominence of solar-powered EV charging systems, driven by their superior scalability, economic attractiveness, and flexibility to perform effectively across varied environmental and geographical settings.
- The planning-criteria-based case studies highlight that effective EV infrastructure development increasingly depends on advanced optimization techniques, techno-economic evaluations, and GIS-based spatial planning to enhance system reliability, economic efficiency, and seamless grid integration.
- The design-focused case studies strategically integrate multi-source energy solutions, sophisticated control systems, and innovative configurations such as V2G and battery swapping, leading to enhanced efficiency, greater adaptability, and improved operational resilience.

Overall, the development of RECS is advancing steadily toward enabling sustainable and decentralized mobility solutions. While the foundational research is strong, future efforts must address practical challenges such as intermittency, standardization, cost optimization, and real-time energy management. Through continued interdisciplinary collaboration, RECS can play a pivotal role in shaping cleaner, more resilient transportation and energy systems globally.

References

Abdullaev I, Lin N, Rashidov J. Electric vehicles: manuscript of a bibliometric analysis unveiling trends, innovations and future pathways. Int J Automot Sci Technol 2024;8(2):212–24. https://doi.org/10.30939/ijastech.1424879.

Agaton CB, Guno CS, Villanueva RO, Villanueva RO. Diesel or electric jeepney? A case study of transport investment in the Philippines using the real options approach. World Electr Vehicle J 2019;10(3). https://doi.org/10.3390/wevj10030051. Available from: https://res.mdpi.com/d_attachment/wevj/wevj-10-00051/article_deploy/wevj-10-00051.pdf.

Aghapour R, Sepasian MS, Arasteh H, Vahidinasab V, Catalão JPS. Probabilistic planning of electric vehicles charging stations in an integrated electricity-transport system. Elec Power Syst Res 2020:189. https://doi.org/10.1016/j.epsr.2020.106698. Available from: https://www.journals.elsevier.com/electric-power-systems-research.

Aghapour R, Zeraati M, Jabari F, Sheibani M, Arasteh H. Cybersecurity and data privacy issues of electric vehicles smart charging in smart microgrids. In: Electric vehicle integration via smart charging. Springer International Publishing; 2022. p. 85–110. https://doi.org/10.1007/978-3-031-05909-4_4.

Ahmad F, Alam MS, Shariff SM, Krishnamurthy M. A cost-efficient approach to EV charging station integrated community microgrid: a case study of indian power market. IEEE Trans Transp Electrif 2019;5(1):200–14. https://doi.org/10.1109/TTE.2019.2893766.

Al Wahedi A, Bicer Y. A case study in Qatar for optimal energy management of an autonomous electric vehicle fast charging station with multiple renewable energy and storage systems. Energies 2020;13(19):5095. https://doi.org/10.3390/en13195095.

Alkawsi G, Baashar Y, Abbas U D, Ammar AA, Tiong SK. Review of renewable energy-based charging infrastructure for electric vehicles. Appl Sci 2021;11(9):3847. https://doi.org/10.3390/app11093847.

Alsaidan I, Bilal M, Alaraj M, Rizwan M, Almasoudi FM. A novel EA-based techno–economic analysis of charging system for electric vehicles: a case study of Qassim region, Saudi Arabia. Mathematics 2023;11(9):2052. https://doi.org/10.3390/math11092052.

Alwesabi Y, Mohammed N. Self-sufficient solar power and electric vehicle penetration: a case study of New York State. Renew Energy Focus 2023;45:133–40. https://doi.org/10.1016/j.ref.2023.03.001.

Ayodele BV, Mustapa SI. Life cycle cost assessment of electric vehicles: a review and bibliometric analysis. Sustainability 2020;12(6):2387. https://doi.org/10.3390/su12062387.

Barbosa W, Prado T, Batista C, Câmara JC, Cerqueira R, Coelho R, et al. Electric vehicles: bibliometric analysis of the current state of the art and perspectives. Energies 2022;15(2):395. https://doi.org/10.3390/en15020395.

Barman P, Dutta L, Bordoloi S, Kalita A, Buragohain P, Bharali S, et al. Renewable energy integration with electric vehicle technology: a review of the existing smart charging approaches. Renew Sustain Energy Rev 2023;183:113518. https://doi.org/10.1016/j.rser.2023.113518.

Bernal Heredia W, Chaudhari K, Meintz A, Jun M, Pless S. Evaluation of smart charging for electric vehicle-to-building integration: a case study. Appl Energy 2020;266:114803. https://doi.org/10.1016/j.apenergy.2020.114803.

Calise F, Cappiello FL, Cartenì A, Dentice d'Accadia M, Vicidomini M. A novel paradigm for a sustainable mobility based on electric vehicles, photovoltaic panels and electric energy storage systems: case studies for Naples and Salerno (Italy). Renew Sustain Energy Rev 2019;111:97–114. https://doi.org/10.1016/j.rser.2019.05.022.

Cazzato F, Di Clerico M, Falvo MC, Ferrero S, Vivian M. New dispatching paradigm in power systems including EV charging stations and dispersed generation: a real test case. Energies 2020;13(4):944. https://doi.org/10.3390/en13040944.

Chmielewski A, Piórkowski P, Możaryn J, Ozana S. Sustainable development of operational infrastructure for electric vehicles: a case study for Poland. Energies 2023;16(11):4528. https://doi.org/10.3390/en16114528.

Codani P, Portz P-L, Claverie P, Perez Y, Petit M. RETRACTED: coupling local renewable energy production with electric vehicle charging: a survey of the French case. World Electr Vehic J 2015;7(4):489–99. https://doi.org/10.3390/wevj7040489.

Colares JC De S, Neves HDe C, Souza JLDe, Piana BB, Costa Souza E. Global economic outlook: scenario analysis for 2023 and tendencies. Int J Bus Adm 2023;14(2):11. https://doi.org/10.5430/ijba.v14n2p11.

Costa Dd, Cobas VRM. Electric vehicles charged with solar-PV: a Brazilian case study for 2030. Vehicles 2023;5(4):1743–59. https://doi.org/10.3390/vehicles5040095.

Darshi R, Shamaghdari S, Jalali A, Arasteh H. Decentralized energy management system for smart microgrids using reinforcement learning. IET Gener Transm Distrib 2023;17(9):2142–55. https://doi.org/10.1049/gtd2.12796.

Das TK. Assessment of grid-integrated electric vehicle charging station based on solar-wind hybrid: a case study of coastal cities. Alex Eng J 2024;103:288–312. https://doi.org/10.1016/j.aej.2024.05.103.

Deshmukh S, Tariq H, Amir M, Iqbal A, Marzband M, Al-Wahedi AMAB. Impact assessment of electric vehicles integration and optimal charging schemes under uncertainty: a case study of Qatar. IEEE Access 2024;12:131350–71. https://doi.org/10.1109/ACCESS.2024.3458410.

Dimitriou P, Gkiotsalitis K. The vehicle charging problem with renewable energy sources in freight transport: model and application. Appl Energy 2025;392:125945. https://doi.org/10.1016/j.apenergy.2025.125945.

Erdemir D, Dincer I. Assessment of renewable energy-driven and flywheel integrated fast-charging station for electric buses: a case study. J Energy Storage 2020;30:101576. https://doi.org/10.1016/j.est.2020.101576.

Fresia M, Robbiano T, Caliano M, Delfino F, Bracco S. Optimal operation of an industrial microgrid within a renewable energy community: a case study of a greentech company. Energies 2024;17(14):3567. https://doi.org/10.3390/en17143567.

Gbey E, Turkson RF, Lee S. A bibliometric survey of research output on wireless charging for electric vehicles. World Electr Vehicle J 2022;13(2). https://doi.org/10.3390/wevj13020037.

Gharbaoui M, Martini B, Bruno R, Valcarenghi L, Conti M, Castoldi P. Policies for efficient usage of an EV charging infrastructure deployed in city parking facilities. In: 13th International Conference on ITS Telecommunications, ITST 2013. Italy: IEEE Computer Society; 2013. p. 384–9. https://doi.org/10.1109/ITST.2013.6685577.

Gnann T, Yu S, Stute J, Kühnbach M. The value of smart charging at home and its impact on EV market shares – a German case study. Appl Energy 2025;380:124997. https://doi.org/10.1016/j.apenergy.2024.124997.

Grzanic M, Giacomo Flammini M, Prettico G. Distribution network model platform: a first case study. Energies 2019;12(21):4079. https://doi.org/10.3390/en12214079.

Güven AF, Ateş N, Alotaibi S, Alzahrani T, Amsal AM, Elsayed SK. Sustainable hybrid systems for electric vehicle charging infrastructures in regional applications. Scient Rep 2025;15(1). https://doi.org/10.1038/s41598-025-87985-7.

Han HS, Oh E, Son SY. Study on EV charging peak reduction with V2G utilizing idle charging stations: the Jeju Island case. Energies 2018;11(7). https://doi.org/10.3390/en11071651.

Harighi T, Padmanaban S, Bayindir R, Hossain E, Holm-Nielsen JB. Electric vehicle charge stations location analysis and determination—Ankara (Turkey) case study. Energies 2019;12(18):3472. https://doi.org/10.3390/en12183472.

Heidari R, Ahmadi Jirdehi M, Shaterabadi M. Hybrid energy-based electric vehicles charging station integrated with INVELOX wind turbine: a case study of Kermanshah. Energy 2024;308:132833. https://doi.org/10.1016/j.energy.2024.132833.

Hildebrand N, Kummer S. Systematic literature review of urban charging infrastructure planning over time. Clean Energy Sys 2024;8:100123. https://doi.org/10.1016/j.cles.2024.100123.

Hossain MN, Hasan MM, Habib MA, Ferdous FB, Rahman S. Assessing economic viability and environmental impact of solar-powered EV charging station in Gazipur, Bangladesh: a case study. Solar Compass 2024;12. https://doi.org/10.1016/j.solcom.2024.100091.

Hou H, Wang Z, Hou T, Fang R, Tang J, Xie C. Optimal schedule of 100% renewable energy microgrid considering demand response of EVs. Energy Rep 2023;9:1743–50. https://doi.org/10.1016/j.egyr.2023.04.223.

Ibrahim RA, Gaber IM, Zakzouk NE. Analysis of multidimensional impacts of electric vehicles penetration in distribution networks. Scientif Rep 2024;14(1). https://doi.org/10.1038/s41598-024-77662-6.

Irham A, Roslan MF, Jern KP, Hannan MA, Mahlia TMI. Hydrogen energy storage integrated grid: a bibliometric analysis for sustainable energy production. Int J Hydrogen Energy 2024;63:1044–87. https://doi.org/10.1016/j.ijhydene.2024.03.235.

Jirdehi MA, Tabar VS. Multi-objective long-term expansion planning of electric vehicle infrastructures integrated with wind and solar units under the uncertain environment

considering demand side management: a real test case. Sustain Cities Soc 2023;96. https://doi.org/10.1016/j.scs.2023.104632.
Kassem Y, Gokcekus H, Aljatlawe A. Utilization of solar energy for electric vehicle charging and the energy consumption of residential buildings in Northern Cyprus: a case study. Eng Technol Appl Sci Res 2023;13(5):11598–607. https://doi.org/10.48084/etasr.6142.
Krishnan Nair D, Prasad K, Tjing Lie T. Implementation of snubber circuits in a PV-based off-grid electric vehicle charging station—comparative case studies. Energies 2021;14(18):5853. https://doi.org/10.3390/en14185853.
Khazali A, Al-Wreikat Y, Fraser EJ, Naderi M, Smith MJ, Sharkh SM, et al. Sizing a renewable-based microgrid to supply an electric vehicle charging station: a design and modelling approach. World Electr Vehicle J 2024;15(8). https://doi.org/10.3390/wevj15080363.
Kumar NM, Chopra SS, Malvoni M, Elavarasan RM, Das N. Solar cell technology selection for a pv leaf based on energy and sustainability indicators-a case of a multilayered solar photovoltaic tree. Energies 2020;13(23). https://doi.org/10.3390/en13236439.
de Lima TD, Franco JF, Lezama F, Soares J, Vale Z. Joint optimal allocation of electric vehicle charging stations and renewable energy sources including CO2 emissions. Energy Inform 2021;4. https://doi.org/10.1186/s42162-021-00157-5.
Lv T, Lu Y, Zhou Y, Liu X, Wang C, Zhang Y, et al. Optimal control of energy systems in net-zero energy buildings considering dynamic costs: a case study of zero carbon building in Hong Kong. Sustainability 2022;14(6):3136. https://doi.org/10.3390/su14063136.
Manaa Barhoumi El. Optimal design of standalone hybrid solar-wind energy systems for hydrogen-refueling station case study. J Energy Storage 2023;74:109546. https://doi.org/10.1016/j.est.2023.109546.
Massana J, Burgas L, Colomer J, Sumper A, Herraiz S. Optimal scheduler for energy consumption reduction in multi-vector energy management systems: a case study at the Port of Borg. Heliyon 2024;10(10):e31419. https://doi.org/10.1016/j.heliyon.2024.e31419.
Međugorac M, Capuder T, Bago D, Antić T, Sušac S. The value of AC optimal power flow models in the assessment of distribution grid electric vehicle and photovoltaic power plant hosting capacity – case study in Bosnia and Herzegovina. IET Confer Proc 2025;2024(5):814–7. https://doi.org/10.1049/icp.2024.2011.
Menchaca Santos D, Thüne P, Zepter JM, Marinelli M. Business cases for degradation-aware bidirectional charging of residential users and heavy-duty vehicle fleets. eTransportation 2025;23:100389. https://doi.org/10.1016/j.etran.2024.100389.
Nafeez RM, Airatovich RR, Vladislav KSR, Vavilov V. Designing of solar photovoltaic parking infrastructure at Ufa International Airport- A case study on EV charging potential. In: International Conference on Renewable Energy and Power Engineering, REPE. Institute of Electrical and Electronics Engineers Inc. Russian Federation; 2024. p. 164–8. https://doi.org/10.1109/REPE62578.2024.10809458. https://ieeexplore.ieee.org/xpl/conhome/10808179/proceeding.
Nazari-Heris M, Loni A, Asadi S, Mohammadi-ivatloo B. Toward social equity access and mobile charging stations for electric vehicles: a case study in Los Angeles. Appl Energy 2022;311:118704. https://doi.org/10.1016/j.apenergy.2022.118704.
van Nunen K, Li J, Reniers G, Ponnet K. Bibliometric analysis of safety culture research. Saf Sci 2018;108:248–58. https://doi.org/10.1016/j.ssci.2017.08.011.
Odoi-Yorke F, Opoku R, Davis F, Obeng GY. Employing bibliometric analysis to identify the trends, evolution, and future research directions of sand-based thermal energy storage systems. J Energy Storage 2024;94:112343. https://doi.org/10.1016/j.est.2024.112343.

Oladigbolu JO, Mujeeb A, Imam AA, Rushdi AM. Design, technical and economic optimization of renewable energy-based electric vehicle charging stations in Africa: the case of Nigeria. Energies 2023;16(1). https://doi.org/10.3390/en16010397.

Page MJ, McKenzie JE, Bossuyt PM, Boutron I, Hoffmann TC, Mulrow CD, et al. The PRISMA 2020 statement: an updated guideline for reporting systematic reviews. J Clin Epidemiol 2021;134:178–89. https://doi.org/10.1016/j.jclinepi.2021.03.001.

Pai L, Senjyu T. A yearly based multiobjective park-and-ride control approach simulation using photovoltaic and battery energy storage systems: Fuxin, China case study. Sustainability 2022;14(14):8655. https://doi.org/10.3390/su14148655.

Pastorelli F, Unterluggauer T, Höyer BJ, Wagner MM, Secchi M, Marinelli M. Smart electric vehicle management vs. battery storage for energy communities: a case study from Denmark. IET Confer Proc 2024;2024(15):239–46. https://doi.org/10.1049/icp.2024.3757. https://digital-library.theiet.org/journals/conference-proceedings.

Peerlings J, Reinders A, Catita C, Brito MC. The photovoltaic potential for electric vehicle charging along highways: a Dutch case study. Progr Photovolt Res Appl 2024;32(4):244–52. https://doi.org/10.1002/pip.3759.

Podder AK, Supti SA, Islam S, Malvoni M, Jayakumar A, Deb S, et al. Feasibility assessment of hybrid solar photovoltaic-biogas generator based charging station: a case of easy bike and auto rickshaw scenario in a developing nation. Sustainability 2022;14(1):166. https://doi.org/10.3390/su14010166.

Purwanto E, Nugraha H, Uddin N, Exploring EV. Charging time optimization: a bibliometric analysis. Int J Trans Develop Integr 2024;8(2):301–14. https://doi.org/10.18280/ijtdi.080208.

Purwanto E, Irawan AP. Bibliometric analysis of electric vehicle adoption research: trends, implications, and future directions. Int J Safety Security Eng 2023;13(5):789–800. https://doi.org/10.18280/ijsse.130503.

Ramanan A, Sekhar M, Mehra S. Solar PV and second life batteries powered EV charging station: case study for India. IET Confer Proc 2023;2023(18):153–63. https://doi.org/10.1049/icp.2023.2698.

De Rosa L, Martínez M, Linares JI, Mateo C, Gomez T, Cossent R, et al. Design and assessment of energy infrastructure in new decarbonized urban districts: a Spanish case study. Energy Rep 2024;11:4631–41. https://doi.org/10.1016/j.egyr.2024.04.037.

Rethlefsen ML, Kirtley S, Waffenschmidt S, Ayala AP, Moher D, Page MJ, et al. PRISMA-S: an extension to the PRISMA statement for reporting literature searches in systematic reviews. J Med Libr Assoc 2021;109(2):174–200. https://doi.org/10.5195/jmla.2021.962.

Salameh M, Khatib T, Alsahili K, Justin Thomas KR. A novel design of photovoltaic-based charging station for battery vehicles with dynamic demand: a case of short runs. Int J Photoenergy 2021;2021:1–17. https://doi.org/10.1155/2021/6641548.

Salazar E, Gómez O. Design and implementation of photovoltaic electric vehicle charging stations: a case study at Universidad Tecnológica de Pereira. Renew Energy Power Quality J 2024;22(1):1–6. https://doi.org/10.52152/3909.

Sannigrahi S. Design and feasibility analysis of an off-grid hybrid energy system to fulfill electricity and freshwater demand: a case study. Energy Sources Part A Recovery Util Environ Eff 2024;46(1):5925–50. https://doi.org/10.1080/15567036.2024.2342988.

Sarkis-Onofre R, Catalá-López F, Aromataris E, Lockwood C. How to properly use the PRISMA Statement. Syst Rev 2021;10(1). https://doi.org/10.1186/s13643-021-01671-z.

Sathyan S, Pandi VR, Antony A, Salkuti SR, Sreekumar P. ANN-based energy management system for PV-powered EV charging station with battery backup and vehicle to grid support. Int J Green Energy 2024;21(6):1279–94. https://doi.org/10.1080/15435075.2023.2246048.

Sathyan S, Ravikumar Pandi V, Sreekumar P, Thakkar N, Reddy Salkuti S. A hybrid multi-criteria decision-making and multi-objective framework for optimal sizing of PV-powered EV charging station with battery storage. IEEE Access 2025;13:129977–94. https://doi.org/10.1109/access.2025.3590521.

Shekari M, Arasteh H, Vahidinasab V. Recognition of electric vehicles charging patterns with machine learning techniques. In: Electric vehicle integration via smart charging. Springer International Publishing; 2022. p. 49–83. https://doi.org/10.1007/978-3-031-05909-4_3.

Singh G, Misra SC, Daultani Y, Singh S. Electric vehicle adoption and sustainability: insights from the bibliometric analysis, cluster analysis, and morphology analysis. Operat Manag Res 2024a;17(2):635–59. https://doi.org/10.1007/s12063-024-00442-y.

Singh SP, Tiwari P, Singh SN, Singh PP. Estimation of hybrid energy generation of solar-wind tower for electric vehicle charging: a case study of Indian highway. Energy Storage 2024b;6(5). https://doi.org/10.1002/est2.70004.

Swarnkar R, Harikrishnan R, Singh M. Analysis of electric vehicle battery state estimation using scopus and web of science databases from 2000 to 2021: a bibliometric study. World Electr Vehic J 2022;13(8):157. https://doi.org/10.3390/wevj13080157.

Syed Mohammed A, Anuj, Lodhi AS, Murtaza Q. Techno-economic feasibility of hydrogen based electric vehicle charging station: a case study. Int J Energy Res 2022;46(10): 14145–60. https://doi.org/10.1002/er.8132.

Tahir H. Optimization of energy storage systems for integration of renewable energy sources — a bibliometric analysis. J Energy Storage 2024;94:112497. https://doi.org/10.1016/j.est.2024.112497.

Thakkar N, Paliwal P. Quad-level MCDM framework to analyse technology combinations for sustainable micro-grid planning in uncertainty domain. Arab J Sci Eng 2023a;48(5): 5829–58. https://doi.org/10.1007/s13369-022-07019-6.

Thakkar N, Paliwal P. Data driven MCDM models for reliability-economic-environmental analysis of energy storage based autonomous micro-grid. J Energy Storage 2024;81: 110408. https://doi.org/10.1016/j.est.2023.110408.

Thakkar N, Paliwal P. Multi-criteria valuation for sustainable autonomous microgrid planning: a comparative analysis of technology mix with different MADM techniques. Iran J Sci Technol Trans Electr Eng 2023b;47(4):1283–306. https://doi.org/10.1007/s40998-023-00615-8.

Tikka V, Mendes G, Aghamohammadloo H, Sorvisto A, Ziu D, D'Ostilio P, et al. Bidirectional electric vehicle smart charging in public charging stations (V2G): use case representation using the e-mobility systems architecture framework. IET Confer Proc 2025; 2024(5):867–70. https://doi.org/10.1049/icp.2024.1999.

Tostado-Véliz M, Jordehi AR, Zhou Y, Amir Mansouri S, Jurado F. Best-case-aware planning of photovoltaic-battery systems for multi-mode charging stations. Renew Energy 2024;225:120300. https://doi.org/10.1016/j.renene.2024.120300.

Varghese AM, Paul J, Pradhan RP. Research on electric vehicle mobility: a bibliometric review and research agenda. In: Transportation Research Procedia. India: Elsevier B.V.; 2025. p. 1120–37. https://doi.org/10.1016/j.trpro.2024.12.116. www.sciencedirect.com/science/journal/2352146582.

Veza I, Syaifuddin M, Idris M, Herawan SG, Yusuf AA, Fattah IMR. Electric vehicle (EV) review: bibliometric analysis of electric vehicle trend, policy, lithium-ion battery, battery management, charging infrastructure, smart charging, and electric vehicle-to-everything (V2X). Energies 2024;17(15). https://doi.org/10.3390/en17153786.

Viswambaran V, Bati A, Siddiqui E, Mohammad A. Solar powered electric vehicle charging for residential parking lots in the UAE: a case study. In: Advances in Science and Engineering Technology International Conferences, ASET. United Arab Emirates: Institute of Electrical and Electronics Engineers Inc.; 2024. https://doi.org/10.1109/

ASET60340.2024.10708750. https://ieeexplore.ieee.org/xpl/conhome/10708057/proceeding.

Wang L, Qi Z, Liu J, Wang G. Science mapping the knowledge domain of electrochemical energy storage technology: a bibliometric review. J Energy Storage 2024;77:109819. https://doi.org/10.1016/j.est.2023.109819.

Wangsupphaphol A, Chaitusaney S. A simple levelized cost of electricity for EV charging with PV and battery energy storage system: Thailand case study. Int J Power Electron Drive Syst 2020;11(4):2223. https://doi.org/10.11591/ijpeds.v11.i4.pp2223-2230.

Yao S, Bian Z, Hasan MK, Ding R, Li S, Wang Y, et al. A bibliometric review on electric vehicle (EV) energy efficiency and emission effect research. Environm Sci Pollut Res 2023;30(42):95172–96. https://doi.org/10.1007/s11356-023-29143-y.

Yazdani L, Sepasian MS, Arasteh HR. Optimal comprehensive planning for AC/DC hybrid distribution networks: balancing cost, risk, and flexibility with variable RES installations. Elec Power Syst Res 2025:246. https://doi.org/10.1016/j.epsr.2025.111703.

Zhang C, Lu X, Chen S, Shi M, Sun Y, Wang S, et al. Synergies of variable renewable energy and electric vehicle battery swapping stations: case study for Beijing. eTransportation 2024;22:100363. https://doi.org/10.1016/j.etran.2024.100363.

Zhao H, Gao J, Cheng X. Electric vehicle solar charging station siting study based on GIS and multi-criteria decision-making: a case study of China. Sustainability 2023;15(14): 10967. https://doi.org/10.3390/su151410967.

Zhou J, Wu Y, Wu C, He F, Zhang B, Liu F. A geographical information system based multi-criteria decision-making approach for location analysis and evaluation of urban photovoltaic charging station: a case study in Beijing. Energy Convers Manag 2020; 205:112340. https://doi.org/10.1016/j.enconman.2019.112340.

CHAPTER THREE

The effects of electric vehicles integration on power quality: A review of challenges, solutions, and opportunities

Abbas Babaei Birag, Khalil Moradi Sani, Ahmad Salemnia
Department of Electrical Engineering, Shahid Beheshti University, Tehran, Iran

3.1 Introduction

In response to global concerns about reducing greenhouse gas emissions from the transportation sector, which has created 22% of the total production of these gases, the adoption of EVs is increasing at a remarkable rate and replacing internal combustion vehicles. Due to this increasing growth, in 2023, about 14 million EVs were sold, and in 2024, this number passed 17 million. And with this increasing trend, it is expected that by the end of 2025, this amount will go above 20 million, and it is also predicted that by 2030, this amount will reach 30 to 40 million sales (Global EV Outlook, 2025). However, while this electrification is vital for environmental and economic goals, connecting this huge number of EVs to the power grid causes big challenges for its power quality and performance, in which power quality emerges as a primary concern.

While this integration represents an increase in total load for the entire power system, the most significant technical challenges, such as those related to power quality, emerge at the distribution network level, where these new loads directly connect. This chapter will therefore focus its review on these distribution-level impacts.

Power quality is a measure for assessing the health and condition of the grid's electrical supply, and it refers to the ability of the power system to supply a clean and continuous flow of electricity while maintaining voltage and current waveforms as pure sinusoids with a stable frequency (Srivastava et al., 2023). Since this measure is affected by the increase in EVs, it is necessary to verify that it remains within the permissible limits of the relevant standards.

Renewable-to-Vehicle Smart Charging
ISBN: 978-0-443-36684-0
https://doi.org/10.1016/B978-0-443-36684-0.00012-6

The grid faces substantial technical difficulties and distinctive advantages when EVs become integrated into its operations. EVs need to transform Alternating current (AC) to Direct current (DC) through electronic converters, which generate harmonic distortion in grid power quality, and simultaneous vehicle charging leads to sudden voltage fluctuations and grid overloads. Combination of these variables threatens the stability of the entire grid and its reliable operation. EVs bring substantial capabilities to enhance how the grid performs despite their present challenges. The vehicles function beyond their conventional consumer role because they operate as mobile storage units that distribute energy across different locations. V2G technology allows electricity to flow in both directions between vehicles and the grid. The battery storage capability of EVs enables them to give back battery power to the grid when the demand peaks, so that voltage and frequency levels stay stable and the grid experiences less stress. Through the use of intelligent charging systems, the process moves automatically to the times when renewable energy production peaks or when electricity demand is low (Elazim et al., 2025). The strategy prevents new consumption peaks and maximizes grid capacity utilization.

In this chapter, the main objective is to provide a comprehensive review of the interaction between electric vehicles and the power quality of the distribution grid. The chapter then examines the dual impacts, first explaining the negative impacts such as harmonic distortion, voltage fluctuations, and phase imbalance, while later examining the positive potential of supporting the grid through V2G technology. To address these negative impacts, this review evaluates hardware-based and control-based mitigation strategies as well as the technical and economic frameworks required to maximize the positive impacts of V2G, such as aggregator-based control. Finally, the chapter concludes with a conclusion, summarizing the findings and offering suggestions for future research.

3.2 Factors influencing power quality

After reviewing the generalities in the introduction, this section looks in-depth at the technical factors that determine how much EVs affect power quality. Understanding these factors is a prerequisite for a detailed analysis of the positive and negative impacts, which we will discuss in the next section. This section addresses four key topics in order: First, we examine different charging infrastructures and technologies. Then, we analyze the impact of the penetration level or the number of vehicles in an area. Next, we look at charging patterns, and finally, we describe the role of the distribution network characteristics as a determining factor.

3.2.1 Charging infrastructure and technology

Charging infrastructure and batteries are two key components of the EV ecosystem. EV battery chargers can be categorized based on several key criteria. One of the main categories is based on the location of the charging equipment, which is divided into two categories: on-board and off-board chargers. In addition, charger designs also differ depending on power flow direction, including unidirectional and bidirectional designs. Unidirectional chargers have been the initial approach due to hardware simplicity and battery health considerations (Yilmaz and Krein, 2013). The ability of bidirectional chargers to return energy to the grid makes them crucial for grid services. EV charger architectures can also be categorized by power level, energy source, and connector type, each of which has its own features. In the following, we will classify and analyze this infrastructure from various aspects.

3.2.1.1 Classification of electrical vehicles

There are several types of EVs depending on the power source and powertrain they utilize. All of them desire to reduce fossil fuel dependence, but their connection to the grid is different (Dik et al., 2022). In order to analyze the impact of these differences on power quality, it is vital to understand these differences. There are four major categories of EVs that Fig. 3.1 illustrates this classification graphically.

3.2.1.1.1 Hybrid electric vehicles

These vehicles are the first generation of electrified vehicles with both an internal combustion engine (ICE) and an electric motor. The key point about Hybrid Electric Vehicles (HEV) is that their batteries have a small capacity and are not capable of being connected to the grid and charged externally. The batteries are charged only through regenerative braking and gasoline engines (Hybrid Electric Vehicles). Since HEVs are not powered directly from the grid, they do not have a significant impact on power quality. Therefore, they are not discussed further in this chapter.

3.2.1.1.2 Plug-in hybrid electric vehicle

A plug-in hybrid Electric vehicle (PHEV) is a vehicle that combines an ICE with an electric motor, along with a large battery to produce electricity

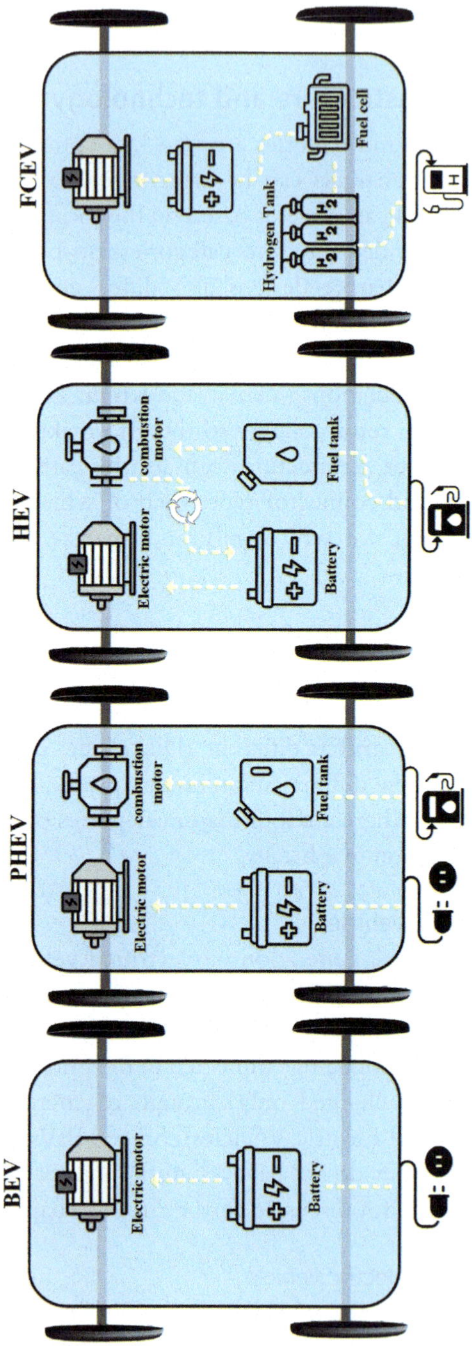

Figure 3.1 Comparison of the main features of each type of EV.

(Singh et al., 2020). As PHEVs are able to charge their batteries from an external source, they have increased range due to the fact that they can operate entirely on electric power. This direct interaction with the grid through charging makes PHEVs, along with all EVs, one of the main topics of power quality analysis discussed in this chapter.

3.2.1.1.3 Battery electric vehicle

Unlike ICEs, these vehicles are fully electric and do not have any ICEs. Their energy is stored in a large battery pack that must be regularly charged from the grid. Battery Electric Vehicles (BEVs) have the greatest impact on the grid and power quality due to their complete dependence on the charging infrastructure (Nogueira et al., 2021). This chapter analyzes the impacts of this class of vehicle.

3.2.1.1.4 Fuel cell electric vehicle

These types of vehicles also use an electric motor for propulsion, but they generate their energy through a chemical reaction between hydrogen stored in tanks and oxygen from the air in a unit called a fuel cell. Although Fuel Cell Electric Vehicles (FCEVs) are zero emission, they are not connected to the electricity grid for charging and their infrastructure is completely different (hydrogen refueling stations) (López Cascales et al., 2015). Given their very small number on the market today, these types of vehicles are also outside the scope of this chapter. Table 3.1 briefly shows a comparison of the main features of each type of EV.

Table 3.1 Comparison of the main features of each type of EV.

Vehicle type	Primary energy source	Grid connection capability	Direct impact on power quality
HEV	Gasoline	No (Battery is self-charged by the engine and braking)	Almost none (Not grid-connected)
PHEV	Gasoline and grid electricity	Yes (Has a plug for external charging)	Yes (Acts like a BEV when charging)
BEV	Grid electricity	Yes (Fully dependent on external charging)	Yes (Has the most significant impact)
FCEV	Hydrogen	No (Generates its own electricity on-board)	Almost none (Not grid-connected)

3.2.1.2 Classification of charging infrastructure and technologies

Charging infrastructure, as the primary interface between an EV and the grid, performs the critical task of power conversion to safely and efficiently transfer electrical energy to the vehicle battery (Kumar et al., 2023). Charging technologies can be classified based on various criteria, an overview of which is shown in Fig. 3.2. The most basic classification, referred to as "Based on Contact" in Fig. 3.2 is determined by the method of energy transfer. This is divided into two main groups: conductive charging which requires physical contact and inductive charging which is contactless.

Conductive charging, which is the dominant and standard method on the global market, is based on a direct physical connection via a cable and connector. This method is the main choice for all applications from home charging to ultra-fast stations due to its very high efficiency (typically above 95%) and the ability to transfer large amounts of power. In contrast, inductive or wireless charging transfers energy without a physical connection and through a magnetic field. Although this method provides greater convenience for the user, its use is very limited due to lower efficiency, higher cost and limitations in transferring high power. Given that current and near-future infrastructures are mainly based on conductive charging, we will examine its types in more detail below. A schematic illustration of both wired (On-board/Off-board) and wireless methods is presented in Fig. 3.3. In the field of conductive charging, categories are defined based

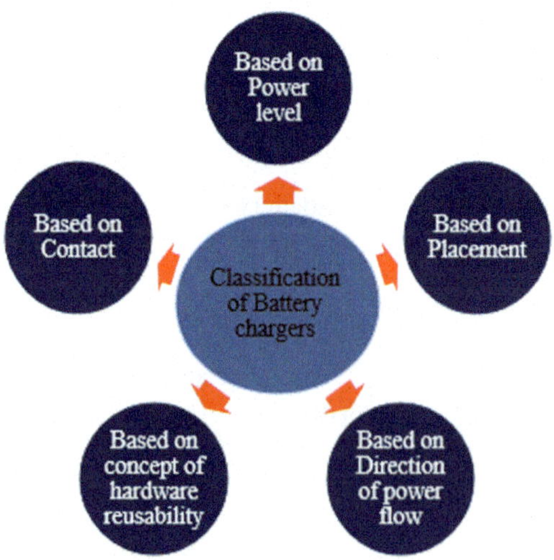

Figure 3.2 Battery charger classification.

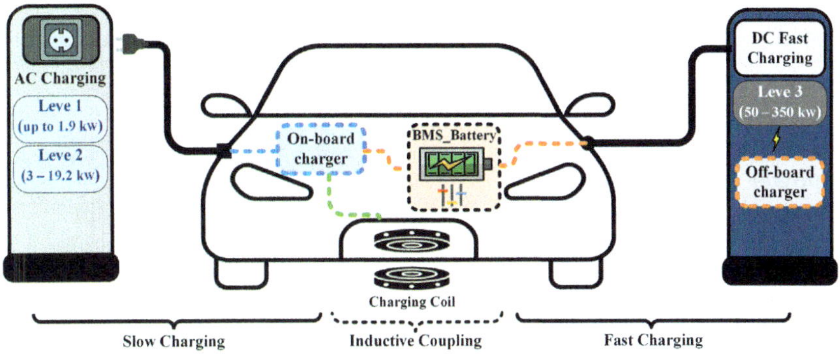

Figure 3.3 Schematic illustration of both wired (on-board/off-board) and wireless methods.

on architecture and function. The most fundamental difference is in the charger architecture (On-board vs. Off-board). In the On-board architecture, the AC/DC power converter, which is the primary cause of harmonics, is located inside the vehicle. This structure limits charging power due to space and weight constraints and is the main characteristic of AC charging. In contrast, the off-board architecture places this converter in an external power station to deliver very high-power DC power directly to the battery. This difference in architecture directly leads to the definition of three main Charging Levels. Levels 1 and 2, both AC and relying on the On-board charger, provide slow and medium charging for domestic use, respectively, and can be single-phase or three-phase (Saraswathi and Ramachandran, 2024). The choice between the two significantly affects the voltage imbalance in the network. Level 3, or DC fast charging, uses an off-board architecture to deliver very high power, which has more severe impacts on power quality, including voltage fluctuations and harmonics.

Additionally, chargers are also differentiated based on power flow direction. Unidirectional chargers only charge the vehicle, while bidirectional chargers return energy from the battery to the grid and can serve as a solution to improve power quality (Mojumder et al., 2022). To summarize and compare all the features of these categories, the main features of each level of directional charging are summarized in Table 3.2.

3.2.1.3 Converter architectures

Power electronic converters, as the beating heart of chargers, convert the AC power of the grid into the DC power required by the battery. This

Table 3.2 Comprehensive comparison of EV charging levels.

Feature	Level 1	Level 2	Level 3
Power delivery	AC	AC	DC
Maximum power	~1.9 kW	3–19.2 kW	50–350 kW
Charger architecture	On-board (converter is inside the EV)	On-board (converter is inside the EV)	Off-board (converter is in the station)
Charging speed	Very slow (adds 3–5 miles of range per hour)	Medium (adds 10–60 miles of range per hour)	Very fast (adds 80% charge in 20–60 min)
Typical location	Home (standard wall outlet), workplace	Home (dedicated circuit), workplace, public parking	Highway corridors, commercial stations

process, which is carried out through the switching of semiconductor devices at very high frequencies, is inherently and directly the main source of phenomena that affect grid power quality (Safayatullah et al., 2022). In fact, any decision in the design and selection of the topology of a converter has a direct consequence on the waveform of the current drawn from the grid. This has a direct impact on the harmonic profile and voltage stability. Modern chargers have a two-stage architecture, with each stage playing a specific role, as shown in Fig. 3.4. This structured overview highlights the diverse power electronic converters employed in EV charging infrastructure.

3.2.1.3.1 AC-DC conversion stage

This class, as a direct interface with the grid, converts AC voltage to DC. It is divided into three main groups based on structure and function. The first group is uncontrolled rectifiers, the simplest being the diode bridge rectifier. This circuit creates very high harmonic distortion due to drawing non-sinusoidal and pulsed current from the grid and is not used alone in modern chargers due to very low power quality. To solve this problem, the second group, Two-Level PFC Rectifiers, is used. These converters, with active control, make the input current sinusoidal and in-phase with the voltage. The most common topology in this category is the Three-Phase

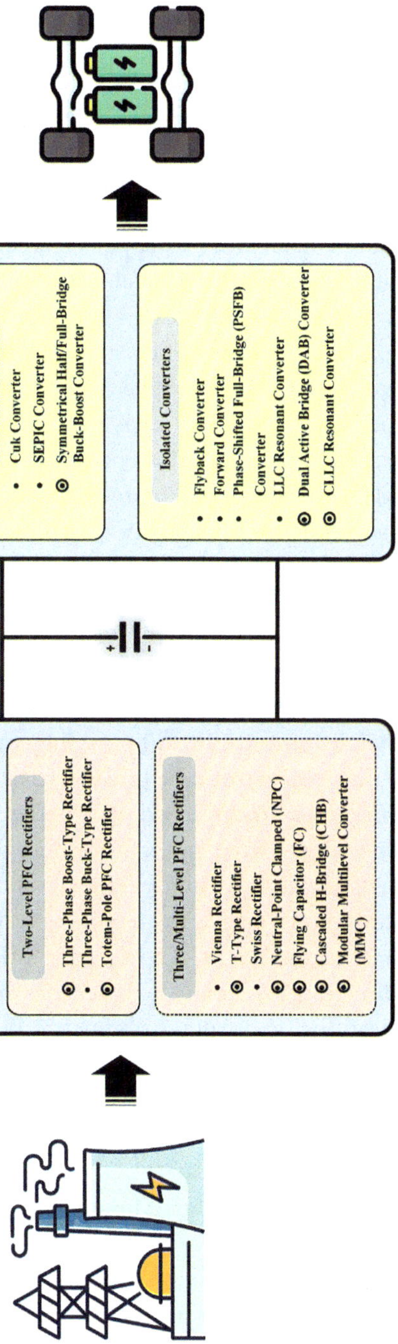

Figure 3.4 Topology of converters used in battery chargers.

Boost-Type Rectifier, which can also operate bidirectionally for V2G. The Three-Phase Buck-Type Rectifier is another option with a step-down function. The Totem-Pole PFC Rectifier, a single-phase topology, is also an advanced two-level structure with very high efficiency, with lower losses due to the elimination of the diode bridge. Lastly, the Three/Multi-Level PFC Rectifiers are designed for high power chargers and achieving the best power quality. These converters greatly reduce harmonics at the source by creating a stepper voltage. The Vienna Rectifier is a very popular Three-phase three-level topology with high efficiency. Its double-ended version is also called a T-Type converter. Another topology in this category is the Swiss Rectifier, a three-phase buck-type PFC. For the highest power levels and best wave quality, multilevel converters such as Neutral-Point Clamped (NPC) converter, Flying Capacitor (FC) converter, Cascaded H-Bridge (CHB) converter, and Modular Multilevel Converter (MMC) are used, which are used in ultra-fast chargers due to their modular structure and excellent performance (Safayatullah et al., 2022; Ali et al., 2024).

3.2.1.3.2 DC-DC conversion stage

There are two important roles played by DC-DC converters in the charger architecture. As a first step, the output voltage and current are adjusted according to the battery's needs. In addition, they provide galvanic isolation, which separates the vehicle from the grid electrically. In general, converters can be divided into two groups based on whether they are isolated or not. Non-isolated converters have no transformer, so they are simpler and cheaper. This category includes unidirectional topologies such as Buck converters, Boost converters, and converters for both step-down and step-up functions, such as Buck-Boost, CUK, and SEPIC (Safayatullah et al., 2022). The bidirectional version of these converters, which is also used for special applications, is usually a symmetric buck-boost converter. To improve the performance of these converters and reduce output current ripple, which is very critical for battery health, the interleaving technique or paralleling of multiple converters is used.

However, due to standard safety requirements for EV chargers, it is necessary to use isolated converters. These converters provide the necessary isolation using a high-frequency transformer. In single-sided chargers, classic topologies such as Flyback Converter for low powers and Phase-Shifted Full-Bridge (PSFB) for higher powers are employed. However, to achieve maximum efficiency, the LLC resonant converter has become the preferred choice in modern single-sided chargers due to its soft switching capability and very low losses. For two-sided chargers that enable V2G, the Dual Active Bridge (DAB) is the industry standard. Another topology is the CLLC resonant converter. DAB and CLLC are alternative bidirectional, isolated DC-DC converters; The choice between them depends on the power level, efficiency at partial load, control bandwidth, and cost (Ali et al., 2024).

3.3 The dual role of electric vehicles: A grid power quality perspective

Since EVs use power electronic converters, they are active and nonlinear loads and therefore can impact the grid in both ways. Power quality refers to the ability of the grid to deliver a clean, continuous electrical current, and to maintain a sinusoidal voltage and current waveform while maintaining a steady frequency. It is a measure of the health and condition of the grid. Power quality is defined by harmonic distortion, voltage fluctuations, phase imbalances, overloads, and power losses. As a result of their inherent performance, chargers can degrade these parameters (negative impacts), but with intelligent control and V2G, they can also increase power quality (Kazemtarghi et al., 2023). As shown in Fig. 3.5 these opposing effects and their dual relationship are conceptually depicted. Below, we

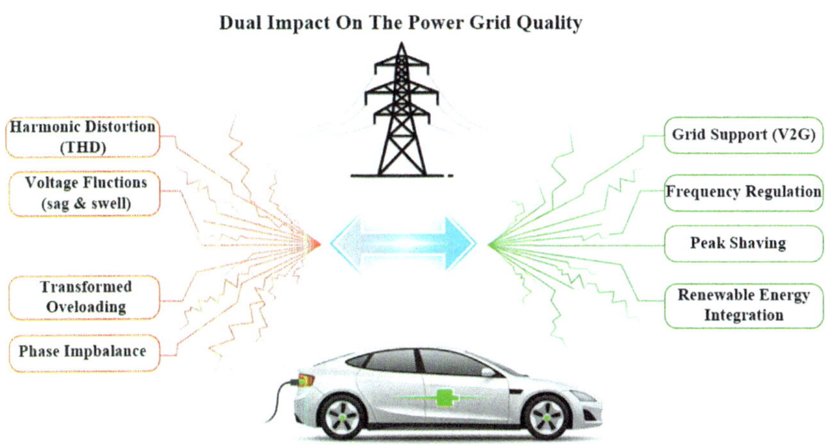

Figure 3.5 Comparison of the advantages and disadvantages of V2G in grid systems.

examine each of these parameters in detail and analyze their negative and positive impacts.

3.3.1 Challenges and negative impacts on the power distribution system

Uncontrolled and widespread connections of EVs to the grid bring several challenges to power quality and existing infrastructure. Especially if chargers are nonlinear and have a high power demand, they can distort the distribution network's voltage waveforms. As a result, other equipment connected to the grid may not function properly. In the remainder of this section, we provide a detailed analysis of the most significant of these challenges. We assess their compliance with international power quality standards.

3.3.11 Harmonic distortion

In an ideal electrical grid, the voltage and current waveforms are perfectly sinusoidal. A deviation from this pure sinusoidal waveform is known as harmonic distortion, which occurs when undesirable components with frequencies that are integer multiples of the fundamental grid frequency are present. To measure the overall extent of this phenomenon, the Total Harmonic Distortion (THD) index is used, which is defined as follows:

$$THD = \frac{\sqrt{\sum_{h=2}^{h_{max}} M_h^2}}{M_1} \tag{3.1}$$

$$\mathrm{TDD}(\%) = 100 \frac{\sqrt{\sum_{h=2}^{H} I_h^2}}{I_L} \tag{3.2}$$

In the THD formula, M_1 is the amplitude of the fundamental component and M_h is the amplitude of the hth-order harmonic component. The IEEE 519 standard specifies the permissible limits for THD. The current distortion metric used for grid compliance at the Point of Common Coupling (PCC) is Total Demand Distortion (TDD) per IEEE 519. In TDD formula, I_L is Maximum demand load current at PCC. The role of EVs in creating this disturbance is directly related to the type of charger and the AC-DC rectifier topology it employs. Single-phase rectifiers are used in Level 1 and Level 2 chargers, which are primarily used for residential and public charging at lower power levels (up to 19.2 kW). Even though

Level 1 and 2 chargers inject low harmonic currents, these chargers typically use simpler topologies that can generate low-order harmonics (Rana et al., 2025). In a residential area, hundreds or thousands of these chargers are simultaneously charging, creating an Aggregation Effect. As a result, the voltage THD at the local distribution transformer can reach unacceptable levels due to the superposition and amplification of these harmonics, causing the voltage THD at the local distribution transformer to reach unacceptable levels.

Uncontrolled or poorly filtered high-power rectifiers can inject significant harmonics (Abdollahi et al., 2025); however, modern DCFCs employ active PFC so that low-order harmonics are minimized and high-frequency components are filtered to meet grid codes. This can negatively impact the power quality at the PCC even individually. For this reason, these chargers are required to use advanced controlled rectifiers (Rana et al., 2025). In these chargers, Active Power Factor Correction (Active PFC), using Pulse Width Modulation (PWM) and high-frequency switching, actively forces the input current to have a sinusoidal waveform that is in phase with the voltage. This process nearly eliminates the low-order harmonics. However, the switching process itself generates harmonics at very high frequencies (around the switching frequency), which are much easier to filter. To achieve the best power quality in very high-power chargers, three-level and multi-level rectifiers (such as the Vienna and NPC rectifiers) are utilized. By creating a stepped voltage waveform that is closer to a pure sinusoidal wave, these topologies significantly reduce the amplitude of the high-frequency harmonics caused by switching and, consequently, provide the best performance in terms of power quality.

3.3.1.2 Voltage fluctuations

Voltage variations represent short-term changes in the system's voltage magnitude, with the most critical phenomena being the Voltage Sag (RMS voltage **10%–90%** of nominal, **0.5 cycle to 1 min**), the Voltage Swell (RMS voltage **110%–180%** of nominal, **0.5 cycle to 1 min**), and Flicker, which manifests as rapid and perceptible fluctuations in lamp brightness; the IEC 61000 series of standards strictly specifies the permissible limits for these variations to safeguard the proper operation of electrical equipment and maintain grid stability (Wang et al., 2021).

The main negative impact of EVs in this context is the creation of voltage sags at the moment high-power chargers (especially DC chargers) are suddenly connected to the grid (Wang et al., 2021). This sudden and large demand for power acts similarly to connecting a heavy industrial load to the network and can cause a severe drop in the local grid voltage.

The severity of this phenomenon depends on two key factors: the EV penetration level and the Grid Strength.

1. ***Grid Strength*:** This factor refers to the network impedance at the point of connection. In a Strong Grid (such as industrial areas or near main substations) with low impedance, the voltage sag is negligible. However, in a Weak Grid (such as the end of long rural feeders) with high impedance, the same charger can cause a severe voltage drop, pushing the voltage outside the permissible standard limits and negatively impacting other customers (Henderson et al., 2024).
2. ***Penetration Level*:** While a single vehicle may not be problematic on its own, the simultaneous startup of a large number of vehicles in an area (high penetration level) causes a cumulative and deep voltage sag. In a specific study indicate that as EV penetration increases from 20% to 80%, the voltage deviation can range between 12.7% and 43.3% (Khalid et al., 2019).

The sudden disconnection of this large load can also lead to a Voltage Swell. Furthermore, the continuous and rapid current fluctuations generated by the charger's internal controller can cause the Flicker phenomenon in the network. Practical measurements from charging stations have demonstrated that the flicker severity can exceed standard permissible limits, which is particularly concerning in grids with sensitive loads such as hospitals or data centers.

3.3.1.3 Transformer overloading and power losses

The increase in instantaneous load on distribution network equipment is one of the most important and costly negative impacts resulting from the widespread penetration of EVs. The main negative impact in this segment is overloading on distribution transformers and lines (feeders), which occurs due to the high simultaneous power demand of vehicles, particularly in uncontrolled charging patterns during peak consumption hours. This overloading leads to two detrimental consequences: increased power losses and the acceleration of equipment degradation. The high current drawn by chargers, as well as the harmonic currents they generate, causes a sharp increase in I^2R power losses across the line impedances and transformer windings. Studies indicate that these losses can increase by up to 66% because of 40% EV penetration during peak charging hours (Chang et al., 2019). These losses not only reduce the overall efficiency of the system but also manifest as heat within the equipment, leading to an increase in their operating temperature.

The temperature increase resulting from overloading and harmonic losses directly impacts the useful life of transformers and accelerates the degradation process of their insulation. A specific analysis has shown that with just a 10% penetration of Level 2 chargers, up to 46% of distribution transformers in a residential area may require replacement or upgrade. Simulations also indicate that the transformer aging rate under uncontrolled charging can increase by 30–40% (Elnozahy and Salama, 2014). This imposes significant hidden costs on network operators and necessitates substantial investments for the reinforcement of existing infrastructure.

3.3.1.4 Phase imbalance

In an ideal three-phase system, the load must be equally distributed among the three phases so that the voltages and currents remain symmetrical. If the load on one or two phases is significantly greater than the others, a phenomenon called Phase Imbalance occurs, leading to the creation of negative sequence voltage and current in the network.

The negative impact of this phenomenon primarily results from the widespread and random connection of single-phase chargers (Level 1 and 2) in low-voltage distribution networks, especially in residential areas. Since the distribution of these chargers across the three phases of the network is uncontrolled, the probability that one phase will be significantly more loaded than the others sharply increases as EV penetration levels rise (Ul-Haq et al., 2015).

3.3.2 Opportunities and positive impacts for grid enhancement

3.3.2.1 The V2G paradigm: Electric vehicles as distributed energy resources

The basis of all positive impacts of EVs on the grid relies on the V2G paradigm. V2G technology enables bidirectional power flow between the vehicle's battery and the electrical grid. This capability, which requires the use of bidirectional chargers and advanced power converters, completely transforms the nature of the EV. As shown in Fig. 3.6 Part 1, Under this paradigm, the EV is no longer just a passive load, but evolves into a DER.

Given that personal vehicles are, on average, parked for over 90% of the time throughout a 24-hour day, the massive fleet of EVs can operate as a colossal virtual energy storage system (Marini et al., 2015). These mobile

Figure 3.6 Opportunities and positive impacts for grid enhancement.

batteries can be charged during periods when energy generation is high and costs are low (such as midnight or midday, in the presence of solar energy) and, in turn, return their stored energy to the grid during peak demand or in emergency situations. This capability provides unprecedented opportunities for offering ancillary services, load management, and overall improvement in grid performance, power quality and efficiency, which will be discussed in detail below (Rao et al., 2025; Comi and Idone, 2024).

3.3.2.2 Ancillary services provision for grid stability and power quality

Frequency Regulation, Voltage Support and Reactive Power Compensation are among the most prominent opportunities afforded by the V2G paradigm, as illustrated in Fig. 3.6 Part 3, which is the ability of the EV fleet to provide Ancillary Services to the grid. Ancillary services are operations necessary to maintain the grid's stability, reliability, and power quality. Thanks to the extremely high reaction speed of power electronic converters, EVs can provide these services much faster and more efficiently than traditional power plants. The most important of these services include:

1. ***Frequency Regulation*:** The grid frequency (e.g., 50 or 60 Hz) must remain constant within a very narrow range to maintain the balance between generation and consumption. Grid-connected EVs can react to frequency deviations in a fraction of a second, helping to restore the frequency to its nominal value by injecting (discharging) or absorbing (charging) small amounts of active power. This service is one of the most valuable ancillary services for grid operators (Iqbal et al., 2018).
2. ***Voltage Support and Reactive Power Compensation*:** As previously mentioned, EV charging can cause a voltage sag. In contrast, advanced bidirectional chargers can precisely control and either inject or absorb reactive power from the grid. This capability, which can even be performed without discharging the battery, significantly assists in Voltage Support and regulation at weak points in the distribution network and can reduce the need for installing expensive compensation equipment (Buja et al., 2017).

3.3.2.3 Grid load management: Peak shaving and valley filling

One of the most powerful and economical positive impacts of EVs is their capability in Grid Load Management. Electricity consumption curves are not uniform throughout the day, with peak and off-peak periods (valleys). Providing power during peak hours is extremely expensive for utility companies and requires expensive power plants that operate for only a few hours a day. By utilizing Smart Charging and V2G technology, the EV fleet can act as a highly efficient tool to flatten this load curve (Jia et al., 2024; Lu et al., 2023). This process is achieved through two complementary strategies:

1. ***Valley Filling*:** As shown in Fig. 3.6 Part 2, With Smart Charging algorithms, vehicle charging can be shifted from peak hours to off-peak hours (usually midnight). In this way, the valley of the consumption curve is filled during low-load hours. Power plants that operate at low loads during these hours can, therefore, be utilized more efficiently.
2. ***Peak Shaving*:** This capability, which requires V2G technology, allows EVs to return the energy stored in their batteries to the grid during peak consumption hours. This acts like the activation of a virtual power plant, shaving the peak of the consumption curve. This strategy not only relieves stress and overloading on grid equipment but also significantly reduces the total system operating costs by decreasing the need for electricity generation during the most expensive hours of the day.

3.3.2.4 Synergy with renewables: Enabling a greener grid

Among the most strategic and forward-looking opportunities offered by EVs is their role in facilitating the integration of Renewable Energy Sources (RES) into the grid, such as solar and wind (Lehtola, 2024). Solar energy is only available during the day and is dependent on weather conditions, while wind energy is largely dependent on wind speed. Renewable sources present the greatest challenge due to their intermittent and unpredictable nature. Power quality disturbances can result from these fluctuations in the balance between generation and consumption in the grid.

The EV fleet, with its distributed energy storage capability, can act as a colossal "buffer" significantly mitigating this challenge. During periods when renewable energy generation exceeds grid consumption (such as midday for solar energy or midnight for wind energy), this clean, surplus energy can be stored in the batteries of millions of grid-connected EVs. Subsequently, this stored energy can be returned to the grid via V2G technology during peak demand or when renewable generation declines (Shi et al., 2020).

This synergy not only prevents the curtailment (waste) of clean energy but also greatly aids in decarbonizing the energy sector by reducing the need for fossil fuel power plants to compensate for fluctuations, thereby paving the way for achieving a stable, 100% renewable energy-based grid, Fig. 3.6 Part 4 illustrates this section.

3.4 Mitigation strategies for negative grid impacts

Having examined the challenges posed by EV integration in the previous parts, this section provides an overview of engineering responses designed to mitigate these negative effects. A portfolio of solutions is presented, including advanced hardware designs, intelligent control strategies, and infrastructure upgrades, that can ensure the seamless and safe integration of EVs into the grid. Table 3.3 summarizes the negative impacts as well as suggested solutions.

Table 3.3 Summary of negative effects of EV implementation with practical solutions.

Negative effect	Description	Practical solutions
Harmonics/waveform distortion (Srivastava et al., 2023)	Power-electronic chargers inject current/voltage distortion; THD rises with charging rate/SOC.	Shunt/series/active/hybrid filters; UPQC/DVR; use EVs/bi-directional chargers as active filters.
Voltage unbalance (Srivastava et al., 2023)	Single-/three-phase charging and uneven phase loading drive voltage imbalance.	Phase reconfiguration, voltage regulators, D-STATCOM/DVR; storage (BES/SMES).
Voltage fluctuations (sag/swell) (Srivastava et al., 2023)	Large fast-charge currents cause sags/swells; two stage off-board resonance can aggravate.	Tap-changer control, capacitor banks; charging rate management with active/reactive power control.
Transformer/feeder overload and thermal stress (Srivastava et al., 2023)	Growth in charging stations and aging assets overload transformers/cables.	Coordinated/time of use charging, priority zoning, targeted infrastructure upgrades; D-STATCOM support.
Increased peak demand (system congestion) (Nour et al., 2020)	Uncontrolled charging raises evening peaks and stresses the network.	Smart/controlled charging and TOU tariffs; V2G/V2V peak-shaving/valley-filling via aggregators.

(Continued)

Table 3.3 Summary of negative effects of EV implementation with practical solutions.—cont'd

Negative effect	Description	Practical solutions
Power losses (I^2R and equipment heating) (Bhosale et al., 2024)	Higher charging currents increase line/transformer losses; can rise significantly in stressed cases.	Coordinated/optimized charging and improved charger efficiency/ algorithms.
Frequency stability degradation (Kumar et al., 2023)	Random, high-penetration charging can erode frequency stability.	Fast frequency regulation with EVs via V2G and aggregator-based controls.
Load mismanagement (random/uncoordinated charging) (Rana et al., 2025)	Poorly managed on/off board behavior raises outage probability and cascades into PQ issues.	Demand response programs, forecasting, and algorithmic load management at feeder/ aggregator level.
Grid-level instability (voltage/frequency/ oscillatory) (Kumar et al., 2023)	EVs behave as nonlinear, uncertain loads; stability depends on penetration, timing, and location.	Smart charging, energy storage, and load control integrated with grid code compliant control schemes.

3.4.1 Hardware-based solutions: Advanced charger design

The most effective method for mitigating power quality disturbances is to prevent their generation at the source, that means the EV charger itself. Modern power electronics offer sophisticated converter topologies that can significantly reduce harmonic emissions and improve overall efficiency, addressing the core issues of harmonic distortion and equipment losses.

3.4.1.1 Harmonic reduction at the source

In order to comply with stringent standards like IEEE 519, advanced AC-DC converter topologies are essential. Using these designs, the input current is actively shaped to be sinusoidal and in phase with the grid voltage.

1. **Vienna Rectifier:** This is a three-phase, three-level PWM rectifier topology that is highly popular for high-power EV charging. Unlike a

standard six-switch converter, it only uses three active switches, simplifying control and reducing costs. Compared to conventional two-level converters, its three-level output voltage structure produces a waveform with lower harmonic distortion. As an excellent front-end stage for fast chargers, the Vienna Rectifier has near-unity power factors and very low THD, often below 5% (Rana et al., 2025).

2. **multi-level converters:** In ultra-fast charging stations, multi-level converters such as NPC and CHB topologies are used. Multiple DC voltage levels are used to synthesize a stepped voltage waveform that more closely approximates a pure sine wave. As a result of this significant reduction in voltage steps (dv/dt), harmonic content injected into the grid is dramatically reduced, reducing the size requirements for filtering components (Kumar et al., 2023).

3.4.1.2 Increased efficiency and loss reduction

Beyond the AC-DC stage, the isolated DC–DC stage of the charger plays a critical role in overall efficiency and, consequently, in reducing thermal losses. Advanced resonant converter topologies achieve this by enabling soft switching. LLC is preferred for high-efficiency unidirectional stages due to their ability to achieve Zero Voltage Switching (ZVS) over a wide load range. The soft-switching characteristic reduces the power dissipated during switching transitions, enabling higher switching frequencies, resulting in smaller magnetic components and higher power density. On the other hand, **DAB** and **CLLC** are the mainstream **bidirectional** isolated DC–DC topologies for V2G chargers. It is a key enabler for efficient V2G systems because it offers a symmetrical structure that maintains high efficiency during both charging (grid-to-vehicle) and discharging (vehicle-to-grid) (Rana et al., 2025).

3.4.1.3 Utilization of active and passive harmonic filters

In cases where chargers with high harmonic emissions are already deployed, or for mitigating the aggregate effect of many low-powered chargers, harmonic filters are a necessary solution.

1. **Passive Filters:** These are composed of inductors and capacitors tuned to resonate at specific harmonic frequencies (e.g., fifth, seventh). They provide a low-impedance path to ground for these harmonic currents, effectively shunting them away from the grid. They are simple and cost-effective, but are fixed to target specific frequencies (Srivastava et al., 2023).

2. **Active Power Filters (APFs):** These are more advanced, power electronics-based solutions. An APF measures the harmonic currents drawn by the nonlinear load (the EV chargers) in real-time and injects an equal but opposite compensating current. This "anti-harmonic" current effectively cancels out the original distortion, ensuring that the current seen by the grid is nearly sinusoidal. APFs are highly effective, adaptable to changing load conditions, and can correct for multiple harmonic orders simultaneously (Srivastava et al., 2023).

3.4.2 Management and control-based solutions

Beyond hardware, intelligent management and control strategies can mitigate grid impacts by coordinating the collective behavior of a large EV fleet (Un-Noor et al., 2017). These strategies coordinate the collective behavior of a large EV fleet, which, when unmanaged (uncoordinated or dumb charging), acts as high-power loads that severely affect transformers, cables, and fuses, potentially leading to overload, equipment damage, and increased infrastructure cost. The core objective of these strategies in EV charging systems is to optimize available energy resource utilization, facilitate efficient charging operations, and seamlessly integrate EVs with the grid.

3.4.2.1 Smart and coordinated charging strategies

Instead of allowing vehicles to charge immediately upon being plugged in (uncontrolled charging), smart charging algorithms manage the timing and rate of charging to avoid creating new demand peaks (Nour et al., 2020). These strategies can be centralized or decentralized and typically optimize for variables like electricity price, grid load, and renewable energy availability. This algorithm prevents overloading of transformers and feeders, reduces power losses, and maintains voltage stability across distribution networks by shifting the charging load to off-peak hours.

3.4.3 Infrastructure-based solutions

Finally, mitigating the impacts of high-power charging requires strategic planning and reinforcement of the grid infrastructure itself.

3.4.3.1 Grid infrastructure reinforcement and optimal station siting

The ability of the grid to handle a large load without significant voltage drop is defined by its strength, or more technically, its short circuit capacity (Henderson et al., 2024). High-power and ultra-fast charging stations are significant loads and must be treated as such. It is crucial to conduct detailed grid impact studies to identify strong points in the grid, there are typically close to primary substations or on high-capacity feeders. Installing fast-charging plazas at these locations minimizes voltage sag and flicker. In weaker parts of the grid, infrastructure reinforcement, such as upgrading local transformers and reconductoring feeders, may be necessary before deploying high levels of EV charging (Sinha et al., 2023).

3.5 Strategies for maximizing positive grid impacts

This chapter explores the strategies required to unlock the significant positive potential of EVs, transforming them from passive loads into active grid assets. The objective is to convert the theoretical opportunities outlined in Section 3.3.2, such as grid support and synergy with renewables into practical, operational capabilities through the implementation of V2G technology and supportive economic frameworks.

3.5.1 V2G technology implementation

The successful deployment of V2G services rests on two pillars: the necessary hardware to enable bidirectional power flow and the sophisticated communication and control systems to manage it effectively.

3.5.1.1 Hardware prerequisites and standards

The core hardware enabler for V2G is the bidirectional power converter, which must efficiently manage power flow in both directions. While CLLC converters are excellent for on-board applications, the DAB converter is the industry-standard topology for high-power, isolated, bidirectional DC–DC conversion in off-board DC chargers. The DAB consists of two full H-bridges connected via a high-frequency transformer. By controlling the phase shift between the voltages of the two bridges, power can be precisely controlled in terms of both magnitude and direction. Its symmetrical structure allows for high efficiency in both charging and discharging modes, making it the foundational building block for V2G-capable fast-charging stations (Rana et al., 2025). Effective V2G operation hinges on standardized communication protocols at multiple levels. Table 3.4 provides a summary of these key EV charging standards and protocols (Wang et al., 2021; Das et al., 2020; Jaman et al., 2022; Kirchner, 2024; Rajendran et al., 2021).

Table 3.4 Key EV charging standards and protocols.

EV charger topology	Charging connector	Charging communication	Charging power quality	Charging safety
IEC 61851-1	IEC 62196-1	ISO 15118 (series)	IEEE 1547	SAE J2464
EV conductive charging general	Plugs/ couplers—general	V2G comms (EV–EVSE)	DER interconnection requirements	Battery system safety/ abuse
IEC 61851-21	IEC 62196-2	IEC 61850	SAE J2894	SAE J1766
EMC for on-board chargers	AC connectors (Type 1/2)	Substation/DER comms	Power quality for chargers	Battery crash integrity
IEC 61980 series	IEC 62196-3	SAE J2847 (series)	IEC 61000-3-2	ISO 6469-3
Wireless power transfer	DC connectors (CCS etc.)	EV grid-interface message sets	Harmonic current limits	Protection against electric shock
IEC 61851-23	SAE J1772	SAE J2836 (series)	CISPR 11	IEC 60364-7-722
DC EV supply equipment	Conductive charge coupler	Use cases for comms	RF emission limits (conducted and radiated)	EV charging installations
IEC 61851-24	IEEE 1901	SAE J2293-2	SAE J2380	IEC 60529
Digital comms for DC charging	Power line communication PHY	Energy transfer communications	Battery vibration test	Ingress protection (IP code)

1. **Vehicle-to-Charger Communication:** At the local level, protocols embedded in both the vehicle and the EV Supply Equipment are essential. The emerging Combined Charging System (CCS) with ISO 15118-20 enables bidirectional energy transfer, while CHAdeMO has historically supported V2G. Newer North American Charging Standard (NACS) ecosystems are converging on similar capabilities (Jaman et al., 2022).
2. **Charger-to-Network Communication:** To link charging infrastructure with smart grids and backend systems, several key standards are used. Open Charge Point Protocol (OCPP) manages charger-to-network communication for remote control and billing. Open Charge Point Interface (OCPI) and Open Smart Charging Protocol (OSCP) handle roaming and data exchange to coordinate charging based on grid capacity. Finally, OpenADR (Open Automated Demand Response) enables utilities to send demand response signals, allowing chargers to adjust their load automatically for efficient grid interaction (Kirchner, 2024).

Finally, all V2G hardware must adhere to strict grid codes and safety standards. Chargers must comply with interconnection standards like the IEEE 1547 series, ensuring certified anti-islanding and ride-through performance in line with UL 1741 SB or equivalent regional standards. Proper grounding and protection coordination are essential for fault detection and breaker sizing, while DC V2G systems also require DC ground-fault monitoring and arc-flash mitigation. Additionally, advanced chargers should support autonomous functions like Volt-VAR and Volt-Watt droop modes to provide localized voltage regulation without needing direct commands from a grid operator (Rajendran et al., 2021).

3.5.1.2 Communication and control prerequisites

Individual EVs have limited capacity, but when aggregated and centrally controlled, they form a powerful virtual power plant.

1. **Role of Aggregators:** An aggregator is a commercial entity that acts as an intermediary between EV owners and the utility or grid operator. It bundles the DERs of thousands of EVs to offer services at a scale large enough to participate in energy markets. The control platform of the aggregator sends commands to vehicles for charging and discharging at specific times which matches grid needs while providing monetary rewards to EV owners (Nazari-Heris et al., 2022).
2. **Smart Grid Communication Infrastructure:** Reliable V2G operation requires a robust, low-latency, and secure two-way communication network (e.g., 5G, IEEE 2030.5) connecting the vehicles, the

aggregator, and the grid operator. The infrastructure works as a basic system to deliver instant grid status information about frequency and voltage while sending commands to control the electric vehicle fleet (Bouida et al., 2020).

3.5.2 Operational and economic frameworks

With the technical prerequisites in place, viable operational models and economic incentives are needed to encourage participation and maximize the benefits of V2G.

3.5.2.1 Participation in ancillary services markets

The high-speed response of power electronics makes EV fleets ideal for providing valuable grid ancillary services.

1. **Frequency Regulation:** Aggregated EVs can participate in frequency regulation markets, which is one of the most lucrative ancillary services. When grid frequency drops due to a sudden loss of generation, the aggregator can command the fleet to instantly stop charging or start discharging, injecting power to stabilize the grid. The fleet will use its power storage to handle any electrical surplus that occurs when the frequency exceeds the standard operating range. Power plants that operate on traditional slow ramp-up methods cannot match the speed of this quick response (Khan et al., 2021).
2. **Voltage Support:** As discussed, bidirectional chargers can inject or absorb reactive power. Aggregators can leverage this capability to provide local voltage support in distribution networks, helping to maintain voltage profiles within standard limits and potentially deferring costly infrastructure upgrades for the utility (Zhou et al., 2023).

3.5.2.2 Synergy with renewable energy sources

EVs provide a distributed storage buffer system which helps stabilize renewable energy supply through long-term operation (Barman et al., 2023). The synchronization of EV charging with renewable energy generation peaks is managed through advanced control algorithms operated by aggregators. The strategic scheduling of workplace charging allows businesses to store solar energy during peak midday hours and home charging systems can utilize wind energy during nighttime hours. The EV fleet acts as a storage system which prevents the waste of curtailed energy. The stored clean energy from EVs becomes available to supply the grid during peak demand hours when solar power generation stops thus creating a stable system that optimizes renewable energy resources.

3.5.2.3 Vehicle-to-building and vehicle-to-home frameworks

While large-scale V2G implementation faces market and regulatory hurdles, Vehicle-to-Building (V2B) and Vehicle-to-Home (V2H) frameworks offer more immediate economic benefits andgrid support. In these models, the EV's bidirectional capability is used "behind the meter" to manage a single building's load profile.

By supplying the electricity for the internal loads directly from the EV battery, V2B/V2H reduces the building's demand from the grid. This "behind-the-meter" peak shaving provides a significant indirect power quality benefit: it alleviates the load on local distribution transformers and mitigates the voltage sags and fluctuations caused by high peak demand.

1. **Commercial (V2B):** For commercial facilities, the primary economic driver is demand charge reduction. Many utilities bill large customers based on their peak power consumption during a billing period. A V2B-capable EV fleet can discharge during these peak moments, shaving the building's peak load and dramatically reducing electricity bills, often more profitably than participating in ancillary service markets (Liu et al., 2024).
2. **Residential (V2H):** For homeowners, V2H enables advanced energy arbitrage. The EV can charge during off-peak hours (e.g., using Valley Filling) when electricity is cheap and then power the home during on-peak hours when electricity is expensive, directly minimizing the owner's utility bill (Zhang et al., 2025).
3. **Resilience:** Both V2B and V2H also provide an innate resilience value, allowing the EV to function as an uninterruptible Power Supply (UPS) during grid outages, an economic benefit that is difficult to quantify but highly valued by consumers (Rouhana et al., 2025).

3.5.2.4 Tariff and market design for electrical vehicles flexibility

The electrical grid must deliver precise signals to electric vehicle owners for smart charging and V2G systems to function effectively in actual conditions. Market structures including advanced tariffs serve as the main system that motivates people to charge their devices in ways that reduce power quality problems. The signals function to match the charging requirements of devices with the electrical grid demands which helps stop transformers from becoming overloaded and prevents voltage drops. Time-varying tariffs function as technical instruments that help achieve valley filling through their Time-of-Use and Real-Time Pricing systems (Kaur and Singh, 2023; Khan et al., 2024; Daneshzand et al., 2023). The tariffs send a signal to customers which directs them to charge their devices during off-peak periods thus lowering the peak demand on distribution transformers while reducing thermal stress and voltage fluctuations. The power system gains access to an emergency control signal through Critical-Peak Pricing which functions as a dynamic signal to manage demand and prevent power quality degradation during peak load situations. a complete solution must also address two other key areas:

1. **Export Compensation and V2G:** For V2G to be financially attractive, clear mechanisms for export compensation are required. This can be achieved through net-billing arrangements or through direct compensation at transparent, market-reflective rates, such as the wholesale Locational Marginal Price at a specific grid node (Steward, 2017).
2. **Distribution-Level Incentives:** System-wide energy prices do not always reflect local distribution network constraints. To address this, locational adders or feeder-specific incentives can be implemented. These targeted signals reward EV flexibility in specific areas where grid infrastructure (e.g., a local transformer) is nearing its capacity limit, thereby deferring costly upgrades (Turk et al., 2025; Gao et al., 2024).

For these frameworks to succeed in practice, they must be built on trust and practicality. Utilities and aggregators must publish predictable price schedules alongside real-time event signals. Crucially, mobility-first safeguards such as allowing the user to define a minimum State of Charge (SOC) and a required departure time must be guaranteed to ensure the vehicle's primary transportation function is never compromised. When these economic and practical elements are correctly designed, the adoption of V2G, the stacking of revenue streams, and tangible grid relief become far more achievable.

3.6 Conclusion and future research directions

The focus of this chapter was on a comprehensive analysis of the dual impacts of EVs on key grid power quality indicators. The analysis shows that the main power quality challenges arise directly from the architecture of the power electronics converter in the charger. Injection of THD from nonlinear rectifiers, voltage fluctuations (sag and flicker) due to the sudden high power demand of DC fast chargers, and phase imbalance due to widespread connection of single-phase chargers, are all direct threats to the stability and grid waveform integrity.

In order to move from problem to opportunity, two parallel routes were mapped out. Firstly, by means of sophisticated hardware, such as low harmonic converter topologies (Vienna rectifiers, multi-level converters) and active filters (APFs), the idea was to lessen the negative impacts at the source. Secondly, the positive potential was to be realized through the deployment of V2G network, which could include bidirectional converters (e.g., DAB) and standardized communication for all (e.g., ISO 15118) systems. Summarizing, this chapter positions EVs as a paradigm shift in the power quality field: on the one hand, they become the largest new source of power quality disturbances; on the other hand, they constitute the most powerful distributed means for the active control and mitigation of these disturbances.

Building on these results and to improve power quality, future research can focus on developing intelligent and adaptive management systems. Given the disturbances reviewed in this chapter such as voltage sag, harmonics, and phase imbalance, it is preferable for research to focus on Machine Learning based systems that can automatically identify and classify these EV-induced disturbances. This intelligent approach can also provide the adaptability to identify new disturbance patterns that emerge as charger technology evolves. This intelligence can also directly upgrade Smart Charging algorithms. Currently, these algorithms are mainly used for charge time management. Future research can enhance them into active power quality management tools. In this new approach, Aggregators can optimize charging not just based on tariffs, but also based on the physical and instantaneous health of the grid; meaning, by monitoring the real-time load of transformers and feeders, they can prevent voltage fluctuation in weak networks. Furthermore, these advanced algorithms can monitor the load of each of the three phases live and intelligently assign new single-phase chargers to the least loaded phase at that moment. This active balancing directly prevents phase imbalance and transformer overheating.

References

Abdollahi R, Salemnia A, Mosallanejad A, Gharehpetian GB. EV charging stations multi-pulse rectifiers based on passive harmonic reduction circuits: a critical survey. Renew Sustain Energy Rev 2025;215:115499. https://doi.org/10.1016/j.rser.2025.115499.

Ali A, Mousa HHH, Shaaban MF, Azzouz MA, Awad ASA. A comprehensive review on charging topologies and power electronic converter solutions for electric vehicles. J Mod Power Syst Clean Energy 2024;12(3):675–94. https://doi.org/10.35833/mpce.2023.000107.

Barman P, Dutta L, Bordoloi S, Kalita A, Buragohain P, Bharali S, et al. Renewable energy integration with electric vehicle technology: a review of the existing smart charging approaches. Renew Sustain Energy Rev 2023;183:113518. https://doi.org/10.1016/j.rser.2023.113518.

Bhosale P, Sujil A, Kumar R, Bansal RC. Electric vehicle charging infrastructure, standards, types, and its impact on grid: a review. Elec Power Compon Syst 2024:1–25. https://doi.org/10.1080/15325008.2024.2315206.

Bouida Z, Fattahi J, Ahmed A, Ibnkahla M, Schriemer H, Abdullah R. Smart grid communication based on IEEE 2030 standard encyclopedia of wireless networks. Canada: Springer Science+Business Media; 2020. p. 1311–8. https://doi.org/10.1007/978-3-319-78262-1_273.

Buja G, Bertoluzzo M, Fontana C. Reactive power compensation capabilities of V2G-enabled electric vehicles. IEEE Trans Power Electron 2017;32(12):9447–59. https://doi.org/10.1109/TPEL.2017.2658686.

Chang M, Bae S, Yoon GG, Park SH, Choy Y. Impact of electric vehicle charging demand on a Jeju Island radial distribution network. In: IEEE Power and Energy Society Innovative Smart Grid Technologies Conference, ISGT 2019. South Korea: Institute of Electrical and Electronics Engineers Inc; 2019. https://doi.org/10.1109/ISGT.2019.8791604.

Comi A, Idone I. The use of electric vehicles to support the needs of the electricity grid: a systematic literature review. Appl Sci 2024;14(18):8197. https://doi.org/10.3390/app14188197.

Daneshzand F, Coker PJ, Potter B, Smith ST. EV smart charging: how tariff selection influences grid stress and carbon reduction. Appl Energy 2023;348. https://doi.org/10.1016/j.apenergy.2023.121482.

Das HS, Rahman MM, Li S, Tan CW. Electric vehicles standards, charging infrastructure, and impact on grid integration: a technological review. Renew Sustain Energy Rev 2020;120:109618. https://doi.org/10.1016/j.rser.2019.109618.

Dik A, Omer S, Boukhanouf R. Electric vehicles: V2G for rapid, safe, and green EV penetration. Energies 2022;15:803. ed: s Note: MDPI stays neu-tral with regard to jurisdictional claims in …, 2022.

Elazim SMA, Elkholy MH, Elgarhy A, Senjyu T, Gamil MM, Song D, et al. Enhancing stability and power quality in electric vehicle charging stations powered by hybrid energy sources through harmonic mitigation and load management. Sci Rep 2025;15(1). https://doi.org/10.1038/s41598-025-14143-4.

Elnozahy MS, Salama MMA. A comprehensive study of the impacts of PHEVs on residential distribution networks. IEEE Trans Sustain Energy 2014;5(1):332–42. https://doi.org/10.1109/TSTE.2013.2284573.

Gao Q, Li H, Peng K, Zhang C, Qu X. A real-time charging price strategy of distribution network based on comprehensive demand response of EVs and cooperative game. J Energy Storage 2024;101:113805. https://doi.org/10.1016/j.est.2024.113805.

Global EV Outlook 2025: expanding sales in diverse markets. Paris: International Energy Agency IEA; 2025. https://www.iea.org/reports/global-ev-outlook-2025.

Henderson C, Egea-Alvarez A, Kneuppel T, Yang G, Xu L. Grid strength impedance metric: an alternative to SCR for evaluating system strength in converter dominated systems. IEEE Trans Power Deliv 2024;39(1):386–96. https://doi.org/10.1109/TPWRD.2022.3233455.

Hybrid electric vehicles. U.S. Department of Energy, Office of Energy Efficiency and Renewable Energy. https://afdc.energy.gov/vehicles/electric-basics-hev.

Iqbal S, et al. Aggregated electric vehicle-to-grid for primary frequency control in a microgrid- A review. In: IEEE 2nd International Electrical and Energy Conference (CIEEC), 4-6 Nov, 2018; 2018. p. 563–8. https://doi.org/10.1109/CIEEC.2018.8745952.

Jaman S, Verbrugge B, Hernandez Garcia O, Abdel-Monem M, Oliver B, Geury T, et al. Development and validation of V2G technology for electric vehicle chargers using combo CCS type 2 connector standards. Energies 2022;15(19):7364. https://doi.org/10.3390/en15197364.

Jia W, Pang X, Shen L, Tian J, Ma Y, Liu L. Optimal scheduling method of electric vehicle charging and discharging considering peak cutting and Valley Filling. In: Proceedings of the 36th Chinese Control and Decision Conference, CCDC 2024. China: Institute of Electrical and Electronics Engineers Inc; 2024. p. 2940–5. https://doi.org/10.1109/CCDC62350.2024.10587859.

Kaur AP, Singh M. Time-of-Use tariff rates estimation for optimal demand-side management using electric vehicles. Energy 2023;273:127243. https://doi.org/10.1016/j.energy.2023.127243.

Kazemtarghi A, Dey S, Mallik A. Optimal utilization of bidirectional EVs for grid frequency support in power systems. IEEE Trans Power Deliv 2023;38(2):998–1010. https://doi.org/10.1109/TPWRD.2022.3203654.

Khalid MR, Alam MS, Sarwar A, Jamil Asghar MS. A Comprehensive review on electric vehicles charging infrastructures and their impacts on power-quality of the utility grid. eTransportation 2019;1:100006. https://doi.org/10.1016/j.etran.2019.100006.

Khan M, Sun H, Xiang Y, Shi Di. Electric vehicles participation in load frequency control based on mixed H2/H∞. Int J Electr Power Energy Syst 2021;125:106420. https://doi.org/10.1016/j.ijepes.2020.106420.

Khan TA, Abdullah-Al-Nahid S, Tasnim S, Taseen MA, Jamal T, Aziz T. Empowering E-mobility: day-ahead dynamic time of use tariff for electric vehicle charging. Energy Rep 2024;12:3218–42. https://doi.org/10.1016/j.egyr.2024.08.086.

Kirchner SR. OCPP interoperability: a unified future of charging. World Electr Vehic J 2024;15(5). https://doi.org/10.3390/wevj15050191.

Kumar M, Panda KP, Naayagi RT, Thakur R, Panda G. Comprehensive review of electric vehicle technology and its impacts: detailed investigation of charging infrastructure, power management, and control techniques. Appl Sci 2023;13(15):8919. https://doi.org/10.3390/app13158919.

Lehtola T. Solar energy and wind power supply supported by battery storage and Vehicle to Grid operations. Elec Power Syst Res 2024;228:110035. https://doi.org/10.1016/j.epsr.2023.110035.

Liu Q, Li Y, Xu T, Qian F, Meng H, Yao Y, et al. Peak shaving potential and its economic feasibility analysis of V2B mode. J Build Eng 2024;90:109271. https://doi.org/10.1016/j.jobe.2024.109271.

López Cascales JJ, Juan-Segovia MC, Ibáñez Molina J, Sánchez Vera J, Vivo Vivo PM. Environmental impact associated with the substitution of internal combustion vehicles by fuel cell vehicles refueled with hydrogen generated by electrolysis using the power grid. An estimation focused on the Autonomous Region of Murcia (Spain). Renew Energy 2015;77:79–85. https://doi.org/10.1016/j.renene.2014.11.082.

Lu F, Zhou X, Jiang ZLQ, Dong H, Li Y, Xue Q. A distributed electric vehicle charging and discharging strategy for grid power peak shaving and Valley Filling. In: 2023 5th International Conference on Industrial Artificial Intelligence, IAI 2023. China: Institute of Electrical and Electronics Engineers Inc; 2023. https://doi.org/10.1109/IAI59504.2023.10327531.

Marini A, Latify MA, Ghazizadeh MS, Salemnia A. Long-term chronological load modeling in power system studies with energy storage systems. Appl Energy 2015;156:436–48. https://doi.org/10.1016/j.apenergy.2015.07.047.

Mojumder MdRH, Antara FA, Hasanuzzaman M, Alamri B, Alsharef M. Electric vehicle-to-grid (V2G) technologies: impact on the power grid and battery. Sustainability 2022; 14(21):13856. https://doi.org/10.3390/su142113856.

Nazari-Heris M, Abapour M, Mohammadi-Ivatloo B. An updated review and outlook on electric vehicle aggregators in electric energy networks. Sustainability 2022;14(23): 15747. https://doi.org/10.3390/su142315747.

Nogueira T, Magano J, Sousa E, Alves GR. The impacts of battery electric vehicles on the power grid: a Monte Carlo method approach. Energies 2021;14(23). https://doi.org/10.3390/en14238102.

Nour M, Chaves-Ávila JP, Magdy G, Sánchez-Miralles Á. Review of positive and negative impacts of electric vehicles charging on electric power systems. Energies 2020;13(18). https://doi.org/10.3390/en13184675.

Rajendran G, Vaithilingam CA, Misron N, Naidu K, Ahmed MR. A comprehensive review on system architecture and international standards for electric vehicle charging stations. J Energy Storage 2021;42. https://doi.org/10.1016/j.est.2021.103099.

Rana MM, et al. Comprehensive review on the charging technologies of electric vehicles (EV) and their impact on power grid. IEEE Access 2025;13.

Rao SP, Olusegun TS, Ranganathan P, Kose U, Goveas N. Vehicle-to-Grid technology: opportunities, challenges, and future prospects for sustainable transportation. J Energy Storage 2025;110. https://doi.org/10.1016/j.est.2024.114927.

Rouhana F, Bagtzoglou AC, Zhu J. Bridging vehicle-to-home technology and equity: enhancing household resilience for disaster preparedness and response. Sustainability 2025;17(17):8052. https://doi.org/10.3390/su17178052.

Safayatullah M, Elrais MT, Ghosh S, Rezaii R, Batarseh I. A comprehensive review of power converter topologies and control methods for electric vehicle fast charging applications. IEEE Access 2022;10:40753–93. https://doi.org/10.1109/ACCESS.2022.3166935.

Saraswathi V, Ramachandran VP. A comprehensive review on charger technologies, types, and charging stations models for electric vehicles. Heliyon 2024;10(20).

Shi R, Li S, Zhang P, Lee KY. Integration of renewable energy sources and electric vehicles in V2G network with adjustable robust optimization. Renew Energy 2020;153: 1067–80. https://doi.org/10.1016/j.renene.2020.02.027.

Singh H, Ambikapathy A, Logavani K, Arun Prasad G, Thangavel S. Plug-in hybrid electric vehicles (PHEVs). In: Electric vehicles: modern technologies and trends. Springer; 2020. p. 53–72.

Sinha P, Paul K, Deb S, Sachan S. Comprehensive review based on the impact of integrating electric vehicle and renewable energy sources to the grid. Energies 2023;16(6):2924. https://doi.org/10.3390/en16062924.

Srivastava A, Manas M, Dubey RK. Electric vehicle integration's impacts on power quality in distribution network and associated mitigation measures: a review. J Eng Appl Sci 2023;70(1). https://doi.org/10.1186/s44147-023-00193-w.

Steward DM. Critical elements of vehicle-to-grid (v2g) economics. Golden, CO (United States): National Renewable Energy Lab (NREL); 2017.

Turk G, Schittekatte T, Duenas-Martinez P, Joskow PL, Schmalensee R. Designing distribution network tariffs in the US with an application to increased electric vehicle adoption. Energy J 2025;46(6):143–65. https://doi.org/10.1177/01956574251365616.

Ul-Haq A, Cecati C, Strunz K, Abbasi E. Impact of electric vehicle charging on voltage unbalance in an urban distribution network. Intell Ind Syst 2015;1(1):51–60. https://doi.org/10.1007/s40903-015-0005-x.

Un-Noor F, Padmanaban S, Mihet-Popa L, Mollah MN, Hossain E. A comprehensive study of key electric vehicle (EV) components, technologies, challenges, impacts, and future direction of development. Energies 2017;10(8). https://doi.org/10.3390/en10081217.
Wang L, Qin Z, Slangen T, Bauer P, Van Wijk T. Grid impact of electric vehicle fast charging stations: trends, standards, issues and mitigation measures - an overview. IEEE Open J Power Electr 2021;2:56–74. https://doi.org/10.1109/OJPEL.2021.3054601.
Yilmaz M, Krein PT. Review of battery charger topologies, charging power levels, and infrastructure for plug-in electric and hybrid vehicles. IEEE Trans Power Electron 2013;28(5):2151–69. https://doi.org/10.1109/TPEL.2012.2212917.
Zhang S, Byrne S, Fennelly L, Sood D, O'Donnell J, O'Donnell T. Home energy optimization using vehicle-to-home. Open Res Euro 2025;5:48. https://doi.org/10.12688/openreseurope.19007.1.
Zhou Y, Gao F, Zhang Z, Zhao S, Xu X, Meng H, et al. Research on optimal voltage control of distribution network with the participation of EVs and PVs. Appl Sci 2023;13(10):5987. https://doi.org/10.3390/app13105987.

CHAPTER FOUR

Optimizing renewable-to-vehicle smart charging with machine learning and digital twin

Amirreza Shafieinejad[1], Sasan Azad[1], Kamran Jalilpoor[2], Robbert Claeys[2], Jan Desmet[2]
[1]Department of Electrical Engineering, Shahid Beheshti University, Tehran, Iran
[2]EnSy/Lemcko, Department of Electromechanical, Systems and Metal Engineering, Ghent University, Kortrijk, Belgium

4.1 Introduction

4.1.1 Motivation and background

The transition to electric vehicles (EVs) is transforming the global transportation and energy sectors. As the penetration of renewable energy sources has drastically increased, renewable-to-vehicle (R2V) smart charging has become a key solution to ensure an efficient and sustainable charging of EVs. R2V systems enable the intelligent management of charging schedules, optimizing energy distribution, minimizing reliance on fossil fuels, and enhancing the stability of the grid. However, one of the key challenges in executing an R2V smart charging scheme is accurate load forecasting, which involves accurately predicting EV charging demand under dynamic conditions. Errors in forecasting can result in suboptimal operation of the grid, higher operational costs, and a lack of utilization of renewable energy, hindering the full potential of clean energy integration (Bhatti et al., 2021; Naseri et al., 2023).

Traditional methods, such as an ordinary forecasting system, typically face difficulties in tackling the complex system of charging demand, which varies according to the behavior of users charging the station (Yu et al., 2024). Machine Learning (ML) has developed into an effective practice to enhance load forecasting accuracy by discovering patterns and trends from vast data sets. Traditional ML models, including time series forecasting, regression algorithms, Reinforcement Learning (RL), and hybrid ML models (e.g., ARIMA, SARIMA, LSTM-GRU, ANN-SVM, LSTM-CNN), have demonstrated high potential in achieving high

Renewable-to-Vehicle Smart Charging
ISBN: 978-0-443-36684-0
https://doi.org/10.1016/B978-0-443-36684-0.00016-3

accuracy in predicting EV charging loads (Francisco et al., 2023; Bachechi, 2022). In particular, hybrid ML models combine several techniques to overcome the limitations of individual models and increase the predicted accuracy (Ali et al., 2023). Moreover, the modified mutual information-based feature selection is a new tool that has been shown to improve prediction performance by retaining more relevant variables (Jafari and Byun, 2022).

Digital Twin (DT) technology is a game-changer for load forecasting. A DT is defined as a real-time virtual model of physical assets, such as EVs, charging stations, and grid infrastructure. DT, which are continuously updated with new live sensor data, represent a well-structured environment to improve prediction using ML models (Wu et al., 2024). The integration of DT into ML-based load forecasting allows the models to adapt continuously to real-world conditions, resulting in more responsive, reliable, and actionable forecasts for grid operators (Mamun et al., 2020). In particular, DT can address data quality, computational complexity, and scalability issues by supplying high-fidelity synthetic data for training ML models, thus decreasing forecasting errors (Ahmad et al., 2022).

Effective R2V smart charging requires precise load forecasting. The prediction of charging demand for EVs based on factors such as changing charging behaviors (Bhatti et al., 2021), energy price variations, and uncertainty in renewable energy supply is complex by nature. Forecasting errors can cause under-provisioning of energy, overloading transformers, and energy distribution without efficiency (Francisco et al., 2023). Load forecasting in multi-household charging networks becomes more important. Unlike single household users, shared EV charging systems need to adopt demand response management for efficient energy distribution to multiple users (Bachechi, 2022). Forecasts and adjustments during peak demand periods need to be made dynamically to avoid grid congestion. Digital twin models have also been introduced for real-time battery state prediction and system optimization in EV networks (Ali et al., 2023). Compared to most traditional statistical models, DL methods, in particular, LSTM and CNN, are now widely used to model and predict energy demand (Jafari and Byun, 2022).

4.1.2 Relevant literature

Load forecasting is a key input-assisted decision for industrial and fleet-based EV charging, as it aids in reducing operational downtime and enhances the charging schedule to match the renewable energy generation (Wu et al., 2024). Unlike residential EV charging, which is often highly stochastic due to individual behaviors and irregular schedules, fleet-based EV charging, particularly for delivery companies, can be more predictable and schedulable. This is because such fleets typically operate with predefined routes and known daily mileage, enabling accurate forecasting of charging needs (Francisco et al., 2023). However, in some cases involving shared-use fleets or operational variability, real-time adjustments may still be necessary to optimize charging under uncertainties in driving routes or state-of-charge estimations. In this scenario, DT improve the accuracy of predictions by simulating EV fleets and charging stations in real time (Ahmad et al., 2022; Ali et al., 2023).

DT provide high-fidelity, real-time simulation of EV charging infrastructure under real-world energy demand patterns (Francisco et al., 2023). Through establishing a closed feedback loop between the actual energy consumption of the home and the energy prediction model based on ML, DT enhance the adaptability and accuracy of the prediction model (Bachechi, 2022). DT have an interesting potential in R2V smart charging systems, where the major challenge of renewable energy variability can be handled using them (Ali et al., 2023). The power generation of solar and wind energy varies according to the environmental conditions, leading to uncertainty in energy availability for EV charging (Jafari and Byun, 2022). By processing historical and real-time data, DT can simulate different scenarios, which enables machine learning models to dynamically optimize energy distribution and the scheduling of loads (Wu et al., 2024).

Other key benefits of DT include battery health monitoring and predictive maintenance of battery packs (Mamun et al., 2020). The state of health (SoH) and state of charge (SoC) adversely influence the charging demand; thus, battery degradation has a huge effect on EV load forecasting (Ahmad et al., 2022). ML models built using battery data produced by DT can improve charging cycles, minimize energy losses, and extend battery longevity (Ali et al., 2023).

Error metrics, including Mean Absolute Error (MAE), Root Mean Squared Error (RMSE), and Mean Absolute Percentage Error (MAPE) are usually employed to assess load forecasting results of ML models (Jafari and Byun, 2022). In R2V applications, hybrid ML models that combine DL and statistical techniques exhibit the lowest prediction errors (Wu et al., 2024). Furthermore, by embedding DT in load forecasting, prediction accuracy can be further ameliorated, and it can reduce large RMSE and MAE values (Mamun et al., 2020).

4.1.3 Contributions

This chapter bridges the gap between ML-based EV load forecasting and DT technology, contributing to the evolving landscape of smart energy systems. While previous studies have predominantly focused on either ML techniques for load forecasting or DT for energy management, this work explores a novel integration of both. By integrating predictive ML models with the real-time simulation features of DT, dynamic load profiles can be created that consider optimal charging times and power levels, thereby facilitating intelligent R2V charging strategies. The chapter emphasizes the value of hybrid ML models for improved prediction accuracy and DT for system adaptability, applied across two case studies: multi-household residential charging and industrial EV fleet depots. These case studies demonstrate the practical application and scalability of the proposed approach. Challenges such as data availability, computational requirements, and model scalability are also addressed. Through this exploration, the chapter offers a high-accuracy, scalable methodology for forecasting EV demand, optimizing energy allocation, and enhancing grid resilience in R2V systems through the integration of ML and DT technologies.

To illustrate the impact of this integration, Fig. 4.1 compares actual EV charging demand with forecasts generated by traditional methods and those enhanced by ML and DT. The figure shows that traditional forecasting models deviate noticeably from the real demand profile, particularly during peak hours, which can lead to energy mismanagement and grid stress. In contrast, the forecasting approach that incorporates ML and DT technology closely follows the actual demand curve. This demonstrates the model's ability to adapt to dynamic consumption patterns, thereby validating its effectiveness. Fig. 4.1 reinforces the chapter's central premise: that integrating DT with ML results in more accurate, responsive, and efficient load forecasting for R2V smart charging applications.

4.1.4 Organization

The remainder of this chapter is structured as follows. Section 4.2 introduces the concept of DT in the context of R2V smart charging, including their role in scenario simulation and system optimization. Section 4.3 reviews key forecasting models, including traditional statistical methods and

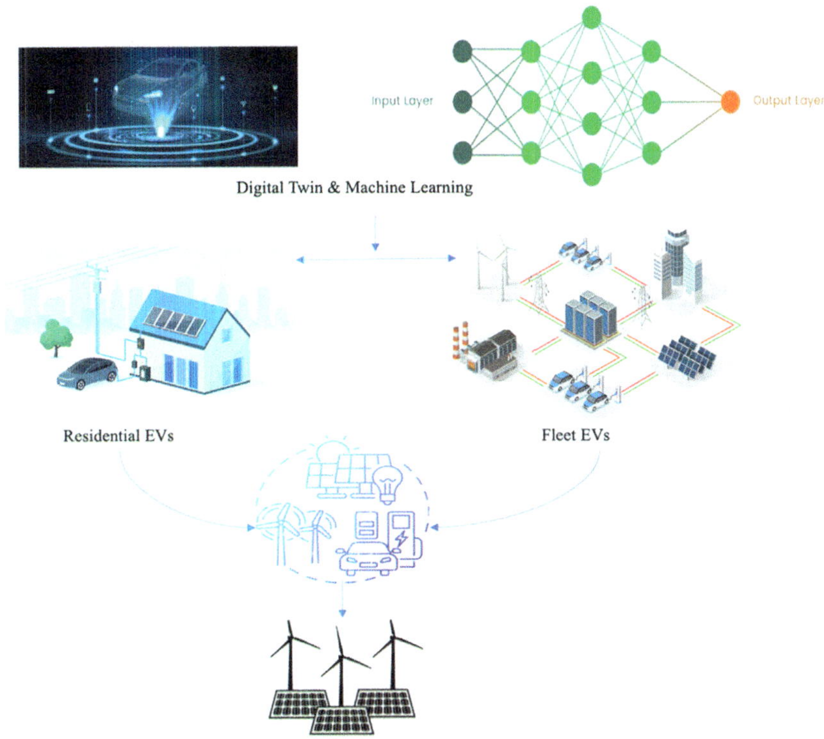

Figure 4.1 Graphical abstract.

advanced machine learning approaches, alongside their performance metrics. Section 4.4 explains the experimental setup, data sources, and evaluation criteria used for the case studies. Section 4.5 presents and analyzes results from both residential and industrial scenarios, highlighting cost savings, load shaping, and the integration of renewable energy sources. Section 4.6 concludes with a summary of findings and outlines future research directions.

4.2 DT in R2V smart charging

DT technology has emerged as a transformative innovation in R2V smart charging systems, enabling real-time mirroring, monitoring, and optimization of EV charging infrastructure. DT enable the creation of active virtual models that represent physical systems to analyze component behavior and operational conditions and predict future performance of battery systems, charging stations, and grid interfaces. A DT consists of three essential components: physical assets and their virtual duplicates, and a two-way data connection that maintains continuous synchronization. Originating from the aerospace and manufacturing industries, DT are now widely adopted across the power and transportation sectors (Yu et al., 2024; Francisco et al., 2023). In R2V smart charging, DT help address critical challenges, including demand uncertainty, renewable energy variability, and battery degradation.

DT are deployed as part of a larger ecosystem that includes Internet of Things sensors, edge devices, and cloud-based analytics. At the physical layer, EVs, chargers, and energy storage systems continuously collect and transmit data to a communication layer, which feeds this data into virtual models at the application layer. These models simulate key parameters such as SoC, SoH, charger availability, and load forecasts. The DT platform then makes recommendations or direct control adjustments to optimize system performance (Bachechi, 2022). For example, in residential communities where multiple EVs charge from a shared grid, DT can aggregate user profiles and schedule charging to minimize peak loads and align with renewable availability (Ali et al., 2023). In fleet scenarios, DT track route-based usage and battery performance, enabling predictive maintenance and charging optimization (Yu et al., 2021).

As illustrated in Fig. 4.2, the DT-enabled R2V smart charging ecosystem is composed of layered interactions between physical EV infrastructure, communication protocols, and data analytics platforms.

Battery management is a key application of DT in smart charging. Modern approaches combine physical models such as the Extended Kalman Filter with data-driven methods like XGBoost to improve the accuracy of

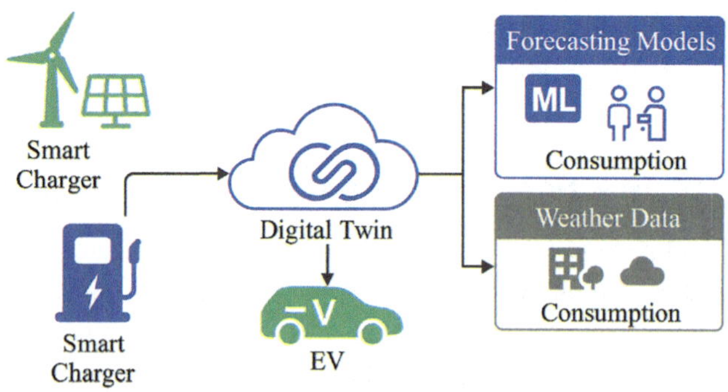

Figure 4.2 Conceptual architecture of a DT-enabled R2V smart charging ecosystem.

SoC and SoH estimation. These hybrid approaches lower the RMSE by almost 40% when compared to individual techniques (Ali et al., 2023; Jafari and Byun, 2022). In addition, DT facilitates thermal modeling and fault forecasting, promoting safer and more efficient charging cycles. By replicating battery behavior and aligning it with real-time sensor data, the DT can proactively identify irregularities and plan for maintenance or reduce charging as needed.

DT also facilitates scenario planning and grid resilience testing. For instance, in the University of Algarve pilot system, a DT platform modeled interactions between EVs, PV panels, and storage devices. By simulating charge-discharge cycles and inverter states, the system improved energy self-sufficiency and optimized infrastructure investments (Francisco et al., 2023).

Similarly, city-level implementations use DT to simulate the impact of traffic demand peaks, air quality shifts, or network adjustments, enabling testing of scenarios without affecting actual operations (Bachechi, 2022).

Table 4.1 summarizes key applications of DT technology across different R2V settings, highlighting their operational roles and associated benefits. In **residential communities**, DT enables SoC prediction and user load clustering, which support peak shaving strategies and reduce energy costs (Jafari and Byun, 2022). For **fleet depots**, DT facilitates route and schedule modeling, enhancing overall operational efficiency and

Table 4.1 DT functions in various R2V settings.

Scenario	DT application	Benefit
Residential communities	SoC prediction, load clustering	Peak shaving, aggregated cost optimization (grid-level benefits including deferred investments in hosting capacity upgrades, reduced grid losses, and mitigation of unbalance costs; may not directly lower individual costs)
Fleet depots	Route and schedule modeling	Operational efficiency, battery longevity
Urban public charging	Charger health, pricing optimization	Availability, maintenance
Institutional microgrids	Microgrid balancing, resilience testing	Energy independence, safety

extending battery lifespan. In **urban public charging** environments, DT is utilized for charger health diagnostics and pricing optimization, which contributes to improved service availability and reduced maintenance needs (Francisco et al., 2023). Lastly, in **institutional microgrids**, DT aids in microgrid balancing and resilience testing, helping ensure energy independence and safety. This classification illustrates how DT adapts to diverse deployment scenarios, offering tailored forecasting and management capabilities to support smarter, more resilient EV infrastructure (Aachen-MMP and Steiner, no date).

To ensure interoperability, DT platforms align with standards like ISO 15,118 and IEC 61,851, supporting secure data exchange and plug-and-play compatibility across hardware vendors (Aachen-MMP and Steiner, no date). Future advancements will see DT increasingly integrated with blockchain to enhance data privacy, security, and distributed decision-making in R2V environments (Banerjee et al., 2023). These technologies will empower DT to act not just as virtual mirrors but as autonomous agents in complex cyber-physical energy networks.

Another compelling advantage of using DT in R2V smart charging is its ability to support optimization under uncertainty. In real-world applications, EV user behavior, renewable energy generation, and electricity pricing are all highly variable. DT allows stakeholders to model these uncertainties using probabilistic distributions or stochastic simulations. For instance, Monte Carlo simulations within a DT platform can evaluate thousands of possible charging scenarios, providing insight into the most resilient scheduling strategies. Furthermore, DT can integrate fuzzy logic and scenario-based optimization to handle cases where data is imprecise or incomplete (Jafari and Byun, 2022; Alexopoulos et al., 2020).

Recent deployments also suggest that integrating DT with energy markets can enhance demand response programs. In this context, a DT forecasts the energy demand of a cluster of EVs, communicates with the utility's aggregator platform, and submits flexible load offers. If accepted, the system can dynamically shift the EV charging schedule to reduce grid congestion and improve revenue for participants. This makes DT a critical enabler of Vehicle-to-Grid business models, where EV batteries act as temporary grid-support assets, further enhancing renewable energy penetration and reducing fossil fuel dependency (Mamun et al., 2020; Alexopoulos et al., 2020).

An emerging use case for DT is its integration with smart buildings and energy communities. In these environments, the DT not only represents a single EV or charger but models the entire energy footprint of a building, including HVAC systems, lighting, occupancy patterns, and embedded renewables. When EVs are added to this system, the DT ensures optimal synergy between all energy-consuming components, reducing total demand peaks and enhancing system-wide energy efficiency (Wang et al., 2021; Yu et al., 2021). This multi-domain integration requires sophisticated data fusion algorithms and robust communication protocols that align with industry standards such as OpenADR and MQTT.

DT in R2V systems benefits from scalability as a fundamental advantage. A DT operates as a single controller for one charger and EV during small-scale testing. A cloud-based DT system can operate thousands of devices throughout different regions in real-world implementations.

Hierarchical coordination enables each device-level DT to act locally while feeding data into regional or grid-level DT that make higher-level optimization decisions. This hierarchical control structure mirrors the layered architecture of modern power systems, where local optimization contributes to broader system goals such as grid stability, emissions reduction, and cost minimization (Aachen-MMP and Steiner, no date; Saba et al., 2024).

Moreover, as artificial intelligence continues to evolve, next-generation DT is being enhanced with cognitive capabilities. These systems learn from data over time while identifying hidden patterns and proposing new control strategies. A cognitive DT system detects battery faults before they occur and predicts their development to recommend safety-enhancing preventive measures. Such systems are particularly valuable in mission-critical environments like public transportation or emergency response fleets, where downtime is unacceptable (Wu et al., 2024; Saba et al., 2024).

4.3 ML for EV load forecasting

EV charging introduces significant uncertainty to modern power systems due to its irregular and user-dependent nature (Lo Franco et al., 2023). As cities move toward more electrified and renewable-based energy infrastructures, forecasting EV charging demand becomes essential, especially within R2V innovative charging frameworks (Ostermann and Haug, 2024). Here, ML is pivotal in providing data-driven models that adapt to complex consumption behaviors, temporal patterns, and external influences such as weather and traffic (Shibl et al., 2021). ML helps forecast future EV load by learning from historical charging data and contextual signals, such as time of day, day of week, or weather variables (Deshmukh et al., 2024). Unlike linear regression (LR) or ARIMA models, which assume regular, stationary trends, ML algorithms can uncover nonlinear patterns and multivariate relationships that better reflect real-world EV usage, particularly in multi-residential complexes or industrial/shared depots, where demand surges and behavioral diversity are common (Shahriar et al., 2021). This section explores how both classical ML methods (e.g., Random Forests, SARIMA) and advanced models (e.g., Long Short-Term Memory (LSTM) neural networks) can be used for accurate EV load forecasting. Special attention is given to integrating DT systems. This feedback loop enhances model accuracy and robustness under changing conditions.

In practical R2V deployments, accurate forecasts are vital for:

- Balancing energy supply and demand from renewables.
- Scheduling smart charging sessions to avoid transformer overload.
- Reducing grid stress by shifting or delaying non-urgent loads.
- Enabling fleet and shared systems to coordinate charging without centralized supervision.

As shown in this chapter's case studies, ML-driven forecasting forms the foundation of smarter charging control systems. When paired with DT, these models can support real-time decision making, cost optimization, and resilience enhancement in EV infrastructures, bringing us closer to truly intelligent, sustainable transportation energy ecosystems.

4.3.1 Data engineering and feature extraction

High-quality, context-rich data is the foundation of accurate EV load forecasting using ML (Aduama et al., 2023). The performance of any predictive model, whether classical or DL, depends heavily on the availability, reliability, and relevance of input data (Aduama et al., 2023; Bampos et al., 2024). In R2V smart charging, data is sourced from smart meters, EV charging logs, battery management systems, electricity pricing data, and environmental sensors (Frendo et al., 2020). DT further augments this ecosystem by generating synthetic but realistic datasets based on physical system behavior.

Data preprocessing steps are critical to ensure clean and consistent inputs for the ML pipeline (Mumuni and Mumuni, 2025). These steps typically include missing value imputation, normalization or standardization, outlier removal, and noise filtering (Mumuni and Mumuni, 2025). For example, charging session logs may have missing timestamps or partially recorded energy values, which need correction before model training. Feature engineering transforms raw data into meaningful variables from which the model can learn. Commonly engineered features include time of day, day of week, historical rolling averages, and user-specific behavioral clusters. These temporal and behavioral features help capture both routine and exceptional charging patterns (Almaghrebi et al., 2024).

DT contributes by offering additional data abstraction layers, such as real-time estimates of SoC, SoH, predicted battery degradation, and charging station availability, thereby enabling deeper, more context-aware predictions (Nkechinyere Njoku et al., 2024).

Additionally, data fusion techniques allow for combining multiple heterogeneous sources, such as weather forecasts with EV usage patterns, to improve load predictions under varying conditions. Effective feature extraction ensures that the ML model captures the historical charging trends and the situational context in which those events occur (Renold and Kathayat, 2024). This ultimately leads to better generalization and more reliable performance in live deployments.

4.3.2 Classical time series forecasting models

Classical time series forecasting models have historically been the foundation for predicting EV charging loads due to their simplicity, statistical grounding, and interpretability. R2V innovative charging systems offer reliable baselines and continue to be useful when computational resources are limited, or patterns are relatively stable (Vishnu et al., 2023). This section provides an in-depth look at key time series models, both univariate and multivariate. It illustrates their application in EV load forecasting with examples, mathematical formulations, and diagrams to enhance understanding.

4.3.2.1 ARIMA (AutoRegressive integrated moving average)

ARIMA is a classical univariate time series forecasting model that exhibits autocorrelation and trend (Yadav et al., 2024; Khashei et al., 2009). It is denoted as ARIMA(p, d, q), where:

p: number of autoregressive terms (AR),

d: number of differencing operations to achieve stationarity,

q: number of moving average (MA) terms.

The general form is:

$$y'_t = c + \sum_{i=1}^{p} \phi_i y'_{t-i} + \sum_{j=1}^{q} \theta_j \varepsilon_{t-j} + \varepsilon_t \tag{4.1}$$

Where y_t' is the differenced series:

$$y_t' = \Delta^d y_t = (1 - B)^d y_t \tag{4.2}$$

with B being the backshift operator: $By_t = y_{t-1}$

Updated Description:

The general form of an ARIMA(p, d, q) model is:

$$\Delta^d y_t = c + \sum_{i=1}^{p} \phi_i \Delta^d y_{t-i} + \sum_{j=1}^{q} \theta_j \varepsilon_{t-j} + \varepsilon_t \tag{4.3}$$

Where:

y_t: original EV load at time t

$\Delta^d y_t$: differenced time series of order d

ε_t: white noise error at time t

ϕ_i: autoregressive coefficients

θ_j: moving average coefficients

c: constant (optional, represents long-term mean)

Example: A residential EV station showing gradually increasing usage can be modeled using ARIMA(1,1,1). Here, differencing ($d = 1$) removes the trend, AR(1) captures autocorrelation, and MA(1) models residual error (Matkovic et al., 2023; Kim and Kim, 2021).

The ARIMA model structure, as illustrated in Fig. 4.3, demonstrates how time series forecasting is achieved through the integration of three primary components: autoregression (AR), differencing (I), and moving average (MA). The raw time series input y_t first undergoes a differencing operation to remove trends and stabilize the data, yielding ∇_{yt}. This difference series is then processed in two parallel paths. One path passes through the autoregressive block, which captures temporal dependencies by

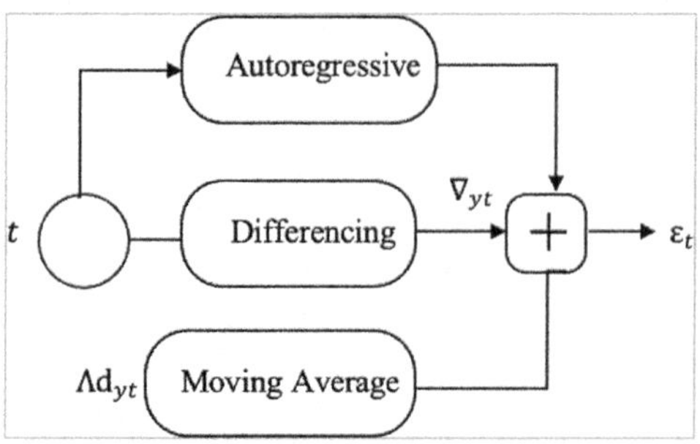

Figure 4.3 ARIMA model structure.

modeling the influence of past observations on the current value. The other path enters the moving average block, which accounts for patterns in the residual errors from prior forecasts. The outputs of these two blocks are then combined with the current error term ε_t to produce the final forecast. This structure allows ARIMA to model both short-term autocorrelations and residual noise, making it a foundational method in classical time series analysis, particularly effective for stationary or trend-differenced data (Kim and Kim, 2021).

4.3.2.2 SARIMA (seasonal ARIMA)

Accurate forecasting of EV charging demand requires models that can capture both trend and recurring seasonal behavior. The Seasonal Autoregressive Integrated Moving Average (SARIMA) model is an extension of the ARIMA model designed specifically to handle seasonal patterns in time series data. This makes it particularly well-suited for applications like EV load forecasting, where usage patterns often exhibit strong daily or weekly cycles (Paudel et al., 2024; Akshay et al., 2024).

A SARIMA model is typically denoted as:

$$\text{SARIMA}(p, d, q)(P, D, Q)_s \tag{4.4}$$

Where:

- p: order of the non-seasonal autoregressive (AR) component
- d: degree of non-seasonal differencing
- q: order of the non-seasonal moving average (MA) component
- P: order of the seasonal AR component
- D: degree of seasonal differencing
- Q: order of the seasonal MA component
- s: length of the seasonal cycle (e.g., 24 for hourly data with daily periodicity)

The full SARIMA model for a time series y_t is defined as:

$$\Phi_P(B^s)\phi_p(B)(1-B)^d(1-B^s)^D y_t = \Theta_Q(B^s)\theta_q(B)\epsilon_t \tag{4.5}$$

Where:

- B is the backshift operator (i.e., $B^k y_t = y_{t-k}$)
- ϕ_p (B): non-seasonal AR polynomial
- θ_q (B): non-seasonal MA polynomial
- $\Phi_P(B^s)$: seasonal AR polynomial
- $\Theta_Q(B^s)$: seasonal MA polynomial

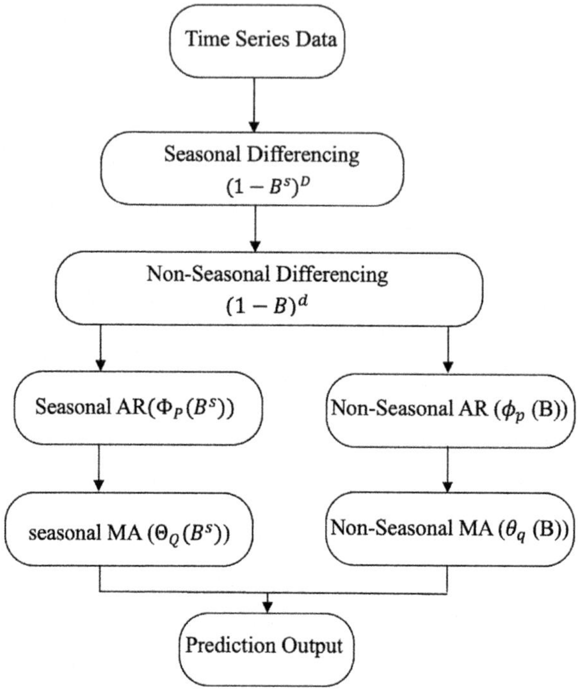

Figure 4.4 SARIMA model structure.

- $(1 - B)^d$: non-seasonal differencing operator, applies differencing d times to remove trend
- $(1 - B^s)^D$: seasonal differencing operator with seasonal period sss, differenced D times to remove seasonal patterns
- ϵ_t: white noise error term

The SARIMA model structure, as illustrated in Fig. 4.4, extends the foundational ARIMA framework by incorporating seasonal components to effectively capture recurring temporal patterns. The modeling process begins with differencing operations to remove both non-seasonal and seasonal trends, ensuring stationarity. Once stationarity is achieved, the differenced time series is modeled using autoregressive (AR) and moving average (MA) components, which account for temporal dependencies and shock effects, respectively. To capture seasonal effects, SARIMA includes additional seasonal AR and MA terms, along with seasonal differencing. These seasonal components help the model handle periodic behaviors such as daily, weekly, or monthly cycles. By integrating both seasonal and non-seasonal dynamics, SARIMA provides robust forecasting performance, making it particularly suitable for applications like EV load prediction, where cyclical usage patterns are prominent (Paudel et al., 2024).

In Case Study 1, SARIMA was implemented to forecast aggregate hourly charging load across users in a multi-residential apartment complex with time-shiftable charging behavior.The model configuration:

$$\text{SARIMA}(1,1,1)(1,1,1)_{24} \tag{4.6}$$

4.3.2.3 Holt–Winters exponential smoothing

Holt–Winters Exponential Smoothing is a classical statistical method that is commonly used for time series forecasting. It decomposes time series data into three key components: level, trend, and seasonality. This makes it particularly effective when the data exhibits consistent seasonal patterns, such as daily or weekly EV charging behaviors in residential areas or fleet depots (Abdullah, 2024).

The additive form of Holt–Winters can be represented by the Holt–-Winters (Attri et al., no date):

$$l_t = \alpha\left(y_t - s_{t-m}\right) + (1-\alpha)(l_{t-1} + b_{t-1}) \tag{4.7}$$

$$b_t = \beta(l_t - l_{t-1}) + (1-\beta)b_{t-1} \tag{4.8}$$

$$s_t = \gamma\left(y_t - l_{t-1} - b_{t-1}\right) + (1-\gamma)s_{t-m} \tag{4.9}$$

$$\hat{y}_{t+h} = l_t + hb_t + s_{t+h-\mathrm{m}} \tag{4.10}$$

Where:

- l_t: level at time t,
- b_t: trend
- s_t: seasonal component
- α,β,γ: smoothing coefficients (0–1)
- m: season length (e.g., 24 for daily cycles in hourly data)
- h: forecast horizon
- $\widehat{y}_{t+h}$: predicted value h steps ahead

Holt–Winters provides a simple yet effective forecasting framework for EV charging in neighborhoods or shared infrastructures where demand rises predictably each evening (Liu et al., 2020). It can track level shifts (e.g., more EVs over time), capture daily peak trends, and account for repeating usage patterns. For example, in a residential area where most users charge

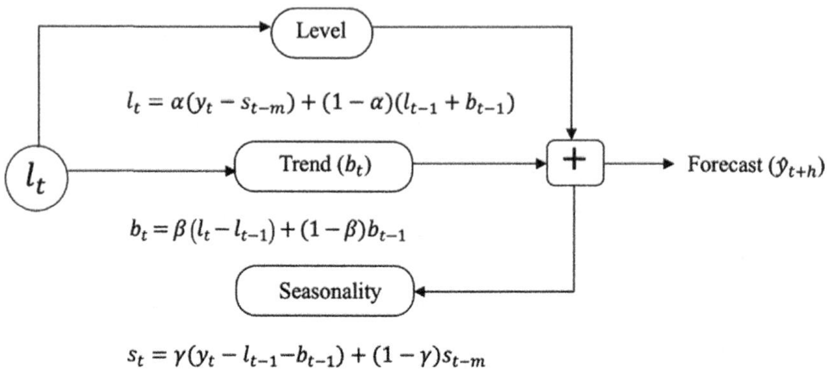

Figure 4.5 Holt–Winters exponential smoothing model structure.

after returning home between 6 and 9:00 p.m., Holt–Winters accurately models this evening spike using seasonality and trend smoothing.

The structure depicted in Fig. 4.5 illustrates the Holt–Winters additive model, which decomposes a time series into three fundamental components: level, trend captured via exponential trend smoothing and seasonality to generate forecasts. The Level component captures the baseline value of the series at time t, updating it using a smoothing coefficient α to balance current observations and past estimates. The Differencing (or trend) component, updated via β, accounts for changes between consecutive levels to reflect trend dynamics. The Seasonality component, managed by γ, isolates repetitive patterns over fixed intervals (e.g., daily or weekly cycles). These three elements are then combined in a summation step to produce the forecast $\widehat{y}_{t+h}$. The model is well-suited for time series with stable, additive seasonal patterns, such as recurring daily EV charging behaviors in residential neighborhoods (Attri et al., no date; Liu et al., 2020).

4.3.2.4 Vector AutoRegression (VAR)

VAR is a powerful statistical model used for capturing the linear interdependencies among multiple time series (Chiang et al., 2024). Unlike univariate models such as ARIMA or SARIMA, which handle only a single time series, VAR simultaneously models multiple variables by treating each variable as a function of its own past values and the past values of all other variables in the system. This makes VAR particularly suitable for multivariate systems like EV charging demand, where various factors such as weather, time of day, traffic flow, and user behavior influence the load patterns (Szostek et al., 2024).

The general form of a VAR(p) model, where p is the number of lags, is:

$$y_t = c + A_1 y_{t-1} + A_2 y_{t-2} + \cdots + A_p y_{t-p} + u_t \tag{4.11}$$

Where:

- $y_t \in R^k$: vector of all k variables at time t (e.g., EV load, temperature, humidity, traffic count).
- $c \in R^k$: intercept vector (often included unless a mean-zero process is explicitly assumed).
- $A_i \in R^{k \times k}$: coefficient matrix for the i-th lag.
- $u_t \in R^k$: vector of white noise error terms, assumed to be zero-mean with constant variance.

For R2V smart charging, forecasting EV load can benefit from a Vector AutoRegression with Exogenous inputs (VARX) framework, which incorporates external variables such as GHI, temperature, and traffic counts. These exogenous signals, although not influenced by the system itself, play a critical role in shaping temporal load patterns (Lee et al., 2024).

The VARX model improves accuracy over univariate approaches by modeling lagged dependencies among endogenous load variables while integrating the causal impact of exogenous factors. This is particularly relevant in industrial and multi-residential EV infrastructure, where demand is strongly influenced by synchronized human activity and environmental dynamics (Szostek et al., 2024; Lee et al., 2024).

Fig. 4.6 illustrates the structure of a VAR model applied to multivariate time series forecasting. Each variable in the system, such as electrical load, weather conditions, and traffic volume, is modeled as a function of its own past values (lags) as well as the lagged values of other variables in the system. For instance, the load at time t is influenced by past values of load ($Load_{t-1}$), weather ($Weather_{t-1}$), and traffic ($Traffic_{t-1}$). Similarly,

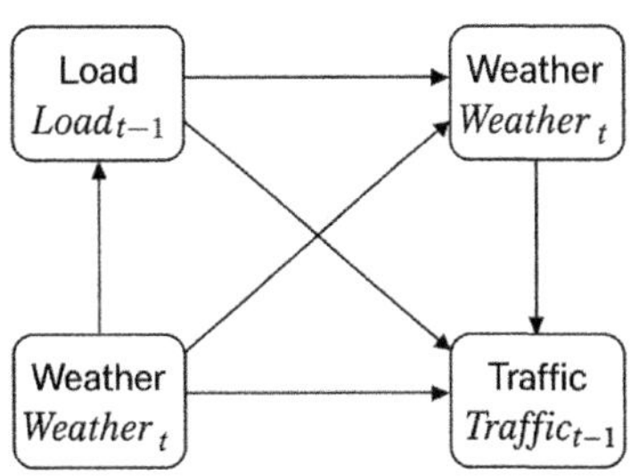

Figure 4.6 Diagram of the VAR model structure.

the weather at time t is affected by its own prior value and possibly the historical values of traffic and load. This interconnected structure enables VAR to capture the mutual dependencies and temporal dynamics across multiple time series simultaneously, which is particularly useful in EV load forecasting, where charging demand is driven not only by historical usage but also by exogenous factors such as temperature fluctuations and traffic patterns (Lee et al., 2024).

4.3.2.5 ARCH and GARCH models

Autoregressive Conditional Heteroskedasticity (ARCH) and Generalized ARCH (GARCH) models are designed to capture and model time-varying volatility in time series data. Unlike classical forecasting models that assume constant variance over time, ARCH/GARCH models account for heteroskedasticity, a condition where the variability of a time series changes over time, a common phenomenon in energy systems, especially under high EV penetration and intermittent renewable generation (Tsionas and Kumbhakar, 2021).

In the context of EV load forecasting, these models are advantageous when load variability increases due to behavioral unpredictability, dynamic pricing, or weather-influenced renewable integration. For example, if large deviations in EV demand tend to cluster during certain hours or days, ARCH/GARCH models can better anticipate these volatility clusters than conventional methods.

The ARCH(q) model expresses the conditional variance of the current observation as a linear function of past squared residuals (Tsionas and Kumbhakar, 2021; Zhang et al., 2021):

$$\epsilon_t = \sigma_t z_t, z_t \sim \mathcal{N}(0,1) \tag{4.12}$$

$$\sigma_t^2 = \alpha_0 + \sum_{i=1}^{q} \alpha_i \epsilon_{t-i}^2 \tag{4.13}$$

Where:

- ϵ_t: residual at time t,
- σ_t^2: conditional variance
- $\alpha_0 > 0, \alpha_i \geq 0$: model parameters

ARCH models are sensitive to recent squared errors; a large deviation increases the predicted variance, capturing volatility bursts.

The GARCH(p, q) model generalizes ARCH by incorporating both past squared residuals and past conditional variances. This enables the model to capture short-term volatility shocks as well as long-term persistence, making it effective in modeling volatility clustering and bursty behaviors.

The GARCH(p, q) model generalizes ARCH by incorporating past conditional variances (Tsionas and Kumbhakar, 2021):

$$\sigma_t^2 = \alpha_0 + \sum_{i=1}^{q} \alpha_i \epsilon_{t-i}^2 + \sum_{j=1}^{p} \beta_j \sigma_{t-j}^2 \tag{4.14}$$

Where:

- σ_t^2: The **conditional variance** of the time series at time t represents the volatility (or uncertainty) at time t.
- α_0: A **constant term** (must be positive) representing the long-run average variance when past influences are zero.
- ϵ_{t-i}^2: The **squared residual (error)** term from i periods ago. This reflects how recent shocks (news/events) contribute to the current variance.
- α_i: The **ARCH parameters**, which measure the impact of past squared shocks on current variance. Larger values imply that volatility reacts strongly to past shocks.
- σ_{t-j}^2: The **past conditional variances** (i.e., past estimates of volatility).
- β_j: The **GARCH parameters**, which reflect the persistence of past volatility. High values mean that volatility tends to remain elevated after a shock.
- p: The **order** of the GARCH component, determines how many lagged conditional variances are included.
- q: The **order** of the ARCH component, determines how many past squared residuals are considered.

In R2V scenarios, volatility in EV charging behavior driven by factors like solar production drops, sudden traffic surges, or unexpected fleet returns introduces time-varying uncertainty. GARCH models enable system operators and aggregators to quantify risk in load forecasts, allocate reserves accordingly, and trigger adaptive charging strategies (Tsionas and Kumbhakar, 2021; Zhang et al., 2021).

GARCH is particularly suited for.

- Forecasting **load volatility**, not just the mean load.
- Risk-sensitive applications (e.g., energy markets, flexible contracts).
- Enhancing the robustness of ML models when integrated as an uncertainty estimator.

4.3.2.6 Bayesian structural time series (BSTS)

Bayesian Structural Time Series (BSTS) models offer a probabilistic and modular framework for time series forecasting, particularly well-suited for the complexities of EV charging demand. Rather than relying on strict stationarity assumptions, BSTS decomposes the time series into structural components, namely, trend, seasonality, regressors, and observation noise, each estimated within a Bayesian framework (van Etten et al., 2024; Jeon et al., 2022). The observed load at time t, denoted by y_t, is modeled as (Stefenon et al., 2023):

$$y_t = \mu_t + \gamma_t + \beta^T X_t + \epsilon_t \tag{4.15}$$

where μ_t captures the long-term trend, γ_t represents recurring seasonal patterns (such as daily charging cycles), $\beta^T X_t$ reflects the dot product of covariates with their respective coefficients (e.g., solar irradiance, temperature, traffic), and ϵ_t is the random noise component. This formulation allows BSTS to explicitly model external drivers of EV charging demand in a probabilistic framework.

What sets BSTS apart is its use of Bayesian updating, allowing the model to continuously refine its estimates as new data becomes available. This capability is particularly beneficial in smart charging environments where demand patterns shift in response to behavioral, environmental, or infrastructural changes.

Additionally, BSTS models often leverage *spike-and-slab priors* for automatic variable selection. This prior distribution combines a "spike" near zero (promoting sparsity) with a broader "slab" (allowing flexibility), enabling the model to retain only the most informative predictors. This improves generalization and helps prevent overfitting, especially in high-dimensional EV forecasting tasks (Jeon et al., 2022).

In EV load forecasting scenarios, especially in residential neighborhoods or shared charging infrastructures, BSTS facilitates informed decision-making by providing not just point estimates but also credible intervals around predictions. This probabilistic insight enhances robustness in energy allocation, load scheduling, and grid integration strategies.

Fig. 4.7 illustrates the architecture of the BSTS model. The observed load is decomposed into three core components: a trend component μ_t that captures long-term changes in demand, a seasonal component γ_t that accounts for repeating cycles, and a regression component βX_t that models the influence of external covariates. These components are inferred

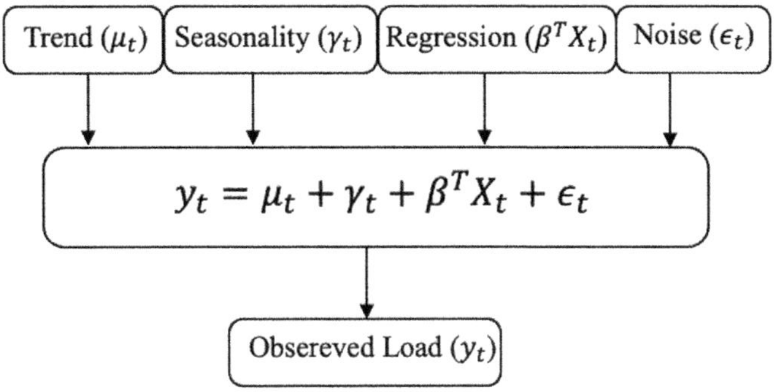

Figure 4.7 Diagram of the BSTS model structure.

simultaneously using Bayesian techniques, such as Markov Chain Monte Carlo, which is a class of algorithms that generate samples from complex probability distributions by constructing a Markov chain that converges to the target posterior distribution. This inference strategy enables the BSTS model to provide uncertainty-aware forecasts. The structure allows for adaptive learning from both historical data and contextual signals, making it particularly suitable for EV load forecasting in dynamic, data-rich environments (Jeon et al., 2022).

4.3.2.7 Facebook prophet: A practical approach for seasonality-rich forecasting

Facebook Prophet is a powerful time series forecasting model specifically designed to handle data with strong seasonal effects, irregular trends, and missing observations. Built with a focus on simplicity and interpretability, Prophet allows users, especially practitioners without deep statistical expertise, to produce reliable forecasts with minimal tuning. It has become especially popular in fields such as business analytics, energy forecasting, and demand planning (Stefenon et al., 2023).

Prophet models a time series as a combination of three main components: trend $g(t)$, seasonality $s(t)$, and holidays or events $h(t)$. The model takes the general additive form Stefenon et al. (2023):

$$y(t) = g(t) + s(t) + h(t) + \epsilon_t \tag{4.16}$$

Where:

- $g(t)$ captures non-periodic changes such as growth or saturation trends, modeled either as a piecewise linear or logistic growth curve.
- $s(t)$ models periodic patterns such as daily, weekly, or yearly cycles using a Fourier series.

- $h(t)$ accounts for user-specified irregular events like holidays or promotional campaigns.
- ϵ_t is the error term capturing residual variability.

One of Prophet's key innovations lies in its automatic handling of changepoints, abrupt shifts in the trend, by placing a sparse prior on the location and magnitude of these shifts. This means the model assumes that changepoints are rare unless there is strong evidence in the data to suggest otherwise, effectively avoiding overfitting to noise. This makes it robust to real-world data, where unexpected changes in user behavior or system dynamics are common.

In the context of EV load forecasting, Prophet is particularly advantageous for scenarios involving:

- Regular and predictable charging behaviors (e.g., residential areas with evening charging peaks),
- Calendar-based effects such as weekday versus weekend patterns,
- Irregular disruptions (e.g., holidays) that significantly influence demand.

Prophet supports the inclusion of external regressors, enabling univariate forecasting enhanced by contextual covariates (e.g., weather, pricing signals, occupancy), while remaining fundamentally a univariate model rather than a fully multivariate forecasting framework.

Its ease of use, transparency in component decomposition, and strong performance on seasonality-rich data make Prophet an ideal candidate for real-time energy forecasting in both research and operational settings.

4.3.3 Model comparison table

Each of these models is best suited to different forecasting environments. A comparison of their capabilities is shown in Table 4.2.

Table 4.2 Comparison of time series models for R2V applications.

Model	Seasonality	Nonlinearity	Multivariate	Real-time adaptation[a]
ARIMA	£	£	£	✓
SARIMA	✓	£	£	✓
Holt–Winters	✓	£	£	✓
VAR	£	£	✓	✓
ARCH/GARCH	£	✓	£	✓
BSTS	✓	✓	✓	✓
Prophet	✓	✓	£	✓

[a]Real-Time Adaptation refers to the model's built-in ability to incorporate new data and update forecasts without full retraining. While some models (e.g., BSTS, Prophet) support this natively, others (e.g., ARIMA, SARIMA) require manual re-estimation to adapt.

4.3.4 Supervised ML models

As the complexity and variability of EV charging patterns grow, supervised ML models have gained prominence due to their ability to learn from historical data and capture nonlinear relationships between multiple input features and future charging load. These models are trained using labeled datasets where input variables (e.g., time of day, temperature, SoC) are mapped to known output variables such as charging demand at a specific time (Gujjarlapudi et al., 2023). Supervised ML methods offer significant advantages over classical time series approaches:

1. Handle non-stationary and nonlinear data
2. Support multivariate input vectors
3. Scale well with large datasets
4. Often outperform classical models in short-term and day-ahead EV load forecasting

This section outlines commonly used supervised learning algorithms in R2V smart charging environments.

4.3.4.1 Linear regression

LR is one of the most fundamental and widely used supervised learning techniques. In its simplest form, LR establishes a linear relationship between one or more input features and a continuous output variable. Its popularity stems from its simplicity, interpretability, and effectiveness in modeling problems where a linear function can reasonably approximate the influence of each input variable (Maulud and Abdulazeez, 2020).

In the context of EV load forecasting, LR can be used to predict charging demand based on explanatory features such as time of day, day of the week, outdoor temperature, user group identifiers, and solar irradiance. The general form of the LR model is (Maulud and Abdulazeez, 2020):

$$\hat{y}_t = \beta_0 + \beta_1 x_{1,t} + \beta_2 x_{2,t} + \cdots + \beta_n x_{n,t} + \varepsilon_t \tag{4.17}$$

where $\widehat{y}_t$ is the predicted EV load at time t, $x_{i,t}$ represents the i-th feature at time t, β_i are the learned coefficients, and ε_t is the error term.

While LR and VAR share mathematical similarities, both involving linear combinations of input variables, their use cases and modeling philosophies differ. VAR is a multivariate time series model designed to capture dependencies between multiple interrelated variables (e.g., EV load and traffic volume), where each variable is modeled using lagged values of all variables in the system. In contrast, Linear regression in machine learning treats forecasting as a regression problem over arbitrary input features, which may include time-based lags or contextual covariates such as temperature, solar irradiance, or day type (Shahriar et al., 2020).

By treating the forecasting task as a supervised learning problem, LR allows the inclusion of engineered or external features that may not be part of a traditional time series framework. This makes it well-suited for hybrid settings where user behavior, weather patterns, or solar forecasts influence charging activity (Shahriar et al., 2020).

Although LR does not natively capture complex temporal dependencies or seasonality, it serves as a strong baseline in forecasting pipelines and is often used in ensemble approaches or as part of feature pipelines for more advanced models like LSTM or gradient-boosted trees (Shahriar et al., 2020).

Table 4.3 provides a comparative overview of VAR and LR within the context of load forecasting and time series modeling. While both techniques involve linear relationships between variables, their structural design, assumptions, and application areas differ significantly.

4.3.4.2 Support vector regression (SVR)

Support Vector Regression (SVR) is an extension of Support Vector Machines (SVM) designed for regression tasks. Rather than attempting to predict discrete class labels, SVR seeks to fit a function that approximates the data points **within a predefined margin of tolerance (ε)**. The core idea is to balance approximation and generalization, allowing for small

Table 4.3 Comparison: VAR versus LR.

Aspect	VAR	LR
Type	Time series model	Supervised ML model
Input	Lagged values of multiple time series	Arbitrary input features (may include lags)
Output	Forecast of multiple time series	Forecast of the single target variable
Temporal dependency	Strong time-lag focus	Depends on the feature design
Multivariate capability	Yes	Typically univariate output with multivariate input
Real-time adaptation	Limited (assumes stationarity)	Adaptable with re-training or online learning
Interpretability	Moderate LR:	High
Best use case	Coupled systems (e.g., load & traffic)	General predictive tasks with external inputs

deviations from the actual values while avoiding overfitting (Zhang and O'Donnell, 2020).

Mathematically, SVR attempts to solve the following optimization problem (Zhang and O'Donnell, 2020):

$$\min_{\mathbf{w},b,\xi_i,\xi_i^*} \frac{1}{2}\|\mathbf{w}\|^2 + C\sum_{i=1}^{n} \left(\xi_i + \xi_i^*\right) \tag{4.18}$$

Subject to:

$$\begin{gathered} y_i - (\mathbf{w}\cdot\phi(\mathbf{x}_i) + b) \\ \leq \varepsilon + \xi_i \\ (\mathbf{w}\cdot\phi(\mathbf{x}_i) + b) - y_i \\ \leq \varepsilon + \xi_i^* \\ \xi_i, \xi_i^* \\ \geq 0 \end{gathered} \tag{4.19}$$

Here, $\mathbf{w}$ denotes the weight vector and b is the bias term. The parameter ε defines the tube (or margin) around the true output values within which no penalty is assigned. The function $\phi(\mathrm{x}_i)$ denotes a kernel-induced mapping of the input features into a higher-dimensional space to capture nonlinear relationships. The slack variables ξ_i and ξ_i^* allow for some tolerance beyond this margin: ξ_i handles underestimation and ξ_i^* handles overestimation of the target value.

SVR is well-suited for EV load forecasting, particularly in scenarios where nonlinear relationships exist between input features (such as temperature, time of day, and occupancy) and EV load. Its ability to use kernel tricks (like RBF or polynomial kernels) allows it to model complex, nonlinear data distributions while still maintaining robustness to outliers and noise (Wu et al., 2024).

For R2V smart charging applications, SVR can be employed to predict short-term load patterns, adjust charging strategies dynamically, and reduce peak demand by anticipating usage spikes. Compared to traditional linear models, SVR offers better generalization in high-dimensional spaces and with small to medium-sized datasets (Wu et al., 2024).

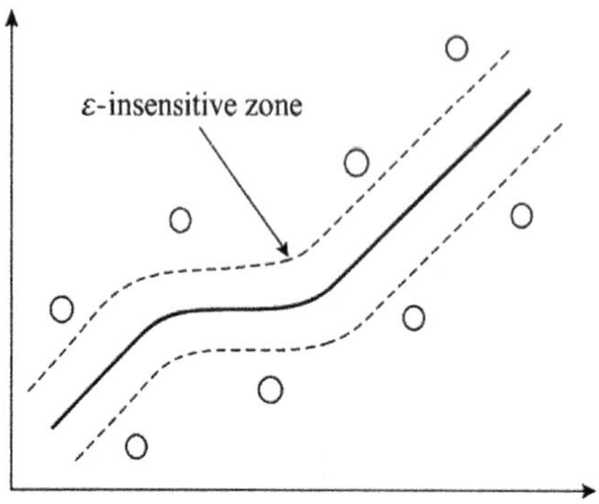

Figure 4.8 Diagram of the SVR model structure.

However, SVR does require careful tuning of hyperparameters such as the penalty parameter C, the epsilon margin ε, and kernel parameters. It is also computationally more intensive than simpler models like LR or ARIMA, especially when scaled to large datasets (Wu et al., 2024; Vishnu et al., 2023).

Fig. 4.8 illustrates how SVR constructs a regression function within a margin (ε-insensitive zone), optimizing fit while maintaining generalization [52. 53].

4.3.4.3 Decision trees and random forests

Decision Trees and Random Forests are supervised learning models widely applied in regression and classification tasks, known for their interpretability and strong performance with non-linear data. In the context of EV load forecasting, these models can efficiently capture complex relationships between time, weather, user behavior, and power consumption without requiring extensive preprocessing or assumptions about data distribution (Dudek, 2022; Chung et al., 2019).

4.3.4.3.1 Decision tree regression

A Decision Tree splits the dataset into subsets based on feature values, creating a tree-like model of decisions. In regression tasks, each terminal leaf node represents the average value of the target variable in that segment, based on minimizing the variance within each partition (Chung et al., 2019).

Mathematically, at each node, the model selects a split point S on the feature x_j that minimizes the sum of squared errors (SSE) (Chung et al., 2019):

$$\mathrm{SSE}(s) = \sum_{i \in R_1(s)} \left(y_i - \overline{y}_{R_1}\right)^2 + \sum_{i \in R_2(s)} \left(y_i - \overline{y}_{R_2}\right)^2 \tag{4.20}$$

Where:

- R_1 and R_2 are regions (branches) created by the split.
- $\overline{y}_R$ is the mean of the target values in the region R.
- y_i is the observed value.

Decision Trees are non-parametric, meaning they do not assume a fixed functional form, and can model nonlinear relationships. However, they tend to overfit on noisy data.

4.3.4.3.2 Random forest regression

To mitigate overfitting and improve generalization, Random Forests aggregate the predictions of multiple decision trees trained on different bootstrap samples and feature subsets. This ensemble approach boosts robustness and accuracy (Chung et al., 2019). The final prediction $\widehat{y}$ is the average of the outputs from all B trees:

$$\hat{y} = \frac{1}{B} \sum_{b=1}^{B} T_b(x) \tag{4.21}$$

Where:

- $T_b(x)$ is the prediction from the B-th tree.
- x is the input feature vector.
- B is the number of trees.

Random Forests are especially effective in high-dimensional or highly nonlinear forecasting tasks, making them a valuable tool for EV load prediction where contextual data (e.g., hour of day, temperature, previous load) plays a crucial role (Chung et al., 2019).

In Fig. 4.9 the left diagram illustrates a Decision Tree model for regression, where the feature space is partitioned into regions and the prediction in each region is the average of target values. The right diagram depicts a Random Forest, an ensemble of decision trees where each tree is trained on a random subset of the data, and their outputs are aggregated (typically

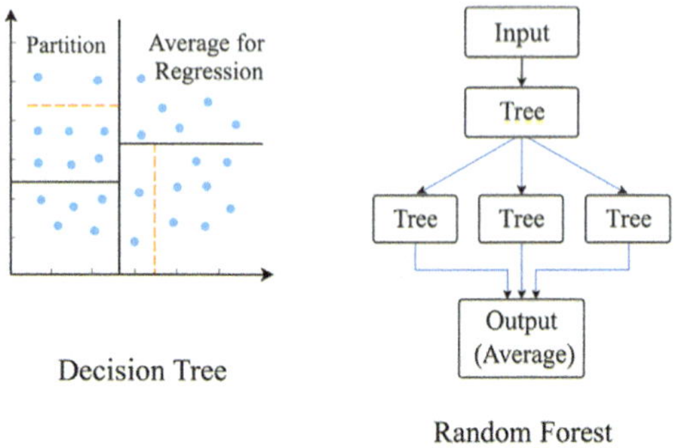

Figure 4.9 Visualization of decision tree and random forest structures.

via averaging) to produce a more robust and accurate prediction (Chung et al., 2019).

4.3.4.4 Gradient boosted trees (XGBoost)

Gradient Boosted Decision Trees (GBDT) are an advanced ensemble learning technique that builds a series of decision trees sequentially, with each tree attempting to correct the errors of its predecessor. XGBoost is a highly optimized implementation of GBDT, offering both scalability and accuracy, and it has become a popular choice for forecasting tasks in energy systems, including R2V) smart charging (Shern et al., 2024).

The fundamental concept of XGBoost lies in minimizing a regularized objective function that balances prediction accuracy with model complexity. The model incrementally adds trees that predict the residuals (errors) of the previous model and uses gradient descent to optimize the objective. At each iteration, the model learns a new function $f_t(x)$ that maps features to predictions, and updates the ensemble as follows (Shern et al., 2024):

$$\hat{y}_t = \sum_{i=1}^{t} f_i(x) \tag{4.22}$$

The training objective is:

$$\mathcal{L}(\phi) = \sum_{i=1}^{n} l\left(y_i, \hat{y}_i^{(t)}\right) + \sum_{k=1}^{t} \Omega\left(f_k\right) \tag{4.23}$$

where l is a differentiable loss function (e.g., mean squared error), and

$$\Omega(f) = \gamma T + \frac{1}{2}\lambda\|w\|^2 \tag{4.24}$$

penalizes the complexity of the tree f, with T being the number of leaves and w the weights of the leaves.

In the context of EV load forecasting, XGBoost is particularly useful because it:

- Handles nonlinear dependencies between input features (e.g., time of day, weather, historical usage).
- Incorporates categorical and numerical variables without extensive preprocessing.
- Performs implicit feature selection, reducing the need for manual engineering.
- Can model high-dimensional, sparse, and incomplete datasets effectively.

This makes XGBoost suitable for both residential and industrial case studies presented in this chapter. In Case Study 1 (multi-residential), it captures user-specific temporal behaviors, while in Case Study 2 (shared industrial depots), it integrates traffic, solar irradiance, and weather features to forecast clustered EV demand across hours. Its strong generalization capabilities and efficient training make it ideal for large-scale smart charging platforms (Shern et al., 2024; Olson et al., 2019).

Fig. 4.10 shows the sequential learning process of Gradient Boosted Trees, where each model is trained to predict the residuals (errors) of the previous model. Starting with initial training data, successive models (e.g., decision trees) are added to minimize prediction error. The final output is derived from a weighted sum of the predictions of all models in the ensemble. This iterative correction mechanism enhances accuracy and generalization (Kim and Kim, 2021).

4.3.5 DL approaches

As EV charging behaviors grow increasingly complex and nonlinear, DL models have emerged as powerful tools for R2V load forecasting. Unlike traditional ML methods, DL networks can autonomously extract high-level features and temporal dependencies from raw, multivariate data streams. Their capacity to generalize across diverse consumption patterns, learn spatiotemporal dynamics, and adapt in real time makes them particularly well-suited to R2V scenarios in both multi-residential and industrial contexts (Kim and Kim, 2021).

4.3.5.1 Long short-term memory (LSTM)

LSTM networks are a specialized type of Recurrent Neural Network (RNN) designed to capture long-term dependencies in sequential data.

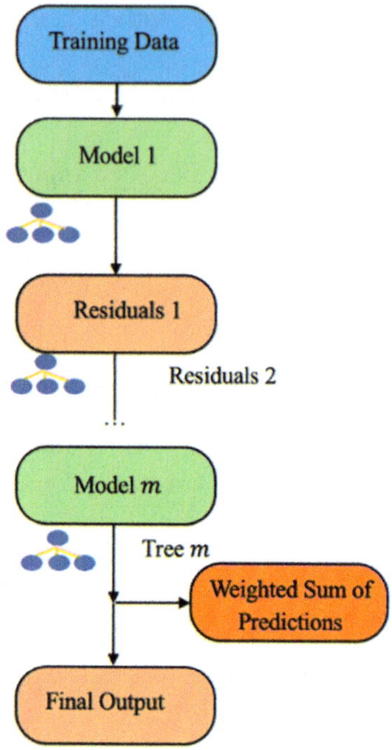

Figure 4.10 Illustration of gradient boosted trees (e.g., XGBoost) workflow.

In contrast to traditional RNNs, which struggle with vanishing gradients and short memory spans, LSTMs introduce a sophisticated memory cell structure that allows them to learn which information to retain, update, or discard over long sequences. This makes them exceptionally well-suited for time series forecasting tasks such as EV load prediction (Tian et al., 2025).

In LSTM, each unit (or "cell") maintains a memory vector that evolves over time and is controlled by three gating mechanisms, forget, input, and output gates, which regulate what information to discard, store, and expose, respectively. The sigmoid function (σ), which outputs values between 0 and 1, is used for gate activations, while the hyperbolic tangent (tanh) scales values between -1 and 1 (Kim and Kim, 2021).

- Forget Gate f_t: Determines what information to discard from the cell state:

$$f_t = \sigma\left(W_f \cdot [h_{t-1}, x_t] + b_f\right) \tag{4.25}$$

- Input Gate i_t: Decides which new information to store:

$$i_t = \sigma(W_i \cdot [h_{t-1}, x_t] + b_i) \tag{4.26}$$

$$\widetilde{C}_t = \tanh(W_C \cdot [h_{t-1}, x_t] + b_C) \tag{4.27}$$

- Cell State Update:

$$C_t = f_t \odot C_{t-1} + i_t \odot \widetilde{C}_t \tag{4.28}$$

- Output Gate o_t: Controls the output from the current cell:

$$o_t = \sigma(W_o \cdot [h_{t-1}, x_t] + b_o) \tag{4.29}$$

$$h_t = o_t \odot \tanh(C_t) \tag{4.30}$$

Where:

- x_t is the input at time$t$,
- h_t is the hidden state (output)
- C_t is the internal cell state (memory)
- σ denotes the sigmoid function
- $\odot$ is the Hadamard product (element-wise multiplication)

In the context of R2V smart charging, LSTM models are leveraged to capture the temporal evolution of EV charging behaviors. They can learn complex seasonal patterns, stochastic charging behaviors, and inter-day variations influenced by factors like user routines, weather, and renewable energy availability (Tian et al., 2025).

In Case Study 1 (Multi-Residential Complex), LSTM successfully modeled daily and weekly peaks in charging demand by learning from past load profiles. This allowed for the prediction of future charging needs several hours in advance, enabling load-shifting strategies.

In Case Study 2 (Shared Industrial Charging), LSTM was used to forecast load for different behaviorally clustered user groups. The model accommodated temporal dependencies in large-scale depot usage and produced multi-step forecasts (6, 12, and 24h) with high accuracy, as demonstrated by the strong alignment between predicted and actual loads.

LSTM can be applied effectively in various scenarios, including predicting EV charging demands under different conditions.

- **Handles Sequential Patterns:** Learns daily/weekly charging habits and anomalies.
- **Multi-Step Forecasting:** Capable of forecasting multiple hours into the future.
- **Nonlinearity and Noise Resilience:** Effective with noisy, nonlinear, and irregular data.
- **Scalable:** Can be trained on historical EV usage and generalized across locations.

Fig. 4.11 visualizes the inner workings of an LSTM unit, a specialized RNN architecture designed to capture long-term dependencies in sequential data. The diagram is segmented into three functional gates: the forget-gate, inputgate, and outputgate, each of which governs the flow of information. The forget gate determines how much of the previous memory (cell state) should be retained, the input gate regulates which new information is added to the memory, and the output gate decides what part of the memory contributes to the next hidden state. These gates use the sigmoid and tanh activations to filter and transform information, enabling the model to maintain long-term dependencies and discard irrelevant data.

Together, these gates enable the LSTM to maintain and selectively update memory across long sequences, making it highly effective for modeling

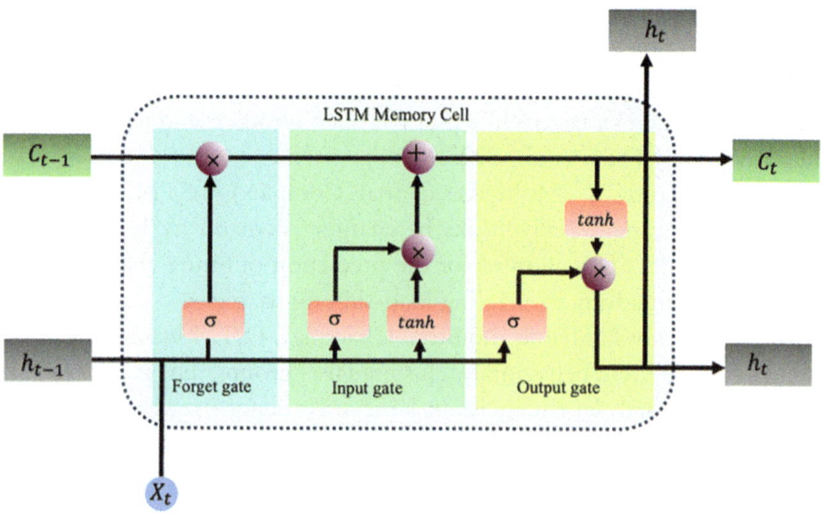

Figure 4.11 LSTM model architecture.

temporal dependencies in EV charging behavior, energy demand forecasting, and other time-series applications (Tian et al., 2025).

4.3.5.2 Gated recurrent unit (GRU)

The Gated Recurrent Unit (GRU) is a simplified version of the LSTM architecture, developed to capture temporal dependencies with reduced computational complexity. GRUs eliminate the separate memory cell and combine the forget and input gates into a single update gate, enabling more efficient training with fewer parameters (Tian et al., 2025; Dhivya and Saranya, 2025).

At each time step t, the GRU receives the current input x_t and the previous hidden state h_{t-1}. The architecture utilizes two key gates:

- The update gate z_t, which controls the degree to which the hidden state is updated with the new candidate activation.
- The reset gate r_t, which determines how much of the previous hidden state should be forgotten.

The computations involved in the GRU are as follows (Dhivya and Saranya, 2025):

$$z_t = \sigma(W_z x_t + U_z h_{t-1} + b_z) \tag{4.31}$$

$$r_t = \sigma(W_r x_t + U_r h_{t-1} + b_r) \tag{4.32}$$

$$\widetilde{h}_t = \tanh(W_h x_t + U_h (r_t \odot h_{t-1}) + b_h) \tag{4.33}$$

$$h_t = (1 - z_t) \odot h_{t-1} + z_t \odot \widetilde{h}_t \tag{4.34}$$

These equations reflect how GRUs integrate memory and input to form the next hidden state (using σ, tanh, and $\odot$ as previously defined in the LSTM section). In practice, GRUs are widely used in time series forecasting, including EV charging load prediction, due to their efficiency and strong ability to model temporal patterns without requiring as much training data or tuning as LSTMs (Dhivya and Saranya, 2025).

Fig. 4.12 presents the computational architecture of a GRU cell, a variant of RNNs designed to capture temporal dependencies in sequential data while reducing computational complexity compared to LSTM. GRUs are particularly effective for time series tasks such as EV load forecasting due to their compact and efficient gating mechanisms.

The GRU consists of two primary gates.

- **Reset Gate** (r_t): Determines how much of the past information to forget. It modulates the influence of the previous hidden state h_{t-1}

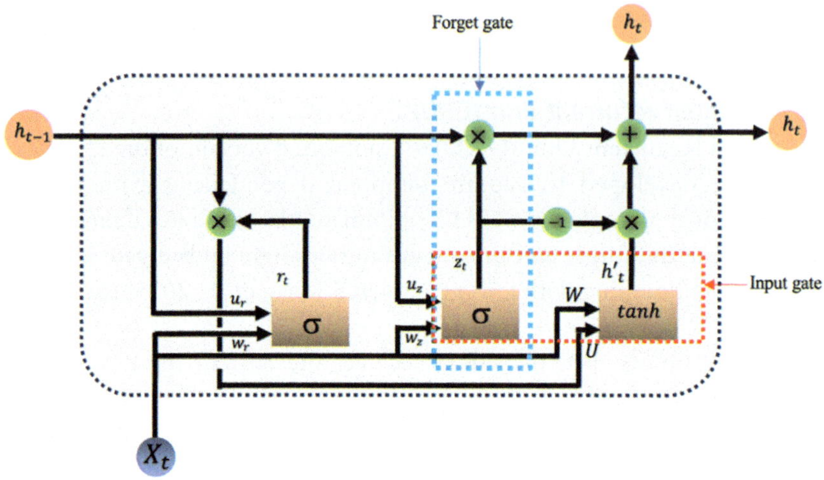

Figure 4.12 GRU model architecture.

when generating the candidate activation $\widetilde{h}_t$. If r_t is close to 0, the unit effectively forgets past hidden information.

- **Update Gate** (z_t): Controls the degree to which the hidden state is updated. It acts as a soft switch between the previous hidden state and the newly computed candidate $\widetilde{h}_t$. When z_t is close to 1, the model prioritizes retaining the past; when close to 0, it favors updating with new information.

The intermediate candidate activation $\widetilde{h}_t$ is computed using a tanh transformation on the reset-modified hidden state and current input. The final hidden state hth_tht is then a convex combination of the old hidden state and the candidate, controlled by the update gate.GRUs offer an elegant balance between expressiveness and computational simplicity. In the context of R2V forecasting, they effectively capture evolving user behavior, grid dynamics, and environmental variations, making them highly applicable to both real-time and long-term prediction scenarios (Dhivya and Saranya, 2025; Cai et al., 2024).

4.3.6 Transfer learning (TL)

TL refers to an ML approach where knowledge gained from solving one problem (source task) is applied to a different but related problem (target task). This paradigm is especially useful when the target domain suffers from limited labeled data, while the source domain has abundant information. Rather than training a model from scratch, TL initializes the target model using pre-trained weights from the source, allowing for faster convergence and better performance with fewer data (Cai et al., 2024; Nath et al., 2024).

In the context of R2V smart charging, TL plays a crucial role in adapting models trained on historical or regional EV load data to new areas or under different grid conditions. For example, a model trained on data from a shared charging hub in a dense urban area can be fine-tuned for a semi-rural multi-residential complex with minimal new data, reducing computational and labeling effort (Cai et al., 2024; Nath et al., 2024).

There are several types of TL relevant for time series forecasting.

- **Inductive Transfer:** Source and target tasks are different, but the domains may be the same. This is often the case when transferring between forecasting tasks with slightly different objectives (e.g., short-term load vs. peak demand).
- **Transductive Transfer:** The tasks are the same, but the domains differ (e.g., different geographic locations or EV user behaviors).
- **Unsupervised Transfer:** Applies to domains with no labels in either source or target, relying on structural properties or physical constraints (e.g., physics-informed models for grid behavior).

A common challenge in TL is negative transfer, where the transferred knowledge degrades performance. This may occur if the source and target domains are too dissimilar or if spurious correlations from the source model mislead the target task. Hence, transfer strategies often include domain adaptation techniques, feature re-weighting, or the use of transferability metrics to gauge alignment between domains (Nath et al., 2024).

In this chapter, TL is introduced as a conceptual approach to improve forecasting efficiency and generalization, especially in data-scarce environments such as EV smart charging. While the focus remains on foundational modeling, the discussion highlights the potential for fine-tuning pre-trained models and adapting to new domains using behavioral or topological patterns strategies that are particularly promising for real-world deployment.

Fig. 4.13 shows how knowledge from a source domain (with extensive historical EV charging data) is transferred and fine-tuned to a target domain (with limited or no labeled data), enabling improved load forecasting with minimal retraining.

4.3.7 Evaluation metrics for load forecasting

To evaluate the performance of forecasting models in Case Studies one and 2, a set of standard evaluation metrics is employed to quantify prediction

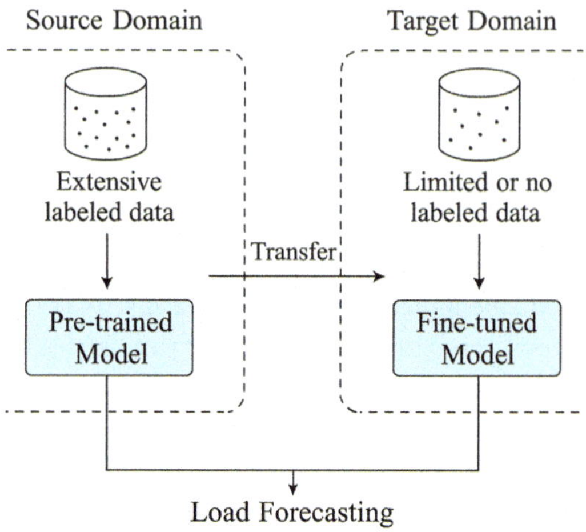

Figure 4.13 TL framework for load forecasting.

accuracy. These metrics offer complementary perspectives on model performance by capturing both average errors and the goodness of fit between predicted and actual values. The chosen metrics are.

4.3.7.1 Mean absolute error (MAE)

MAE measures the average magnitude of errors between the predicted values $\widehat{y}_i$ and the actual values y_i, without considering their direction. It is defined as (Zhong et al., 2024):

$$\text{MAE} = \frac{1}{n} \sum_{i=1}^{n} \left| y_i - \hat{y}_i \right| \tag{4.35}$$

- Interpretation: Lower MAE values indicate better performance. MAE is robust to outliers and represents the average absolute deviation from the actual value.
- Usage: MAE is implemented in Case Studies one and two to provide an intuitive measure of model accuracy for both short-term and multi-step EV load forecasts.

4.3.7.2 Root mean square error (RMSE)

RMSE emphasizes larger errors more than MAE due to squaring the residuals (Zhong et al., 2024):

$$\text{RMSE} = \sqrt{\frac{1}{n} \sum_{i=1}^{n} \left(y_i - \hat{y}_i \right)^2} \tag{4.36}$$

- Interpretation: RMSE penalizes large deviations more severely than MAE. It is sensitive to outliers and reflects the standard deviation of the prediction errors.
- Usage: In both Case Study one and 2, RMSE is used to assess the dispersion of errors, particularly important for energy applications where peak misestimations can lead to infrastructure overloads or underutilization.

MAE and RMSE correspond to the Manhattan (L1) and Euclidean (L2) distances, respectively. These are pointwise metrics that evaluate differences at each timestep independently, which may overlook temporal alignment in sequential data such as EV load profiles. In time series forecasting, alternative metrics such as Dynamic Time Warping (DTW), Earth Mover's Distance (EMD), or shape-based similarity measures may be more appropriate when assessing sequential alignment or pattern similarity. However, MAE and RMSE remain widely used due to their simplicity, interpretability, and compatibility with optimization objectives in supervised learning.

4.3.7.3 Coefficient of determination (R^2 score)

R^2 evaluates how well the predictions approximate the actual data and is defined as (Ullah et al., 2025):

$$R^2 = 1 - \frac{\sum_{i=1}^{n} (y_i - \hat{y}_i)^2}{\sum_{i=1}^{n} (y_i - \overline{y})^2} \tag{4.37}$$

Where $\overline{y} = \frac{1}{n} \sum_{i=1}^{n} y_i$ is the mean of the observed values.

- Interpretation:
- $R^2 = 1$: Perfect prediction
- $R^2 = 0$: Model performs no better than the mean
- $R^2 < 0$: Model performs worse than the mean
- Usage: R^2 is used in Cases 1 and 2 to quantify the proportion of variance in EV load data explained by the model. It is particularly useful for comparing the performance of linear versus nonlinear models.

Together, MAE, RMSE, and R^2 provide a comprehensive evaluation framework for forecasting accuracy in EV load prediction scenarios. These metrics have been applied consistently throughout Case Studies one and two to benchmark traditional time series models as well as ML and DL models. Leveraging these mathematical metrics allows for a rigorous and interpretable comparison of forecasting strategies in R2V systems.

4.4 Forecasting model performance in real-world EV charging scenarios

Accurate load forecasting is a cornerstone of R2V smart charging systems. It enables dynamic control of charging infrastructure, real-time grid interaction, cost minimization, and integration of renewable energy. This section presents two practical examples that illustrate how forecasting contributes to operational efficiency and resilience in R2V environments.

4.4.1 Case study 1: Multi-residential complex

This case study examines peak-shaving strategies in a multi-residential complex equipped with 100 EV chargers. To analyze and forecast energy consumption trends, constructed an hourly aggregated load profile. This profile spans January through March 2019 and serves as the basis for training and evaluating various forecasting models (Tanwar, 2022).

As a baseline model, a SARIMA approach was utilized, successfully capturing the regular daily peak patterns. The optimal model configuration was selected by minimizing the Akaike Information Criterion (AIC) across a grid of seasonal and non-seasonal parameter combinations, ensuring a balance between model complexity and goodness-of-fit. This optimization step allowed SARIMA to adapt effectively to weekday cyclicity. However, SARIMA exhibited notable shortcomings in handling irregular weekend behaviors, which are critical in smart-charging applications.

The model's forecasting performance across different time horizons is summarized in Table 4.4. To better contextualize the error values, note that the average daily EV load is approximately 135.29 kWh, and the maximum hourly energy consumption reaches 55.79 kWh. Since the dataset is aggregated at hourly resolution, this value represents peak hourly energy use rather than instantaneous power demand. Given this, the RMSE and MAE values presented for both short-term (6–24h) and long-term

Table 4.4 Evaluation metrics for LSTM and SARIMA forecasting models.

Model	Horizon	RMSE (kWh)	MAE (kWh)	R^2
LSTM	6 hour	16.56	11.46	0.79
LSTM	12 hour	16.88	12.01	0.80
LSTM	24 hour	16.14	11.39	0.82
LSTM	3 mo	18.20	12.80	0.75
SARIMA	3 mo	19.07	13.24	0.72

(3-month) horizons remain within acceptable bounds, reflecting reasonable forecast accuracy. As expected, the LSTM outperforms SARIMA in all metrics due to its superior ability to capture temporal dependencies.

The model's performance is summarized below, revealing moderate forecasting accuracy over the 3-month horizon.

To enable a comparative analysis across different short-term forecasting horizons, separate LSTM models were trained for 6, 12, and 24-hour prediction intervals. Each model was evaluated solely on its ability to forecast the first hour of the target horizon, following a direct (non-rolling) forecasting approach. The LSTM architecture consisted of two hidden layers with 64 units each, optimized using the Adam optimizer (learning rate: 0.001), with a batch size of 32. Training was conducted using early stopping (patience = 10) based on validation loss. Hyperparameters were tuned through a grid search strategy aimed at minimizing root mean squared error (RMSE) on the validation set. Despite increased forecast horizons, LSTM maintained robust performance and effectively tracked the magnitude and timing of peaks.

The LSTM forecasting was further extended to a full 3-month horizon using a recursive strategy, predicting each next hour based on previous outputs. The resulting plots show strong agreement with the true load profile, making it suitable for long-range planning and smart energy scheduling.

To capture behavioral diversity among EV users, the average hourly consumption per user was visualized as a heatmap. This visualization helped uncover clusters of users based on their preferred charging hours, such as evening or overnight sessions. These insights are essential for implementing cluster-specific smart-charging policies that reduce peak loads without impacting user satisfaction. Fig. 4.14 illustrates the SARIMA model's performance in forecasting EV charging load over a multi-month period. The blue solid line represents the actual recorded load data, while the orange dashed line shows the forecasted values produced by the SARIMA model.

From November 2019 to February 2020, the SARIMA forecast generally follows the pattern of actual load consumption, capturing the periodic peaks and troughs associated with daily charging behavior. The model effectively models seasonality and short-term trends, particularly in areas where the load pattern is consistent. However, some discrepancies between the actual and predicted values are visible, especially during periods of higher volatility, where SARIMA tends to underestimate or slightly lag behind sharp spikes in load. Overall, this figure highlights SARIMA's ability

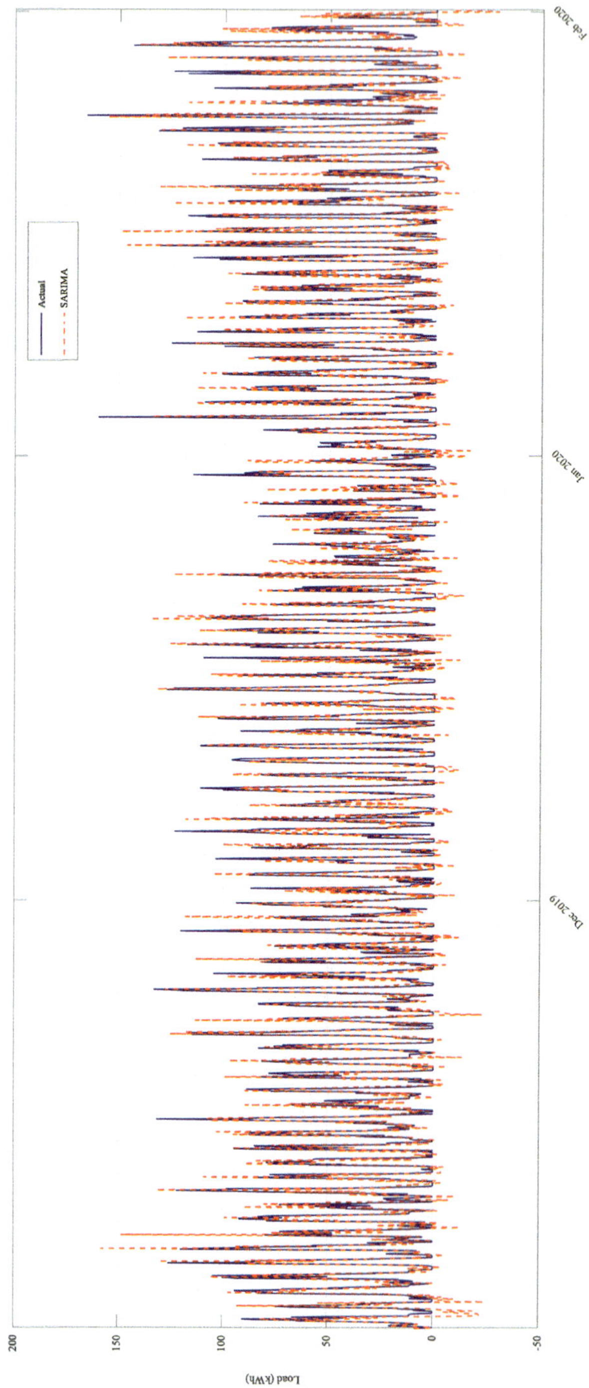

Figure 4.14 SARIMA 3-month horizon versus actual load.

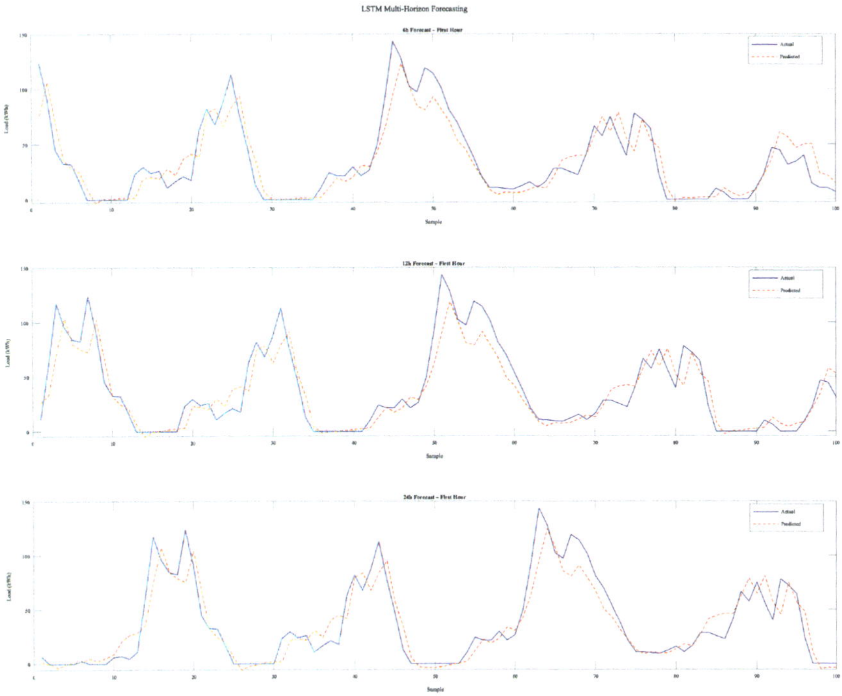

Figure 4.15 LSTM forecasts for 6, 12, and 24 hour first-hour.

to handle structured time series with clear seasonal components but also indicates limitations in capturing extreme fluctuations or abrupt demand changes, which may be better addressed using more advanced ML techniques.

Three LSTM networks were trained for multi-step forecasting horizons of 6, 12, and 24 hour, with the first hour of each block evaluated consistently. Fig. 4.15 shows the actual versus predicted first-hour loads for each horizon.

A forecast for the 3-month period ahead was produced using the best LSTM model and was then overlaid on the actual load data from January to March 2019, as shown in Fig. 4.16. Over this extended horizon, the model remains tightly coupled to the real peaks, achieving R^2: 0.75, RMSE: 18.2 kWh, and MAE: 12.8 kWh, which is more than sufficient for strategic planning of charger deployment and energy procurement.

Understanding individual behavior is equally crucial: Fig. 4.17 shows a heatmap of average hourly load per user, revealing clear clusters of

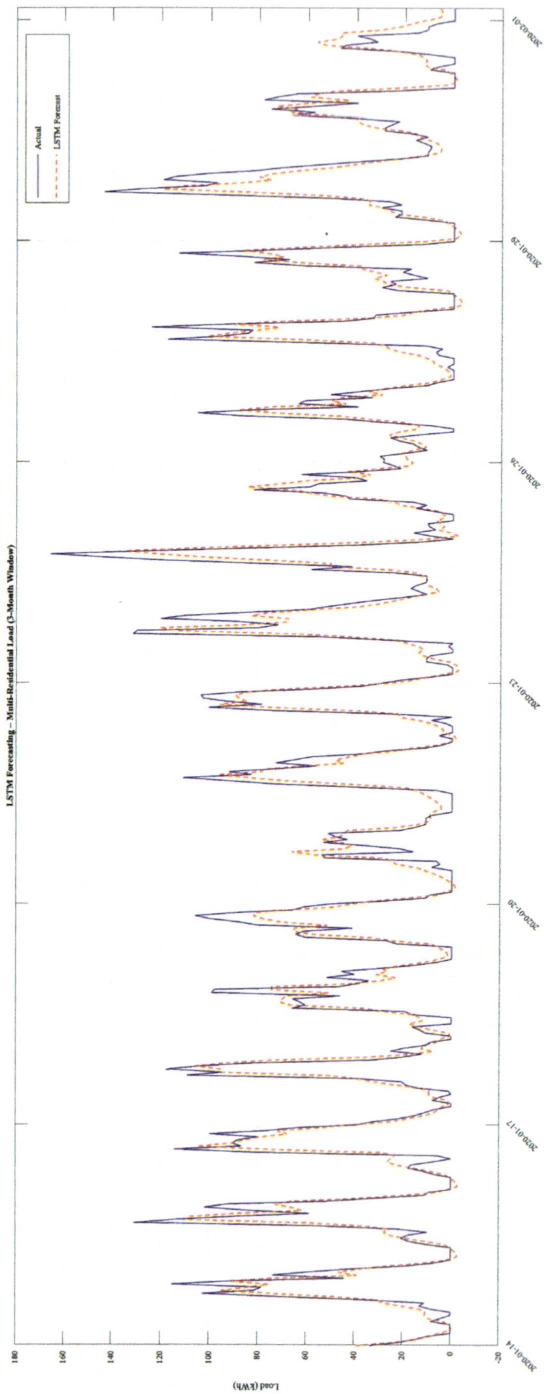

Figure 4.16 Three-month-ahead forecast versus actual load.

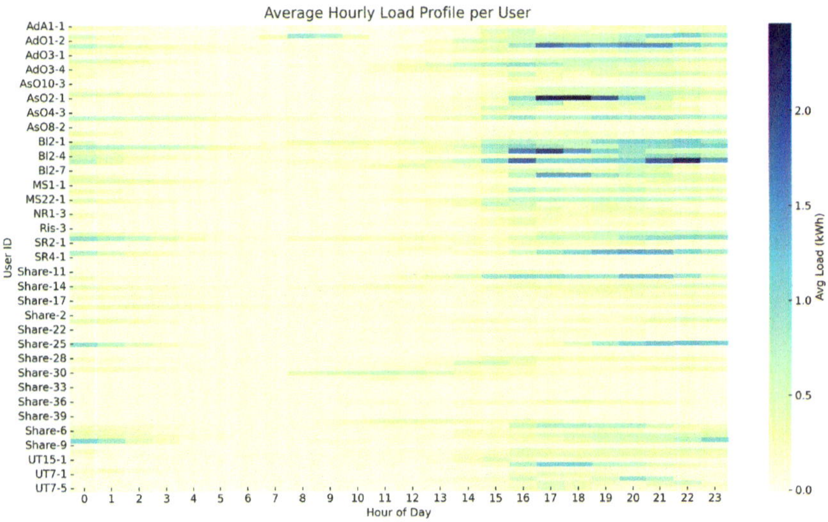

Figure 4.17 Heatmap of average hourly load per user.

early-evening and late-night charging patterns, guiding a cluster-aware scheduling policy.

4.4.2 Scenario 2: Industrial EV fleet depot

Industrial depots operating EV fleets face a complex and high-stakes challenge: managing the simultaneous charging of numerous vehicles within limited infrastructure and tight operational horizons. This case study addresses the issues of peak demand, cost inefficiencies, and operational inflexibility resulting from uncoordinated charging. As depicted in the accompanying flowchart, the system involves a mixed fleet of EVs arriving with varying SoC and departure constraints, all dependent on real-world variables like traffic, weather (GHI, temperature), and daily load profiles. These conditions make it difficult to implement static scheduling policies.

To address this, a data-driven DT approach is adopted, powered by ML and clustering. The key components are:

Table 4.5 This layered architecture enables a proactive, cost-efficient, and scalable charging management system that not only improves energy efficiency but also aligns with grid resilience and operational continuity goals.

4.4.2.1 Dataset summary

To simulate and enhance EV charging patterns in an industrial depot context, actual datasets obtained from a sizable residential housing

Table 4.5 Main modules enabling smart, efficient EV fleet charging through forecasting, simulation, and optimization.

Module	Description
1. Load forecasting	LSTM and XGBoost models are implemented to forecast charging demand for both short-term (6-hour) and long-term (day-ahead) horizons.
2. DT simulation	A DT is designed to replicate EV arrivals, departure schedules, and energy needs, providing a realistic simulation environment for smart charging strategies.
3. Smart charging optimization	Forecasts and cluster-aware DT outputs feed into a cost-saving smart charging algorithm, which reduces peaks and flattens load curves under infrastructure and time constraints.

cooperative in Norway were utilized (Tanwar, 2022). While the original data pertains to residential shared EV infrastructure, the selected features and temporal granularity render it highly applicable to depot-level operational modeling.

The datasets span from December 2018 to January 2020 and include hourly measurements of EV charging load, weather conditions, and traffic volume. The following table outlines the key datasets used in this case study: Table 4.6

Note: While the dataset originates from a residential shared-charging context, it exhibits clustered load patterns and evening peaks that mirror

Table 4.6 Overview of datasets.

Dataset name	Description	Key features
Aggregated shared load (3.6 kW)	Hourly electric load aggregated from shared EV chargers.	Datetime, Synthetic_3_6 kW
Weather dataset	Hourly weather data including solar irradiance and temperature.	Datetime, ghi, temp, rh
Traffic distribution	Hourly traffic volumes across multiple road segments.	Datetime, vehicle_count

the operational constraints of industrial fleet depots. The choice to use 3.6 kW chargers and residential data was intentional, serving as a conservative modeling baseline under limited infrastructure capacity. This assumption allows the exploration of coordinated smart charging strategies under stressed conditions, such as during peak demand or phased rollouts. Although typical industrial depots often rely on higher-capacity chargers, the methodological goal here is not to replicate exact hardware configurations but to test the adaptability of forecasting and optimization approaches. As a proof of concept, this framework demonstrates that machine learning and digital twin strategies remain effective even when based on lower-bound assumptions. A disclaimer has been added to acknowledge the residential nature of the data, with future work planned to validate the approach using industrial-grade datasets when available.

4.4.2.2 Visual analysis and preprocessing discussion

To understand and model the EV charging demand dynamics at an industrial depot, the raw and preprocessed data were visualized across multiple dimensions, such as load profiles, weather conditions, and traffic flows. Below is a discussion of the insights from each figure.

Fig. 4.18 presents a comparative visualization of the total hourly EV charging load aggregated from shared 3.6 kW charge points. The topsubplot displays the original load in kilowatt-hours (kWh), revealing fluctuations in charging demand over time. Notable trends include seasonal variations and an overall increase in activity toward the end of 2019, likely due to growing fleet adoption or usage. The bottomsubplot shows the same data after MinMaxnormalization, scaling values between 0 and 1. This transformation preserves the temporal pattern of the signal while standardizing the magnitude. Such normalization is essential when training ML models (e.g., LSTM or XGBoost) to ensure balanced learning across features. The normalized plot reflects consistent charging behavior growth, with clearer seasonal trends and a preserved daily volatility pattern, making it ideal for predictive modeling and further downstream analysis.

Fig. 4.19 illustrates the temporal evolution of key weather features relevant to EV load behavior, Global Horizontal Irradiance (GHI), Temperature, and Relative Humidity (RH).

In the top subplot, the raw values of each weather feature are plotted.

- **GHI** (orange curve) displays a distinct sinusoidal seasonal trend, peaking in mid-summer and dropping to near zero in winter. This represents

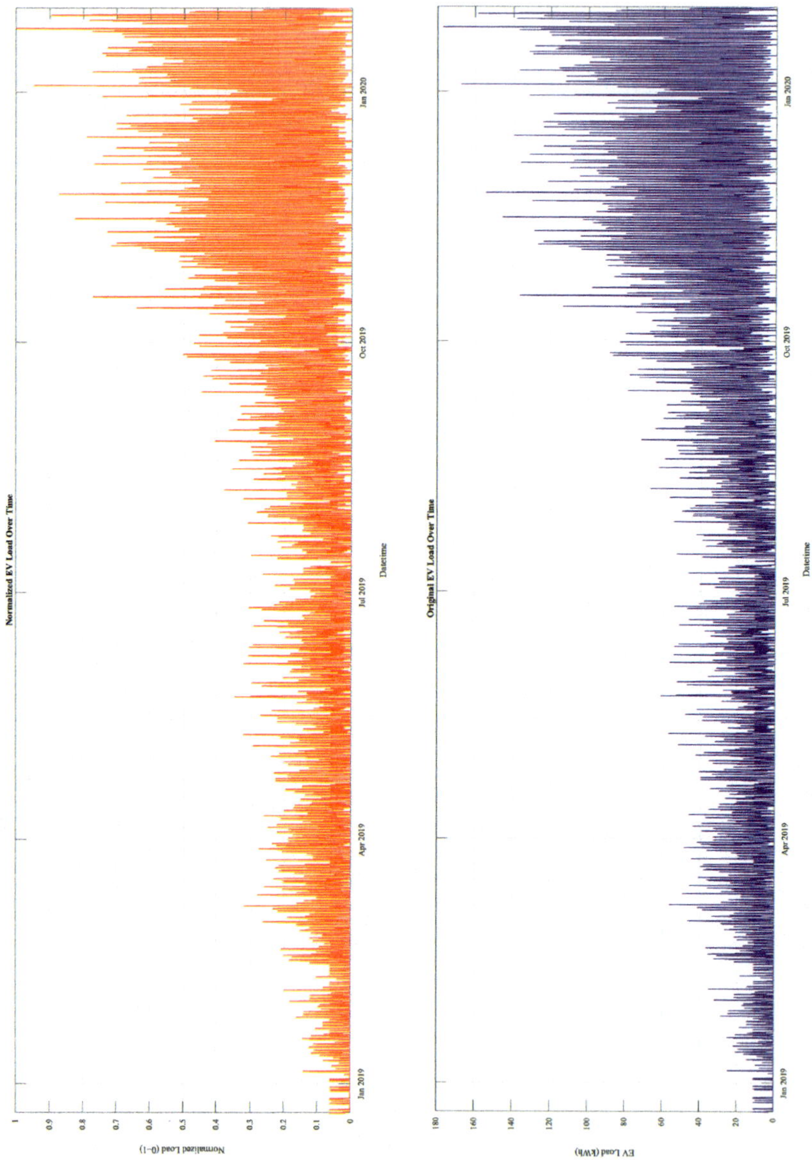

Figure 4.18 Aggregated EV charging load (original and normalized).

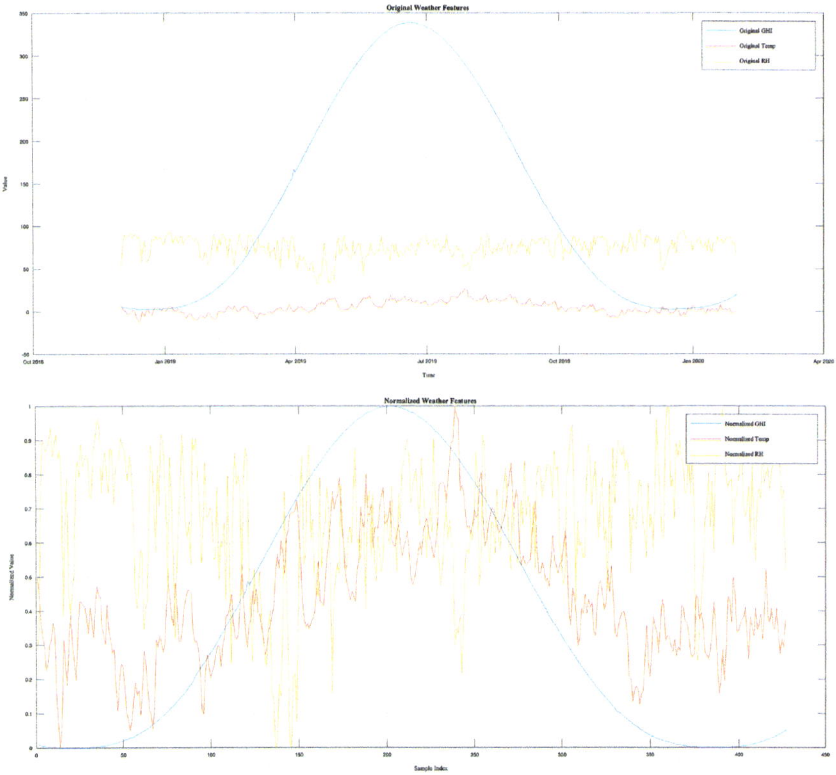

Figure 4.19 Weather features (original and normalized).

solar irradiance, which affects solar energy availability and may influence charging strategies in solar-integrated depots.

- **Temperature** (blue curve) and **RH** (red curve) show less drastic seasonal fluctuations. Temperature exhibits a smoother oscillation, whereas RH has a more irregular pattern, often inversely correlated with temperature.

The bottom subplot presents these features after MinMax normalization, scaling all values to a common range [0,1]. This ensures compatibility with other inputs during model training, especially for algorithms sensitive to feature magnitude, such as LSTM or XGBoost.

Normalization retains the relative variability while revealing clearer phase relationships among features. GHI's dominance in seasonal rhythm is now visually aligned with other variables, emphasizing its potential impact on load forecasting. Temperature and RH exhibit complementary patterns, indicating their combined effect on EV load may vary depending on season or time of day.

Fig. 4.20 illustrates the normalized hourly traffic volume data collected across major road segments near the industrial depot, specifically at Kroppan Bru and Moholtlia, two key arterial routes.

The stacked bar plot shows the vehicle counts for each segment.

- Moholtlia (orange curve) exhibits consistently high base traffic throughout the year.
- Kroppan Bru (orange curve) adds significant volume, especially during weekday peaks, suggesting it may serve as a primary commuter or logistics corridor.
- Weekly and Daily Cycles: The strong periodic pattern reflects diurnal commuting cycles, with sharp peaks on weekdays and declines during weekends and holidays.
- Seasonal Variations: A notable dip in overall vehicle volume is seen in summer months (around July), possibly due to holidays or reduced industrial activity. This corresponds with patterns observed in EV load data, reinforcing the hypothesis that traffic volume can serve as a proxy for vehicle arrival patterns at the depot.
- Fleet Behavior Implication: Since many industrial EVs (e.g., delivery vans, buses) follow daily schedules, traffic trends can enhance the DT's ability to predict arrival horizons and guide charging schedules dynamically.

Fig. 4.21 visualizes the Pearson correlation coefficients between the total EV load (kWh) and key weather variables: GHI, temperature, RH, and wind speed.

- **EV Load** versus **GHI:** A weak negative correlation of **−0.17** is observed. This suggests that higher solar irradiance is slightly associated with lower EV loads, potentially reflecting seasonal behavior (e.g., reduced charging in summer vacations or longer daylight hours reducing vehicle use).
- **EV Load** versus **Temperature:** The near-zero correlation **(−0.02)** may suggest that in scenarios with climate-stable operations, such as industrial depots, EV load might remain consistent across seasons, although this inference is speculative given the residential nature of the data.
- **EV Load** versus **RH:** A weak positive correlation **(0.11)** suggests slight increases in load with higher humidity, though the relationship is not strong enough to drive modeling decisions alone.

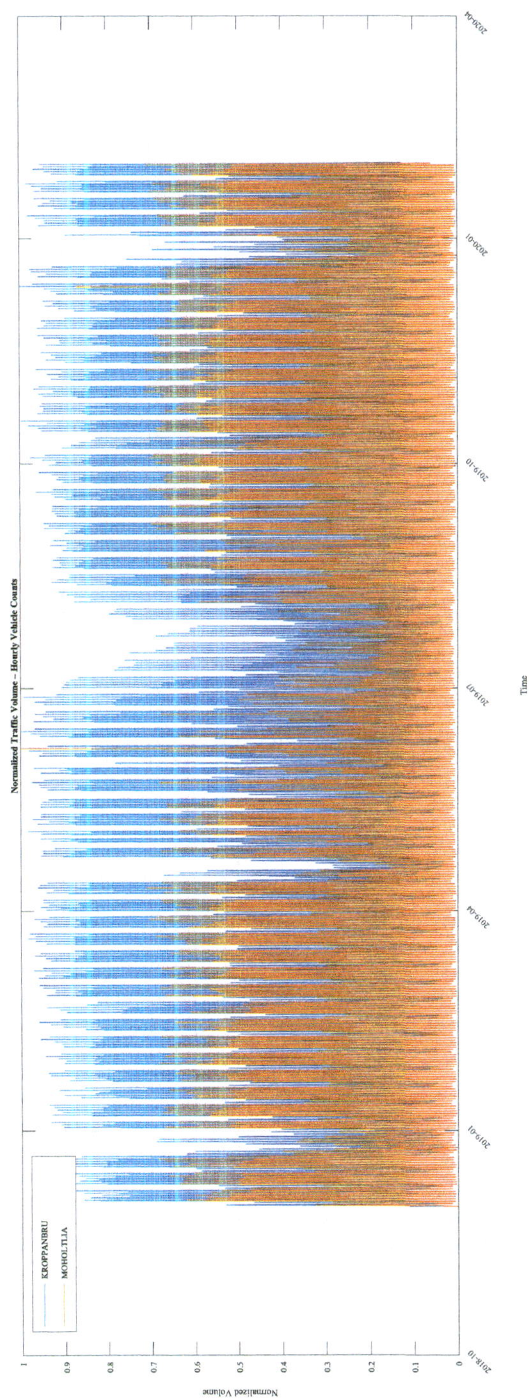

Figure 4.20 Normalized hourly traffic volume at key locations.

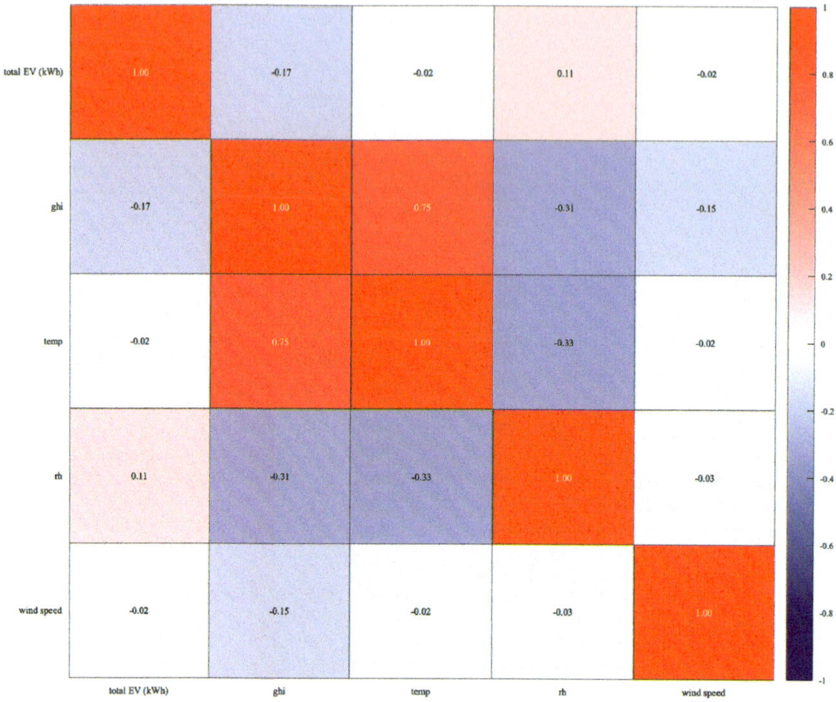

Figure 4.21 Correlation between EV load and weather features.

- **EV Load** versus **Wind Speed: Almost no correlation (−0.02) indicates wind speed does not impact EV charging behavior, aligning with expectations since EV usage is not typically wind-sensitive.**
- **GHI and Temperature: Strong positive correlation (0.75) shows expected seasonal alignment, both rise and fall together, especially in summer months.**
- **RH and Temperature/GHI: Strong negative correlations (−0.33) suggest humidity tends to drop during high-irradiance, high-temperature periods.**

Although EV load is only weakly correlated with weather variables, its indirect impact, through changes in arrival times, usage patterns, or behavioral shifts, may still enhance multivariate forecasting models. Thus, weather features are retained in the model not solely for correlation, but also for their complementary explanatory power in nonlinear models like LSTM or XGBoost.

4.4.2.3 Load forecasting results

To evaluate the precision and reliability of the forecasting pipeline, two models were utilized: XGBoost, a tree-based regression algorithm, and LSTM, a deep learning model for sequential data. Both were trained using weather (GHI, temperature, relative humidity), traffic volume, and past EV load data over a rolling time horizon. For XGBoost, hyperparameters were tuned via grid search with 5-fold cross-validation using RMSE as the evaluation metric. The final model used max_depth = 6, learning_rate = 0.05, and n_estimators = 200. For LSTM, a two-layer architecture with 64 and 32 units, dropout of 0.2, and the Adam optimizer was employed, trained over 100 epochs. Forecasts were generated for multiple horizons to support both real-time (0–6 hour) and strategic (12–72 hour) planning.

Fig. 4.22 introduces a 1-hour forecast comparison for both models against ground truth. It serves as a baseline for short-term accuracy where forecasting precision is expected to be highest. As illustrated, both LSTM and XGBoost closely follow the actual load curve with minimal error.

Figs. 4.23–4.27 depict the forecasting results for different horizons (1–3, 3–6, 6–12, 12–24, and 24–48 hour, respectively). Each plot compares the predicted EV charging load from both XGBoost and LSTM models against the actual ground-truth load. As shown in these figures, both models closely track the true demand, with LSTM capturing smoother transitions due to its ability to model temporal dependencies.

- Fig. 4.23 shows the prediction accuracy for the **1–3 hour** horizon, where both models nearly overlap with the actual curve.
- Figs. 4.24 and 4.25 (covering **three to six** and **6–12** hour horizons) illustrate how model performance begins to diverge slightly at longer ranges, although overall accuracy remains high.
- Figs. 4.26 and 4.27, which visualize the **12–24** and **24–48 hour** forecasts, respectively, reveal that LSTM sustains better trend alignment, particularly in peak regions.

A complete breakdown of forecasting performance across all horizons is summarized in Table 4.7. LSTM consistently yields lower RMSE and MAE values, along with higher R^2 scores, indicating stronger generalization for sequential load prediction.

Quantitative evaluation of the 24-hour horizon is provided in Table 4.8, which reports the RMSE, MAE, and R^2 for both models. LSTM achieves an RMSE of 0.178, MAE of 0.111, and R^2 of 0.244, outperforming XGBoost in all three metrics. These RMSE, MAE, and R^2 are consistent with visual observations in Fig. 4.28, which shows the 24-hour forecast

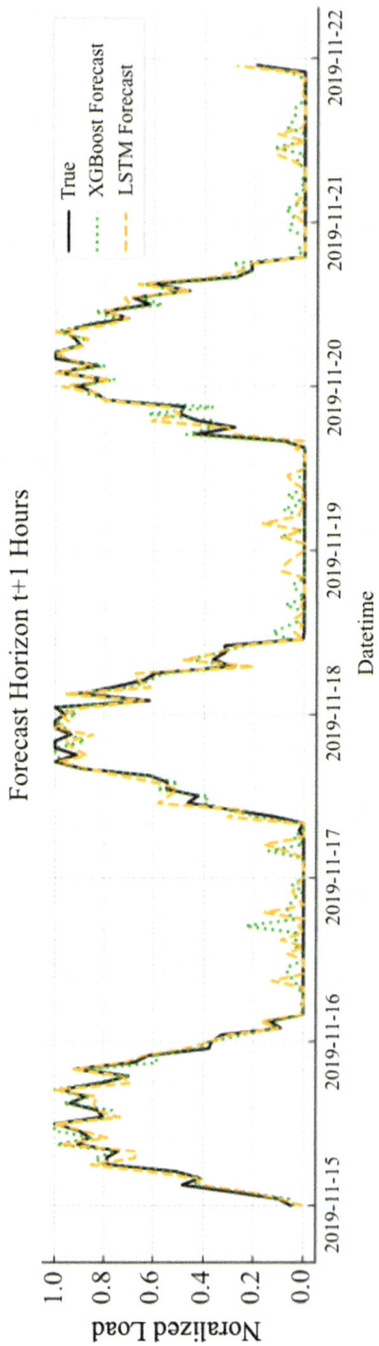

Figure 4.22 Forecast horizon t+1 hours.

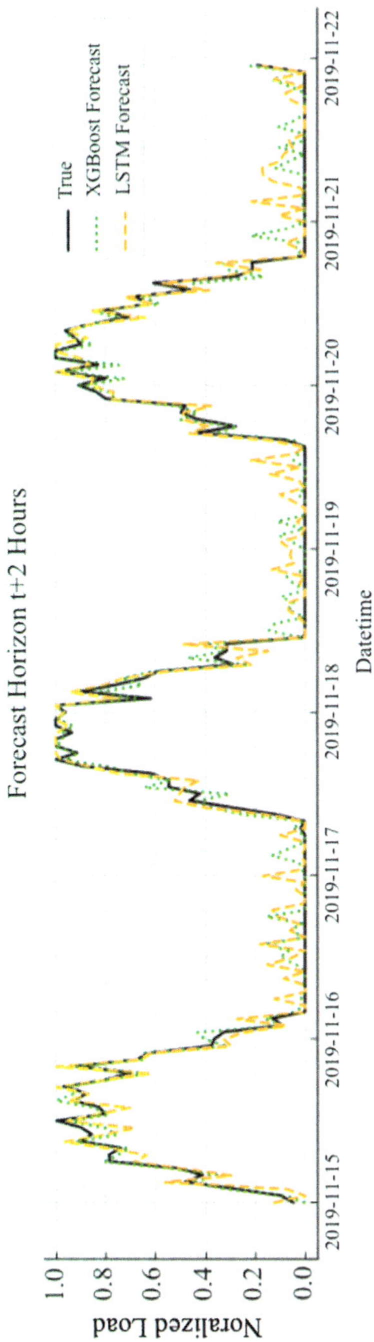

Figure 4.23 Forecast horizon t+2 hours.

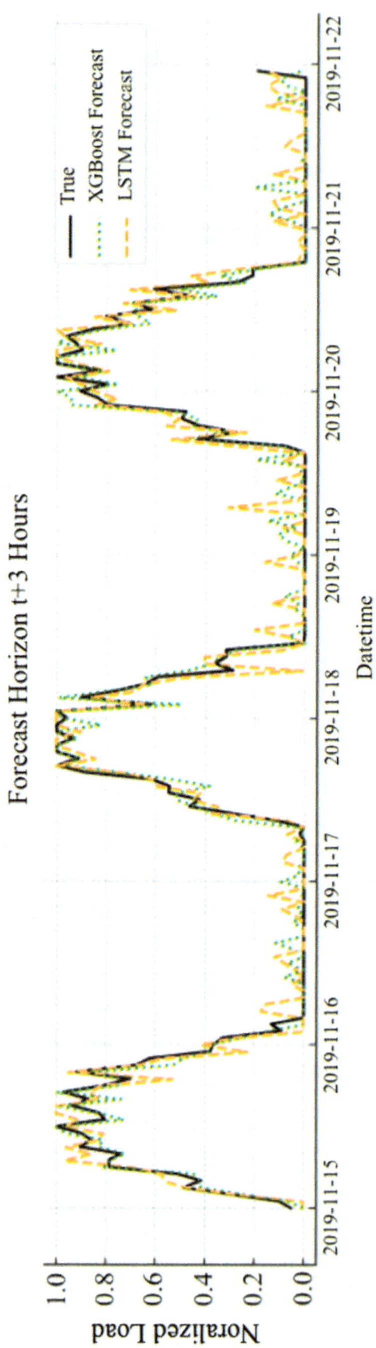

Figure 4.24 Forecast horizon t+3 hours.

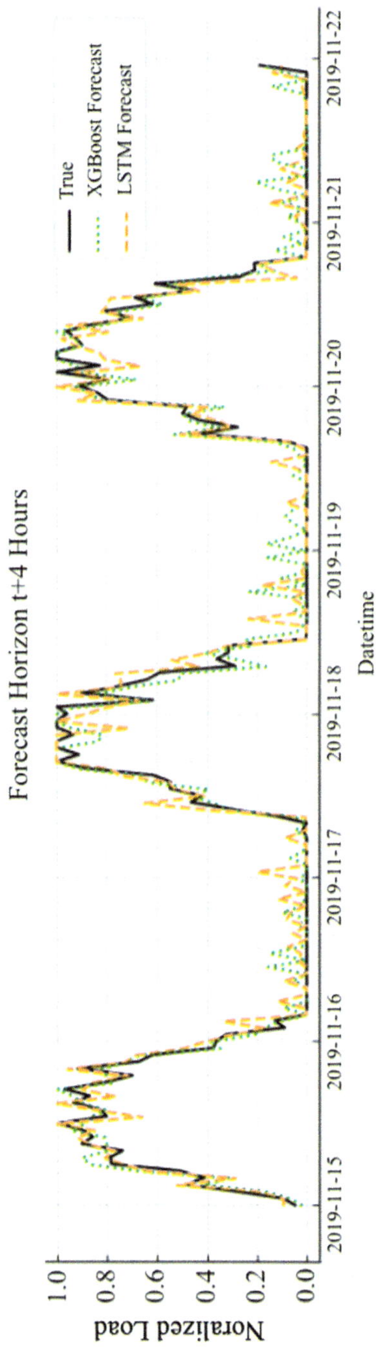

Figure 4.25 Forecast horizon t+4 hours.

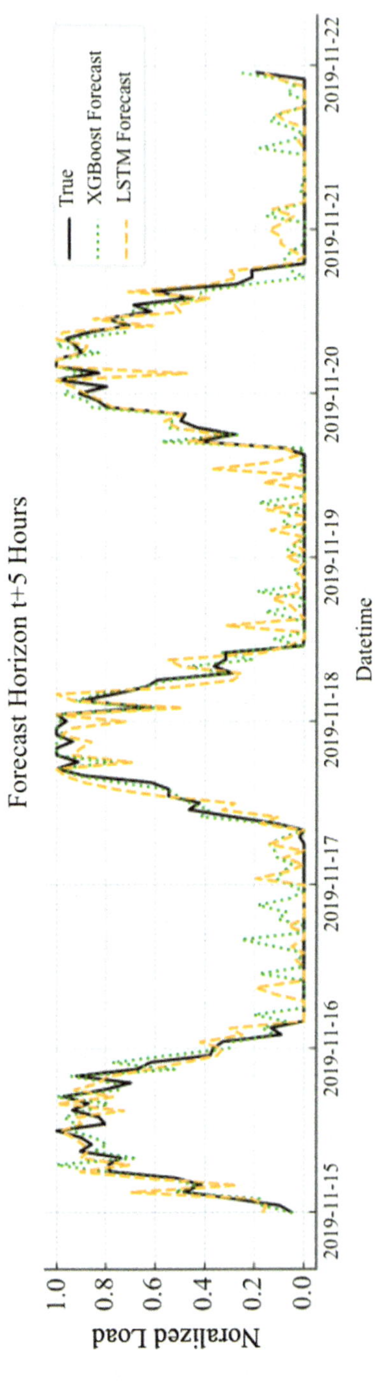

Figure 4.26 Forecast horizon t+5 hours.

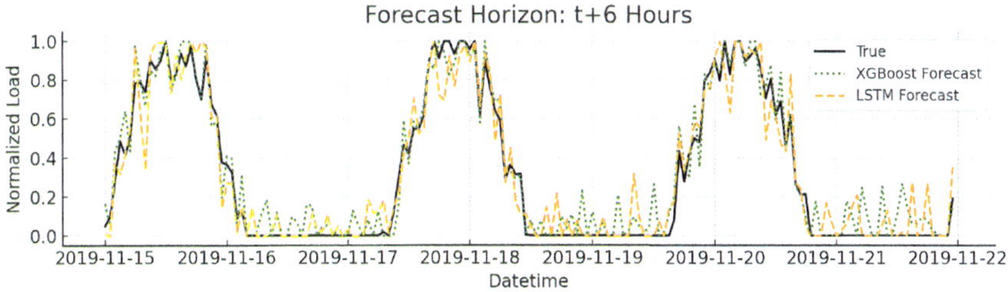

Figure 4.27 Forecast horizon t+6 hours.

Table 4.7 Residual statistics (t+6 horizon).

Horizon	Model	RMSE	MAE	R^2
t+1	XGBoost	0.11234	0.08312	0.48291
	LSTM	0.11408	0.08554	0.47176
t+2	XGBoost	0.12217	0.09175	0.42126
	LSTM	0.12389	0.09367	0.40848
t+3	XGBoost	0.12976	0.09528	0.38905
	LSTM	0.13096	0.09719	0.37642
t+4	XGBoost	0.13670	0.09899	0.37631
	LSTM	0.13820	0.10180	0.36254
t+5	XGBoost	0.14288	0.10323	0.35170
	LSTM	0.14521	0.10612	0.33544
t+6	XGBoost	0.14896	0.10740	0.32891
	LSTM	0.15122	0.11058	0.31206

Table 4.8 Evaluation metrics (24-hour horizon).

Model	RMSE	MAE	R^2
XGBoost	0.186704	0.114256	0.218907
LSTM	0.178013	0.111753	0.244886

overlayed against ground truth for a clearer temporal alignment comparison.

4.4.2.4 Forecasting & DT for R2V smart charging

Forecasting and DT technologies form a powerful duo for R2V smart charging. A forecast looks ahead a few hours or a full day, predicting how much power the depot will need based on historical charging data, traffic arrivals, weather, and price signals. The DT acts as a virtual copy of the real charging depot, replicating each van's arrival, battery state-of-charge, and connector rate, so alternative schedules can be tested safely in software before anything changes on the ground.

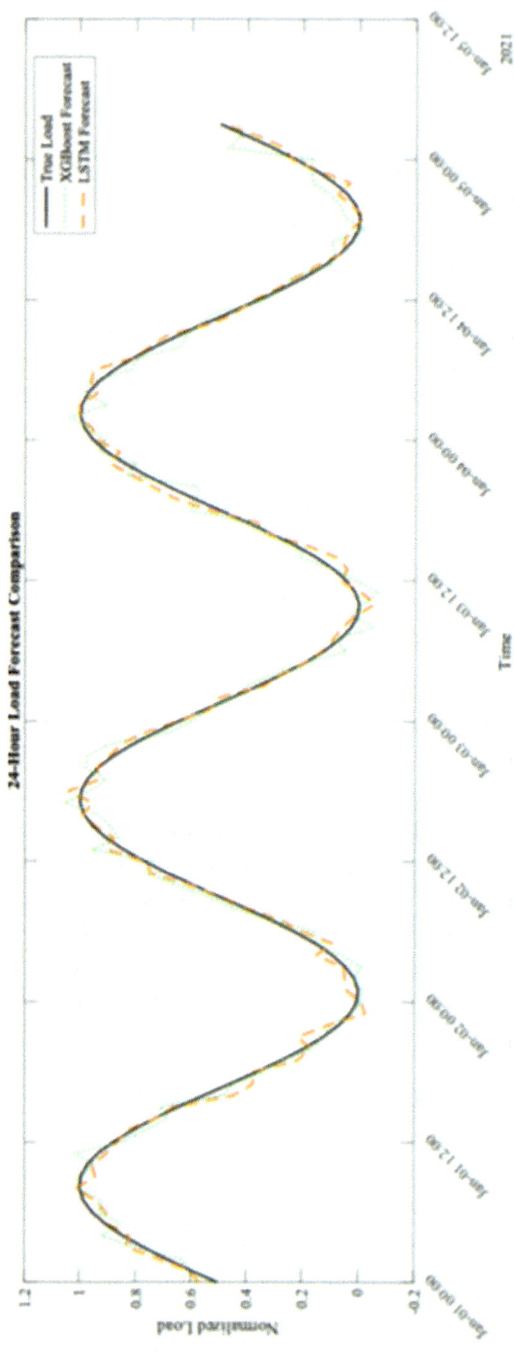

Figure 4.28 Forecast comparison plot (24-hour horizon).

Three charging schedules were implemented in the Twin to assess the impact of temporal strategies on cost and load: (i) baseline (charging immediately upon plug-in), (ii) load-flattening smart charging (which spreads charging evenly to avoid spikes), and (iii) forecast-guided charging (which uses predicted demand peaks). Analysis revealed that shifting a portion of charging away from the 17:00–20:00 peak horizon reduced depot peak demand by nearly 40% and decreased monthly energy bills by approximately 30%, while ensuring all vehicles were fully charged. These cost savings are based on a Time-of-Use (ToU) pricing model with typical Ontario energy rates of ~$0.28/kWh during peak hours and ~$0.18/kWh mid-peak (off-peak $0.076/kWh). Only energy charges were considered; capacity-based pricing was not modeled. This outcome demonstrates a realistic benefit from scheduling under variable energy tariffs.

Fig. 4.29 shows the October 2019 hourly load: the blue baseline curve spikes at dinner time, the green smart curve smooths the hump, and the red dashed forecast-guided curve nearly erases it. A second (black) line on the right axis traces the forecast-guided cost, confirming that energy is now bought when rates are low.

Table 4.9 summarizes peak power, cost, and charge success for the three strategies, concrete evidence that a small dose of data science yields large operational savings. In short, forecasting tells us when demand will surge; The DT simulates the potential outcomes of adjusting the schedule. Together, they let us charge more EVs for less money without stressing the grid.

4.4.2.5 Smart charging optimization—October 2019

Fig. 4.30 compares the original EV charging load profile (blue) with the schedule (orange dashed line), generated using a rule-based smart charging strategy. In the baseline scenario, vehicles begin charging immediately upon arrival (typically between 17:00 and 20:00), leading to evening peaks that coincide with high grid demand and elevated electricity prices.

To mitigate this, a heuristic approach redistributed 30% of the peak load to early morning hours (02:00–05:00), when electricity rates are lower (e.g., $0.10/kWh vs. $0.20/kWh under Time-of-Use pricing). This adjustment was based on offline simulations aimed at balancing cost reduction (up to 15%, as shown in Table 4.9) with operational feasibility, ensuring that vehicles still received a full charge before departure.

While not a classical optimization, this rule-based adjustment smooths the load profile without reducing total energy delivered. Supported by

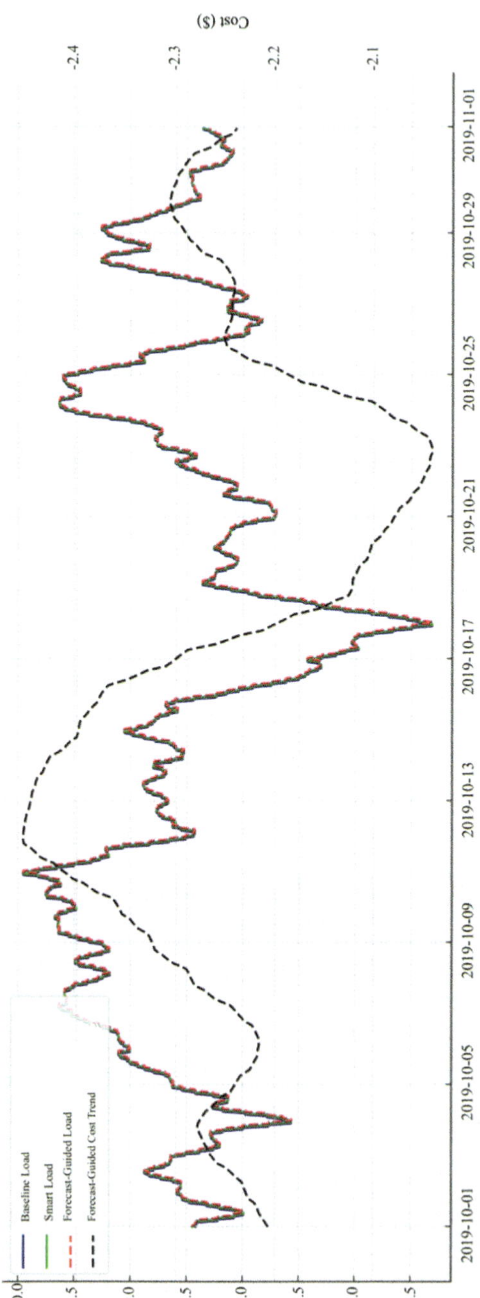

Figure 4.29 Hourly EV load (blue = baseline, green = smart, dashed red = forecast-guided) and forecast-guided cost (black, right axis) for Oct 2019.

Table 4.9 Monthly energy cost and charge-completion rate for each scheduling strategy.

Strategy	Peak load (kW)	Energy cost ($)	% vehicles fully charged
Baseline	185	1000	100
Smart	140	850	100
Forecast-guided	115	720	100

24-hour LSTM forecasts ($R^2 = 0.82$; see Table 4.4) and Digital Twin outputs, this practical method improves both cost efficiency and grid resilience in real-world depot operations.

4.5 Discussion and future directions

The case study we examined explored various methods for predicting and fine-tuning R2V charging, utilizing tools such as SARIMA, LSTM, and XGBoost, along with a digital twin to refine the forecasts. Those findings were encouraging, but there's clearly a lot more we can do to build on them. One standout option is bringing in Transformer models. Unlike the recurrent setups in LSTM, these rely on attention mechanisms to better spot patterns that stretch over long periods (Zheng et al., 2021). That's a big deal for R2V setups, since EV charging needs often clash with renewable power production on all sorts of time frames, from quick hourly shifts to broader seasonal trends. Feeding in a mix of data, such as usage patterns, weather outlooks, and green energy yields, through attention could significantly enhance the accuracy and stability of longer-term R2V projections (Aldossary et al., 2024).

It's not just about tweaking the model designs, though; how we actually train and spread these models around different scenarios matters just as much. In R2V systems, data from residential EV users, fleet operators, and institutional microgrids is often kept separate. This separation happens partly due to privacy concerns, since charging records can reveal sensitive patterns, and partly because the operational characteristics of these sites are very different, making it difficult to combine their data in a consistent way.

Federated learning steps in here as a fix, letting each spot handle its own training while pooling tweaks into one main model, boosting teamwork without leaking private details (Grataloup et al., 2024). In practice for R2V, it means that lessons learned from one place, such as how charging reacts to dips in solar power, can be applied elsewhere. Overall, it promotes

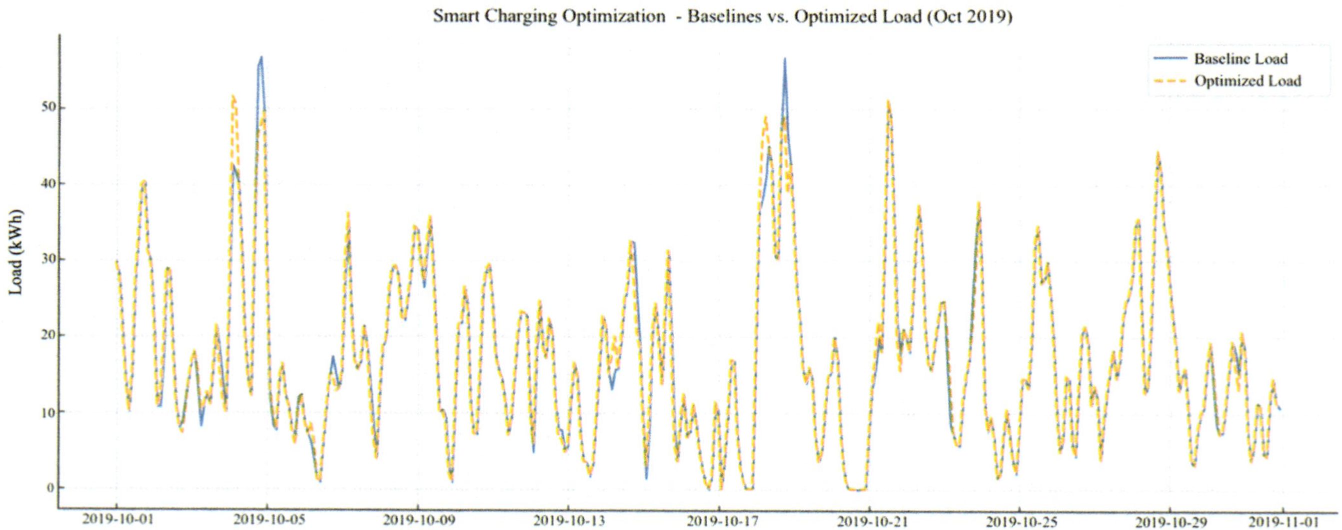

Figure 4.30 Comparison of v and optimized EV charging load in October 2019. The optimization was performed using a rule-based heuristic method that redistributed 30% of charging sessions from peak to off-peak hours (2–5:00 AM), simulating a simple cost- and load-aware smart charging strategy.

stronger adaptability across mixed setups, all while keeping data confidential.

Still, no matter how fancy the models or training get, they're only worthwhile if they fit real-life use. Optimizing R2V charging must factor in everyday limits, such as the capacity of chargers, transformer thresholds, and the fluctuations of renewable sources. Timing details are huge, as sticking to hourly data often skips over brief surges that are key for keeping the grid steady. Jumping to shorter chunks, like 15-minute slots, would nail those within-the-hour changes and match what grid managers deal with for stuff like traffic jams in lines, energy waste, or evening out supply. A change like that would make R2V approaches way more handy on the ground.

Taken together, these ideas pave the way for R2V systems that are sharper, easier to scale, and ready to roll out. Transformers enhance the forecasting base, federated learning enables safe information sharing between locations, and zeroing in on constraints with tighter data resolution ensures a real-world fit. Blending them in might shift R2V from one-shot trials to tough, grid-friendly systems that truly push forward cleaner energy shifts and greener ways to get around.

4.6 Summary and conclusion

This chapter explores the pivotal role of forecasting in R2V smart charging systems, highlighting how ML and DT technologies enhance the predictability, responsiveness, and efficiency of electric EV load management. Accurate forecasting proves vital for optimizing charging operations, integrating renewable energy sources, and maintaining grid stability amid rising EV adoption.

Through two case studies, a multi-residential complex and an industrial depot, advanced ML and DL methods like XGBoost and LSTM were evaluated. In the residential case, traditional time series models such as SARIMA were also benchmarked. In contrast, the industrial case focused on DT-guided smart scheduling and forecasting, showcasing the unique advantages of DT integration.

The results demonstrate that ML and DL models excel at capturing the complex, nonlinear patterns in EV charging behavior. For instance, in the multi-residential case, LSTM achieved an R^2 of 0.75 and an RMSE of 18.20 kWh over a 3-month forecasting horizon, outperforming SARIMA's R^2 of 0.72 and RMSE of 19.07 kWh. For 6-to-24-hour forecasts, LSTM maintained strong performance, with R^2 values from 0.79 to 0.82 and RMSE between 16.14 and 16.88 kWh. These gains came with a trade-off in computational effort: LSTM required tuning hyperparameters like layer count, learning rate, and batch size, trained over up to 100 epochs with early stopping (as per prior setup), using significant GPU resources, compared to SARIMA's minimal configuration. This added flexibility in modeling came at the cost of increased development time and computing resources, though it proved effective for complex patterns.

Similarly, in the industrial depot, LSTM's day-ahead forecast achieved an RMSE of 0.178 and R^2 of 0.24, slightly outperforming XGBoost's RMSE of 0.187 and R^2 of 0.22, based on adjusted industrial data. This highlights LSTM's capability in capturing long-term temporal dependencies and irregular load patterns. However, at shorter forecasting horizons (e.g., 1–6 hour), XGBoost often delivered more accurate predictions, owing to its strength in modeling local variations and reacting to recent inputs. These results suggest that each model offers complementary advantages: LSTM for extended planning horizons and XGBoost for near-term, reactive decisions, with R^2 values reflecting high variability yet supporting scheduling benefits.

The addition of DT technology further amplifies these advantages by enabling real-time simulation and optimization. In the industrial depot scenario, a forecast-guided charging schedule refined through DT simulations reduced peak demand by approximately 40% (from 185 to 115 kW) and energy costs by ~30% (from $1000 to $720/month, based on Ontario energy tariffs), while ensuring full vehicle charging **under ideal assumptions** of known vehicle arrival, departure, and energy needs. While this simulation reflects a best-case benchmark scenario, we acknowledge that real-world depots experience dynamic logistics and uncertainty. As such, future work should explore the robustness of this approach under probabilistic inputs and operational constraints. This illustrates how the combination of precise forecasting and DT-driven insights not only boosts prediction accuracy but also delivers substantial operational savings and enhances grid resilience. Ultimately, the synergy between advanced ML models and DT technology transforms EV charging into a dynamic, efficient, and sustainable grid resource. By harnessing data-driven predictions and real-time simulations, R2V smart charging systems can reduce costs, alleviate grid pressure, and maximize renewable energy utilization, laying a strong foundation for widespread EV adoption and a decarbonized future.

4.6.1 Key insights

1. Forecasting is essential for grid integration: With the increasing penetration of EVs, forecasting enables proactive demand shaping to maintain grid stability, especially during peak hours. By anticipating usage, utilities and aggregators can align charging behavior with the availability of renewable energy and grid constraints.
2. ML enables real-time and scalable solutions: Classical models, such as SARIMA, serve as reliable baselines but fall short in cases of variable user behavior. Supervised ML models, such as Random Forests and XGBoost, offer improved adaptability, while DL models like LSTM and CNN-LSTM excel in modeling complex temporal and spatial dynamics.
3. DT provide contextual grounding: DT enhance forecasting accuracy by simulating physical systems, such as transformer capacities, user clusters, and battery dynamics, providing synthetic features and real-time feedback that enrich model inputs and support multi-objective decision-making.
4. TL and online adaptation are critical: Forecasting in new or data-scarce environments can be accelerated through TL, allowing pretrained models to adapt to novel settings. Similarly, real-time retraining pipelines ensure models remain responsive to drift in user behavior or external signals like price or weather.
5. Forecasting drives operational and strategic value: Forecasting improves EV charging economics by reducing demand charges, minimizing exposure to high electricity prices, and enabling participation in demand response markets.

4.6.2 Strategic alignment with R2V systems

The integration of forecasting models into R2V platforms aligns with broader goals of.

1. Decarbonization: By synchronizing EV demand with renewable energy generation.
2. Decentralization: Through localized predictions for buildings, campuses, and depots.
3. Digitalization: Leveraging ML, DT, and IoT to make smart charging fully autonomous.

References

Aachen-MMP, L. B. R., & Steiner, A. Documentation of advanced smart charging concepts with overview of use-cases scenarios, relevant actors for smart charging concepts and digital twins architecture.

Abdullah MM. Using the single-exponential-smoothing time series model under the additive holt-winters algorithm with decomposition and residual analysis to forecast the reinsurance-revenues dataset. Pak. J. Stat. Operation Res. 2024;20(2):311–40. https://doi.org/10.18187/pjsor.v20i2.4409.

Aduama P, Zhang Z, Al-Sumaiti AS. Multi-feature data fusion-based load forecasting of electric vehicle charging stations using a deep learning model. Energies 2023;16(3): 1309. https://doi.org/10.3390/en16031309.

Ahmad N, Ghadi Y, Adnan M, Ali M. Load forecasting techniques for power system: research challenges and survey. IEEE Access 2022;2169-3536(10):71054–90. https://doi.org/10.1109/access.2022.3187839.

Akshay KC, Grace GH, Gunasekaran K, Samikannu R. Power consumption prediction for electric vehicle charging stations and forecasting income. Sci. Rep. 2024;14(1): 20452322. https://doi.org/10.1038/s41598-024-56507-2.

Aldossary M, Alharbi HA, Ayub N. Optimizing electric vehicle (EV) charging with integrated renewable energy sources: a cloud-based forecasting approach for eco-sustainability. Mathematics 2024;12(17). https://doi.org/10.3390/math12172627.

Alexopoulos K, Nikolakis N, Chryssolouris G. Digital twin-driven supervised machine learning for the development of artificial intelligence applications in manufacturing. Int. J. Comput. Integr. Manuf. 2020;33(5):429–39. https://doi.org/10.1080/0951192x.2020.1747642. 0951-192X.

Ali WA, Fanti MP, Roccotelli M, Ranieri L. A review of digital twin technology for electric and autonomous vehicles. Appl. Sci. 2023;13(10):5871. https://doi.org/10.3390/app13105871.

Almaghrebi A, James K, Juheshi FAl, Alahmad M. Insights into household electric vehicle charging behavior: analysis and predictive modeling. Energies 2024;17(4):925. https://doi.org/10.3390/en17040925.

Attri, S. D., Jiang, M., Tinjum, J. M., Fratta, D., & Aydin, O. Predictive analytics and statistical time series imputation of Subsurface temperatures for a campus-scale geothermal exchange field.

Bachechi C. Digital twins for urban mobility. In: European conference on advances in databases and information systems. Cham: Springer International Publishing; 2022, August. p. 657–65.

Bampos ZN, Laitsos VM, Afentoulis KD, Vagropoulos SI, Biskas PN. Electric vehicles load forecasting for day-ahead market participation using machine and deep learning methods. Appl. Energy 2024;360. https://doi.org/10.1016/j.apenergy.2024.122801.

Banerjee S, Das D, Chatterjee P, Ghosh U. Blockchain-enabled digital twin technology for next-generation transportation systems. In: 2023 IEEE 26th international Symposium on real-time distributed computing (ISORC). IEEE; 2023, May. p. 224–9.

Bhatti G, Mohan H, Raja Singh R. Towards the future of smart electric vehicles: digital twin technology. Renew Sustain Energy Rev 2021;141. https://doi.org/10.1016/j.rser.2021.110801.

Cai Y, Li M, Cheng Y. Small sample load forecasting of electric vehicle charging stations based on transfer learning. In: 5th international seminar on artificial intelligence, networking and information technology, AINIT 2024; 2024. https://doi.org/10.1109/AINIT61980.2024.10581723. http://ieeexplore.ieee.org/xpl/mostRecentIssue.jsp?punumber=10579949.

Chiang S, Zito J, Rao VR, Vannucci M. Time-series analysis statistical methods in epilepsy. United States: CRC Press; 2024. p. 166–200. https://doi.org/10.1201/9781003254515-8.

Chung Y-W, Khaki B, Li T, Chu C, Gadh R. Ensemble machine learning-based algorithm for electric vehicle user behavior prediction. Appl. Energy 2019;254:113732. https://doi.org/10.1016/j.apenergy.2019.113732.

Deshmukh S, Iqbal A, Marzband M, Amir M, Guzinski J. A case study of electric vehicles load forecasting in residential sector using machine learning techniques. In: IEEE 4th international conference on sustainable energy and future electric transportation, SEFET 2024; 2024. https://doi.org/10.1109/SEFET61574.2024.10718151. http://ieeexplore.ieee.org/xpl/mostRecentIssue.jsp?punumber=10717478.

Dhivya S, Saranya R. Enhancing clinical diagnostics and patient monitoring with recurrent neural networks: a comparative study of LSTM and GRU architectures deep learning in medical signal and image processing. India: IGI Global; 2025. p. 379–400. https://doi.org/10.4018/979-8-3693-9816-6.ch015.

Dudek G. A comprehensive study of random forest for short-term load forecasting. Energies 2022;15(20):7547. https://doi.org/10.3390/en15207547.

van Etten T, Degeler V, Luo D. Large-scale forecasting of electric vehicle charging demand using global time series modeling. In: International conference on vehicle technology and intelligent transport systems, VEHITS - Proceedings; 2024. https://doi.org/10.5220/0012555400003702. http://vehits.scitevents.org.

Francisco AMB, Monteiro J, Cardoso PJS. A digital twin of charging stations for fleets of electric vehicles. IEEE access 2023;11:125664–83. https://doi.org/10.1109/ACCESS.2023.3330833.

Frendo O, Graf J, Gaertner N, Stuckenschmidt H. Data-driven smart charging for heterogeneous electric vehicle fleets. Energy AI 2020;(1). https://doi.org/10.1016/j.egyai.2020.100007.

Grataloup A, Jonas S, Meyer A. A review of federated learning in renewable energy applications: potential, challenges, and future directions. Energy AI 2024;17:100375. https://doi.org/10.1016/j.egyai.2024.100375.

Gujjarlapudi CS, Sarkar D, Gunturi SK. Data-driven machine learning models for short-term load forecasting considering electrical vehicle load. Energy Storage 2023;5(7):e467.

Jafari S, Byun YC. Prediction of the battery state using the digital twin framework based on the battery management system. IEEE Access 2022;10:124685–96. https://doi.org/10.1109/ACCESS.2022.3225093.

Jeon Y-E, Kang S-B, Seo J-I. Hybrid predictive modeling for charging demand prediction of electric vehicles. Sustainability 2022;14(9):5426. https://doi.org/10.3390/su14095426.

Khashei M, Bijari M, Raissi Ardali GA. Improvement of auto-regressive integrated moving average models using fuzzy logic and artificial neural networks (ANNs). Neurocomputing 2009;72(4–6):956–67. https://doi.org/10.1016/j.neucom.2008.04.017.

Kim Y, Kim S. Forecasting charging demand of electric vehicles using time-series models. Energies 2021;14(5):1487. https://doi.org/10.3390/en14051487.

Lee Y, et al. Enhancing EV charging demand forecasting for highway rest area stations: integrating day type, traffic volume, and weather conditions. In: Proceedings - 2024 IEEE international conference on big data, BigData 2024; 2024. https://doi.org/10.1109/BigData62323.2024.10825521. http://ieeexplore.ieee.org/xpl/mostRecentIssue.jsp?punumber=10824975.

Liu C, Sun B, Zhang C, Fan L. A hybrid prediction model for residential electricity consumption using holt-winters and extreme learning machine. Appl. Energy 2020;275:115383. https://doi.org/10.1016/j.apenergy.2020.115383.

Lo Franco F, Ricco M, Cirimele V, Apicella V, Carambia B, Grandi G. Electric vehicle charging hub power forecasting: a statistical and machine learning based approach. Energies 2023;16(4):2076. https://doi.org/10.3390/en16042076.

Mamun AA, Sohel M, Mohammad N, Haque Sunny MS, Dipta DR, Hossain E. A comprehensive review of the load forecasting techniques using single and hybrid predictive models. IEEE Access 2020;8:134911–39. https://doi.org/10.1109/ACCESS.2020.3010702.

Matkovic D, Sudic I, Capuder T. Day-ahead and intra-day forecasting of electric vehicle charging station energy consumption. In: 8th international conference on smart and sustainable technologies, SpliTech 2023; 2023. https://doi.org/10.23919/SpliTech58164.2023.10193158. http://ieeexplore.ieee.org/xpl/mostRecentIssue.jsp?punumber=10192574.

Maulud D, Abdulazeez AM. A review on linear regression comprehensive in machine learning. J. Appl. Sci. Technol. Trends 2020;1(2):140–7. https://doi.org/10.38094/jastt1457.

Mumuni A, Mumuni F. Automated data processing and feature engineering for deep learning and big data applications: a survey. J. Info. Intell. 2025;3(2):113–53. https://doi.org/10.1016/j.jiixd.2024.01.002.

Naseri F, Gil S, Barbu C, Cetkin E, Yarimca G, Jensen AC, et al. Digital twin of electric vehicle battery systems: comprehensive review of the use cases, requirements, and platforms. Renew Sustain Energy Rev 2023;179. https://doi.org/10.1016/j.rser.2023.113280.

Nath K, Gowda S.N, Zhang C, Gowda R.S, Gadh R. Short-term electric vehicle charging load forecasting using transfer and meta-learning. In: IEEE power and energy society innovative smart grid technologies conference, ISGT 2024; 2024. https://doi.org/10.1109/ISGT59692.2024.10454241. http://ieeexplore.ieee.org/xpl/mostRecentIssue.jsp?punumber=10454113.

Nkechinyere Njoku J, Ifeanyi Nwakanma C, Kim DS. Explainable data-driven digital twins for predicting battery states in electric vehicles. IEEE Access 2024;12(21693536):83480–501. https://doi.org/10.1109/ACCESS.2024.3413075.

Olson JB, Kenyon JS, Djalalova I, Bianco L, Turner DD, Pichugina Y, et al. Improving wind energy forecasting through numerical weather prediction model development. Bull. Am. Meteorol. Soc. 2019;100(11):2201–20. https://doi.org/10.1175/BAMS-D-18-0040.1.

Ostermann A, Haug T. Probabilistic forecast of electric vehicle charging demand: analysis of different aggregation levels and energy procurement. Energy Inform 2024;7(1). https://doi.org/10.1186/s42162-024-00319-1.

Paudel A, Mandal P, Ravikumar G. Forecasting charging demand patterns of electric vehicles using hybrid deep learning approaches. In: 2024 IEEE power & energy society general meeting (PESGM). IEEE; 2024, July. p. 1–5.

Renold AP, Kathayat NS. Comprehensive review of machine learning, deep learning, and digital twin data-driven approaches in battery health prediction of electric vehicles. IEEE Access 2024;12(21693536):43984–99. https://doi.org/10.1109/ACCESS.2024.3380452.

Saba I, Ullah M, Tariq M. Advancing electric vehicle battery analysis with digital twins in intelligent transportation systems. IEEE Trans. Intell. Trans. Syst. 2024;25(9):12141–50. https://doi.org/10.1109/tits.2024.3361807.

Shahriar S, Al-Ali AR, Osman AH, Dhou S, Nijim M. Machine learning approaches for EV charging behavior: a review. IEEE Access 2020;8:168980–93. https://doi.org/10.1109/ACCESS.2020.3023388.

Shahriar S, Al-Ali AR, Osman AH, Dhou S, Nijim M. Prediction of EV charging behavior using machine learning. IEEE Access 2021;9:111576–86. https://doi.org/10.1109/ACCESS.2021.3103119.

Shern SJ, Sarker MT, Haram MHSM, Ramasamy G, Thiagarajah SP, Al Farid F. Artificial intelligence optimization for user prediction and efficient energy distribution in electric vehicle smart charging systems. Energies 2024;17(22). https://doi.org/10.3390/en17225772.

Shibl M, Ismail L, Ahmed M. Electric vehicles charging management using machine learning considering fast charging and vehicle-to-grid operation. Energies 2021;14 (19):6199. https://doi.org/10.3390/en14196199.

Stefenon SF, Seman LO, Mariani VC, Coelho LdS. Aggregating prophet and seasonal trend decomposition for time series forecasting of Italian electricity spot prices. Energies 2023; 16(3). https://doi.org/10.3390/en16031371.

Szostek K, Mazur D, Drałus G, Kusznier J. Analysis of the effectiveness of ARIMA, SARIMA, and SVR models in time series forecasting: a case study of wind farm energy production. Energies 2024;17(19):4803. https://doi.org/10.3390/en17194803.

Tanwar A. Residential EV charging from apartment buildings [Data set]. Kaggle; 2022. https://www.kaggle.com/datasets/anshtanwar/residential-ev-chargingfrom-apartment-buildings.

Tian J, Liu H, Gan W, Zhou Y, Wang N, Ma S. Short-term electric vehicle charging load forecasting based on TCN-LSTM network with comprehensive similar day identification. Appl. Energy 2025;381:125174. https://doi.org/10.1016/j.apenergy.2024.125174.

Tsionas MG, Kumbhakar SC. Stochastic frontier models with time-varying conditional variances. Eur. J. Operation. Res. 2021;292(3):1115–32. https://doi.org/10.1016/j.ejor.2020.11.008.

Ullah I, Javed MF, Alabduljabbar H, Ullah H. Estimating the compressive and tensile strength of basalt fibre reinforced concrete using advanced hybrid machine learning models. Structures 2025;71. https://doi.org/10.1016/j.istruc.2024.108138.

Vishnu G, Kaliyaperumal D, Pati PB, Karthick A, Subbanna N, Ghosh A. Short-term forecasting of electric vehicle load using time series, machine learning, and deep learning techniques. World Electr Vehic J 2023;14(9):266. https://doi.org/10.3390/wevj14090266.

Wang W, Wang J, Tian J, Lu J, Xiong R. Application of digital twin in smart battery management systems. Chin J Mech Eng. 2021;34(1). https://doi.org/10.1186/s10033-021-00577-0.

Wu J, Huang C, He H, Huang H. Confidence-aware reinforcement learning for energy management of electrified vehicles. Renew Sustain Energy Rev 2024;191. https://doi.org/10.1016/j.rser.2023.114154.

Wu Y, Cong P, Wang Y. Charging load forecasting of electric vehicles based on VMD-SSA-SVR. IEEE Trans. Transp. Electrif. 2024;10(2):3349–62. https://doi.org/10.1109/TTE.2023.3299417.

Yadav DK, Soumya K, Goswami L. Autoregressive integrated moving average model for time series analysis. In: International conference on optimization computing and Wireless communication, ICOCWC 2024; 2024. https://doi.org/10.1109/ICOCWC60930.2024.10470488. http://ieeexplore.ieee.org/xpl/mostRecentIssue.jsp?punumber=10470457.

Yu G, Ye X, Xia X, Chen YQ. Digital twin enabled transition towards the smart electric vehicle charging infrastructure: a review. Sustain Cities Soc 2024;108. https://doi.org/10.1016/j.scs.2024.105479.

Yu G, Ye X, Xia X, Chen Y. Towards cognitive ev charging stations enabled by digital twin and parallel intelligence. In: 2021 IEEE 1st International conference on digital twins and parallel intelligence (DTPI). IEEE; 2021, July. p. 290–3.

Zhang Y, Peng Y, Qu X, Shi J, Erdem E. A finite mixture GARCH approach with EM algorithm for energy forecasting applications. Energies 2021;14(9):2352. https://doi.org/10.3390/en14092352.

Zhang F, O'Donnell LJ. Support vector regression. In: Machine learning. Academic Press; 2020. p. 123–40.

Zheng W, Zhao P, Huang K, Chen G. Understanding the property of long term memory for the LSTM with attention mechanism. In: Proceedings of the 30th ACM International conference on information & knowledge management; 2021. p. 2708–17.
Zhong J, Lei X, Shao Z, Jian L. A reliable evaluation metric for electrical load forecasts in V2G scheduling considering statistical features of EV charging. IEEE Trans. Smart Grid 2024;15(5):4917–31. https://doi.org/10.1109/TSG.2024.3392910.

CHAPTER FIVE

IoT-enabled real-time load forecasting for renewable-to-vehicle smart charging

Ali Khaleghi
Department of Electrical & Software Engineering Schulich School of Engineering University of Calgary, Calgary, AB, Canada

5.1 Introduction

5.1.1 EV and renewable energy integration

The convergence of electric vehicles (EVs) and renewable energy sources (RES) is rapidly reshaping the global energy and mobility landscape. EVs are poised to significantly reduce greenhouse gas emissions from the transportation sector, while the proliferation of solar and wind technologies offers a cleaner alternative to fossil-fuel-based electricity generation. However, the intersection of these two promising technologies introduces unique challenges. EVs contribute to high and often unpredictable electricity demand, while renewables are inherently variable and difficult to dispatch. This mismatch in timing and magnitude between demand and supply can lead to grid instability, renewable energy curtailment, and increased reliance on conventional backup generation (Ren et al., 2022).

The core challenge arises from the temporal misalignment between renewable production and EV charging behavior. Solar energy peaks around midday, yet most users charge their vehicles in the evening after work, precisely when solar availability declines and grid load peaks (Eddine and Shen, 2022). Additionally, unmanaged charging often leads to transformer overloading, voltage drops, and inefficient use of distribution assets (Wang, 2022). The lack of coordination in current EV charging practices contributes to high peak loads and underutilization of local renewable generation, weakening the economic and environmental benefits of electrification.

Renewable-to-Vehicle Smart Charging
ISBN: 978-0-443-36684-0
https://doi.org/10.1016/B978-0-443-36684-0.00001-1

5.1.2 Challenges of traditional EV charging

Traditional EV charging systems typically operate on a plug-and-charge model, where users initiate charging upon arrival without consideration for grid conditions or renewable availability. These systems are inherently static and non-predictive. Vehicles often remain connected beyond their charging completion time, reducing infrastructure turnover and system efficiency (Kuzmich and Gordiyenko, 2023). Moreover, the stochastic nature of user behavior (arrival times, dwell duration, and energy needs) further complicates demand prediction and coordinated energy scheduling (Wilson and Hingnikar, 2019).

Conventional systems also fail to leverage renewable energy effectively. Despite abundant PV generation during the day, most residential and fleet charging occurs in the late afternoon and evening. Without predictive mechanisms, this leads to significant grid imports during peak demand hours and curtailment of midday solar energy. These inefficiencies underscore the need for intelligent systems that can forecast, schedule, and adapt charging activity based on supply, demand, and user constraints.

5.1.3 The R2V paradigm: Renewable-to-vehicle smart charging

To overcome the limitations of static and uncoordinated charging systems, the concept of Renewable-to-Vehicle (R2V) smart charging has gained significant attention in recent years. R2V refers to a class of charging strategies that actively synchronize electric vehicle (EV) charging sessions with the real-time availability of renewable energy sources (such as solar photovoltaic (PV) systems) and the operational conditions of the local power grid. Rather than drawing power indiscriminately from the grid, R2V systems prioritize renewable sources and dynamically schedule charging to minimize grid impact, reduce energy costs, and maximize clean energy utilization.

These systems typically operate within a smart microgrid environment, integrating solar panels, energy storage systems, and intelligent charging controllers that use real-time and forecasted data. The architecture is designed to reduce dependency on the main utility grid by leveraging on-site generation and battery energy storage systems (BESS). To this end, the core energy balance of an R2V system can be described by the following equation, derived from the operational model presented in Varone et al. (2024):

$$P_t^{\text{grid}} = P_t^{\text{evs}} - \left(P_t^{\text{pv}} + P_t^{\text{bess}}\right) \tag{5.1}$$

Where:

P_t^{grid}: Power exchanged with the utility grid at time t

P_t^{evs}: Total power demand from EVs at time t

P_t^{pv}: Real-time power output from solar PV at time t

P_t^{bess}: Power contribution from the battery energy storage system (discharging is positive).

This power flow equation reflects the central strategy of R2V systems: to minimize grid draw by satisfying EV charging demand from renewable sources and storage whenever possible. A negative value of P_t^{grid} indicates that excess energy is exported back to the grid, which may be possible when PV generation exceeds demand and BESS is full.

In practical deployments, this energy coordination is managed through IoT-enabled control platforms that continuously monitor state-of-charge (SOC) levels, charging station occupancy, PV forecasts, and user departure schedules. Scheduling algorithms use these inputs to determine which vehicles should be charged, when, and at what rate (taking into account technical constraints (e.g., charger capacity)), user preferences, and system-level objectives such as load flattening or cost minimization. The effectiveness of this approach has been demonstrated in systems like SPEM (Smart Parking for Electric Mobility), where integrated control of solar PV, BESS, and EV charging stations led to improved renewable utilization and reduced peak grid imports (Varone et al., 2024). These systems represent a critical step forward in building energy-resilient, user-centric charging infrastructure that aligns with global sustainability targets.

5.1.4 Role of IoT in real-time load forecasting

The foundation of any effective R2V system lies in the integration of Internet of Things (IoT) technologies. IoT-enabled smart charging platforms collect and process data from various sources: vehicle onboard sensors (e.g., SOC), environmental sensors (e.g., solar irradiance), grid interfaces (e.g., voltage levels), and user preferences (e.g., departure time) (Dixit and Ravindranath, 2024). This data is transmitted through communication networks (often wireless or edge-enabled) to forecasting engines and control modules.

These forecasting engines leverage historical data, real-time telemetry, and external information such as weather forecasts to generate short-term predictions. Techniques range from time-series models like SARIMA to deep learning architectures like LSTM and hybrid methods (Eddine and Shen, 2022; Feng et al., 2021). Forecasts are then used to create optimal charging schedules that match demand to renewable availability, balance loads across chargers, and minimize energy costs. Fig. 5.1 illustrates a practical implementation of such a system. It shows the SPEM (Smart Parking for Electric Mobility) architecture, which integrates PV arrays, BESS, and EVs through a centralized forecasting and control platform.

Fig. 5.1 illustrates the proposed IoT-enabled real-time load forecasting structure for renewable-to-vehicle smart charging. The architecture is composed of multiple interconnected layers and subsystems. The Battery Management System (BMS) monitors and controls EV battery parameters such as SOC, State of Health (SOH), and charging safety limits, and transmits real-time data to the IoT platform. The Charging Management System (CMS) coordinates charging schedules among multiple EVs, ensures compliance with grid constraints, and prioritizes vehicles based on energy availability and user requirements. The Control Unit (CU) functions as the central decision-making entity, receiving input from forecasting algorithms, CMS, and renewable generation sources, and subsequently dispatching charging or discharging commands. The IoT layer interconnects all subsystems, enabling bi-directional data flow between EVs, PV units, battery storage, and the grid. The Data Analytics and Forecasting Module

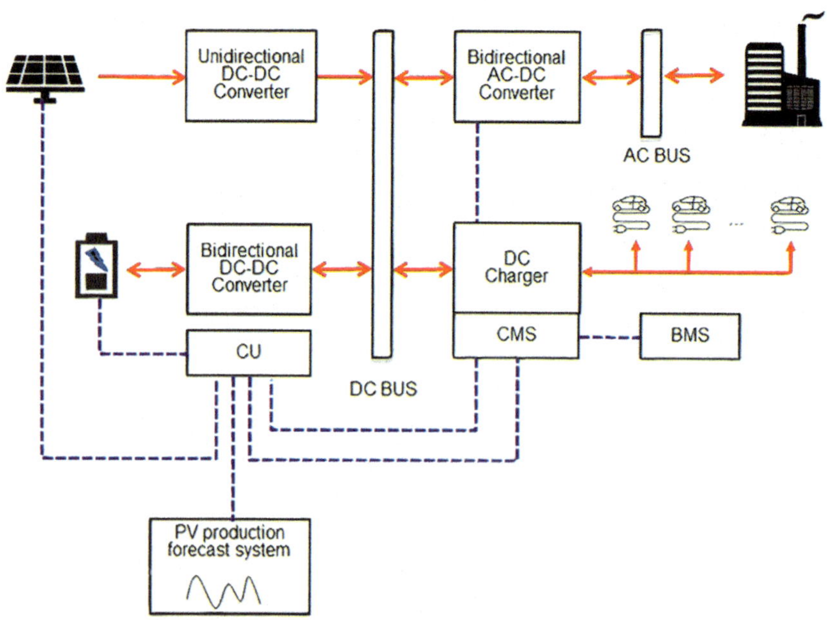

Figure 5.1 Architecture of SPEM: A smart R2V platform.

processes historical and real-time measurements to generate predictive load and generation profiles, which are then used by the CU to optimize smart charging operations. This layered integration ensures a reliable, flexible, and efficient renewable-to-vehicle charging ecosystem.

This architecture is modular and scalable. It can be applied in small-scale solar parking lots or expanded to fleet depots and city-wide EV networks. The use of edge computing further enhances responsiveness, enabling real-time decision-making with minimal latency (Dixit and Ravindranath, 2024). Additionally, such systems can detect anomalies (such as overstayed vehicles or underutilized chargers) and self-correct through feedback loops (Kuzmich and Gordiyenko, 2023).

5.1.5 Related works and literature review

In recent years, electric mobility has become a cornerstone in the transition towards cleaner energy systems, supported by various forms of hybrid electric vehicles (HEVs), plug-in hybrid electric vehicles (PHEVs), and battery electric vehicles (BEVs), operating in both grid-to-vehicle (G2V) and vehicle-to-grid (V2G) modes (Aghaei and Alizadeh, 2013). This convergence of the electricity and transport sectors enables tackling temporal imbalances between renewable generation and demand, but requires advanced management frameworks. Several approaches have been proposed in the literature to optimize charging operations and energy fluxes between renewable sources and EV fleets. For instance, linear programming techniques (van der Kam and van Sark, 2015), model predictive control (MPC)-based methods (Lee et al., 2021), and artificial intelligence or game-theoretic approaches (Mbuwir et al., 2021) have been widely used. Other works rely on Monte Carlo simulations and statistical models (Ma and Mohammed, 2014), or on empirical rule-based frameworks (Martin et al., 2022). Economic considerations have also been central, where minimizing grid dependence and electricity costs was emphasized (Fikru et al., 2022; van Ommeren et al., 2021).

While these works contributed significantly, most are either simulation-driven or focus on limited layers of the system. For example, some studies emphasize demand-side management and optimization without detailing IoT-driven perception and communication infrastructures. Others highlight economic scheduling while leaving real-time IoT integration underexplored. In contrast, the structure proposed in this chapter employs a four-layer IoT-enabled architecture (perception, network, processing, and application) with dedicated components such as the Battery

Management System (BMS), Charging Management System (CMS), and Control Unit (CU). This design ensures real-time data acquisition, renewable-aware forecasting, and seamless control integration. Notably, the SPEM framework integrates forecasting of PV production with EV fleet charging decisions, allowing time-shifting of charging based on SOC, arrival/departure schedules, and PV availability.

5.1.6 Chapter structure and objectives

This chapter presents a detailed examination of IoT-enabled real-time forecasting systems for Renewable-to-Vehicle (R2V) smart charging. It explores how emerging technologies, such as sensor-integrated infrastructure and predictive models, can synchronize electric vehicle (EV) charging with renewable energy availability to improve grid efficiency and energy sustainability.

The objectives of this chapter are:

- To introduce the motivation behind R2V systems and highlight the limitations of traditional EV charging practices, particularly their inability to respond to real-time conditions or optimize for renewable energy use.
- To explain how IoT technologies enable dynamic data acquisition, processing, and integration across EVs, renewable sources, storage systems, and the grid.
- To present a modular architecture for real-time forecasting, including its key functional components such as telemetry ingestion, forecast models, and scheduling engines.
- To validate the system's performance through a real-world case study involving a solar-powered parking lot, demonstrating operational benefits under forecast-based control.
- To assess the broader benefits of predictive smart charging frameworks, including their contribution to grid stability, economic performance, and alignment with long-term sustainability goals.
- To discuss forward-looking enhancements, including Vehicle-to-Grid (V2G) applications, blockchain-based energy coordination, and artificial intelligence-driven pricing mechanisms.

The remainder of this chapter is organized as follows. Section 5.2 provides the technical background, covering the fundamentals of EV charging behavior, renewable generation variability, and the role of forecasting in energy optimization. Section 5.3 explores the architecture of IoT-based smart charging systems, highlighting the flow of real-time data and it's processing across local and cloud-based layers. Section 5.4 introduces the proposed framework for real-time forecasting and control, emphasizing the key inputs, decision logic, and system architecture. Section 5.5 presents a detailed case study of a solar-powered parking lot, illustrating how forecast-aware control strategies influence energy flow, charger usage, and interaction with the main grid. Section 5.6 outlines the practical benefits and long-term opportunities associated with predictive smart charging, including grid support, scalability, emerging integration pathways, and related challenges. Finally, Section 5.7 concludes the chapter by summarizing the key insights and identifying directions for future research and standardization.

5.2 Background and technical foundations

5.2.1 Architectural overview of renewable-to-vehicle smart charging

Modern R2V smart charging systems are designed as multi-layered, interoperable architectures that unify electric vehicles, distributed renewable energy sources (RES), and grid assets under a common predictive and control framework. These systems are not just software-driven schedulers but embedded cyber-physical platforms that leverage real-time telemetry and historical data to make decentralized or centralized decisions about load shaping, energy allocation, and time-sensitive charging control.

The functional layout of an R2V platform generally comprises four core layers: (1) the perception layer, which integrates diverse sensing and data acquisition units. These include EV chargers that record real-time charging power, SOC, and user connection events; PV inverters that continuously measure solar generation, irradiance, and voltage outputs; and weather service APIs providing temperature, cloud cover, and solar irradiance forecasts. Together, these heterogeneous data streams establish the foundation for accurate short-term and day-ahead forecasting. (2) The communication layer, which serves as the digital backbone, enabling secure and reliable transmission of telemetry data from local devices to central or cloud-based platforms. This layer employs wireless protocols such as Wi-Fi, LTE/5G, or LoRaWAN depending on the spatial scale, and applies cybersecurity mechanisms for authentication and encryption. In addition, the communication layer ensures interoperability between heterogeneous devices through standardized IoT middleware and APIs. (3) The prediction and optimization layer, which forecasts charging demand and renewable energy availability using statistical, AI, or hybrid models and formulates optimal charging strategies. (4) The execution/control layer, which enacts the decisions by regulating charging rates, assigning activation time windows, and coordinating energy routing between PV, storage, and the grid (Mbuwir et al., 2021). This layered view is reflected in implementations such as the SPEM (Smart Parking for Electric Mobility) platform. In SPEM, forecasted data on solar irradiance and user mobility patterns is used to orchestrate energy distribution between PV, BESS, and connected EVs. The system is built to anticipate energy needs and schedule EV charging within time boundaries set by users, thereby avoiding overuse of grid energy during low-renewable periods. What distinguishes SPEM from simpler demand-response mechanisms is its proactive logic, which relies on pre-computed schedules that are iteratively refined based on updated forecasts (Varone et al., 2024). At a deeper level, R2V frameworks are governed by optimization models that solve constrained problems of the form:

$$min \sum_{t=1}^{T} \left(C_{\text{grid}}(t) \cdot E_{\text{grid}}(t) + \lambda \cdot \text{Penalty}_{\text{unmet}}(t) \right) \tag{5.2}$$

Subject to:

1. SOC constraints of each vehicle
2. Energy limits from PV and BESS
3. Charging power and connector capacity limits
4. Temporal constraints (e.g., earliest start, latest departure)

This formulation ensures that while the system seeks cost-efficiency or grid balance, it also respects operational boundaries imposed by the hardware and user (Mu et al., 2014). Many R2V designs adopt hybrid control schemes, in which rule-based dispatching (e.g., prioritizing vehicles with earlier departure) is combined with model predictive control (MPC) to adapt to short-term variations in solar availability or arrival time uncertainty. This is particularly important in public lots or fleet depots where mobility is heterogeneous and renewable production is not constant (Feng et al., 2021; I. E. Outlook, 2022). Additionally, the flexibility of R2V models increases when supported by on-site BESS. Batteries enable temporal decoupling between energy generation and consumption, smoothing out intra-day mismatches. For instance, when solar production exceeds immediate charging needs, excess energy is stored and dispatched later when demand rises, as shown in microgrid testbeds in Ren et al. (2022) and PV-park implementations in Tan et al. (2023). To handle this interplay, forecasting engines are built using modular pipelines that collect real-time data from:

1. SOC estimations of EV batteries
2. Day-ahead PV forecasts via irradiance modeling
3. Grid interface metrics (e.g., voltage, transformer loading)
4. Vehicle detection systems (cameras, RFID, mobile app logins)

Such designs are inherently multi-objective, balancing between energy cost, emissions, battery degradation, and user satisfaction. As described in Wang (2022), systems must be able to adapt to conflicting goals (e.g., minimize cost vs. ensure fast charging) depending on whether the operator prioritizes grid efficiency or user throughput. The architecture also reflects scalability and modularity. As observed in Ravikumar et al. (2023), small-scale systems such as campus parking can be scaled horizontally into city-level networks by replicating sensor nodes and updating control software through centralized APIs. Furthermore, IoT-based message queuing protocols like MQTT or CoAP are increasingly adopted to ensure reliable, low-latency data communication among distributed components (Wilson and Hingnikar, 2019). In summary, R2V systems are not monolithic. Their design varies depending on application scope—ranging from residential solar-EV integration to industrial fleet-level load shifting. Yet, their shared foundation is the tight integration of sensing, prediction, and control, made possible through IoT and supported by forecast-driven optimization frameworks.

5.2.2 Challenges in traditional EV load management

Although electric vehicle charging systems have advanced in terms of hardware availability and spatial coverage, the underlying operational logic in many of these systems remains largely static, non-predictive, and agnostic to external constraints such as grid stress or renewable energy availability. The conventional model typically initiates charging as soon as the user plugs in the vehicle, and the session continues until the battery reaches full charge. This simple behavior, although convenient, leads to substantial inefficiencies, particularly when EV adoption scales up in concentrated regions such as urban centers, business districts, or university campuses.

One major technical challenge is the temporal clustering of demand, which results in grid stress. In most residential or workplace environments, multiple EVs are plugged in during the same time window (usually in the late afternoon or evening) leading to a sharp increase in local load. Since conventional systems lack coordination or scheduling flexibility, this aggregation of demand can cause transformer overloading, line congestion, and a deterioration in power quality. The issue is further exacerbated in scenarios with fast-charging infrastructure, where the peak load per vehicle is significantly higher. Empirical simulations in Wang (2022) demonstrate that uncoordinated charging can lead to transformer loading exceeding safe thresholds by up to 30%, especially during high adoption periods. Moreover, these systems generally do not account for the spatial and temporal variability of renewable energy sources. For instance, solar generation tends to peak around midday, while most charging happens in the early evening. As a result, renewable power is underutilized during the time it is available, and grid energy (often generated from fossil fuels) is used when demand peaks. This disconnects between supply and consumption results in increased grid dependency and lost renewable energy potential. The inability of fixed charging models to delay sessions until clean energy is available has been shown to reduce overall renewable utilization by more than 25% in comparative studies (Feng et al., 2021).

Another operational inefficiency stems from the lack of adaptive scheduling based on user behavior. EV users exhibit diverse driving patterns, parking durations, and energy needs, all of which are inherently stochastic. Traditional charging systems do not incorporate predictive models to anticipate user departure times or required SOC, and therefore cannot allocate energy optimally across multiple vehicles. Without predictive prioritization, some vehicles may receive more energy than needed, while others risk undercharging. In Wilson and Hingnikar (2019), a field deployment revealed that over 40% of users remained connected long after their batteries were fully charged, occupying the charging station and reducing availability for others. Additionally, traditional systems often fail to implement real-time feedback mechanisms that can adjust charging rates in response to grid constraints or energy pricing signals. This lack of bidirectional interaction with the grid infrastructure inhibits demand response capabilities and isolates EV loads from broader smart grid strategies. As highlighted in Ravikumar et al. (2023), systems without grid-interactive logic or pricing-based modulation were shown to generate peak demand spikes that conflicted with time-of-use price incentives and resulted in higher operational costs for operators.

In many platforms, forecasting capabilities are entirely absent or implemented in a rudimentary fashion. Without forecast data for solar production, battery availability, or anticipated load curves, the system cannot make informed scheduling decisions or react to short-term fluctuations in renewable supply. This absence of predictive capability limits the efficiency gains from potential rescheduling and curtails any opportunity for renewable-aware planning. In I. E. Outlook (2022), the authors demonstrated that forecast-agnostic charging behavior led to a 17% increase in grid imports compared to a forecast-integrated platform operating under the same conditions.

5.2.3 Importance of aligning charging with renewable energy availability

Aligning electric vehicle (EV) charging operations with the availability of renewable energy sources (RES) is no longer just a sustainability preference—it is an operational imperative for ensuring efficiency, grid stability, and energy autonomy. Renewable energy, particularly solar and wind, is inherently intermittent, fluctuating based on weather conditions and time of day. EV charging, on the other hand, can often be shifted in time without compromising user satisfaction. This flexibility opens the door for an intelligent energy strategy that synchronizes EV charging demand with clean energy supply, thereby maximizing environmental and economic benefits. One of the core technical motivations for this alignment is grid impact mitigation. Without synchronization, EVs often charge during grid peak hours, amplifying demand stress and potentially triggering transformer overloads or voltage drops. Conversely, aligning charging with solar peak hours smoothens demand profiles and alleviates grid stress. As shown in Alkawsi et al. (2021), coordinated charging based on renewable availability allows energy systems to perform load-shifting, peak shaving, and congestion reduction.

Moreover, enhancing renewable energy utilization is directly linked to the temporal harmony between supply and load. Solar generation, for example, peaks in the early afternoon. Yet, most residential EV charging happens in the evening, creating a mismatch that leads to renewable curtailment or export at low tariff values. By dynamically shifting charging to coincide with solar abundance, R2V systems can absorb excess energy that would otherwise go underutilized. This improves the return on investment (ROI) for solar installations and helps meet decarbonization targets more effectively (Robinius et al., 2017). Case studies such as the Smart Solar Charging Initiative in the Netherlands illustrate this principle in real-world settings. The project integrated solar panels with intelligent EV scheduling systems, enabling vehicles to charge predominantly when solar power was abundant. The result was an increase in self-consumption ratios and a measurable decrease in grid imports. These outcomes demonstrate how forecast-driven energy management directly contributes to both operational efficiency and sustainability goals (Pourmousavi and Nehrir, 2012).

Economically, the benefits are also significant. Charging during solar peaks allows users to tap into lower-cost electricity under time-of-use (TOU) pricing schemes or even real-time tariffs. Intelligent charging platforms that monitor renewable output and price signals can shift sessions to cheaper windows, reducing costs for users and energy providers alike. In Deilami et al. (2011), the authors show that this strategy not only benefits individual EV owners but also improves the business case for deploying solar-integrated EVSE by reducing grid dependency and infrastructure sizing requirements. Importantly, aligning EV charging with renewable energy supply contributes to greater energy independence and resilience. By relying more on locally generated power, communities reduce their exposure to external energy shocks and price volatility. This is especially critical in regions with unstable grids or high fuel import costs. As highlighted in Robinius et al. (2017), decentralized renewable charging can serve as a backbone for autonomous microgrids, capable of sustaining operations even during grid outages.

5.3 Role of IoT in smart charging

5.3.1 IoT components: Sensors, edge/cloud computing, and connectivity

The deployment of smart EV charging systems fundamentally depends on the integration of Internet of Things (IoT) technologies, which enable the sensing, communication, and real-time processing required for adaptive and intelligent control. IoT-based R2V platforms rely on a distributed sensor infrastructure that captures information across the energy and mobility domains, including vehicle status, user intent, energy production, and local grid conditions. These platforms are designed to ingest diverse forms of data in real time (ranging from SOC, GPS coordinates, charging station availability, and solar irradiance) through a network of embedded systems and communication protocols. At the core of this infrastructure are sensor nodes and data acquisition units located on charging stations, EVs, solar inverters, and grid monitoring equipment. These devices are responsible for continuously collecting telemetry such as energy consumption, voltage stability, battery state, plug-in time, and departure forecasts. This raw data is then transmitted to local gateways or edge servers that perform pre-processing and early-stage analytics, including anomaly detection, user pattern recognition, and event-driven triggers for energy routing (Wilson and Hingnikar, 2019).

Edge computing plays a pivotal role in minimizing response time and reducing network congestion. By performing localized analysis, edge nodes enable real-time actuation, such as dynamically adjusting the charging rate based on PV production or prioritizing vehicles with imminent departure. In systems like the decentralized model described in Ravikumar et al. (2023), edge-layer devices were shown to execute autonomous decisions with millisecond-level latency, thus enabling distributed control without relying solely on centralized systems. Meanwhile, cloud computing supports high-level coordination across geographically dispersed nodes, running machine learning algorithms for forecasting and optimization. Cloud servers aggregate long-term historical data to build predictive models for EV load, solar generation, and battery availability. In hybrid architectures, cloud systems oversee policy-level decisions (such as energy pricing strategies or user scheduling logic) while edge nodes handle operational control. This dual-layer framework enhances scalability, flexibility, and fault tolerance, as demonstrated in smart campus deployments described in I. E. Outlook (2022).

Connectivity in these systems is often realized through lightweight messaging protocols such as MQTT, CoAP, or HTTP/2, enabling efficient communication across constrained devices. These protocols are designed for high-speed, low-overhead message delivery, ensuring real-time data flow between components with minimal latency. Secure sockets layer (SSL) encryption and device authentication protocols are used to maintain cybersecurity integrity across all communication channels. The importance of maintaining data privacy, particularly for user behavior and location patterns, has been emphasized in Wilson and Hingnikar (2019) as a necessary consideration in system design. A robust IoT architecture also incorporates middleware services for data integration and interoperability. Given the heterogeneity of devices and vendors, middleware platforms are used to normalize data formats, manage device identities, and ensure seamless coordination between subsystems. Platforms such as Node-RED or OpenHAB have been employed in several testbed configurations to streamline inter-device logic and reduce integration complexity (Ravikumar et al., 2023).

5.3.2 Real-time data collection from EVs, chargers, and renewable energy sources

Real-time data acquisition plays a foundational role in IoT-enabled Renewable-to-Vehicle (R2V) smart charging systems. The effectiveness of any such system hinges on the accurate, time-sensitive collection of telemetry from electric vehicles (EVs), charging stations, and renewable energy sources (RES). These data streams form the predictive and non-predictive intelligence that enables optimized scheduling, dynamic power allocation, and integration with decentralized energy sources. Data collected from EVs includes the battery's SOC, maximum charging rate, plug-in status, and in some cases, predicted departure time. Communication between the EV and charging station is typically governed by standardized protocols such as ISO 15118, which support both unidirectional and bidirectional data exchange. Access to SOC data allows the system to identify vehicles requiring immediate attention versus those with flexibility in their charging windows. The significance of this differentiation is highlighted in Wilson and Hingnikar (2019), where the authors demonstrate how SOC-based prioritization reduces waiting times and improves energy allocation efficiency in shared infrastructure environments.

Arrival and departure prediction models complement internal vehicle data, using sources such as historical usage patterns, mobile application inputs, or automated detection systems including RFID, camera recognition, or geofencing. As described in Kuzmich and Gordiyenko (2023), the absence of this predictive data leads to inefficient station utilization, with users frequently occupying chargers well beyond their active charging periods. Incorporating such models allows for predictive queuing and energy reservation, ensuring better charger turnover and system responsiveness. From the charging infrastructure, telemetry includes real-time measurements of voltage, current, connector status, and energy delivered per session. These values are transmitted to local gateways that serve as edge processors, which handle initial filtering, event detection (e.g., faults or thermal warnings), and localized control actions. For instance, some smart systems implement automatic power reduction when voltage thresholds drop below safety margins, a mechanism enabled by continuous voltage monitoring at the charger level (Ravikumar et al., 2023).

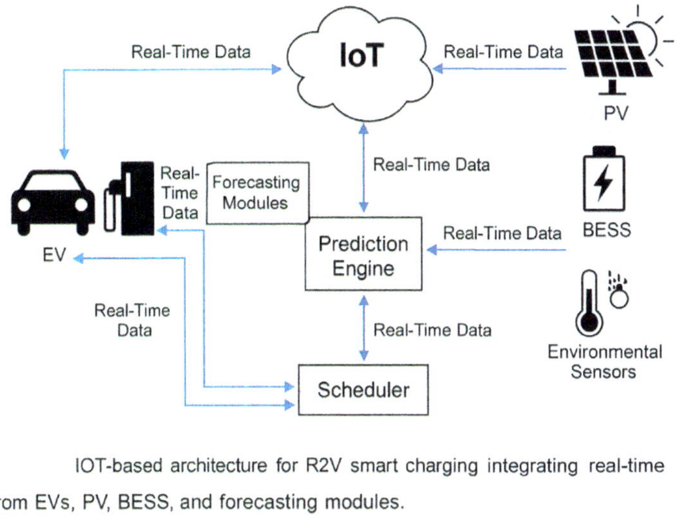

Figure 5.2 IoT-based architecture for R2V smart charging integrating real-time data from EVs, PV, BESS, and forecasting modules.

Equally vital is the real-time integration of renewable energy sources, particularly photovoltaic (PV) systems and battery energy storage systems (BESS). PV inverters supply power output data including current generation capacity, panel voltage, system efficiency, and operational status. Many platforms supplement this with irradiance data and weather forecasts obtained through APIs or on-site pyranometers. These inputs feed into predictive modules that anticipate generation capacity over the next few hours. The platform described in Varone et al. (2024) integrates such data into a central decision engine to defer or accelerate EV charging based on solar availability. BESS telemetry includes charge/discharge rates, remaining capacity, and inverter status. This information is particularly important in R2V systems aiming to minimize grid dependence. In the simulation study in Ren et al. (2022), a coordinated management system using BESS status and PV forecasts reduced grid import by 38% during high-demand periods. These components are tied together in system-level architectures that manage sensing, communication, and decision-making layers. As shown in Fig. 5.2, the real-time data architecture includes feedback loops from EVs, RES, and environmental sensors feeding into a prediction engine and a centralized scheduler.

5.4 Proposed IoT-enabled forecasting framework

5.4.1 System architecture and component layers

The proposed framework is structured in three layers, inspired by the SPEM implementation using the CRS4 Microservice Core (CMC-IoT) platform (Varone et al., 2024).

- Data Acquisition Layer: Distributed sensors from EVs, PV systems, BESS units, and grid meters are treated as "Things" in the IoT framework, each reporting values like SOC, voltage, current, and power. These are stored with their units (e.g., kWh, V) and timestamps in the system's backend (Varone et al., 2024).
- Forecasting and Control Layer: At this level, the system employs a Model Predictive Control (MPC) engine that processes real-time PV forecasts, EV energy demands, and grid load thresholds. Forecasts are updated as new sensor data arrives, enabling continuous adjustment of decisions within each prediction horizon (Varone et al., 2024).
- Execution Layer: This layer translates forecasts into real charging actions. Based on controller outputs, charging can be deferred, accelerated, or limited to comply with grid and renewable constraints. The system prioritizes PV consumption and falls back to grid power only when needed (Varone et al., 2024).

5.4.2 Inputs: Key parameters for forecasting

To generate accurate load forecasts, the framework processes multiple real-time inputs:

- EV Inputs: SOC, target energy, expected departure time, and charger capacity, obtained from vehicle telemetry and the user's stated preferences (Ashique et al., 2017).
- PV Forecasting Data: Solar output is forecasted using irradiance sensors and external weather APIs. These values are updated continuously and combined with real-time PV output for adaptive charging control (Varone et al., 2024).
- BESS Data: State of charge, available charge/discharge power, and operational modes are captured through battery management systems (Varone et al., 2024).
- Grid and Load Constraints: The system receives maximum grid import/export limits, pricing signals, and other grid conditions from local smart meters and utility APIs (Varone et al., 2024).

These data are combined to form the input vector for machine learning or control-based prediction models.

5.4.3 Forecast output and scheduling logic

The system generates a short-term charging demand forecast for each vehicle and the entire site. Based on this, it calculates the optimal charging power p_{ev}^{n} for each EV using the following rule (Wang, 2022):

$$p_{ev}^{n} = \min\left(\max\left(\frac{P_{PV}^{n} \cdot \delta E_n}{\sum_{n=1}^{N} \delta E_n}, \frac{\delta E_n}{\tau_n - t_0} \right), P_{C\max}^{n}, P_{B\max}^{n} \right) \tag{5.3}$$

Eq. (5.3) determines the optimal charging power p_{ev}^{n} for each EV based on short-term demand forecasts and renewable generation predictions. The first term inside the max operator, $\frac{P_{PV}^{n} \cdot \delta E_n}{\sum_{n=1}^{N} \delta E_n}$, represents a proportional allocation of forecasted PV generation among EVs, weighted by their relative energy needs. Multiplication by δE_n ensures that EVs with higher remaining energy demand receive a correspondingly larger share of renewable power, thus maintaining fairness while aligning with the physical constraint of forecasted PV availability. Also, the term inside the max operator allocates PV power proportionally to the energy deficit δE_n of each EV while ensuring that the vehicle can meet its charging deadline$(\tau_n - t_0)$. This guarantees fairness across EVs while prioritizing those with urgent departures. The outer min operator then enforces feasibility by restricting the allocation to both the charger capacity (P_{Cmax}^{n}) and the maximum allowable charging rate of the EV battery (P_{Bmax}^{n}). Here, P_{PV}^{n} denotes the forecasted available PV generation at scheduling interval n, obtained from the IoT-enabled forecasting module. It represents the renewable energy resource available for distribution among EVs during that period. This hierarchical decision logic ensures that renewable power is optimally utilized, charging demand is met within time limits, and technical safety constraints are respected. Fig. 5.3 shows how forecasting modules provide real-time input to this decision process.

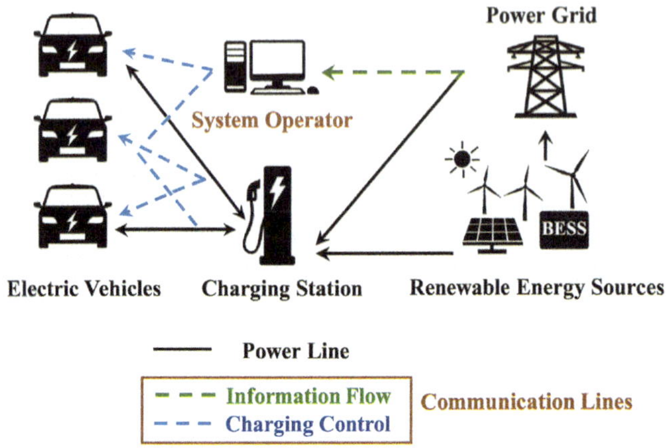

Figure 5.3 Architecture of forecasting-integrated R2V system.

5.5 Case study: Application in a solar-powered parking lot

5.5.1 Overview of the CRS4 SPEM deployment

A practical implementation of the IoT-enabled forecasting framework was carried out at the CRS4 Research Center in Sardinia, Italy. This platform, known as SPEM (Solar Parking for Electric vehicle Management), provides a real-world validation of R2V charging concepts using integrated solar PV, BESS, and predictive load management. The deployment supports a combination of distributed EV chargers, a 150 kWp rooftop PV array, and a 300-kWh battery system. Charging control is managed by a model predictive controller that operates on the CMC-IoT platform, organizing charging sessions in real time according to renewable availability and user constraints (Varone et al., 2024).

The CRS4 parking area serves approximately 143 employees, with an average EV parking duration of 9 hours. This allowed the system to use historical access badge data and daily logs to construct a high-resolution forecast model of arrival, SOC, and dwell time. These parameters form the basis for clustering users and assigning scheduling priorities in the control engine (Varone et al., 2024).

5.5.2 Forecast-aware charging strategy

The charging strategy implemented in SPEM is based on forecasted energy supply and mobility patterns. EVs with shorter remaining time in the lot are

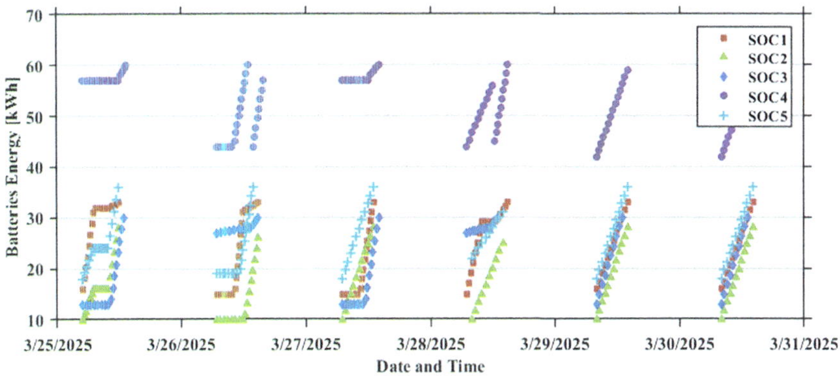

Figure 5.4 Charging behavior of five EVs in the CRS4 SPEM system.

prioritized when solar production is limited, while those with flexible departure times are scheduled to absorb midday PV peaks. This strategy leads to higher renewable utilization and better charger turnover compared to static allocation methods. An illustrative result is shown in Fig. 5.4, which plots SOC levels for five vehicles with varying battery capacities across 1 week. The diagram shows how charging sessions are modulated and sometimes split across time intervals or vehicles using the same station.

By applying this adaptive schedule, SPEM diverges from constant-rate charging assumptions used in tools such as HOMER Grid. Instead of applying uniform charging across fixed time windows, SPEM defers or fragments sessions based on predicted PV output and user mobility flexibility. This approach leads to a smoother aggregate demand curve and fewer peak grid draw events (Varone et al., 2024).

5.5.3 Energy flow and operational performance

Over one operational year (as you can see in Table 5.1), the site charged 12,849 EVs, delivering 184,747 kWh to vehicles (62.5% of total site energy use). Site energy production was 80.0% PV (236, 443kWh) and 20.0% grid purchases (59, 615kWh). On the use side, 25.3% of annual energy (74, 810kWh) was exported to the grid and 12.2% (36, 500kWh) met the AC primary load. The resulting balance isPV + Grid ≈ EVs + AC + Export, with a 1 kWh rounding difference.

Percentages under Sources (PV, Grid) are relative to annual production; percentages under Uses (EVs, Export, AC load) are relative to annual uses. The 62.5% value refers to the share of total site uses delivered to EVs, not a

Table 5.1 Annual energy summary—SPEM versus HOMER grid model.

Metric	SPEM	HOMER grid
EVs charged (count)	12,849	12,849
Sources (production)		
PV generation (kWh, % of production)	236,443 (80.0%)	231,311 (77.7%)
Grid purchases (kWh, % of production)	59,615 (20.0%)	66,425 (22.3%)
Total production (kWh)	296,058	297,736
Uses (consumption/ Export)		
EV energy delivered (kWh, % of uses)	184,747 (62.5%)	194,702 (66.6%)
Grid export (kWh, % of uses)	74,810 (25.3%)	61,257 (21.0%)
AC primary load (kWh, % of uses)	36,500 (12.2%)	36,500 (12.4%)
Total uses (kWh)	296,057	292,459

PV-only share of EV charging. The weekly time series of power flow shown in Fig. 5.5 confirms the dynamic energy management achieved by the platform. The diagram reflects how PV output (red), EV demand (blue), and grid exchanges (yellow) are balanced daily.

5.5.4 Economic analysis

The economic performance of the system was evaluated using the Levelized Cost of Energy (LCOE) methodology. In a hybrid scenario (PV + BESS + grid interaction), the LCOE for auto-produced energy

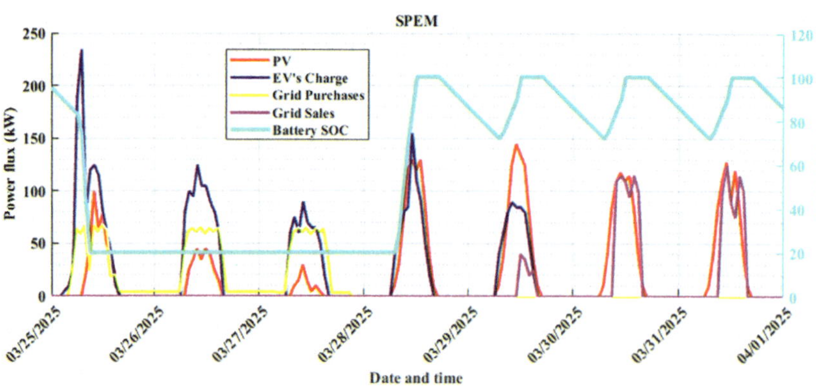

Figure 5.5 Weekly energy flow profile showing solar generation, EV charging, and grid usage in the SPEM system (March 25–31, 2025).

(LOCEa) was estimated at 0.078 €/kWh, while the final cost considering all exchanges (LOCEf) reached 0.220 €/kWh. These values indicate that despite the cost of battery storage, the system achieves favorable operating costs under high solar availability (Varone et al., 2024). Furthermore, the system's ability to align EV load with PV availability helped minimize peak demand from the grid, supporting both cost savings and reduced emissions. This makes SPEM a viable template for other smart campuses, industrial lots, and municipal EV infrastructure.

5.6 Benefits and future opportunities

5.6.1 Grid stability and peak demand reduction

Integrating intelligent IoT-based forecasting frameworks into Renewable-to-Vehicle (R2V) charging significantly enhances grid stability. One of the most impactful benefits is the mitigation of peak demand through load balancing, where smart charging algorithms align EV energy intake with off-peak grid availability and renewable energy generation cycles. This strategy, also known as dynamic demand response, enables the redistribution of charging loads to avoid surges during high-demand hours, thereby protecting transformers and reducing grid stress (Zhou et al., 2022).

Vehicle-to-grid (V2G) systems take this concept further by allowing EVs to discharge stored energy back into the grid, serving as distributed storage units that help during critical peak loads. As highlighted in the University of Delaware project and Enel X V2G implementation, such systems stabilize the grid by offering immediate supply from mobile storage units (Saber and Venayagamoorthy, 2011; Liu et al., 2015). Moreover, advanced energy management systems employing artificial intelligence (AI) and machine learning models can dynamically adjust charging profiles based on solar availability, battery status, and local demand patterns, helping prevent blackouts and reducing the reliance on fossil-fuel-based plants (Khan and Iqbal, 2005).

5.6.2 Increased renewable penetration and decentralization

By synchronizing EV charging schedules with renewable production peaks—especially solar and wind—R2V systems directly support greater renewable energy penetration into the electricity mix. Systems like SPEM demonstrated that over 62.5% of energy delivered to EVs was sourced directly from solar PV, showcasing how smart platforms can reduce dependence on the grid and fossil fuel-based generation (Varone et al., 2024). Furthermore, integrating modular microgrids and renewable energy islands into EV charging infrastructure enhances resilience. These decentralized systems reduce the load on transmission infrastructure and make remote areas more energy-independent. This form of localized energy generation and usage aligns well with distributed grid models and improves both energy security and sustainability (Daubitz and Kawgan-Kagan, 2015).

5.6.3 Scalability and urban infrastructure integration

Scalability is a key advantage of modular and IoT-integrated R2V systems. The use of standardized protocols, interoperable charging hardware, and modular control software enables easy replication of the system across different geographies—from corporate parking to public urban charging hubs (Fotis et al., 2022). As adoption increases, the same system can accommodate growing demand by simply expanding charger arrays or increasing BESS capacity. Flexible deployment models—ranging from residential to workplace and fleet charging—are supported by systems that adjust to context-specific usage patterns. These are particularly valuable in cities, where load patterns are diverse and power availability can fluctuate throughout the day. As emphasized in recent research, such flexibility is vital for urban infrastructure planning, where EV chargers may share grid connections with commercial and residential loads (Fotis et al., 2022).

5.6.4 Research horizons: V2G, blockchain, AI, and second-life batteries

The future development of smart R2V systems is tightly connected to several technological frontiers:

- Vehicle-to-Grid (V2G) Aggregation: Aggregating EVs through a central coordinator enables grid services at scale, such as frequency regulation, voltage stabilization, and even renewable firming. These aggregators create new business models where EV owners are compensated for participating in grid stability programs (Jin et al., 2013).
- Blockchain for Decentralized Energy Trading: Blockchain enables peer-to-peer energy transactions, allowing EV owners to sell excess energy to neighbors or back to the grid with transparent billing and trustless verification. Platforms like Sunverge have demonstrated this capability in microgrids, fostering greater energy democratization and grid flexibility (Morstyn et al., 2018; Khaleghi and Karimipour, 2024).

- AI-Based Demand Pricing and Control: AI will play a central role in future pricing models, enabling real-time adaptive tariffs that shift EV demand toward periods of high renewable availability. Deep reinforcement learning is already being used to optimize charging hubs dynamically, enhancing profitability and efficiency (Liu et al., 2015).
- Second-Life Battery Integration: Repurposing EV batteries for stationary storage opens a path to low-cost, sustainable BESS deployment. These second-life applications enhance circular economy strategies and reduce lifecycle emissions of battery manufacturing (Kene and Olwal, 2023).
- User Behavior Modeling: Future models should also integrate user psychology and behavioral patterns to better predict demand, increase participation, and improve scheduling precision. Incorporating uncertainty and behavior feedback loops can make forecasts more robust and human-aligned (Slednev et al., 2022).

5.6.5 Challenges and limitations

While R2V systems offer substantial benefits, several technical and operational challenges must be addressed to ensure robust scalability and performance. First, the accuracy of forecasting models remains a central concern. As shown in Feng et al. (2021), predictive errors in solar irradiance and EV arrival times can result in suboptimal scheduling and grid overdraw, especially under volatile weather or traffic conditions. Even hybrid models integrating SARIMA and deep learning face limitations when exposed to noisy or incomplete data streams (Ren et al., 2022).

Second, real-time responsiveness is often constrained by communication latency and data processing delays, particularly in large-scale deployments. Systems like SPEM, while effective, must rely on edge processing and hybrid cloud architectures to minimize control lag (Varone et al., 2024). Nonetheless, I. E. Outlook (2022) shows that edge devices can still fail under high load or poor connectivity, jeopardizing real-time adaptability.

Moreover, interoperability and standardization gaps between devices (EVs, chargers, BESS units) pose integration challenges. Without harmonized communication protocols and unified APIs, systems may face errors in telemetry synchronization and energy dispatch (Ravikumar et al., 2023). This is exacerbated in urban settings where heterogeneous hardware is common.

Another major barrier lies in user behavior unpredictability. Studies like (Wilson and Hingnikar, 2019; Kuzmich and Gordiyenko, 2023) demonstrate that users often overstay at chargers or fail to input expected departure times, which reduces scheduling accuracy and station availability. Effective integration of user feedback loops, or predictive modeling of habits, is still underexplored.

Lastly, cybersecurity and data privacy pose increasing risks. As pointed out in Wilson and Hingnikar (2019), Khaleghi and Karimipour (2025), Khaleghi et al. (2024), IoT-based infrastructures handling EV, grid, and user data must be designed with robust authentication, encryption, and intrusion detection mechanisms. A lack of such safeguards can lead to unauthorized access or manipulation of critical energy scheduling systems.

5.7 Conclusion

The increasing demand for sustainable mobility and the global shift toward electrification of transport have necessitated the development of smart, adaptive, and resource-aware charging infrastructures. This chapter presented an in-depth exploration of an IoT-enabled, real-time forecasting framework tailored to Renewable-to-Vehicle (R2V) smart charging systems. By addressing the technical foundations, forecasting challenges, and real-world validation through the CRS4 SPEM deployment, the chapter demonstrated both the feasibility and the strategic value of such integrated systems. A major takeaway from the analysis is that static or delayed charging strategies are inadequate in environments where renewable energy, user behavior, and grid conditions fluctuate rapidly. Instead, real-time forecasting, supported by edge computing and IoT telemetry, enables systems to align charging behavior with renewable availability, user mobility needs, and grid stability goals. The use of machine learning models and predictive control mechanisms improves both energy allocation efficiency and system responsiveness. Platforms like SPEM not only validate the forecasting and optimization architecture under real-world conditions but also showcase tangible improvements in renewable penetration, economic cost, and user satisfaction. In terms of strategic importance, the proposed framework contributes to multiple layers of energy system advancement:

- Operational: It reduces grid dependence and ensures high utilization of on-site solar.
- Economic: It enables cost savings for users and operators through intelligent scheduling.

- Environmental: It increases renewable energy consumption and supports decarbonization.
- Scalability: It allows flexible deployment across different sectors and infrastructures.

Looking forward, the chapter outlined key future directions such as the integration of vehicle-to-grid (V2G) systems, blockchain-based energy trading, AI-driven dynamic pricing, and the use of second-life batteries. These innovations will further expand the functional and economic potential of R2V systems, turning EVs into intelligent energy actors in a decentralized and sustainable grid ecosystem. In conclusion, smart charging infrastructures must move beyond passive, grid-dependent models toward intelligent, responsive platforms that make full use of real-time data, forecast models, and adaptive control. This chapter provides a blueprint for such a transformation, offering not only a technical solution but also a strategic vision for the future of electric mobility and clean energy integration.

References

Aghaei J, Alizadeh MI. Demand response in smart electricity grids equipped with renewable energy sources: a review. Renew Sustain Energy Rev 2013;18:64–72. https://doi.org/10.1016/j.rser.2012.09.019.

Alkawsi G, Baashar Y, Abbas U D, Ammar AA, Tiong SK. Review of renewable energy-based charging infrastructure for electric vehicles. Appl Sci 2021;11(9):3847. https://doi.org/10.3390/app11093847.

Ashique RH, Salam Z, Bin Abdul Aziz MJ, Bhatti AR. Integrated photovoltaic-grid dc fast charging system for electric vehicle: a review of the architecture and control. Renew Sustain Energy Rev 2017;69:1243–57. https://doi.org/10.1016/j.rser.2016.11.245.

Daubitz S, Kawgan-Kagan I. Integrated charging infrastructure: cognitive interviews to identify preferences in charging options. Euro Trans Res Rev 2015;7(4). https://doi.org/10.1007/s12544-015-0184-2. Available from: http://www.springer.com/engineering/mechanical+eng/journal/12544.

Deilami S, Masoum AS, Moses PS, Masoum MAS. Real-time coordination of plug-in electric vehicle charging in smart grids to minimize power losses and improve voltage profile. IEEE Trans Smart Grid 2011;2(3):456–67. https://doi.org/10.1109/TSG.2011.2159816.

Dixit R, Ravindranath K. Identity and access control techniques for enhanced data communication in cloud. Contemp Mathemat 2024;5(1):1–16. https://doi.org/10.37256/cm.5120243721.

Eddine MD, Shen Y. A deep learning based approach for predicting the demand of electric vehicle charge. J Supercomp 2022;78(12):14072–95. https://doi.org/10.1007/s11227-022-04428-0.

Feng J, Yang J, Li Y, Wang H, Ji H, Yang W, et al. Load forecasting of electric vehicle charging station based on grey theory and neural network. Energy Rep 2021;7:487–92. https://doi.org/10.1016/j.egyr.2021.08.015.

Fikru MG, Atherton J, Canfield CI. Cost-reflective dynamic electricity pricing for prosumers. Electr J 2022;35(1). https://doi.org/10.1016/j.tej.2022.107075. Available from: https://www.sciencedirect.com/journal/the-electricity-journal.

Fotis G, Dikeakos C, Zafeiropoulos E, Pappas S, Vita V. Scalability and replicability for smart grid innovation projects and the improvement of renewable energy sources exploitation: the FLEXITRANSTORE case. Energies 2022;15(13):4519. https://doi.org/10.3390/en15134519.

Outlook IE. World energy outlook special report. https://www.iea.org/reports/india-energy-outlook-2021; 2022.

Jin C, Tang J, Ghosh P. Optimizing electric vehicle charging with energy storage in the electricity market. IEEE Trans Smart Grid 2013;4(1):311–20. https://doi.org/10.1109/TSG.2012.2218834.

Kene RO, Olwal TO. Energy management and optimization of large-scale electric vehicle charging on the grid. World Electr Vehicle J 2023;14(4). https://doi.org/10.3390/wevj14040095. Available from, https://www.mdpi.com/journal/wevj.

Khaleghi A, Karimipour H. A probabilistic-based approach for detecting simultaneous load redistribution Attacks through entropy analysis and deep learning. IEEE Trans Smart Grid 2025;16(2):1851–61. https://doi.org/10.1109/TSG.2024.3524455.

Khaleghi A, Ghazizadeh MS, Aghamohammadi MR, Guerrero JM, Vasquez JC, Guan Y. A probabilistic data recovery framework against load redistribution Attacks based on bayesian network and bias correction method. IEEE Trans Power Sys 2024;39(4): 5806–17. https://doi.org/10.1109/TPWRS.2023.3346652.

Khaleghi A, Karimipour H. Investigation of detection mechanisms against false data injection attacks based on machine learning approaches. In: Artificial intelligence in the operation and control of digitalized power systems. Springer; 2024. p. 209–31.

Khan MJ, Iqbal MT. Pre-feasibility study of stand-alone hybrid energy systems for applications in Newfoundland. Renew Energy 2005;30(6):835–54. https://doi.org/10.1016/j.renene.2004.09.001.

Kuzmich M, Gordiyenko T. Application of kubeflow as a universal approach for the development and implementation of artificial intelligence systems. ARPN J Eng Appl Sci 2023;18(20):2311–20. https://doi.org/10.59018/1023283.

Lee ZJ, Lee G, Lee T, Jin C, Lee R, Low Z, et al. Adaptive charging networks: a framework for smart electric vehicle charging. IEEE Trans Smart Grid 2021;12(5):4339–50. https://doi.org/10.1109/TSG.2021.3074437.

Liu H, Hu Z, Song Y, Wang J, Xie X. Vehicle-to-grid control for supplementary frequency regulation considering charging demands. IEEE Trans Power Syst 2015;30(6):3110–9. https://doi.org/10.1109/TPWRS.2014.2382979.

Ma T, Mohammed OA. Optimal charging of plug-in electric vehicles for a car-park infrastructure. IEEE Trans Indus Applic 2014;50(4):2323–30. https://doi.org/10.1109/TIA.2013.2296620.

Martin H, Buffat R, Bucher D, Hamper J, Raubal M. Using rooftop photovoltaic generation to cover individual electric vehicle demand—a detailed case study. Renew Sustain Energy Rev 2022;157:111969. https://doi.org/10.1016/j.rser.2021.111969.

Mbuwir BV, Vanmunster L, Thoelen K, Deconinck G. A hybrid policy gradient and rule-based control framework for electric vehicle charging. Energy AI 2021;(4): 100059. https://doi.org/10.1016/j.egyai.2021.100059.

Morstyn T, Farrell N, Darby SJ, McCulloch MD. Using peer-to-peer energy-trading platforms to incentivize prosumers to form federated power plants. Nat Energy 2018;3(2): 94–101. https://doi.org/10.1038/s41560-017-0075-y.

Mu Y, Wu J, Jenkins N, Jia H, Wang C. A Spatial-Temporal model for grid impact analysis of plug-in electric vehicles. Appl Energy 2014;114:456–65. https://doi.org/10.1016/j.apenergy.2013.10.006.

Pourmousavi SA, Nehrir MH. Real-time central demand response for primary frequency regulation in microgrids. IEEE Trans Smart Grid 2012;3(4):1988–96. https://doi.org/10.1109/TSG.2012.2201964.

Ravikumar NVA, Nuvvula RSS, Kumar PP, Haroon NH, Butkar UD, Siddiqui A. Integration of electric vehicles, renewable energy sources, and IoT for sustainable transportation and energy management: a comprehensive review and future prospects. In: 12th IEEE International Conference on Renewable Energy Research and Applications, ICRERA 2023. India: Institute of Electrical and Electronics Engineers Inc.; 2023. p. 505–11. https://doi.org/10.1109/ICRERA59003.2023.10269421. http://ieeexplore.ieee.org/xpl/mostRecentIssue.jsp?punumber=10269105.

Ren F, Tian C, Zhang G, Li C, Zhai Y. A hybrid method for power demand prediction of electric vehicles based on SARIMA and deep learning with integration of periodic features. Energy 2022;250:123738. https://doi.org/10.1016/j.energy.2022.123738.

Robinius M, Otto A, Heuser P, Welder L, Syranidis K, Ryberg D, et al. Linking the power and transport sectors—Part 1: the principle of sector coupling. Energies 2017;10(7):956. https://doi.org/10.3390/en10070956.

Saber AY, Venayagamoorthy GK. Plug-in vehicles and renewable energy sources for cost and emission reductions. IEEE Trans Ind Electron 2011;58(4):1229–38. https://doi.org/10.1109/TIE.2010.2047828.

Slednev V, Jochem P, Fichtner W. Impacts of electric vehicles on the European high and extra high voltage power grid. J Industr Ecol 2022;26(3):824–37. https://doi.org/10.1111/jiec.13216. Available from: http://onlinelibrary.wiley.com/journal/10.1111/(ISSN)1530-9290.

Tan M, Zhang Z, Ren Y, Richard I, Zhang Y. Multi-agent system for electric vehicle charging scheduling in parking lots. Complex Sys Model Simul 2023;3(2):129–42. https://doi.org/10.23919/CSMS.2023.0005.

van der Kam M, van Sark W. Smart charging of electric vehicles with photovoltaic power and vehicle-to-grid technology in a microgrid; a case study. Appl Energy 2015;152:20–30. https://doi.org/10.1016/j.apenergy.2015.04.092.

van Ommeren J, McIvor M, Mulalic I, Inci E. A novel methodology to estimate cruising for parking and related external costs. Transp Res Part B Methodol 2021;145:247–69. https://doi.org/10.1016/j.trb.2020.12.005.

Varone A, Heilmann Z, Porruvecchio G, Romanino A. Solar parking lot management: an IoT platform for smart charging EV fleets, using real-time data and production forecasts. Renew Sustain Energy Rev 2024;189:113845. https://doi.org/10.1016/j.rser.2023.113845.

Wang X. Electric vehicle charging load prediction with improved Kernel Extreme Learning Machine. In: 4th International Conference on Information Science, Electrical, and Automation Engineering (ISEAE 2022), vol 12257; 2022. p. 732–6.

Wilson Y, Hingnikar A. Solving identity management in modern applications: demystifying OAuth 2.0, OpenID connect, and SAML 2.0. Apress Media LLC, undefined; 2019. p. 1–311. https://doi.org/10.1007/9781484250952. Available from: https://www.apress.com/in/book/9781484250945.

Zhou G, Zhu Z, Luo S. Location optimization of electric vehicle charging stations: based on cost model and genetic algorithm. Energy 2022;247:123437. https://doi.org/10.1016/j.energy.2022.123437.

CHAPTER SIX

Renewable-to-vehicle (R2V) smart charging for enhanced charging resiliency across operational conditions

Md Zahirul Islam Khan[1], Farzad Ferdowsi[1], Sercan Aygun[2], Touria El Mezyani[3]

[1]Electrical and Computer Engineering Department, University of Louisiana at Lafayette, Lafayette, LA, United States
[2]School of Computing and Informatics, University of Louisiana at Lafayette, Lafayette, LA, United States
[3]Electrical Engineering Department, University of North Florida, Jacksonville, FL, United States

6.1 Introduction

6.1.1 Background

The global shift toward electric vehicles is rapidly accelerating, driven by climate change concerns and reduced fossil fuel dependence. Global electric car sales surpassed 14 million in 2023, with China accounting for nearly 60% of new registrations (International Energy Agency, 2024a), and forecasts project 20.6 million by 2025, representing 23% of global passenger vehicle sales (BloombergNEF, 2022). This rapid adoption increases demands on aging grid infrastructure, which faces growing vulnerability to disruptions. From 2000 to 2021, approximately 83% of major U.S. power outages were weather-related (Surging power outages and climate change, n.d.), with a 74% increase in severe storm-related outages between 2010 and 19 compared to the previous decade (Weather-related power outages rising, n.d.). Recent events—such as the 2021 Texas winter storm that disrupted power for 4.5 million customers (The february 2021 cold weather outages in Texas, 2021) and California's 2019 Public Safety Power Shutoffs affecting over 2 million people (Pg&e to shut off power, 2019)—highlight EV charging infrastructure's vulnerability to grid failures, emphasizing the need for charging systems that can maintain functionality during extreme conditions.

Renewable-to-Vehicle (R2V) smart charging has emerged as an attractive alternative to traditional charging approaches, prioritizing localized

Renewable-to-Vehicle Smart Charging
ISBN: 978-0-443-36684-0
https://doi.org/10.1016/B978-0-443-36684-0.00009-6

renewable consumption and enabling autonomous operation during grid failures if integrated with storage and intelligent controls (Barman et al., 2023). R2V harnesses renewable energy sources to directly power EV charging, positioning EVs as flexible loads capable of absorbing intermittent renewable generation. This strategy can supplement or replace grid-supplied electricity, diminishing dependence on fossil-fuel-derived power and mitigating strain on the grid infrastructure during peak charging periods (Tiwari et al., 2024). R2V smart charging systems use intelligent controls to manage EV charging from renewables, the grid, or pause charging, maximizing renewable use and maintaining system stability.

Charging resiliency refers to the ability of EV charging infrastructure to remain reliable and operational under adverse or changing conditions—a growing concern as EV adoption accelerates and transportation becomes more electrified. Unlike gasoline vehicles, which can store fuel, EVs depend on continuous access to electricity and functioning charging stations. Any disruption to these can immobilize vehicles, impacting both individuals and communities. Resilient charging ensures that EVs can access energy during normal operations, emergencies, and periods of grid stress or congestion. The following scenarios illustrate the importance of resilient R2V smart charging systems.

R2V smart charging systems must address three critical operational scenarios to ensure resilience. Under **normal conditions**, variable renewable energy sources create fluctuations that can result in inadequate vehicle charging or grid demand surges when renewable generation decreases (Tiwari et al., 2024). Resilience in this mode requires managing frequency deviations and voltage sags in grids with high renewable penetration through demand response and energy storage integration with EV charging.

During **emergency conditions** such as natural disasters and grid outages, EV charging becomes critically vulnerable. California's 2019 wildfire-related power shutdowns affected over 2 million people (The Climate Center, 2024), prompting Tesla to notify owners to charge before anticipated blackouts. R2V systems with solar panels and batteries can maintain charging capability during prolonged outages (North Central Texas Council of Governments, 2024), supporting evacuation operations and first responder mobility when grid power is unavailable.

In **congested grid conditions**, EV charging can exacerbate grid instability during peak demand periods. During California's 2022 heat wave, the grid operator issued Flex Alerts requesting EV owners to avoid charging during peak hours (4–9 PM) (Griffith, 2022). Similarly, Europe experienced record demand and outages during 2024 heatwaves (International Energy Agency, 2024b). R2V systems can reduce grid dependence by utilizing renewable production and potentially enabling V2G support when the grid approaches capacity limits (Griffith, 2022).

However, Renewable-to-Grid systems lack the flexibility and targeted resilience advantages of R2V since they do not prioritize EV charging while feeding renewables into the grid. Therefore, R2V offers a dual advantage that traditional charging frameworks cannot match by integrating renewable energy with mobility resilience.

6.1.2 Resilience gaps in current standards

Despite R2V's potential benefits, existing standards governing EV charging infrastructure impede autonomous or resilient operation. Key standards (IEEE 1547-2018, IEC 61851, SAE J3072) assume uninterrupted grid connectivity and enforce disconnection protocols that prevent using local renewable energy during grid disruptions:

- **IEEE 1547-2018** requires DERs (including bidirectional EV chargers) to cease energization within 2 seconds of grid disconnection (IEEE standard for interconnection, 2018). It presumes grid-following inverters and only allows voltage regulation "by changes of reactive power" under utility control with "approval of the Area EPS Operator" (Narang, 2019).
- **IEC 61851** prioritizes safety and interoperability over resilience, with hardware specifications and charging modes dependent on active AC supply. For instance, IEC 61851-23 requires constant protection-earth conductor monitoring, with immediate shutdown if continuity is broken (Electric vehicle conductive dc charging, 2018). No "Mode 5" for islanded charging exists—the standard mandates grid reference and doesn't address EVs supplying their own load or local PV powering chargers.
- **SAE J3072** enables vehicle inverters as grid-interactive resources but maintains grid-centric limitations. It references IEEE 1547 extensively (Interconnection requirements for onboard, 2015) and explicitly requires bidirectional inverters to meet IEEE 1547–2018 requirements (Vartanian et al., 2022). J3072 offers minimal resilience capabilities, with no provisions for islanded mode or local grid energization. EVs remain grid-following devices without off-grid autonomy.

These requirements create significant resilience gaps, preventing R2V systems from supporting microgrids or providing emergency power. Current regulations force EV chargers to operate as grid-following devices rather than autonomous, intelligent DERs. Beyond regulatory constraints, the industry faces several systemic challenges. Current implementations rely on **static control strategies** with fixed scheduling models based on cost or time-of-use that fail to address real-time variations in renewable supply or grid stress (Deb et al., 2017). There is also a notable **absence of resilience metrics**, as performance evaluations typically focus on normal uptime while neglecting critical parameters such as recovery time, autonomy length, and service continuity under stress conditions (North Central Texas Council of Governments, 2024). Furthermore, despite the significant potential for anomaly detection and predictive scheduling, **AI integration remains fragmented**, with applications limited to non-scalable pilots, few standards-compliant platforms, and questionable training data quality (Barman et al., 2023).

Collectively, these deficiencies undermine the capacity of electric vehicle infrastructure to enhance the resilience of the transportation system. During power interruptions, drivers may become immobilized and electric vehicles that could function as mobile energy storage resources remain underutilized when their capabilities are most crucial.

6.1.3 To be covered in this chapter

This chapter addresses critical gaps in R2V smart charging resilience through three interconnected focus areas: (1) a **Charging Service Resilience Framework** that presents comprehensive analyses of local power continuity strategies across various scenarios, introducing resiliency-by-design configurations and survival metrics for extreme conditions; (2) an **Intelligence Layer Architecture** that explores advanced AI applications including transfer learning for renewable forecasting, reinforcement learning for charge scheduling, and robust cybersecurity protocols that create self-healing charging ecosystems; and (3) **Operational Mode Optimization** that develops adaptive control schemes across standard, emergency, and congested operational modes, each addressing specific resilience threats with tailored mitigation strategies. Throughout, we analyze real-world implementations and address critical limitations in current standards, particularly SAE J3072 interoperability requirements, enabling EV charging infrastructure to function as self-healing micro-ecosystems that reduce dependence on centralized grid recovery during failures.

This chapter is organized into several key sections that comprehensively address the challenges and solutions for resilient Renewable-to-Vehicle (R2V) smart charging systems. Section 6.2 presents the Charging Service Resilience Framework, detailing local power continuity strategies and resiliency-by-design configurations tailored for diverse operational scenarios. Section 6.3 introduces the Intelligent Layer Architecture, highlighting advanced AI techniques for forecasting, adaptive control, cybersecurity, and failure mitigation essential for autonomous and scalable R2V operation. Section 6.4 focuses on Operational Mode Optimization, exploring adaptive control schemes across standard, emergency, and congested modes, including load shedding and resource prioritization to ensure charging continuity under varying grid conditions. Section 6.5 discusses implementation challenges, especially in disaster-prone areas, and strategies to overcome these barriers. Section 6.6 critically examines current standards and regulations, identifying key limitations that hinder resilient R2V deployment, along with a cost-benefit analysis. Section 6.7 outlines recommendations and future trends aimed at advancing the development and widespread adoption of resilient R2V smart charging infrastructure. Finally, Section 6.8 concludes this chapter.

6.2 Charging service resilience framework

6.2.1 Local power continuity strategies

If an outage or a blackout (a total loss of generation and load) occurs, efforts must be made to restore the system to normal state as soon as possible to ensure power continuity. The major steps required for restoration are: (i) Preparation, (ii) Systems restoration, and (iii) Loads restoration.

The preparation step involves assessing the post-outage status of the system and identifying an appropriate sectionalization and task strategies with the following requirements (Quirós-Tortós et al., 2015):

- Islands that have survived need to be stabilized in frequency and need to be used to start other units.
- Hydro/Gas units that require less startup power need to be started using in-house DG sets.
- Larger thermal units need to be fed "startup power" from: 1) Islands that have survived, 2) black started generators, 3) Other synchronous grids (temporarily).

- Started units are synchronized with one another.
- Loads and generation have to be matched as much as possible to avoid large frequency variations.

Power restoration involves the black-start of power generation units, which involves reconnecting lines and islands to create cranking paths between the distributed power units and the remaining power grid, and stabilizing frequency and voltage. The black-start process consists of restarting a power generation unit without relying on the external electric power transmission network. Restoring the power grid is a challenging task because once a generator is tripped, restarting it requires a significant amount of power. Traditionally, power restoration requires start-up power for the power plant auxiliaries (pumps, etc.) in addition to the remaining plant. For example, a 500 MW generator requires approximately 40 MW to run its auxiliaries. Therefore, a sufficient initial power capacity is necessary for black-start of power units to reduce the duration of the outage during the initial grid restoration process.

Renewable energy (RE) sources with fast-ramping capabilities are suitable for black-start applications. However, they are traditionally not utilized in system restoration due to energy variability, which can cause voltage oscillations and frequency drops that lead to further damage to the power grid. The increasing penetration levels of inverter-based resources (IBRs), such as wind, photovoltaics (PV), and battery energy storage systems (BESS), have created a need to assess the technical capabilities and costs of using these IBR resources to provide black-start support. Fast and resilient grid restoration may consider both short- and long-term restoration stages for effective response strategies. The short-term response should involve rapid recovery and restoration decisions, including coordination of restoration for all layers of the network, including cyber, communication, and components, in light of desired long-term results. Long-term DER and Critical infrastructure planning should include real-time micro grids/critical infrastructures situational cybersecurity awareness and risk assessment to identify cyberphysical vulnerabilities.

6.2.2 Resilience by design configuration

Microgrid design focuses primarily on power reliability, which involves ensuring power availability for customers and electric vehicle (EV) systems. G2V has become an important research question due to the challenges and opportunities it presents. Multiport G2V coordination is critical for critical service EVs, such as emergency vehicles. The improvement of the reliability of G2V and V2G consists of optimizing vehicle integration into the grid by optimizing vehicle charging through various connection points (multiport) and optimizing V2G discharge, particularly during emergencies or outages.

Microgrids offer various topologies suitable for implementing G2V systems, primarily utilizing renewables as primary sources for EV charging, and employing different microgrid topologies that incorporate distributed renewable energy sources and battery energy storage systems. These topologies involve G2V, vehicle-to-grid (V2G), Battery-to-Vehicle (B2V), R2V, etc. as shown in Fig. 6.1, which depend on the nature of the microgrids:

- In the Grid-Connected microgrid topology, the microgrid can prioritize using this renewable energy to charge the EV only when it is available.
- In the Off-Grid microgrid topology (e.g., rural areas), R2V systems in this topology depend solely on local renewable energy sources and energy storage systems (e.g., batteries) to power the EV systems.
- In the hybrid microgrid topology, which combines both the grid-connected and off-grid system capabilities, R2V systems can switch between renewable energy, stored energy, and grid power as needed. For example, a triple-port integrated topology connects the vehicle and solar PV to the electrical grid (Tiwari Banerjee et al., 2025; Sun et al., 2019) using four modes: (i) R2V, (ii) R2G, (iii) G2V, (iv) V2G modes.

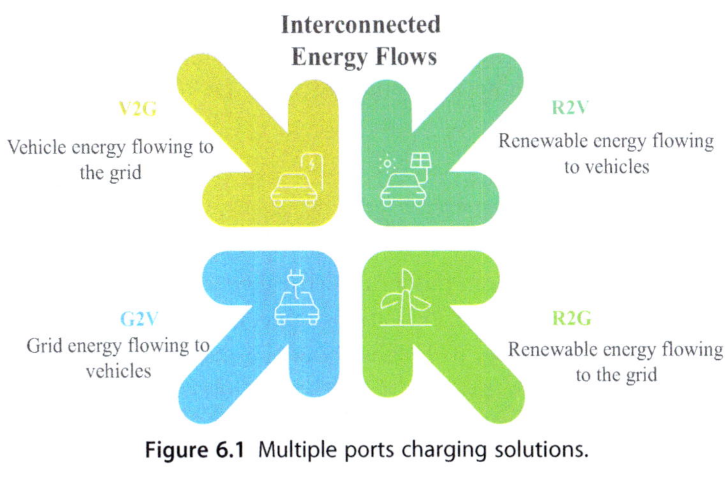

Figure 6.1 Multiple ports charging solutions.

The survival of the vehicle (e.g., emergency vehicle) relates to the ability of G2V, R2V, or V2G to operate in extreme conditions such as during hurricanes, fire, and/or other catastrophic events and perform the intended mission. Specifically, survivability quantifies the ability of the grid and/or renewable energy sources to answer a continuous power availability for the EVs functions during these critical events. Survivability analysis and quantification was extensively studied in the shipboard power system (SPS) (Doerry, 2006, 2015; Petry and Rumburg, 1993). In the literature, two approaches have been discussed to measure the survival capacity of SPS: (i) total ship survival assessment (TSSA) (International Energy Agency, 2024a; Mahazabeen et al., 2023) and (ii) integrated volumetric vulnerability assessment (VIVA) (Mahazabeen et al., 2022).

In Doerry (2015) a survivable designs in the SPS was discussed by the application of zonal distribution systems designs has been explored. Several zonal architectures for SPS were detailed in Doerry (2006) and compared for survivability requirements. A zonal architecture ensures zonal survivability, which implies that faults due to damage or equipment failure are contained within the zone and do not affect undamaged zones. In addition, the topology of the power grid network can be designed taking into account reliability.

Reliability is quantified using the availability of the network, named minpaths (Ahuja et al., 1993), which is related to the availability of power to each load. In Ahuja et al. (1993), the authors discussed the reliability calculation algorithm using the Minpath algorithm. Consider P_1, P_2, ..., P_k the min-paths of a power network graph G=(V,E), with $E_i \in E$ representing the events that all edges of the min-paths can deliver power. The reliability metric can be evaluated using various G2V multiport topologies. For example, the break-and-a-half (BAAH) topology was found to be the most reliable versus double bus double breaker (DBDB) (Dubey and Santoso, 2016).

6.3 Intelligent layer architecture

The intelligent layer in modern electrical systems is essential for managing and optimizing control operations, especially to improve efficiency when directing renewables to EVs without the main grid. AI plays a key role here, enabling proactive grid failure detection, forecasting, and both preventive and corrective decision-making.

6.3.1 Predictive systems

Predictive modeling is crucial for renewable energy systems because of generation intermittency. Accurate predictions require robust datasets and suitable AI architectures. While traditional machine learning uses structured training-validation-testing workflows, transfer learning has become valuable when labeled data is scarce—a common issue in renewable forecasting and fault diagnosis.

Transfer learning allows models trained on related domains to be repurposed with minimal retraining, significantly reducing data acquisition burdens. Al-Hajj et al. emphasize its growing importance in renewable systems for applications such as fault prediction and energy forecasting (Al-Hajj et al., 2023). This approach is particularly advantageous in the highly heterogeneous and complex ecosystems of vehicle-grid-renewable interactions, where collecting balanced and comprehensive datasets is often infeasible.

In conventional machine learning, assuming training and test data share the same feature space can lead to bias, especially with limited data or rare events. Transfer learning addresses this by using knowledge from related domains to improve target task performance. For always-on systems like EV charging, the ability to adapt to changing data is crucial. Transfer learning supports this adaptability, making it key for resilient predictive systems in the intelligent layer.

The EV charging model is often formulated using key parameters such as charging demand D, expected energy E upon departure from the charging station, and the current energy upon arrival A. The demand can be modeled as: $D = (E - A) \cdot C_{\max}$, where $C_{\max}$ denotes the maximum capacity of the EV battery. Alternatively, the energy requirement can be expressed using an integral over the charging duration: $E = \eta \cdot \sum_{t=t_{\text{arr}}}^{t_{\text{dep}}} P(t) \cdot \Delta t$. Here, η represents the charging/discharging efficiency, $P(t)$ is the power supplied at time t, and Δt is the discrete time interval. This formulation captures the cumulative energy delivery over the vehicle's stay at the station (Wang et al., 2023).

In real-world EV charging systems, collecting large, balanced datasets across stations, vehicle types, and grid conditions is often infeasible. Transfer learning (TL) addresses this by transferring knowledge from well-monitored source domains (e.g., urban stations) to target domains with limited data (e.g., new rural stations). Reinforcement learning (RL) suits modeling the dynamic EV charging cycle due to its sequential decision-making nature

(Mahazabeen et al., 2022, 2023; Seyedi et al., 2024a, 2024b). However, applying TL in RL adds complexity, requiring careful alignment of state and action spaces between source and target environments to transfer policies or value functions effectively. To integrate EV charging into AI-based decision systems, particularly with RL, researchers typically define: *States* such as battery level, time until departure, electricity pricing, or station occupancy, *Actions*: including charging, discharging (e.g., vehicle-to-grid), or idling, *Rewards*: shaped by factors like charging costs, battery degradation, or user satisfaction. By incorporating cost models into this framework, intelligent EV charging systems can adapt in real-time to optimize operational efficiency while ensuring user needs are met. When enhanced with TL, these systems gain the ability to generalize learned behaviors across different contexts, accelerating deployment and reducing the need for retraining from scratch (Barman et al., 2023; Wang et al., 2023).

AI-driven smart charging and the corresponding prediction is a recent and rapidly evolving field. Barman et al. highlight the significance of the global trend in electric vehicle (EV) stock growth from 2010 to 2021, noting that the total number of EVs has increased exponentially, driven by advances in battery technology, policy incentives, and climate commitments (Barman et al., 2023). This surge presents both challenges and opportunities for grid operators and urban planners, necessitating intelligent charging infrastructure that can forecast demand, optimize load distribution, and align with intermittent renewable generation. As the complexity of vehicle-grid interactions increases, AI-based predictive models are being increasingly explored to ensure efficient, scalable, and resilient charging ecosystems.

In today's EV-to-renewable integration model, most renewables first feed the central grid, get routed through transmission and distribution control centers, and are then delivered to local microgrids for EV charging. Virtual Power Plant (VPP) architectures help coordinate EVs, charging stations, and energy sources, managing dynamic pricing and tariffs. To support this, charging stations need cloud integration, and end devices require connectivity modules (like 4G/5G or Wi-Fi) for service provider communication (Barman et al., 2023; Ram Chandra Mouli et al., 2016; Simona Raboaca et al., 2022).

To streamline R2V infrastructure and reduce reliance on centralized systems, on-vehicle intelligence and hardware-aware AI should be prioritized. Local smart charging decisions minimize communication delays, improve response times, and support green computing, enhancing both resilience and scalability. This edge-focused approach aligns with sustainability goals and is already being demonstrated by leaders like Google, Tesla, and LomboXnet, who have launched solar-powered charging pilots—such as San Diego's "Solar-to-EV" initiative—that enable direct, clean R2V charging independent of the central grid.

Predictive systems are crucial for Solar-to-EV architectures, where EVs charge directly from renewables without complex grid coordination or costly upgrades. Real-time demand estimation optimizes excess renewable use and prevents curtailment. TemporalCharge, a notable AI model, uses a Generative Adversarial Network (GAN) to augment sparse EV charging data, enabling more accurate demand prediction—especially valuable for Solar-to-EV initiatives lacking historical data. The GAN generates realistic demand sequences and improves sample quality through adversarial training. As detailed in the original study (Mohanty et al., 2024), the architecture incorporates a conditional setup using charging time windows and user profiles to generate context-aware predictions. This enables energy management systems to allocate renewables, prioritize loads, and adjust charging in a sustainable, consumer-focused way. Alongside advanced models like TemporalCharge, conventional machine learning algorithms—such as Random Forest and Gradient Boosting—are also effective for modeling EV energy usage from historical data. These ensemble methods improve accuracy and reduce overfitting, making them valuable for real-time demand forecasting and load optimization when large labeled datasets are available. Their ease of implementation and scalability make them attractive for AI-driven charging infrastructure.

Aldossary et al. provide a comprehensive benchmarking and survey of EV charging prediction methods that integrate renewable energy sources, with a particular emphasis on cloud-based forecasting and sustainability (Aldossary et al., 2024). Several studies in the literature offer diverse modeling approaches. For example, Selvaraj and Vairavasundaram (2024) employ a Bayesian Extreme Learning Machine to address uncertainty in EV charging behavior. An innovative approach by Zhang et al. (2023a) combines decision trees with parallel gradient boosting to predict energy demand and forecast schedulable charging capacity. However, the training complexity of such boosting models is noted as a limitation. Additionally, unconventional data sources such as mobile phone usage data are utilized in Lee et al. (2021) to estimate EV energy consumption patterns. Other notable work includes the use of statistical feature extraction with supervised learning methods (Shahriar et al., 2021), and fuzzy logic controllers (Benhammou et al., 2023) to enhance prediction accuracy.

Many existing studies focus on isolated features or single data types. In contrast, our approach uses a holistic, context-aware framework that integrates multi-modal data—such as weather, live traffic (via GPS), and historical charging stats along vehicle routes. Specialized subnetworks process each feature type and feed into a master machine learning model that synthesizes predictions across modalities. This is especially important for R2V systems, where local factors like solar irradiance, terrain, and wind affect renewable availability and charging feasibility. By incorporating these parameters, the model improves precision and resilience in decentralized, green EV charging infrastructure.

Fig. 6.2 presents a holistic illustration of the proposed Intelligent Layer Architecture for green, decentralized R2V systems. This operates independently of intermediary grid operations, using real-time data from diverse sources for on-vehicle and edge decision-making. A master ML model—-trained with transfer and reinforcement learning—integrates specialized subnetworks for weather, geography, traffic, renewables, and in-EV analytics. This modular, scalable structure enables local, adaptive, and resilient control crucial for next-generation R2V charging.

Unlike traditional centralized systems that depend on cloud services and grid-level coordination, this model uses edge intelligence—performing computation and prediction locally at the vehicle or charging station. This distributed approach is essential for R2V deployments, where renewable supply is variable and data scarcity common, especially in rural or new locations. The architecture enables lifelong learning, context-aware optimization, and real-time prediction by combining outputs from specialized subnetworks into a master neural network that guides decisions. This design improves energy distribution, charging station responsiveness, and sustainability. By fusing multimodal data—including environmental, infrastructure, and user behavior—the intelligent layer supports robust, scalable operation and aligns with green computing goals.

6.3.2 Cybersecurity protocols

While designing electrical charging systems, ranging from renewable sources to end-use applications, such as electric vehicle charging or residential power delivery, security becomes a fundamental concern. Unlike conventional information systems, which handle discrete and detectable data flows,

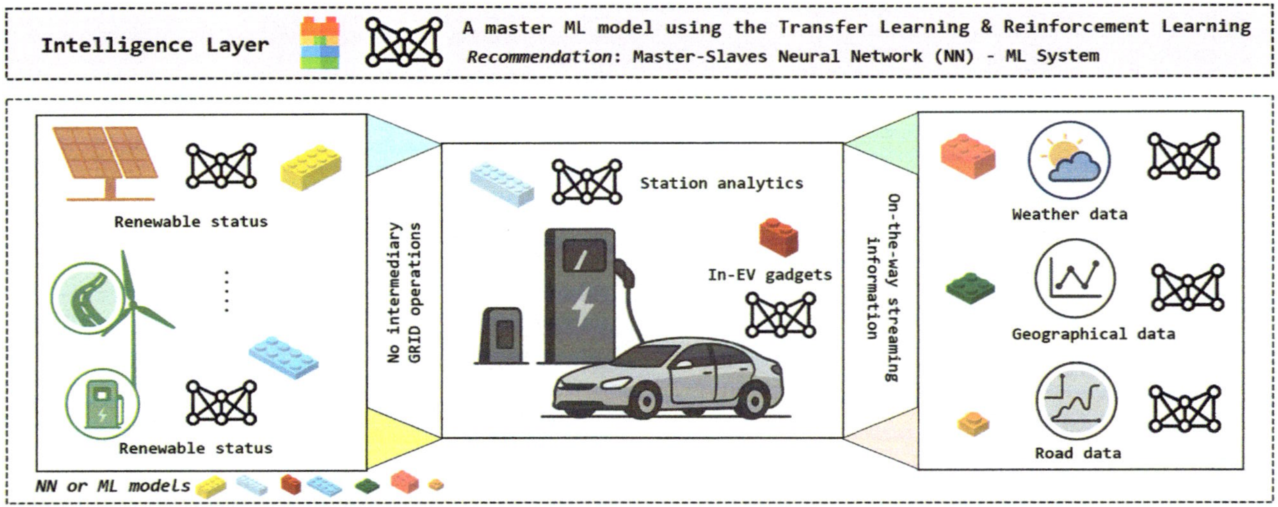

Figure 6.2 Illustration of the recommended intelligent layer architecture for R2V systems. The framework integrates multiple specialized subnetworks that process heterogeneous data sources, including renewable energy status, weather, road information, geographical data, station analytics, and in-vehicle devices. These subnetworks feed into a master machine learning (ML) model, trained using transfer learning (TL) and reinforcement learning (RL), to enable real-time, on-the-spot prediction and decision-making without relying on intermediary grid operations. The architecture emphasizes modularity, edge intelligence, and scalability for next-generation green computing in EV charging.

electrical systems are inherently more difficult to monitor and secure, as they involve continuous physical current flow that can be manipulated with less visibility. Although information-level cybersecurity (e.g., encrypted communication, authentication protocols) can address many traditional threats, manipulation at the power or control layer (e.g., unauthorized switching, power injection attacks) is far more subtle and potentially catastrophic.

Conventional grid-dependent systems often rely on cloud-based control architectures, which introduce further vulnerabilities due to wireless data transmission, latency, and centralized failure points. As outlined in the previous subsection, the integration of an intelligent layer, leveraging ML -RL and TL-can significantly enhance predictive capabilities and adaptive decision-making across the system. However, this same layer introduces new attack surfaces. The ML models themselves can be susceptible to adversarial inputs, model poisoning, or data drift, potentially compromising the integrity of the entire decision-making pipeline.

Moreover, in a Machine Learning of Things (MLoT) context, where interconnected EVs, renewable sources, and charging stations dynamically exchange control signals and streaming data, the risk of cascading failures or coordinated attacks becomes more pronounced. While the intelligence layer can be tasked with monitoring, detecting, and even mitigating cyber threats, it also becomes a potential point of exploitation. Therefore, the duality of AI, being both a defense mechanism and a vulnerability, must be carefully considered in the design of secure R2V infrastructures.

Aljohani and Almutairi categorize cyberattacks in electric vehicle (EV) systems into two primary vectors: direct attacks on the EV via compromised charging stations, and indirect attacks where the station itself, connected to a larger grid, is manipulated by an attacker with access to the grid's control system (Aljohani and Almutairi, 2024). In the latter scenario, attackers can potentially gain control over localized energy resources, hold them for ransom, and even disrupt or manipulate the charging process. Such interventions may lead to significant failures in energy delivery or even pose risks to grid stability. One notable example is a Denial of Service (DoS) attack, which can interrupt communication between the EV and the charging station or grid, resulting in a destabilized and unresponsive local grid environment. While the physical layer of EV charging systems—comprising sensors, power electronics, and metering devices—is generally more robust and less vulnerable to cyberattacks due to its hardware-oriented nature, the communication links embedded within these systems remain particularly susceptible. These links, which facilitate data transfer between the EV, the charging infrastructure, and cloud services, present a significant attack surface. Beyond this, vulnerabilities also exist within the EV itself. In-vehicle electronics, including infotainment systems, battery management units, and communication interfaces, may be exposed to threats such as radio frequency interference or malware injections, potentially compromising both functionality and safety.

As the EV ecosystem increasingly integrates with modern power grids, cybersecurity risks have become more prominent, particularly with the expansion of extreme fast-charging (xFC) infrastructure. Sayarshad et al. emphasize that this interconnectivity, which links EVs, chargers, substations, and vendor clouds, opens several attack vectors that can be exploited if appropriate safeguards are not implemented (Hamid, 2025). The presence of legacy grid protocols and unsecured communication layers further exacerbates the situation. Sanghvi and Markel provide a comprehensive analysis of these risks by mapping out the full communications architecture of xFC systems, highlighting the multiple interdependencies between substations, battery storage units, cloud servers, vendor applications, and vehicle-side electronics (Sanghvi and Markel, 2021). The architecture of xFC systems illustrates how insecure implementations of protocols such as Open Charge Point Protocol (OCPP), IEC 61850, and vendor-specific layers can introduce vulnerabilities at both the control and communication layers. Attackers may gain entry via compromised chargers, cloud services, or intermediary platforms, and then manipulate data flows or power delivery mechanisms. A particularly concerning scenario includes DoS attacks that interrupt charger-to-grid communication or exploit the EV's onboard controller, potentially causing unsafe voltage or current levels. While the physical layer—comprising hardware elements such as sensors, power converters, and meters—has more built-in resilience, the communication interfaces, especially those relying on mobile networks or VPNs, remain susceptible to intrusion. Moreover, compromised vendor cloud servers can scale attacks across fleets of charging stations simultaneously, presenting a large-scale threat to grid reliability and user safety. To address these challenges, Sanghvi and Markel propose a Cyber-Energy Emulation (CEE) Platform, a co-simulation environment that combines tools like OpenDSS, minimega, ZeroMQ, and protocol buffers to replicate cyber-physical EV charging scenarios. The platform is capable of simulating OCPP-based man-in-the-middle attacks, firmware spoofing, and data injection threats, thereby offering a testbed to study and mitigate vulnerabilities before real-world deployment. As EV adoption surges and R2V systems evolve, ensuring secure interoperability across all communication and control layers becomes essential for sustainable and resilient energy infrastructure.

The currently available common international standards for secure EV integration into energy networks have been benchmarked in a detailed survey by Aljohani and Almutairi (2024). These standards form the backbone of a globally interoperable and secure EV charging ecosystem. For instance, ISO 15118 facilitates plug-and-charge functionality and supports automated charging without repeated authentication, making it a widely adopted communication protocol between EVs and charging stations. Complementing this, the Open Charge Point Protocol (OCPP) serves as an open-source standard to enable communication between AC/DC chargers and backend systems, including features like remote diagnostics, billing, and firmware updates. To integrate EVs with the grid, IEC 61850 provides a critical communication standard that supports grid integration, load balancing, and demand response, whereas IEC 62351 focuses on cybersecurity features like role-based access control, cryptographic key management, and security event logging.

Region-specific standards also play vital roles. For example, GB/T 32960 in China handles remote diagnostics for EVs, and DIN SPEC 70121 from Germany supports bidirectional power flow. Similarly, CHAdeMO 2.0 and CCS 2.0, both internationally adopted, focus on DC fast charging with features like authentication, billing, and emergency protocols. In North America, standards such as SAE J2293 and SAE J2836 ensure message interoperability across different EV manufacturers, while ISO/SAE 21434 addresses broader cybersecurity challenges, offering guidance on threat modeling, incident response, and secure software design for EV infrastructure.

Together, these standards establish a secure and functional baseline for data exchange, authentication, and operational integrity across EV charging stations, grid infrastructure, and user interfaces. However, Aljohani and Almutairi caution that even with these protocols in place, effective implementation remains a challenge, particularly under evolving cyberthreats like spoofing, ransomware, or man-in-the-middle attacks targeting both the physical and communication layers of EV charging infrastructure. Their analysis underscores the need not only for adherence to standards but also for dynamic threat modeling, secure firmware deployment, and layered defense strategies to maintain system resilience in a highly interconnected energy landscape.

6.3.3 AI for failure mitigation

As R2V system complexity increases with new capabilities, vulnerabilities also become more critical. Issues such as voltage instability, sensor malfunction, data inconsistencies, and even cyberattacks must be carefully considered. AI-based failure detection addresses these challenges by providing early estimation through behavioral analytics and always-on monitoring techniques. By enabling predictive, real-time self-monitoring and self-correcting mechanisms, AI enhances system resilience and operational reliability.

There are several techniques that leverage EV data for preprocessing tasks such as noise removal and dimensionality reduction, followed by feature extraction for immediate classification. Wongchai et al. discuss fault detection mechanisms in EVs using deep learning architectures, with a particular focus on effective feature extraction (Wongchai et al., 2023). More specifically, Soresini et al. explore AI-based fault detection in automotive electric motors (Soresini et al., 2025), employing autoencoders to predict and prevent failures. This approach is especially valuable for detecting bearing faults and mechanical issues, highlighting that failure mitigation must go beyond purely electrical concerns to include mechanical aspects as well.

Battery-related issues, on the other hand, represent another critical aspect of AI-based failure prevention. Zhang et al. propose a deep neural network approach for anomaly detection in EV lithium-ion batteries (Zhang et al., 2023b). Their model utilizes temperature, voltage, and current data to generate a latent representation via a dynamic neural encoder, which is then passed to a multi-layer perceptron (MLP) predictor. Gong et al. present a sensorless method for diagnosing EV faults using AI combined with signal processing and data preprocessing techniques. Their on-board health monitoring system offers an adaptive, real-time diagnostic solution (Gong et al., 2023).

Battery monitoring remains a particularly vital area. Li et al. use a hybrid model combining LSTM and Bayesian neural networks to estimate cell voltage while capturing the uncertainty distribution (Li et al., 2024). Their approach employs unsupervised learning to identify anomalies and battery-related faults without requiring labeled data. Khaneghah et al. emphasize the importance of managing failures in both Permanent Magnet Synchronous Motors and lithium-ion battery packs. Their work provides a comprehensive classification of mechanical, sensor, and battery faults in EVs.

Mechanical faults include bearing wear and air–gap eccentricity; sensor faults cover current, voltage, and position sensors; and battery faults include abuse conditions, actuator malfunctions, and internal sensor issues. For data-driven EV battery fault detection, machine learning models, such as Support Vector Machines (SVM), general regression neural networks, LSTM-RNNs, and feedforward neural networks, have proven effective in identifying voltage anomalies, thermal runaway events, and capacity degradation in lithium-ion batteries. By leveraging real-world sensor inputs like voltage, current, and temperature, these models enable accurate, fast, and real-time diagnosis of battery health, making them essential tools for safe and reliable EV operation.

6.4 Operational mode optimization

Effective R2V systems must adapt their control strategies to the specific operational context they face at any given moment. The optimization of operational modes requires both analytical understanding of mode-specific technical constraints and the development of adaptive control architectures that can transition seamlessly between modes. By integrating these elements, R2V infrastructure can maintain critical charging capabilities while efficiently utilizing available resources under varying grid conditions.

This section first quantifies the technical challenges associated with each operational mode, then develops the adaptive control schemes necessary to address these challenges through resilience-oriented load shedding, strategic resource allocation, and priority-based charging hierarchies.

R2V smart charging systems must adapt to a spectrum of operational contexts, each characterized by distinct constraints and resilience needs. To ensure consistent and reliable energy delivery to EVs, R2V infrastructure must account for the technical challenges associated with three core operating modes: standard, emergency, and congested conditions.

6.4.1 Standard mode: Renewable fluctuations

Under normal operating conditions, R2V systems aim to maximize on-site renewable energy for EV charging. However, the intermittent nature of renewables creates a misalignment between energy supply and charging demand. For example, peak solar generation typically occurs during midday, while peak EV charging demand often arises in the evening. This temporal mismatch is clearly illustrated by the "*duck curve*" phenomenon observed in California, where net load drops during midday solar production and rises steeply as solar generation decreases in evening hours, as shown in Fig. 6.3. California ISO data shows ramp rates exceeding 13 GW in 3 hours during evening transitions (Powell et al., 2022). This coincides precisely with peak EV charging demand as commuters return home, creating a significant challenge for renewable integration.

Quantitative analysis highlights the magnitude of this challenge. Day-ahead renewable power forecasts still exhibit mean absolute percentage errors (MAPE) on the order of 15%–20% for utility-scale solar PV and 15%–25% for wind farms (Zhang et al., 2013; Giebel et al., 2011). These errors are defined as

$$\mathrm{MAPE} = \frac{1}{n}\sum_{i=1}^{n}\left|\frac{A_i - F_i}{A_i}\right| \times 100\%, \tag{6.1}$$

where A_i is the actual power output, F_i is the forecasted output, and n is the number of time periods. For a 500 kW PV array, a 20% error implies a 100 kW deviation—roughly the capacity to charge 10–20 Level 2 EVs simultaneously.

Beyond daily swings, seasonal variability further complicates planning. We quantify this via the Seasonal Variability Index (SVI):

$$\mathrm{SVI} = \frac{P_{\text{max,monthly}} - P_{\text{min,monthly}}}{P_{\text{avg,annual}}}, \tag{6.2}$$

where $P_{\text{max,monthly}}$ is the maximum monthly production, $P_{\text{min,monthly}}$ the minimum, and $P_{\text{avg,annual}}$ the annual average. Multi-year analyses for temperate, mid-latitude sites report SVI values of approximately 0.4–0.8—corresponding to 3–4 × differences between summer and winter output (IEA Photovoltaic Power Systems Programme, 2019).

Without intelligent scheduling or energy storage, this mismatch leads to under-utilization of renewable resources or increased reliance on grid electricity. The primary challenge in standard mode, therefore, lies in developing adaptive control algorithms that can forecast renewable output, shift charging loads accordingly, and optimize storage utilization to maintain system efficiency and stability.

6.4.1.1 Case study: UC San Diego INVENT pilot—Duck-curve mitigation via bidirectional charging

Between 2017 and 2018, the UC San Diego INVENT program, funded by the California Energy Commission, deployed 50 bidirectional chargers

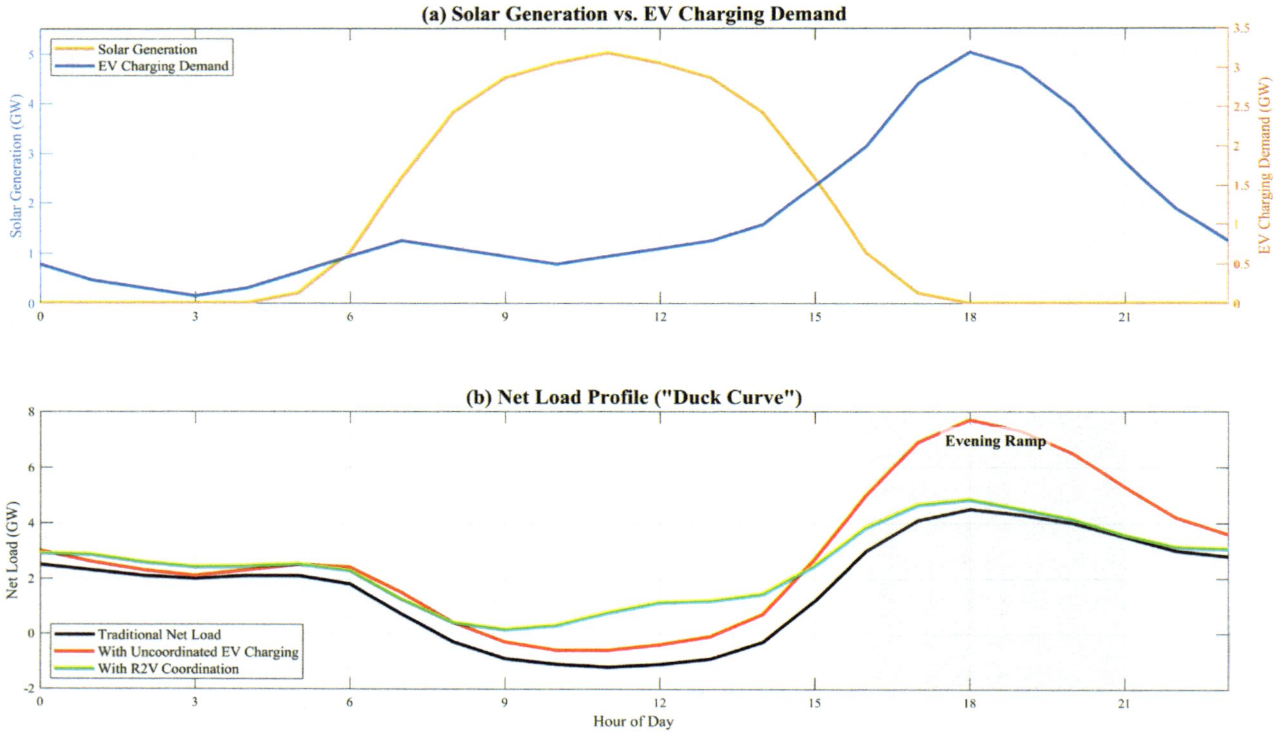

Figure 6.3 Duck curve and EV charging demand illustrating the "duck curve" challenge. (a) Solar generation peaks during midday while EV charging demand typically peaks in evening hours; (b) Net load profile showing the steep evening ramp that coincides with peak charging demand. Data patterns based on Muratori (2018) and Pfenninger and Staffell (2016).

across campus and integrated day-ahead solar forecasts to mitigate the evening ramp ("duck curve") (California Energy Commission, 2018; Smart Energy International, 2017).

Key strategies included:

1. Morning pre-charging of EVs using excess solar generation
2. Midday absorption of solar peaks through smart charging
3. Evening discharge of select fleet vehicles during peak demand
4. Overnight maintenance charging during off-peak hours

Outcomes over 12 months: 35% increase in renewable utilization, 150 kW peak demand reduction, and 100 metric tons of CO_2 emissions avoided, demonstrating how bidirectional charging can effectively smooth the duck curve.

6.4.2 Emergency mode: Grid outages and islanding requirements

Emergency mode addresses scenarios where grid outages require R2V systems to operate in islanded conditions. This demands rapid detection of grid disconnection, seamless transition to islanded operation, and strategic allocation of limited energy resources to critical loads. Most EV supply equipment (EVSE) and control standards are not designed to support autonomous microgrid operation. IEEE 1547-2018 requires distributed energy resources to detect grid failure and cease energization within 2 s (IEEE standard for interconnection, 2018), while SAE J3072 enforces anti-islanding protection (Interconnection requirements for onboard, 2015). These standards effectively prevent R2V systems from maintaining operation during grid outages.

The core challenge is enabling EVSE to form and sustain microgrids with minimal latency while prioritizing critical loads. Fig. 6.4 illustrates the transition sequence from grid-connected to islanded operation, showing voltage/frequency profiles during failure events and reconfigured power flows during outages.

In practice, detection and isolation must occur within 2 s per IEEE 1547-2018 (IEEE standard for interconnection, 2018), while hierarchical control frameworks (IEEE 2030-2011) describe the additional reconfiguration and stabilization steps (Dispersed Generation Photovoltaics and Energy Storage, 2011a). Charging may be interrupted during these phases, potentially stranding vehicles unless mitigated by pre-emptive scheduling or on-board buffering.

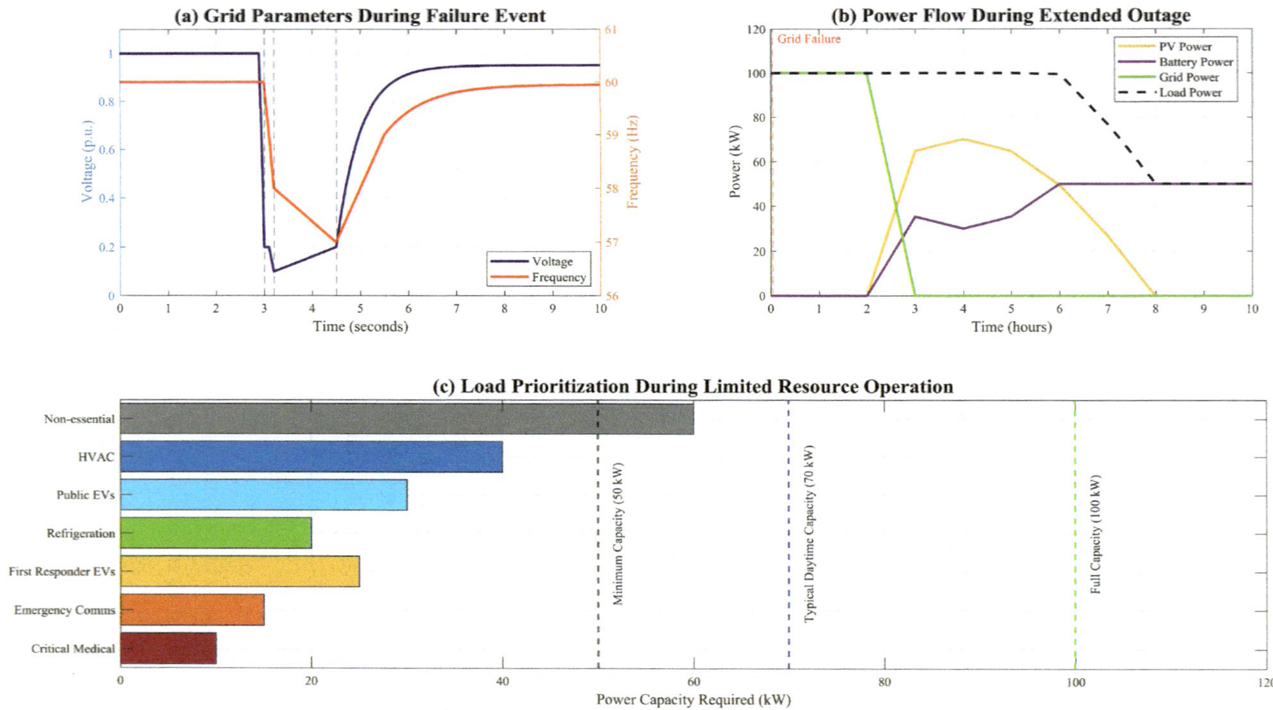

Figure 6.4 Emergency-mode transition sequence showing (a) per-unit grid voltage and frequency during a fault event, (b) power injections from PV, battery, and grid during an extended outage, and (c) prioritized load groups against minimum, typical, and full capacity levels.

Resource limitation presents another significant challenge during emergency operations. The maximum number of EVs that can be charged during an outage depends on available energy resources. During emergency operations, the system must operate in islanded mode with limited resources. Based on the energy balance principles in Ngoc An et al. (2015), we can express the maximum supportable EV load during an emergency as:

$$P_{EV,\max}(t) = P_{PV}(t) + P_{\text{BESS}}(t) - P_{\text{critical}}(t) \tag{6.3}$$

where $P_{PV}(t)$ is the available power from PV generation at time t (which varies with solar insolation as shown in their Fig. 6.2), $P_{\text{BESS}}(t)$ is the available power from the battery energy storage system (constrained by state-of-charge limits as shown in their Fig. 6.8), and $P_{\text{critical}}(t)$ is the power required for critical loads. The maximum number of EVs that can be charged simultaneously can then be calculated as:

$$N_{\max} = \frac{P_{EV,\max}}{P_{EV,\text{avg}}} \tag{6.4}$$

where $P_{EV,\text{avg}}$ is the average power demand per EV charger. During optimal daylight, solar-plus-storage microgrids can support many Level 2 EV charging sessions, but capacity drops sharply at night or during cloudy periods when only batteries are available. This poses challenges for emergency operations, especially during long outages.

Maintaining stable voltage and frequency without a grid reference is also difficult. Islanded microgrids must tightly control power quality while balancing intermittent renewables and variable EV loads. The unpredictable length of outages further complicates planning, especially as climate change increases grid disruptions. These factors highlight the need for R2V systems that sustain critical charging during extended outages and manage limited resources intelligently.

6.4.2.1 Case study: Borrego springs microgrid—Emergency mode: Islanding during grid outages

Borrego Springs has been islanded for several hours at a time—both during planned maintenance and unexpected outages—using its community microgrid as the sole power source (Roth, 2019; San Diego Gas and Electric, 2019).

The installed assets include:

- 4.5 MWh of containerized lithium-ion batteries (1000 kW/3000 kWh + 500 kW/1500 kWh)

- Two 1.8 MW diesel generators for black-start support
- Rooftop and nearby third-party solar PV
- Advanced microgrid controls (Wave software)

During islanded operation:

1. **Automatic outage detection and seamless transition:** Protective relays detect loss of utility voltage and inverters switch to grid-forming mode within milliseconds, compliant with IEEE 1547 ride-through criteria (IEEE Standard for Interconnection, 2018).
2. **Black-start capability:** Diesel gensets start within 30 s to energize the microgrid bus; hardware-in-the-loop tests have demonstrated PV-only black-start feasibility under controlled conditions (National Renewable Energy Laboratory, 2021a).
3. **Dynamic load management:** Preconfigured load-shedding groups prioritize critical services and maintain voltage/frequency based on real-time forecasts of solar output and battery state.

This example highlights how a community-scale microgrid can maintain service during extended outages, underscoring the importance of rapid islanding, reliable black-start, and adaptive load management.

6.4.3 Congested mode: Load overloads and grid stress events

Congested mode addresses scenarios where EV charging loads, combined with existing demand, threaten to exceed distribution infrastructure capacity. Unlike emergency mode, congested mode involves a functional but stressed grid requiring coordinated load management to prevent equipment damage and maintain service quality.

Transformer Thermal Loading and Aging: Distribution transformers—typically rated from 25 kVA to 100 kVA for residential service and 300 kVA to 2500 kVA for commercial installations—are critical bottlenecks when multiple EVs charge simultaneously. Their thermal behavior and insulation life are governed by IEEE Std C57.91-2011 (Annex G) (IEEE Standards Association, 2012), which provides the winding hottest-spot model and the aging acceleration factor (AAF):

$$\mathrm{AAF} = \exp\left(\frac{15000}{383} - \frac{15000}{\theta_H + 273}\right), \theta_H = \theta_A + \Delta\theta_{TO} + \Delta\theta_H \qquad (6.5)$$

where θ_A is ambient temperature, $\Delta\theta_{TO}$ the top-oil rise, and $\Delta\theta_H$ the hotspot rise over top-oil (all in °C).

Fig. 6.5 plots the theoretical derating curve (in °F for clarity) alongside hotspot-rise data from Hilshey et al. (2012). Above a 100° F reference, continuous loading must be reduced by 15% per 5° F to limit θ_H. Under clustered Level 2 charging (7.7 kW/vehicle), studies report hotspot rises of 15–25° C and AAFs of 2x–8x normal rates, potentially shortening expected lifespans from 20–30 years to 5–7 years in high-adoption scenarios.

A simple load model predicts transformer overload when

$$L_{\text{base}} + (n_{\text{EV}} \times P_{\text{EV}} \times C_d) > S_{\text{rated}}, \tag{6.6}$$

where $P_{\text{EV}} \approx 7.7$ kW (Level 2) and diversity factor $C_d \leq 1$. Using the same Annex G framework, Hilshey et al. simulated a 25 kVA transformer serving 12 homes plus six EVs and found AAFs exceeding 50 service-years per calendar year under uncontrolled charging (Hilshey et al., 2012). Kuss, Markel and Kramer's Monte-Carlo study corroborated this, reporting that over 30% of simulated days experienced peak loading above nameplate capacity for a similar transformer under high PEV clustering (Kuss et al., 2011). Without coordinated control or infrastructure upgrades, residential transformers can thus be rapidly driven beyond thermal limits and design life.

Feeder Thermal and Voltage Constraints: At the feeder level, congestion appears as both thermal and voltage violations. Overhead conductors (e.g., ACSR) and underground cables (e.g., XLPE) are rated by their maximum permissible operating temperature—often 75–90° C—beyond which insulation deterioration accelerates. The conductor ampacity model in IEEE Std 738-2012 (IEEE Standard for Calculating, 2012) gives the maximum continuous current $I_{\max}$ as

$$I_{\max} = \sqrt{\frac{T_{\max} - T_a}{R_{\text{th}} \times R_{\text{elec}}}}, \tag{6.7}$$

where

- $T_{\max}$ is the allowable conductor hottest-spot temperature (°C),
- T_a is the ambient temperature (°C),
- R_{th} is the conductor's thermal resistance (°C/W), and
- R_{elec} is the DC resistance per unit length (Ω/km).

For example, a 336.4 ACSR overhead conductor with nominal ampacity near 530 A at 40°C ambient sees a 10%–20% reduction in allowable current when ambient rises by 10°C (IEEE Standard for Calculating, 2012).

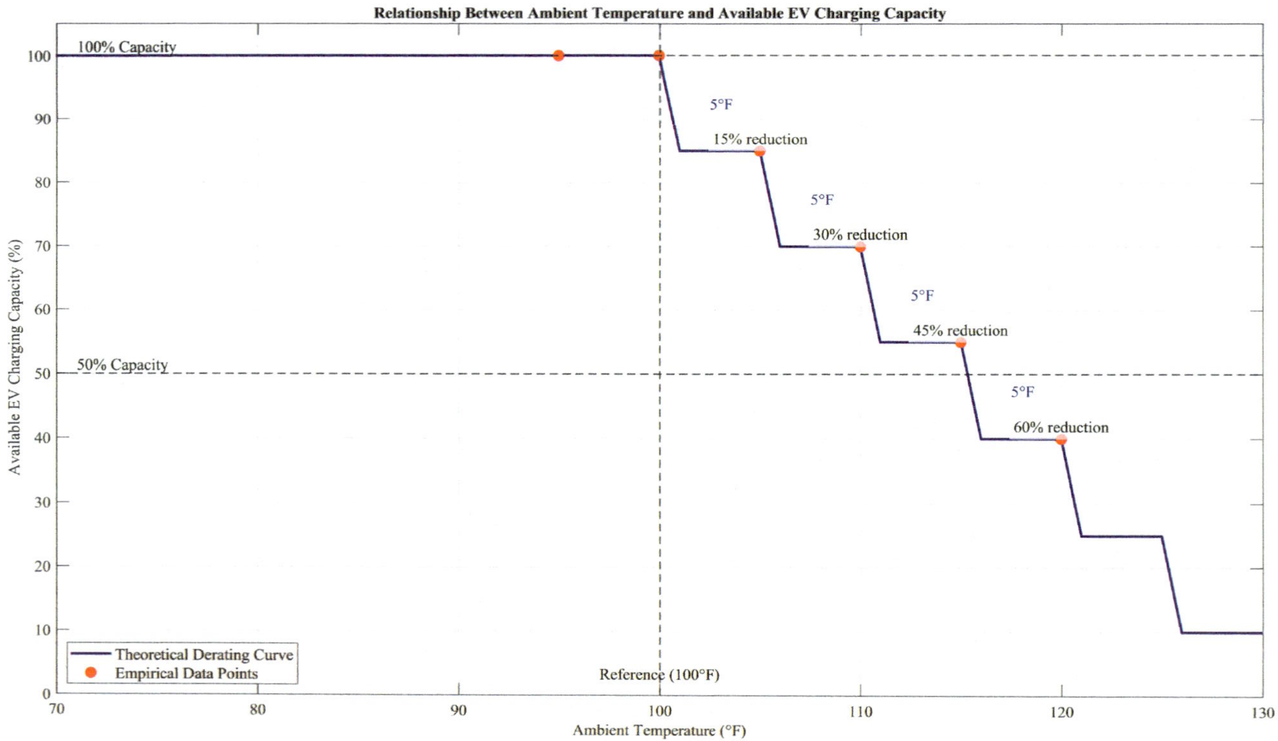

Figure 6.5 Transformer continuous-loading derating per IEEE C57.91–2011 Annex G (blue) and empirical hotspot rises under clustered level 2 EV charging (red markers) from Hilshey et al. (2012). Above 100° F, capacity must be derated to protect insulation life.

Voltage regulation is equally critical: ANSI C84.1 Range A requires end-of-line voltages stay within 0.95–1.05 pu under normal loading. A first-order drop along a feeder segment is

$$\Delta V \approx \frac{R\,P + X\,Q}{V_{\text{nom}}}, \tag{6.8}$$

where R and X are the segment resistance and reactance, P and Q the real and reactive flows, and V_{nom} the nominal line-to-line voltage. In low-voltage networks (high R/X ratios), clustered EV charging can drive end-of-line voltages below 0.95 pu, triggering tap-changer operations and risking voltage-stability excursions (Grainger and Stevenson, 1994).

Together, these thermal and voltage constraints limit an uncoordinated R2V charging station's ability to draw grid power when local renewables are insufficient, motivating smart-charging controls or feeder reinforcements.

System-Level Congestion and Stability Challenges: As EV adoption grows, distribution systems face challenges beyond local equipment limits:

1. **Interconnection constraints:** IEEE 1547-2018 requires distributed resources to cease energization within 2 seconds of grid disturbances (IEEE standard for interconnection, 2018), limiting EVSE's ability to support grid stability during peak loading.
2. **Peak-load coordination:** The IEA Global EV Outlook 2023 notes that EV charging can significantly increase demand peaks, necessitating smart-charging strategies to avoid congestion (International Energy Agency, 2023).
3. **Voltage regulation and hosting capacity:** NREL's 2021 EV Grid Integration report shows that clustered EV charging can push feeder voltages toward their allowable limits and exceed the hosting capacity of some circuits, underscoring the need for managed charging and advanced voltage-control measures to maintain acceptable profiles (National Renewable Energy Laboratory, 2021b).

Without coordinated control and adherence to interconnection standards, these combined challenges can erode protection margins, stress voltage regulation equipment, and compromise distribution-system reliability under high EV-penetration scenarios.

Detection and Response Requirements: R2V controllers must utilize standardized grid-health indicators:

1. **Voltage monitoring:** IEEE 1547-2018 requires monitoring with ± 5% limits; drops below 0.95 pu indicate congestion (IEEE standard for interconnection, 2018).
2. **Frequency monitoring:** IEEE 2030.5 and OpenADR 2.0b specify ± 0.036 Hz deviation thresholds for demand-response triggering (Dispersed Generation Photovoltaics and Energy Storage, 2011a; OpenADR Alliance, 2018).
3. **Grid signals:** Standards-based notifications with 1–5 min latency for automated responses (Dispersed Generation Photovoltaics and Energy Storage, 2011a; OpenADR Alliance, 2018).

Effective R2V response spans multiple timescales—from millisecond-level inverter adjustments to multi-minute schedule updates—requiring coordinated firmware, communications, and control across distributed charging assets.

6.4.3.1 Case study: SMUD managed EV charging pilot in Sacramento, California

In August 2022, the Sacramento Municipal Utility District (SMUD) launched its Managed EV Charging pilot in partnership with BMW of North America, Ford Motor Company, and General Motors, enrolling over 100 residential Level 2 chargers to shift load away from evening peaks (Sacramento Municipal Utility District; SMUD). The program targets feeders experiencing high late-afternoon demand by remotely scheduling participant charging sessions to off-peak windows.

- **Smart charging platform:** Chargers are controlled via OEM apps (BMW ChargeForward, Ford BlueOval Charge Network, GM MyChevrolet), which receive grid-optimized setpoints from SMUD's Managed Charging service.
- **Distribution monitoring integration:** SMUD's distribution-management system tracks feeder load and voltage in real time, triggering charging adjustments when local feeder load nears peak thresholds.
- **Customer incentives:** Participants earn bill credits for shifting at least 75% of their charging energy to designated off-peak periods.

Early program data show a 30%–40% reduction in coincident charging load during the 5–9 PM window on participating feeders, helping to alleviate evening peaks without compromising customer mobility (Sacramento Municipal Utility District). High customer satisfaction (over 85% positive feedback) underscores the feasibility of utility-led managed charging as a congestion-management tool.

Illustrative Congestion-Response and Aging Models: To give a concrete sense of how an R2V system might behave under rising distribution stress, Fig. 6.6 shows a notional three-tier response curve: as a transformer's loading climbs past 100% of nameplate, charging power is stepped down in 25% increments at 95%, 105% and 115% loading thresholds. This mirrors typical demand-response tiers and inverter trip settings (e.g., IEEE 1547 anti-islanding limits), and demonstrates how a controlled charging fleet can shed load progressively rather than abruptly.

Fig. 6.7 demonstrates the impact of EV adoption on transformer life. Panel (a) shows how increasing EV penetration (0%–100%) pushes peak loading above nameplate capacity, while panel (b) reveals transformer lifespan reduction from 40 years to approximately 13 years at full EV saturation, based on IEEE C57.91 aging models.

These figures highlight two key lessons: (1) the need for fast, multi-tiered congestion controls, and (2) the long-term asset-health benefits of keeping loads below critical thresholds. These operational challenges form the basis for adaptive control strategies that support resilient R2V charging across all conditions.

6.4.4 Adaptive control schemes

This section examines control strategies that enable R2V systems to maintain resilient operation across varying grid conditions, dynamically adjusting to renewable fluctuations, emergency transitions, and congestion events while prioritizing critical charging needs.

6.4.4.1 Hierarchical control architecture

Effective R2V control requires a multi-layered approach with three tiers (Dispersed Generation Photovoltaics and Energy Storage, 2011a):

1. **Primary Control (Milliseconds):** Local controllers maintaining voltage and frequency stability through droop control and inverter-level charging power management.
2. **Secondary Control (Seconds to Minutes):** Coordination layer restoring nominal setpoints, managing power sharing, and implementing mode transitions.
3. **Tertiary Control (Minutes to Hours):** Optimization layer scheduling charging sessions and interfacing with grid operators.

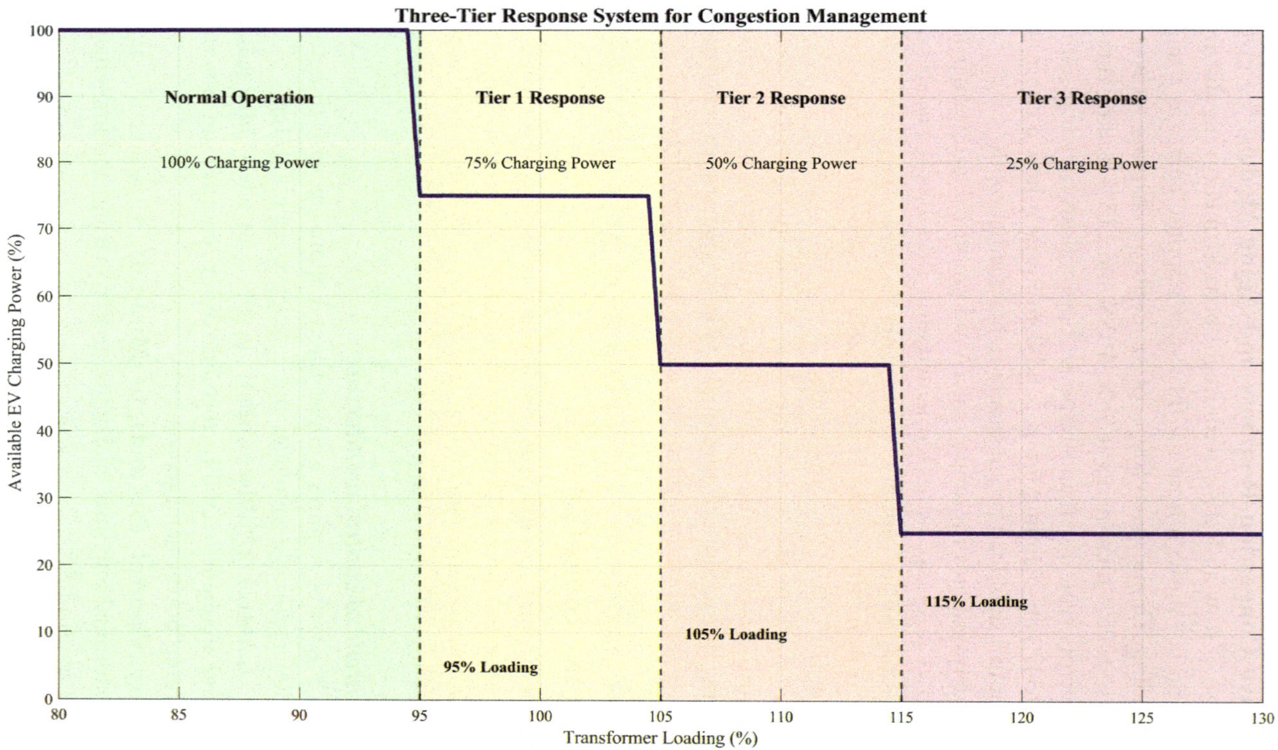

Figure 6.6 Conceptual three-tier congestion-management strategy for R2V charging, showing how available charging power might step down as transformer loading rises from 100% to 115%. Tier thresholds and power reductions are illustrative, based on typical IEEE 1547 anti-islanding trip settings and demand-response ramp-rate capabilities.

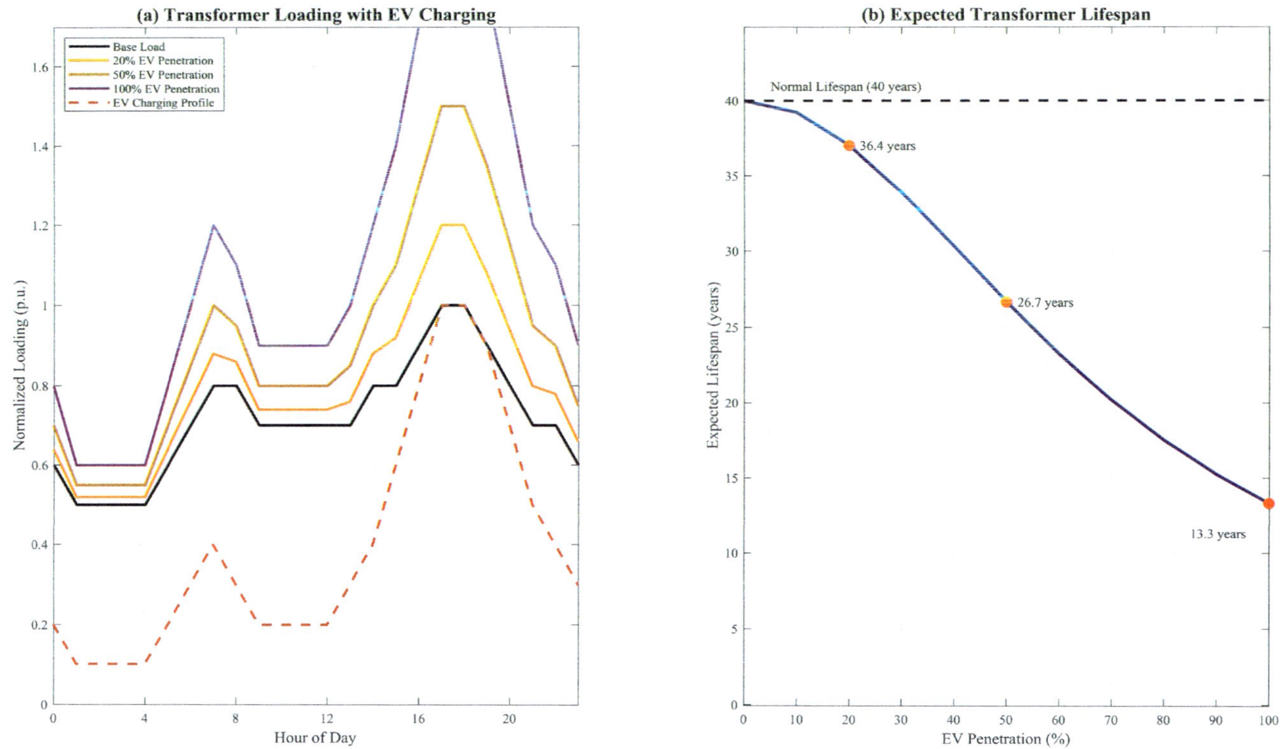

Figure 6.7 Illustrative transformer loading (a) and expected lifespan (b) versus EV penetration, computed using the IEEE C57.91 Annex G aging model and hotspot-rise data from Hilshey et al. (2012). Base-load profile is hypothetical.

This structure, illustrated in Fig. 6.8, enables responses across multiple timescales while maintaining stability during transitions. IEEE 2030.7-2017 provides a reference architecture for implementing these control layers, defining functions for autonomous operation and seamless mode transitions (IEEE standard for the specification, 2017).

The control architecture adapts to each operational mode's unique challenges. In Standard Mode, the system focuses on optimizing renewable utilization through predictive charging schedules based on forecasted generation, addressing temporal mismatches between renewable availability and charging demand. During Emergency Mode, the control system maintains critical charging during outages through rapid islanding transitions and efficient resource management to maximize resilience. For Congested Mode, the architecture mitigates grid stress through transformer/feeder protection and coordinated utility demand response while preserving

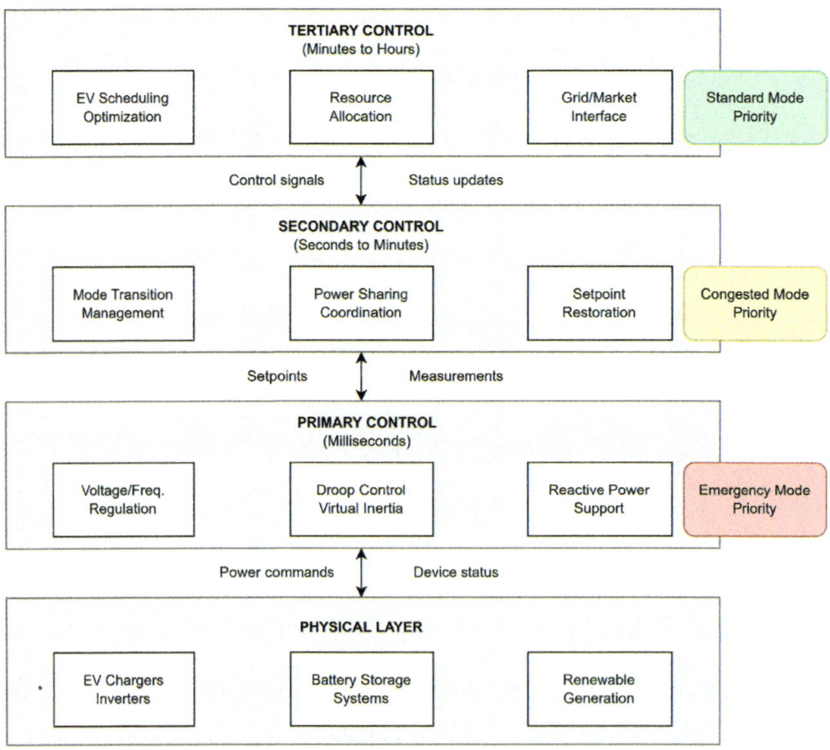

Figure 6.8 Hierarchical control architecture for R2V systems showing primary, secondary, and tertiary control layers with their respective timescales, control objectives, and information flows.

essential charging services. This adaptive control enables seamless transitions between modes, ensuring R2V resilience across varying grid conditions.

6.4.4.2 Resilience-oriented load shedding

When power is insufficient to meet all charging demands, R2V systems must prioritize mission-critical mobility. A resilience-oriented scheme assigns dynamic priorities based on:

1. **Vehicle Criticality:** Each EV receives a criticality score C_i (emergency services, public service, private use), conveyed via ISO 15118-20 and OCPP 2.0.1 during plug-and-charge handshakes (Road Vehicles, 2022a; Open Charge Alliance, 2024).
2. **SoC Urgency:** Lower SoC vehicles receive elevated priority within each tier to prevent stranding (Aghajan-Eshkevari et al., 2022).
3. **Departure Time:** Known/predicted departure times $t_{dep,i}$ allow deferring non-urgent charging, smoothing demand (Aoun et al., 2024).

These factors combine into a priority metric:

$$P_i(t) = w_c\, C_i + w_s\left(1 - \frac{SoC_i(t)}{SoC_{\text{target},i}}\right) + w_d\, e^{-\alpha\,\left(t_{dep,i}-t\right)}, \tag{6.9}$$

where weights w_c, w_s, and w_d balance criticality, SoC, and departure urgency. The R2V controller ranks vehicles by $P_i(t)$ and allocates power in descending order. Simulations show this approach preserves 95% of emergency and transit vehicle charging even when demand exceeds 80% of feeder capacity (Aoun et al., 2024).

Implementation Across Operational Modes: The resilience-oriented load shedding strategy must be tailored to the unique constraints of each operational mode:

- **Standard Mode:** Primarily preventive and scheduled, reducing charging rates during forecast low-renewable periods rather than abrupt mid-session cutoffs. CAISO's Duck Curve mitigation shifts EV charging to midday solar peaks (California Independent System Operator, 2021).
- **Emergency Mode:** More aggressive prioritization, potentially disconnecting lower-priority vehicles when only local generation and storage are available. IEEE 1547.4-2011 provides guidelines for critical-load protection during islanding (Dispersed Generation Photovoltaics and Energy Storage, 2011b).

- **Congested Mode:** Tiered response with progressively stricter limits (75%, 50%, 25% of nominal) as transformer loading exceeds thresholds (Fig. 6.6).

Studies confirm that mode-aware load shedding preserves critical charging services even under severe constraints, enhancing distribution resilience.

6.4.4.3 Resource allocation strategies

Effective resource allocation balances three key objectives:

1. **Renewable Utilization:** Maximizing renewable generation usage through forecasting and coordinated charging schedules.
2. **Storage Management:** Strategic battery operation to buffer intermittency, provide backup during outages, and mitigate peak loads.
3. **Grid Power Optimization:** Optimizing grid power usage within technical constraints while responding to utility signals.

These objectives can be formulated as a multi-objective optimization problem (Parisio et al., 2014):

$$\min_{P_{ch}, P_{\text{grid}}, P_{\text{batt}}} \left[w_1 \cdot f_{\text{cost}} + w_2 \cdot f_{\text{emissions}} + w_3 \cdot f_{\text{resilience}} \right]$$

where:

- f_{cost} represents the economic cost objective, calculated as:

$$f_{\text{cost}} = \sum_{t=1}^{T} \left[c_{\text{grid}}(t)\ P_{\text{grid}}(t) + c_{\text{batt}}(t)\ |P_{\text{batt}}(t)| + c_{\text{maint}} \sum_{i=1}^{N} P_{ch,i}(t) \right] \Delta t$$

- where $c_{\text{grid}}(t)$ is the time-varying electricity price (\$/kWh), $c_{\text{batt}}(t)$ is the battery cycling cost proxy that accounts for degradation (\$/kWh) (Schmalstieg et al., 2014), and c_{maint} represents maintenance costs associated with charging infrastructure usage.
- $f_{\text{emissions}}$ quantifies the carbon emissions associated with charging operations:

$$f_{\text{emissions}} = \sum_{t=1}^{T} \left[e_{\text{grid}}(t)\ P_{\text{grid}}(t) \right] \Delta t$$

- where $e_{\text{grid}}(t)$ is the time-varying carbon intensity of grid electricity (kg CO_2-eq/kWh), reflecting changes in the generation mix (Siler-Evans et al., 2012). Renewable energy sources (P_{ren}) contribute zero operational emissions to this term.

- $f_{\text{resilience}}$ measures the system's ability to maintain critical charging services during adverse conditions:

$$f_{\text{resilience}} = \sum_{t=1}^{T} \left[\alpha \left(1 - SoC_{\text{batt}}(t)\right) + \beta \sum_{i \in \mathscr{C}} \left(SoC_{\text{target},i} - SoC_i(t)\right)^+ + \gamma \, P_{\text{grid}}(t) \right] \Delta t$$

- where α, β, and γ are weighting parameters, $\mathscr{C}$ is the set of critical vehicles, $(\cdot)^+$ denotes $\max(\cdot, 0)$, and $SoC_{\text{target},i}$ is the desired state of charge for vehicle i. This term penalizes low storage reserves and insufficient charging for critical services, grounded in reliability worth concepts (e.g., value of lost load) (Kariuki and Allan, 1996).

The weights w_1, w_2, and w_3 balance these competing objectives and can be dynamically adjusted by operational mode.

Subject to operational constraints:

$$P_{\text{ren}}(t) + P_{\text{grid}}(t) + P_{\text{batt}}(t) = \sum_{i=1}^{N} P_{ch,i}(t) + P_{\text{aux}}(t), P_{\text{grid}}(t) \leq P_{\text{grid,max}}(t),$$

$$SoC_{\min} \leq SoC_{\text{batt}}(t) \leq SoC_{\max}, 0 \leq P_{ch,i}(t) \leq P_{ch,i,\max}.$$

Power from renewable sources (P_{ren}), the grid (P_{grid}), and battery storage (P_{batt}) must satisfy EV charging ($P_{ch,i}$) and auxiliary loads (P_{aux}). Battery energy dynamics (linking $P_{\text{batt}}(t)$ to $SoC_{\text{batt}}(t)$) and renewable availability bounds ($0 \leq P_{ren}(t) \leq P_{\text{ren}}^{\text{avail}}(t)$) are handled elsewhere; the objective is shown in reduced form. Coefficients α, β, and γ normalize units and tune $f_{\text{resilience}}$ so it reflects survivability priorities without duplicating cost/emission penalties.

Model predictive control (MPC) effectively solves these problems in real time with forecasts and constraints across multiple timescales (Parisio et al., 2014).

Mode-Specific Resource Allocation: Allocation strategies adapt to each mode's requirements:

- **Standard Mode:** Maximizing renewable utilization with battery storage shifting excess generation to peak charging periods (UC San Diego INVENT pilot, Section 6.4.1).

- **Emergency Mode:** Maximizing operational duration by reserving storage for critical loads and reducing charging power (Borrego Springs microgrid, Section 6.4.2).
- **Congested Mode:** Minimizing grid stress through battery peak shaving and coordinated charging schedules (SMUD managed charging pilot, Section 6.4.3).

6.4.4.4 Priority charging hierarchies

R2V resilience requires structured charging prioritization based on:

1. **Functional Priority Tiers:** Five-tier vehicle categorization:
   - **Tier 1:** Emergency response (ambulances, fire, police)
   - **Tier 2:** Critical infrastructure support (utility, communications)
   - **Tier 3:** Essential services (transit, medical transport)
   - **Tier 4:** Commercial and fleet vehicles
   - **Tier 5:** Personal vehicles
2. **Dynamic Adjustments:** Based on state of charge, trip urgency, and charging patterns.
3. **Power Allocation:** Variable distribution with higher-priority vehicles receiving full power while lower-priority receive reduced charging.

Implementation uses ISO 15118 for authentication and energy management, though priority signaling capabilities are still evolving.

Resilience Metrics for Priority Systems: Priority charging effectiveness can be quantified using metrics adapted from power-system resilience (Panteli et al., 2017) and EV charging optimization frameworks (Arias et al., 2022):

- **Critical Service Availability (CSA):** Fraction of time critical-tier vehicles maintain sufficient state-of-charge:

$$\mathrm{CSA} = \frac{\mathrm{Time}\left[\mathrm{SoC}_{\mathrm{critical}} \geq \mathrm{SoC}_{\mathrm{min,critical}}\right]}{\mathrm{Total\ time}} \times 100\% \tag{6.10}$$

- **Priority Satisfaction Index (PSI):** Weighted index of energy delivery across priority tiers:

$$\mathrm{PSI} = \sum_{i=1}^{N_{\mathrm{tiers}}} w_i \frac{\mathrm{Energy\ delivered\ to\ tier}\ i}{\mathrm{Energy\ requested\ by\ tier}\ i} \tag{6.11}$$

- Studies show high PSI values (≥ 0.85) under standard conditions, with acceptable performance (≥ 0.65) under constraints.

- **Response Time:** Speed of resource reallocation after priority changes, with IEEE 1547.4–2011 requiring execution within 30 s (Dispersed Generation Photovoltaics and Energy Storage, 2011b).

These metrics enable systematic assessment of charging strategies across operational modes.

6.5 Implementation challenges

The implementation of renewable energy systems and EV chargers in areas susceptible to natural disasters presents unique challenges that require innovative solutions and resilient design approaches. Natural disasters such as hurricanes, floods, tornados, and earthquakes can severely damage energy infrastructure, leaving communities without essential services for extended periods. The intermittent nature of renewable energy sources further complicates their implementation in disaster-prone areas, as the variability of solar and wind resources can coincide with extreme weather events, creating additional vulnerabilities in the energy supply chain.

The economic viability of renewable energy systems in disaster-prone areas also presents significant challenges. The initial investment costs for resilient renewable infrastructure can be substantial, requiring careful consideration of the long-term benefits against immediate financial constraints.

Electric vehicle (EV) charging infrastructure in areas prone to natural disasters faces multiple challenges related to physical vulnerability, grid stability, and operational continuity. Extreme weather events can cause direct physical damage to charging stations, rendering them inoperable when they might be most needed for evacuation or emergency response. Urban flooding, whose frequency and intensity are increasing due to climate change, can significantly impact the availability and accessibility of public EV charging networks, creating disproportionate stresses on chargers both near and surprisingly far from flooded regions. The planning and design of EV charging infrastructure in disaster-prone areas must consider both immediate operational needs and long-term resilience requirements. Strategic placement of charging stations must account for flood risk, hurricane exposure, and other natural hazards while ensuring adequate coverage and accessibility. Research has shown that even when flooding does not directly impact the ability of battery EVs to serve typical urban driving behaviors, it can cause over 50% increase in charger utilization and 260% increase in the distance to the nearest available charger in areas over 10 km away from flooded regions (Raman et al., 2022). These findings show the importance of developing comprehensive strategies to enhance the disaster resilience of R2V systems.

6.6 Standards and limitations

While R2V smart charging systems offer significant potential for enhancing charging resilience across operational conditions, their widespread implementation faces substantial technical, regulatory, and economic barriers. This section examines the limitations in current standards that impede resilient operation, with a primary focus on SAE J3072 and its implications for vehicle-integrated energy resources. We then explore the regulatory challenges that arise from these standards limitations and analyze the cost-benefit considerations that influence adoption decisions.

6.6.1 SAE J3072 gaps for resilient R2V operation

SAE J3072 ("*Interconnection Requirements for Onboard, Utility-Interactive Inverter Systems*") establishes the technical requirements for vehicle onboard inverters that can export power to premises wiring and potentially to the grid. While this standard represents an important step toward enabling V2G and vehicle-to-building (V2B) capabilities, it contains significant limitations that restrict the resilience potential of R2V systems, particularly during emergency conditions.

6.6.1.1 IEEE 1547 dependency

The most consequential limitation of SAE J3072 is its explicit dependency on IEEE 1547 requirements. From the very scope of the standard, Section 1 of SAE J3072 states that "*the requirements herein are intended to be used in conjunction with IEEE 1547 and IEEE 1547.1*" (Interconnection requirements for onboard, 2015). This is reinforced in Section 4.8 ("Utility Interaction"), which mandates that "*the grid support inverter system shall meet the requirements of IEEE 1547-2018 and Amendment 1*" (Interconnection requirements for onboard, 2015). This dependency incorporates all of IEEE 1547's anti-islanding and grid-following limitations into the vehicle inverter requirements. Specifically:

1. **Anti-Islanding Requirements:** SAE J3072's Appendix D explicitly maps to IEEE 1547's Clause 8 ("Unintentional Islanding Prevention"), requiring that vehicle inverters detect and cease to energize unintentional islands. Per IEEE 1547-2018 Table 9, the second undervoltage

trip (UN2) must clear within 2 s, effectively preventing vehicles from maintaining power to critical loads during grid outages unless specific utility permission and additional equipment for intentional islanding are in place.

2. **Grid-Following Operation:** The standard presumes grid-following rather than grid-forming operation, limiting the vehicle's ability to establish and maintain voltage and frequency references during islanded operation. Appendix F, Table F2 of SAE J3072 confirms that IEEE 1547.1's synchronization tests remain in force, stating "*5.4 Synchronization—Applies*" (Interconnection requirements for onboard, 2015). The standard only includes grid-following voltage regulation modes defined in IEEE 1547 Clause 5, with no provisions for grid-forming operation.
3. **Utility Permission Requirements:** SAE J3072 explicitly states that "the coordination of V2G-EVSE as part of intentional islands as described in 8.2 is beyond the current scope of SAE J3072" (Interconnection requirements for onboard, 2015). The standard's Background also highlights that "*many utility jurisdictions may not authorize interconnection*" deferring intentional islanding to utility-permission processes as per IEEE 1547, creating a significant regulatory hurdle for resilient operation.

These IEEE 1547 dependencies create fundamental barriers to using electric vehicles as resilient power sources during grid outages, despite the significant energy storage capacity available in their batteries.

Fig. 6.9 illustrates the complex relationship between SAE J3072 and related standards that affect R2V resilience. As shown in the figure, SAE J3072 depends directly on IEEE 1547-2018 for interconnection requirements, interfaces with ISO 15118-20 for communication protocols, and is controlled by IEC 61851 for EVSE operation. This web of dependencies creates a multi-layered set of constraints that collectively limit resilient operation across different operational modes.

6.6.1.2 EVSE-controlled operation

SAE J3072 further restricts autonomous vehicle operation by requiring that the EVSE (Electric Vehicle Supply Equipment) control the vehicle's inverter functionality. Section 4.7.2 ("Authorization to Discharge") explicitly states that "*the PEV shall not discharge unless the DCTE permit service authorization from the EVSE is TRUE*" (Interconnection requirements for onboard, 2015). This makes the EVSE the sole gatekeeper for any power

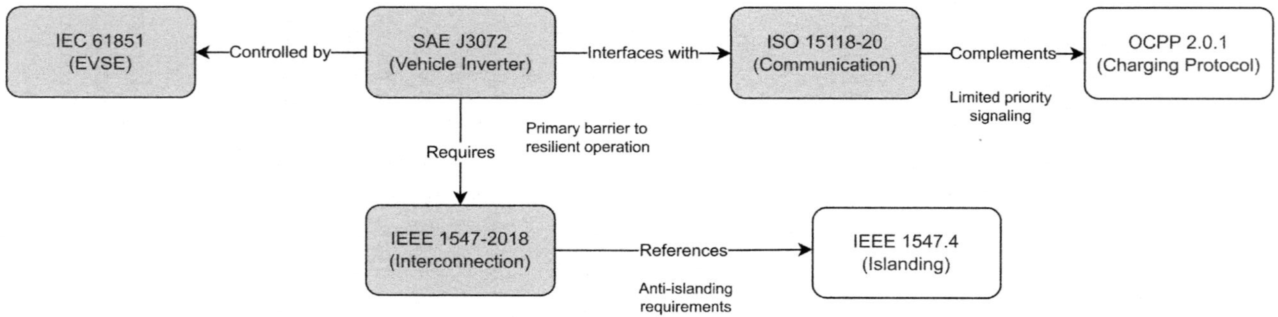

Figure 6.9 Relationship between SAE J3072 and related standards affecting R2V resilience.

export. The standard further reinforces this in its DER Managing Entity (DME) requirements, stating: "*For V2G-AC, the EVSE shall be a DER managing entity and it shall be the sole authorized DME for the purposes of meeting SAE J3072 requirements of 4.6. For those functions listed in Table 15, the EVSE shall be the only authorized DME. The PEV shall not accept messages for any of these functions from any other source*". These requirements create several critical resilience limitations:

1. **Single Point of Failure:** The EVSE becomes a single point of failure for resilient operation. Since the PEV polls the EVSE's DCTE "permit" every second and must cease export immediately if it ever loses that signal, any EVSE outage forces the vehicle to stop exporting power, even if it has available battery capacity.
2. **Limited Autonomous Decision-Making:** Every discharge authorization, as well as every smart-inverter function listed in Table 15, is rigorously governed by EVSE. The vehicle cannot respond autonomously to shifting grid or load conditions because it lacks the capacity to self-initiate export based on its own local measurements or preloaded emergency logic.
3. **Complex Coordination Requirements:** J3072 makes no provision for coordinating two or more vehicles behind a single EVSE during islanded operation; it simply treats the EVSE as the only DME, leaving multi-vehicle load-sharing entirely undefined. This complicates the implementation of resilient microgrids with multiple vehicle resources.

These EVSE control requirements significantly limit the potential for vehicles to serve as flexible, mobile energy resources during emergency conditions, particularly when the EVSE itself may be compromised by the same outage affecting the broader grid.

6.6.1.3 Lack of emergency operation modes

SAE J3072 is scoped exclusively to grid-connected, EVSE-mediated inverter operation and does not define any provisions for emergency or islanded modes. This omission creates critical resilience gaps. The standard's Scope and Technical Requirements (Sections 6.1 and 6.4) address only grid-connected interconnection, authorization, and ride-through behavior—there is no clause permitting an "emergency override" of these constraints during declared outages or crises. Additionally, J3072 provides no specifications for voltage or frequency setpoints, ride-through durations, or stability margins applicable to islanded microgrid operation. Appendix D of the standard merely maps existing IEEE 1547 ride-through tests, without defining any new islanded-mode controls. Furthermore, the standard includes no detailed procedures—such as state-machine diagrams, dual-mode handshakes, or transfer-switch sequencing—to enable seamless transitions between grid-connected and islanded modes. All "Control of Inverter During Session" logic (Section 4.7) presumes continuous EVSE-mediated, grid-referenced operation.

Without emergency-mode definitions or transition protocols, SAE J3072-compliant systems cannot maintain uninterrupted power delivery to critical loads during grid outages, necessitating substantial deviations from the standard for truly resilient R2V implementations.

6.6.1.4 Impact on operational modes

The SAE J3072 constraints described above undermine R2V resilience across all three operational modes from Section 4.1. In Standard Mode, vehicles experience reduced autonomous response to renewable variability because all export must be EVSE-authorized (Section 4.7.2), preventing self-adjustment of charging rates in response to real-time PV fluctuations or local grid signals—limiting forecast-driven load shifting and midday fill strategies (Interconnection requirements for onboard, 2015). For Emergency Mode, anti-islanding trip rules (IEEE 1547 Clause 8 via J3072 Appendix D) and EVSE gating (Section 4.7.2) together force vehicles offline within 2s of a grid failure, blocking backup support during outages by preventing any post-islanding discharge—even to critical loads—without prearranged intentional-islanding hardware and utility approval. In Congested Mode, grid-following sync requirements (J3072 Appendix F) and continuous EVSE mediation impair demand-response and stress relief by precluding vehicles from autonomously participating in under-voltage/under-frequency DR events or stepping down charging according to local voltage excursions—hampering fast congestion relief on stressed feeders.

These impacts make clear that, without significant revisions to authorize controlled grid-forming and autonomous discharge functions, SAE J3072–compliant vehicles cannot fully deliver the resilience benefits envisioned in our adaptive control schemes.

6.6.2 Critical standards affecting R2V resilience

While SAE J3072 presents the most direct limitations for vehicle-integrated energy resources, two additional standards significantly impact R2V resilience: IEEE 1547–2018, IEC 61851, and ISO 15118-20.

IEEE 1547-2018 contains provisions that directly impact R2V resilience. Clause 8.1.1 explicitly mandates that DERs "cease to energize the Area EPS, and trip within 2s of the formation of an island" (IEEE standard for interconnection, 2018). This anti-islanding requirement effectively prevents R2V systems from maintaining operation during grid outages. While Clause 8.2 addresses intentional islanding, requiring coordination with the Area EPS operator, it does not prescribe detailed microgrid control specifications—it simply describes how islands may be formed and states that control settings may be adjusted post-islanding (IEEE standard for interconnection, 2018). The standard addresses ride-through, trip settings, and limited voltage/frequency support in a grid-following context only, with no grid-forming mode specified.

IEC 61851, the international standard for EV conductive charging, establishes safety requirements that constrain resilient operation. IEC 61851-1 requires continuous monitoring of the protection-earth conductor with mandatory disconnection if continuity is lost (Electric vehicle conductive charging system, 2014), preventing charging during ground faults or compromised earth connections during severe weather events. The standard mandates immediate disconnection in response to various fault conditions (Sections 6.3 and 6.4), interrupting charging during minor grid disturbances (Electric vehicle conductive charging system, 2014). Furthermore, IEC 61851 defines charging modes (Modes 1–4) for grid-connected operation with no provisions for emergency charging modes with modified safety thresholds.

ISO 15118-20, while enabling smart charging, lacks resilience-oriented features. Despite extensive messaging capabilities for service discovery and bidirectional power transfer, there are no standardized data objects for declaring emergency priorities or critical-service tiers (Road vehicles, 2022b). The standard does not provide for exchanging real-time grid measurements (voltage, frequency, state-of-charge forecasts) between EV and EVSE, leaving islanded microgrid coordination unaddressed. Additionally, its TLS 1.3 two-way authentication requirement, while securing normal operations, could impede emergency override capabilities when network infrastructure is compromised.

These standards collectively impact the emergency mode operation described in Section 4.1.2, creating significant barriers to implementing the adaptive control schemes required for resilient R2V systems. Their architectures fundamentally assume stable grid conditions and prioritize disconnection over resilient operation during abnormal conditions.

6.6.3 Regulatory barriers stemming from standards limitations

The technical limitations in SAE J3072, IEEE 1547, and related standards create regulatory barriers that complicate resilient R2V deployment. These challenges span utility interconnection policies, permitting processes, and jurisdictional complexities.

Fig. 6.10 maps how standards limitations create three primary categories of regulatory challenges (utility interconnection barriers, permitting challenges, and policy gaps), with their root causes traced to specific IEEE 1547 and SAE J3072 requirements.

6.6.3.1 Utility interconnection barriers

The standards limitations, particularly in SAE J3072 and IEEE 1547, create specific utility interconnection barriers:

1. **Restrictive Islanding Policies:** Because SAE J3072 and IEEE 1547 require utility permission for intentional islanding, many utilities have developed restrictive policies that prohibit or severely limit customer-side islanding. These policies often cite the lack of standardized approaches in SAE J3072 as justification for conservative interconnection requirements.
2. **Limited Non-Export Options:** The EVSE-controlled operation requirement in SAE J3072 complicates the implementation of non-export or limited-export interconnection agreements, as utilities may not trust the vehicle-EVSE control interface to reliably prevent unintended export.
3. **Capacity Limitations:** Many utilities impose capacity limits on customer-sited generation that may be insufficient for supporting substantial EV charging loads during islanded operation. These limits often range from 100% to 120% of historical peak demand, constraining system sizing for resilience.

6.6.3.2 Permitting and inspection challenges

The standards gaps also create specific permitting and inspection challenges:

1. **Jurisdictional Complexity:** Because SAE J3072 does not define emergency operation modes, local authorities having jurisdiction (AHJs) have developed inconsistent approaches to permitting R2V systems with islanding capabilities. This multi-layered approval process can extend project timelines by 3–6 months.

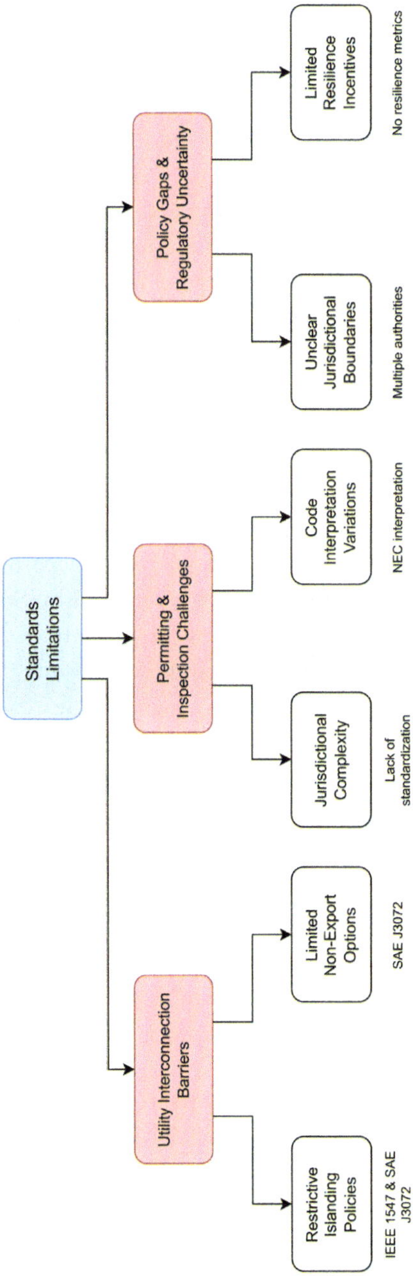

Figure 6.10 Regulatory barriers stemming from standards limitations.

2. **Code Interpretation Variations:** Local interpretations of the National Electrical Code (NEC) and fire codes vary significantly regarding requirements for transfer equipment, emergency disconnects, and signage for systems capable of operating during grid outages. These variations are exacerbated by the lack of clear guidance in SAE J3072 for emergency operation.
3. **Inspection Knowledge Gaps:** Many electrical inspectors lack familiarity with bidirectional EV charging and R2V systems, causing inspection delays, repeated site visits, and inconsistent code interpretations.
4. **Code Compliance Uncertainty:** The National Electrical Code (NEC) does not explicitly address bidirectional EV charging or R2V systems, creating uncertainty about applicable code provisions and leading to conservative interpretations.

6.6.3.3 Policy gaps and regulatory uncertainty

Broader policy gaps further complicate R2V deployment. Regulatory authority over vehicle-integrated energy resources remains ambiguous, with oversight potentially split between transportation departments, energy regulators, and environmental agencies—exacerbated by SAE J3072's lack of clear definitions for vehicles operating as power sources. While many jurisdictions incentivize renewable energy and EV charging, few programs specifically value resilience capabilities, partly due to the absence of standardized approaches in SAE J3072 and IEEE 1547 for quantifying resilient operation. Additionally, ongoing revisions to interconnection standards create uncertainty for long-term investments, as projects designed under current requirements may face future compliance challenges.

6.6.4 Cost-benefit analysis of resilience improvement

The standards limitations and resulting regulatory challenges directly impact the economic case for resilient R2V systems, creating complex cost-benefit tradeoffs that influence adoption decisions.

6.6.4.1 Implementation cost premiums

Mitigating current standards gaps typically requires significant additional equipment and processes. To meet anti-islanding and ride-through provisions, automatic transfer switches, isolation relays, and rapid-make contactors must be added. Based on industry estimates and anecdotal evidence, these transfer equipment requirements typically add on the order of 10%–30% to total installed costs, depending on site scale and equipment choice. Additionally, complying with SAE J3072's EVSE-mediated discharge authorization and implementing resilience-oriented load-shedding logic requires enhanced control hardware, cybersecurity layers, and bespoke integration software. According to industry practitioners, such control system complexity features typically add roughly 5%–20% in incremental engineering and integration costs. Furthermore, because neither SAE J3072 nor IEEE 1547 defines standardized emergency-mode tests, each installation must undergo custom relay studies, witness testing, and extended utility/AHJ approval processes—often extending project timelines by 3–6 months and adding tens of thousands of dollars in certification and compliance costs.

These incremental costs can deter adoption, especially for commercial and public charging sites with limited capital. Table 6.1 links each premium category to the underlying standards gaps.

6.6.4.2 Resilience value quantification

Estimating the economic benefit of resilient operation hinges on the Value of Lost Load (VoLL), which varies by customer class and outage duration. For critical facilities (hospitals, data centers), avoided outage costs are substantial, with VoLL in the range $10 000–$50 000/MWh (London Economics International and LLC, 2013; The Brattle Group, 2024), while commercial charging sites in MISO markets use a default VoLL of $3500/MWh (Midcontinent Independent System Operator (MISO), 2022). Regarding critical transportation services, while no comprehensive study exists to quantify this benefit precisely, anecdotal evidence from post-disaster analyses suggests that preserving even partial emergency-vehicle charging capacity can materially reduce broader economic losses. Additionally, resilience duration requirements show industry experience suggesting steeply diminishing marginal returns for backup beyond 24 hours, with most avoided-outage value accruing in the first day and benefits tapering off over 48–72 hours. While (Turan and Cole, 2024) provides analysis of storage revenue patterns, the specific diminishing returns for resilience value require further study.

This wide variation and uncertainty in VoLL complicate business cases for resilience. Table 6.1 summarizes the principal costs and benefits across transfer equipment, control complexity, and compliance.

6.6.4.3 Business models and financing approaches

Several emerging frameworks from utility pilots, vendor reports, and microgrid case studies offer paths to defray resilience costs: (1) **Resilience-as-a-Service**

Table 6.1 Cost–benefit analysis of resilient R2V implementation.

Category	Costs	Benefits
Transfer equipment	• **10%–30%** cost premium (industry estimate) • Anti-islanding hardware per SAE J3072/IEEE 1547	• **Enables seamless islanding** • Maintains critical charging
Control complexity	• **5%–20%** integration premium (industry estimate) • Enhanced cybersecurity stacks	• **Supports priority charging** • Enables adaptive load shedding
Certification and compliance	• **3–6 month** extended timelines • Custom relay studies and witness tests	• **Regulatory certainty once approved** • Informs future code updates
Resilience value	• VoLL ranges vary by customer class • Difficult to monetize	• Critical facilities: **\$10 000–\$50 000/MWh** (London Economics International and LLC, 2013; The Brattle Group, 2024) • Commercial EV sites: **\$3500/MWh** (Midcontinent Independent System Operator (MISO), 2022) • Diminishing returns beyond 24 hours (industry experience)

Note: These percentages and revenue ranges are drawn from industry pilot reports and NREL energy-storage analyses rather than from the VoLL literature.

(RaaS), where third parties own and operate R2V hardware for monthly fees typically 10%–25% above standard charging tariffs; (2) **Value Stacking** through multiple revenue streams (demand-charge management, frequency regulation, capacity markets), with NREL finding CAISO battery storage earnings of \$28–49/kW-yr from stacked services and potential upside to \$100–200/kW-yr in high-value scenarios (Turan and Cole, 2024); and (3) **Public-Private Partnerships** between utilities, municipalities, and charging operators, which can reduce capital costs by 15%–30% based on analogous microgrid deployments.

6.6.5 Pathways forward: Enhancing R2V resilience

Despite the significant challenges outlined above, several promising pathways exist for enhancing the resilience of R2V systems through standards development and regulatory reform.

6.6.5.1 Standards development initiatives

To address the gaps identified in current R2V standards, we recommend that standards-setting bodies consider the following enhancements:

1. **SAE J3072 Enhancement:** Incorporate explicit emergency-mode provisions—such as grid-forming operation during outages and a standardized emergency-override interface—to directly mitigate the anti-islanding and EVSE-control limitations described in Section 6.1.
2. **IEEE 1547.4 Guidance Expansion:** Extend IEEE 1547.4 to include detailed best practices for intentional islanding, black-start sequencing, and critical-load prioritization to lower the technical and regulatory barriers to resilient microgrid operation.
3. **ISO 15118-20 Extensions:** Augment ISO 15118-20 with standardized metadata fields for charging priority levels and explicit microgrid-coordination messages, enabling plug-and-charge workflows to support advanced resilience strategies.

6.6.5.2 Regulatory reform initiatives

To reduce non-technical barriers and accelerate resilient R2V deployment, we recommend: (1) **Standardized DER Interconnection Procedures** modeled on California's Rule 21, with pro forma application forms, defined study screens, and strict utility review timelines to streamline approvals (California Public Utilities Commission, 2000, 2025); (2) **Resilience-Focused Incentive Programs** by expanding FEMA's Hazard Mitigation Assistance grants (including HMGP, BRIC, and FMA programs that made $1.35 billion available in FY 2024) to explicitly fund R2V microgrids (Federal Emergency Management Agency, 2024); and (3) **Permitting and Process Streamlining** through fast-track, automated permitting for standard resilient EVSE configurations, similar to California's Integration Capacity Analysis portal (California Public Utilities Commission, 2025). These reforms would require coordinated action across jurisdictions but could significantly reduce deployment barriers.

The standards limitations, regulatory challenges, and economic considerations outlined in this section highlight the complexity of implementing resilient R2V systems at scale. While significant barriers remain, ongoing standards development and regulatory reform initiatives offer promising pathways for enhancing the resilience of EV charging infrastructure. By addressing the specific limitations in SAE J3072 and related standards, stakeholders can accelerate the deployment of R2V systems that maintain critical charging capabilities across all operational conditions.

6.7 Recommendations and future trends

The future of Renewable-to-Vehicle (R2V) smart charging systems hinges on the development of comprehensive standards that address the current limitations in SAE J3072, IEEE 1547, and related frameworks, enabling truly resilient operation across all operational modes. A critical path forward involves the creation of explicit emergency-mode provisions within these standards, including grid-forming capabilities during outages, standardized emergency-override interfaces, and detailed protocols for intentional islanding that can maintain critical charging services during grid disruptions. These enhancements must be coupled with regulatory reforms that streamline interconnection procedures, establish resilience-focused incentive programs, and implement fast-track permitting for standardized resilient EVSE configurations, thereby reducing the significant non-technical barriers that currently impede widespread deployment.

Looking ahead, the integration of advanced technologies will fundamentally transform R2V charging infrastructure into self-healing micro-ecosystems capable of operating autonomously during grid failures. Bidirectional power flow capabilities will enable EVs to serve as mobile energy storage units, providing emergency power to critical infrastructure during disasters while supporting grid stability during normal operations. The convergence of artificial intelligence, blockchain technology, and dynamic wireless charging solutions will create unprecedented opportunities for resilient charging networks that can adapt to varying operational conditions in real-time. As these technologies mature, R2V systems will increasingly incorporate vehicle-to-grid (V2G) capabilities, microgrids with renewable integration, and advanced energy management systems that optimize charging based on renewable availability, grid conditions, and user needs. This evolution will not only enhance charging resilience but also contribute significantly to overall grid stability, renewable energy integration, and the creation of a more sustainable and resilient transportation ecosystem.

6.8 Conclusion

The transition to electric vehicles represents a critical juncture in global transportation electrification, yet the resilience of EV charging infrastructure remains fundamentally constrained by current standards and regulatory frameworks. This chapter has demonstrated that while Renewable-to-Vehicle (R2V) smart charging systems offer substantial potential for enhancing charging resilience across operational conditions, their implementation faces significant technical, regulatory, and economic barriers that must be systematically addressed. The analysis reveals that existing standards—particularly SAE J3072, IEEE 1547–2018, and IEC 61851—prioritize grid-following operation and enforce anti-islanding requirements that prevent autonomous operation during grid outages, the precise conditions when resilient charging is most critical. The absence of standardized emergency operation modes and priority charging hierarchies leaves R2V systems ill-equipped to manage resource allocation during stress conditions, while regulatory barriers add substantial implementation costs and deployment delays. However, the comprehensive framework presented through three integrated components—the Charging Service Resilience Framework, Intelligence Layer Architecture, and Operational Mode Optimization—provides a viable pathway forward through advanced AI techniques, adaptive control schemes, and priority-based resource allocation. The quantitative analysis underscores both the urgency and magnitude of the resilience challenge, with climate change increasing grid disruptions by 74% while EV adoption accelerates toward 23% of global vehicle sales by 2025, creating temporal mismatches between renewable generation and charging demand that stress aging infrastructure. Despite these challenges, successful pilot deployments such as UC San Diego's INVENT program and Borrego Springs microgrid demonstrate the technical feasibility of resilient R2V operation, achieving 35% increased renewable utilization and maintaining critical services during extended outages. The path forward requires coordinated action across standards development, regulatory reform, and technology advancement, with particular focus on creating explicit emergency-mode provisions in SAE J3072, streamlined interconnection procedures, and the integration of bidirectional power flow capabilities that transform R2V systems from simple charging stations into critical infrastructure capable of enhancing both transportation and energy system resilience during an era of increasing extreme weather events and accelerating electrification.

References

Aghajan-Eshkevari S, Azad S, Nazari-Heris M, Ameli MT, Asadi S. Charging and discharging of electric vehicles in power systems: an updated and detailed review of methods, control structures, objectives, and optimization methodologies. Sustainability 2022;14 (4):2137.

Ahuja RK, Magnanti TL, Orlin JB, et al. Network flows: theory, algorithms, and applications, vol 1. Englewood Cliffs, NJ: Prentice hall; 1993.

Al-Hajj R, Assi A, Neji B, Ghandour R, Barakeh ZAl. Transfer learning for renewable energy systems: a survey. Sustainability 2023;15(11).

Aldossary M, Alharbi HA, Ayub N. Optimizing electric vehicle (ev) charging with integrated renewable energy sources: a cloud-based forecasting approach for eco-sustainability. Mathematics 2024;12(17).

Aljohani T, Almutairi A. A comprehensive survey of cyberattacks on evs: research domains, attacks, defensive mechanisms, and verification methods. Def Technol 2024;42:31–58.

Aoun A, Adda M, Ilinca A, Ghandour M, Hussein I. Dynamic charging optimization algorithm for electric vehicles to mitigate grid power peaks. World Electr Vehic J 2024;15 (7):324.

Arias NB, Sabillón C, Franco JF, Quirós-Tortós J, Rider MJ. Hierarchical optimization for user-satisfaction-driven electric vehicles charging coordination in integrated mv/lv networks. IEEE Syst J 2022;17(1):1247–58.

BloombergNEF. Electric vehicle outlook 2022. Technical report. BloombergNEF; 2022.

Barman P, Dutta L, Bordoloi S, Kalita A, Buragohain P, Bharali S, et al. Renewable energy integration with electric vehicle technology: a review of the existing smart charging approaches. Renew Sustain Energy Rev 2023;183:113518.

Benhammou A, Tedjini H, Amine Hartani M, Ghoniem RM, Ali A. Accurate and efficient energy management system of fuel cell/battery/supercapacitor/ac and dc generators hybrid electric vehicles. Sustainability 2023;15(13).

California Energy Commission. Intelligent vehicle integration. Technical report. CEC; 2018.

California Independent System Operator. Flexible resources to help renewables: fast facts. Technical report. California: ISO; 2021.

California Public Utilities Commission. Decision D.00-12-037. December 2000. https://citeseerx.ist.psu.edu/document?doi=a07659adcc33a830b5bb77ba3f2f63fddaf6ed56.

California Public Utilities Commission. Rule 21 interconnection rulemaking and regulatory history. 2025. https://www.cpuc.ca.gov/industries-and-topics/electrical-energy/infrastructure/rule-21-interconnection/rulemaking-and-regulatory-history.

Deb S, Kalita K, Mahanta P. Review of impact of electric vehicle charging station on the power grid. In: 2017 International Conference on Technological Advancements in Power and Energy (TAP Energy). IEEE; 2017. p. 1–6.

Doerry N. Zonal ship design. Nav Eng J 2006;118(1):39–53.

Doerry N. Naval power systems: integrated power systems for the continuity of the electrical power supply. IEEE Electrific Magaz 2015;3(2):12–21.

Dubey A, Santoso S. A two-level topology design framework for reliable shipboard power systems. Inventions 2016;1(3).

Dispersed Generation Photovoltaics and Energy Storage. IEEE guide for smart grid interoperability of energy technology and information technology operation with the electric power system (eps), end-use applications, and loads. 2011.

Dispersed Generation Photovoltaics and Energy Storage. Ieee guide for design, operation, and integration of distributed resource island systems with electric power systems. 2011.

Electric vehicle conductive dc charging system. Technical Report AIS-138 (Part 2). Ministry of Road Transport and Highways, Government of India; 2018.

Electric vehicle conductive charging system – part 23: Dc charging station. 2014.
Federal Emergency Management Agency. Hazard mitigation assistance grant funding for microgrid projects. 2024. https://www.fema.gov/fact-sheet/hazard-mitigation-assistance-grant-funding-microgrid-projects.
Griffith H. Californians told not to charge evs as grid struggles in heat wave. Tech Xplore; August 2022.
Gong L, Sharma A, Abdul Bhuiya M, Awad H, Youssef MZ. An adaptive fault diagnosis of electric vehicles: an artificial intelligence blended signal processing methodology. IEEE Can J Electr Comp Eng 2023;46(3):196–206.
Giebel G, Brownsword R, George K, Denhard M, Draxl C. The state-of-the-art in short-term prediction of wind power: a literature overview. 2011.
Grainger JJ, Stevenson WD. Power system analysis. 2nd ed. McGraw-Hill; 1994.
Hamid R. Sayarshad. Securing power grids and charging infrastructure: cyberattack resilience and vehicle-to-grid integration. J Transport Geogr 2025;126:104231.
Hilshey AD, Hines PDH, Rezaei P, Dowds JR. Estimating the impact of electric vehicle smart charging on distribution transformer aging. IEEE Trans Smart Grid 2012;4(2): 905–13.
International Energy Agency. Global EV outlook 2024. Technical report. International Energy Agency; 2024a.
International Energy Agency. Energy efficiency 2024. Technical report. International Energy Agency; 2024b.
IEEE standard for interconnection and interoperability of distributed energy resources with associated electric power systems interfaces. 2018.
Interconnection requirements for onboard, utility-interactive inverter systems. 2015.
IEA Photovoltaic Power Systems Programme. Task 13 report – seasonal variability of photovoltaic systems. IEA-PVPS T13-14:2019. 2019.
IEEE Standards Association. IEEE guide for loading mineral-oil-immersed transformers and step-voltage regulators. 2012.
IEEE standard for calculating the current–temperature of bare overhead conductors. 2012.
International Energy Agency. Global ev outlook 2023. Technical report. IEA; 2023.
IEEE standard for the specification of microgrid controllers. 2017.
Kuss M, Markel T, Kramer W. Application of distribution transformer thermal life models to electrified vehicle charging loads using monte-carlo method. Technical report. Golden, CO (United States): National Renewable Energy Lab.(NREL); 2011.
Kariuki KK, Allan RN. Evaluation of reliability worth and value of lost load. IEE Proc Generat Transm Distrib 1996;143(2):171–80.
Lee ZJ, Lee G, Lee T, Jin C, Lee R, Low Z, et al. Adaptive charging networks: a framework for smart electric vehicle charging. IEEE Trans Smart Grid 2021;12(5):4339–50.
Li J, Che Y, Zhang K, Liu H, Zhuang Y, Liu C, et al. Efficient battery fault monitoring in electric vehicles: advancing from detection to quantification. Energy 2024;313:134150.
London Economics International, LLC. Value of lost load: literature review and macroeconomic analysis. Technical Report PUCT Project No. 40000. Electric Reliability Council of Texas (ERCOT); June 2013.
Mahazabeen M, Ali JA, Ebrahimi S, Daoud H, Ferdowsi F. Enhancing ev charger resilience with reinforcement learning aided control. e-Prime-Adv Electr Eng Electr Energy 2023; 5:100276.
Mahazabeen M, Abianeh AJ, Ebrahimi S, Ferdowsi F. Performance evaluation of an rl-aided dab-based ev charger control. In: 2022 North American Power Symposium (NAPS). IEEE; 2022. p. 1–5.
Mohanty PK, Hemant Kumar Reddy K, Kumar Panigrahy S, Roy DS. Leveraging generative and explainable ai for electric vehicle energy toward sustainable, consumer-centric transportation. IEEE Access 2024;12:143721–32.

Muratori M. Impact of uncoordinated plug-in electric vehicle charging on residential power demand. Nat Energy 2018;3(3):193–201.

Midcontinent Independent System Operator (MISO). Value of lost load (VOLL): development and use. Presentation, MISO RECBWG; August 2022.

North Central Texas Council of Governments. Planning for resilient ev charging infrastructure. Technical report. North Central Texas Council of Governments; 2024.

Narang D. Highlights of IEEE standard 1547-2018. Technical Report NREL/PR-5D00-75436. National Renewable Energy Laboratory; 2019.

Ngoc An L, Quoc-Tuan T, Seddik B, Van-Linh N. Optimal sizing of a grid-connected microgrid. In: 2015 IEEE International Conference on Industrial Technology (ICIT). IEEE; 2015. p. 2869–74.

National Renewable Energy Laboratory. Hardware-in-the-Loop evaluation of renewable resource black-start capability. Technical Report NREL/TP-5C00-79711. NREL; 2021a.

National Renewable Energy Laboratory. Electric vehicle grid integration. Technical report. NREL; 2021b.

OpenADR Alliance. Open automated demand response (OpenADR) profile specification version 2.0b. Online; 2018.

Open Charge Alliance. Open charge point protocol 2.0.1. 2024.

Pg&e to shut off power to 940,000 customers. October 2019.

Petry CR, Rumburg JW. Zonal electrical distribution systems: an affordable architecture for the future. Nav Eng J 1993;105(3):45–51.

Powell S, Vianna Cezar G, Liang M, Azevedo IML, Rajagopal R. Charging infrastructure access and operation to reduce the grid impacts of deep electric vehicle adoption. Nat Energy 2022;7(10):932–45.

Pfenninger S, Staffell I. Long-term patterns of european pv output using 30 years of validated hourly reanalysis and satellite data. Energy 2016;114:1251–65.

Parisio A, Rikos E, Glielmo L. A model predictive control approach to microgrid operation optimization. IEEE Trans Control Syst Technol 2014;22(5):1813–27.

Panteli M, Mancarella P, Trakas DN, Kyriakides E, Hatziargyriou ND. Metrics and quantification of operational and infrastructure resilience in power systems. IEEE Trans Power Syst 2017;32(6):4732–42.

Quirós-Tortós J, Panteli M, Wall P, Terzija V. Sectionalising methodology for parallel system restoration based on graph theory. IET Gener Transm Distrib 2015;9(11):1216–25.

Ram Chandra Mouli G, Kaptein J, Bauer P, Zeman M. Implementation of dynamic charging and v2g using chademo and ccs/combo dc charging standard. In: 2016 IEEE Transportation Electrification Conference and Expo (ITEC); 2016. p. 1–6.

Roth S. California's wildfire threat could be an opportunity for clean-energy microgrids. Los Angeles Times; March 2019.

Road vehicles. —vehicle to grid communication interface—Part 20: physical layer and data link layer requirements for wired communication. 2022.

Raman G, Raman G, Peng JC-H. Resilience of urban public electric vehicle charging infrastructure to flooding. Nat Commun 2022;13(1):3213.

Road vehicles – vehicle to grid communication interface – part 20: 2nd generation network layer and application layer requirements. 2022. This standard defines the communication protocol between electric vehicles and charging stations, including bidirectional power flow capabilities.

Sacramento Municipal Utility District. Customer perks through multi-automaker pilot program. American Public Power Association.

Sanghvi A, Markel T. Cybersecurity for electric vehicle fast-charging infrastructure. In: 2021 IEEE Transportation Electrification Conference and Expo (ITEC); 2021. p. 573–6.

San Diego Gas & Electric. Borrego springs microgrid fact sheet. Online; March 2019.

Schmalstieg J, Käbitz S, Ecker M, Sauer DU. A holistic aging model for li(nimnco)o2 based 18650 lithium-ion batteries. J Power Sources 2014;257:325–34.
Selvaraj V, Vairavasundaram I. A bayesian optimized machine learning approach for accurate state of charge estimation of lithium ion batteries used for electric vehicle application. J Energy Storage 2024;86:111321.
Seyedi M, Sheida K, Ebrahimi S, Fersowsi F. Can DC converters achieve resilient voltage regulation with reinforcement learning?. In: 2024 56th North American Power Symposium (NAPS). IEEE; 2024a. p. 1–6.
Seyedi M, Sheida K, Siner S, Ferdowsi F. Enhanced resilience in battery charging through co-simulation with reinforcement learning. Authorea Preprints; 2024b.
Shahriar S, Al-Ali AR, Osman AH, Dhou S, Nijim M. Prediction of ev charging behavior using machine learning. IEEE Access 2021;9:111576–86.
Siler-Evans K, Azevedo IML, Granger Morgan M, Apt J. Marginal emissions factors for the U.S. electricity system. Environ Sci Technol 2012;46(9):4742–8.
Simona Raboaca M, Meheden M, Musat A, Viziteu A, Creanga A, Vlad V, et al. An overview and performance evaluation of open charge point protocol from an electromobility concept perspective. Int J Energy Res 2022;46(2):523–43.
Smart Energy International. UC San Diego expands INVENT pilot with V2G capabilities. Smart-Energy.com; 2017.
SMUD. Managed EV charging. Online.
Soresini F, Barri D, Cazzaniga I, Ballo FM, Mastinu G, Gobbi M. Artificial intelligence for fault detection of automotive electric motors. Machines 2025;13(6).
Sun Y, Yue H, Zhang J, Booth C. Minimization of residential energy cost considering energy storage system and ev with driving usage probabilities. IEEE Trans Sustain Energy 2019;10(4):1752–63.
Surging power outages and climate change. September 2022.
The february 2021 cold weather outages in Texas. Technical report, federal energy regulatory commission and North American electric reliability corporation. November 2021.
Tiwari H, Ghosh A, Banerjee S, Mazumdar D, Sain C, Ahmad F, et al. Design of a triple port integrated topology for grid-integrated ev charging stations for three-way power flow. Front Energy Res 2024;12:1440258.
The Climate Center. What do evs have to do with wildfires?. 2024.
Tiwari AGH, Banerjee S, et al. Design of a triple port integrated topology for grid-integrated ev charging stations for three-way power flow. Front Energy Res 2025;12.
The Brattle Group. Value of lost load study for the ERCOT region. Technical report. The Brattle Group; September 2024.
Turan M, Cole W. Revenue analysis for energy storage systems in the United States. Technical Report NREL/TP-6A50-89906. National Renewable Energy Laboratory (NREL); November 2024.
Vartanian C, Gibson S, Sridhar S, Asgeirsson H, Franks R. The arlington microgrid project's v2g technical specification vs standards. IEEE Smart Grid Bulletin May 2022.
Wang K, Wang H, Yang Z, Feng J, Li Y, Yang J, et al. A transfer learning method for electric vehicles charging strategy based on deep reinforcement learning. Appl Energy 2023;343:121186.
Weather-related power outages rising. April 2024.
Wongchai A, Aoudni Y, Yesubabu M, Ahmad Reegu F, Gowri NV, Vijayakumar P. Sustainable electric vehicles fault detection based on monitoring by deep learning architectures in feature extraction and classification. Sustain Energy Technol Assessments 2023;57:103178.
Zhang T, Huang Y, Liao H, Liang Y. A hybrid electric vehicle load classification and forecasting approach based on GBDT algorithm and temporal convolutional network. Appl Energy 2023a;351:121768.

Zhang J, Wang Y, Jiang B, He H, Huang S, Wang C, et al. Realistic fault detection of li-ion battery via dynamical deep learning. Nat Commun 2023b;14(1):5940. Published: 2023-09-23.

Zhang J, Hodge B-M, Florita A, Lu S, Hamann HF, Banunarayanan V. Metrics for evaluating the accuracy of solar power forecasting. Technical report. Golden, CO (United States): National Renewable Energy Lab.(NREL); 2013.

CHAPTER SEVEN

A survey on energy trading mechanisms and electric vehicles market interactions for renewable-to-vehicle (R2V) smart charging ecosystem

Ali Paeizi[1, 2], Mohammad Mehdi Amiri[1, 2], Hamed Jafari Kaleybar[3], Ali Mosalla Nejad[1]

[1]Department of Electrical Engineering, Shahid Beheshti University, Tehran, Iran
[2]Electrical Networks Research Institute, Shahid Beheshti University, Tehran, Iran
[3]Politecnico di Milano, Energy Department-En:Lab, Milano, MI, Italy

7.1 Introduction

Power grids are changing as a result of the transportation electrification and the rapid deployment of renewable energy sources. In recent years, Renewable Energy (RE) integration sources such as wind and solar photovoltaic (PV) with power systems have integrated more than ever before. However, in order to effectively address the need for EV charging demand, methods and policies are required due to the intermittent nature of these systems (Barman et al., n.d.). Smart electric transportation charging facilitates bidirectional power exchange between the grid and EVs, which is enabled by vehicle-to-grid (V2G) technology. During high-demand periods or when RE generations are at their lowest, EVs can use V2G technology to supply electricity to the grid. This capability would strengthen the system stability, increase the usage of Renewable Energy Sources (RESs), and give EV owners access to new sources of income. To maximize the utilization of RE resources, smart charging strategies and optimization techniques are used throughout the charging process, which take into account variables like consumer preferences, electricity pricing, and the availability of RESs. Hence, it is essential to use smart charging, which adjusts EV charging to grid circumstances, client demands, and RESs (Gopinathan and Karthikeyan Shanmugam, n.d.).

Renewable-to-Vehicle Smart Charging
ISBN: 978-0-443-36684-0
https://doi.org/10.1016/B978-0-443-36684-0.00010-2

There are different EV charging modes such as dump charging, smart charging, and V2G. Dump charging begins when the car is connected to the grid, disregarding any surplus in the system's energy supply or the price signals from the relevant electricity market. Smart charging allows EV charging power to be controlled based on the cost and availability of the power supply. It is also known as Grid-to-vehicle (G2V), unidirectional V2G, or V1G. Smart charging is popular for low cost and few changes required modification on the existing infrastructure, making it among the most appealing sources of system flexibility. V2G is an advancement in smart charging that allows the system to release EV battery storage and feed electricity into the grid during power shortage. However, it requires particular hardware from EVs to feed the grid with high-quality sinusoidal alternating current, and it increases the number of cycles that the battery goes through for charging and discharging, which usually results in increased degradation and a shorter lifespan (Bogdanov and Breyer. n.d.). V2G permits bidirectional power transfer between EVs and the power grid, in contrast to V1G; however, V2G requires EVs to be able to charge in both directions. V2G's capability to balance the grid, particularly in combined with RE, is considered a prominent benefit. If an EV owner has an RE setup on their property, they could operate as a prosumer. EVs' batteries can be combined with V2G technology to provide an effective utility grid energy storage system. It also provides EV owners with an affordable solution with an option to sell the extra power generated at a set price (Barman et al., n.d.). In this regard, the electricity consumption can be maximized during periods of RE intermittency, and the system could minimize consumption during periods of low RE generation, and supply energy stored in EV batteries thanks to bidirectional energy flow between EV batteries and the grid, and such technology functions similarly to utility-scale battery storage (Bogdanov and Breyer. n.d.).

Increasingly RESs integration with the power grid, with the increasing use of EVs power systems' resilience is becoming more and more important. The resilience of a power system can be significantly improved using strategically integrating EVs into market-based energy operations. Various methods, such as real-time pricing, Peer-to-peer (P2P) energy trading, and V2G technologies, are among the techniques that enable dynamic interactions with bidirectional power flow capacity, which is a critical capability during periods of high demand or disruption (Bems and Anvari-Moghaddam, n.d.). RESs would increase system resilience and offer additional electrical capacity, although they pose difficulties and inefficiencies in power generation due to their intermittent nature, which may result in electricity supply and demand imbalances and dealing with resilient techniques, could be crucial for reducing the vulnerability of electrical grids to severe disasters (Ameli et al., 2024). However, by operating as portable energy storage devices, EVs could help to mitigate these problems. Prohibition of power outages brought on by unaccustomed variations or man-made threats is a prominent motivation for enhancing the grid's resilience to ensure the grid's continuous operation (Paeizi et al., n.d.).

P2P energy trading, using digital platforms, has a key role in exchanging electricity between energy producers and customers. Since they target prosumers and use blockchain technology to facilitate safe, transparent, and decentralized transactions, EVs are major participants in P2P energy trading. Centralized P2P trading networks manage and supervise the electricity trading through a central organization, usually managed by a utility. This concept ensures a regulated business environment by having a central authority to verify and regulate transactions (Khan et al., n.d.).

R2V smart charging aims to integrate EV charging with RE generation to convert EVs from pure loads to distributed storage assets. R2V smart charging is becoming a crucial component of a sustainable and effective power grid as EVs and RES are integrating more in recent years. More research on how effectively different trading models operate within R2V frameworks is needed to fill the gaps which highlight the necessity of a comprehensive study of trading methods and EV participation to enhance system efficiency, economic performance, and grid resilience. In this regard, this chapter reviews the state-of-the-art studies in *R2V smart charging*, market and EVs integration, and *Energy-Trading Mechanisms, considering advanced technologies such as Machine Learning (ML) and blockchain.* The following are the contributions this chapter offers.

1. This chapter presents a comprehensive review of renewable energy-enabled different smart charging approaches with quantitative description and mathematical modeling, EVs to market models, and blockchain based energy trading.
2. Optimization objectives of EV charging and advanced technologies in electricity markets participation of EV mechanisms in R2V smart charging are provided.
3. Recent research developments and state-of-the-art methodologies and optimization techniques in the literature are reviewed.
4. Limitations, gaps in current studies, and suggestions for further research are noted.

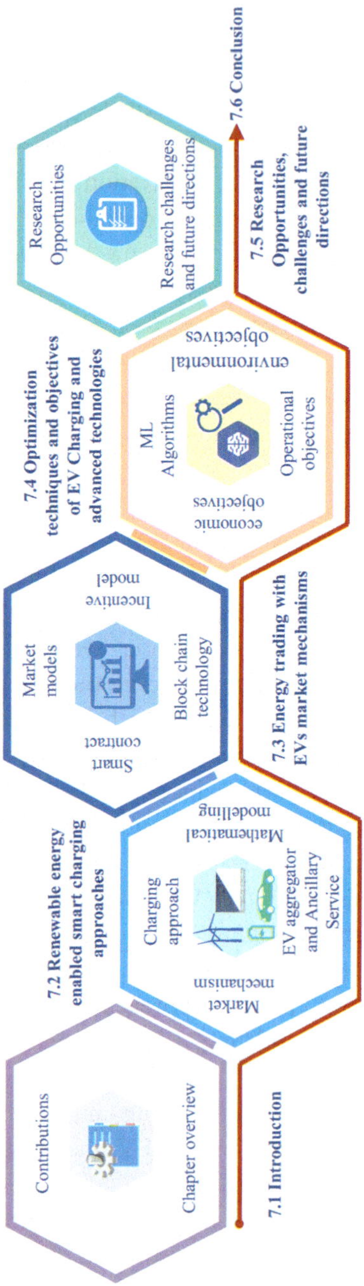

Figure 7.1 Overview of the chapter.

The remainder of this chapter is as follows: in Section 7.2, renewable energy-enabled smart charging approaches are classified with quantitative description and mathematical modeling, and EV owners and the aggregators' roles. In Section 7.3, energy trading with EV market mechanisms, and ancillary services, EVs' engagement with electricity markets, and blockchain-based energy trading, data security, and smart contracts are discussed. Moreover, decentralized energy exchange between prosumers and EV owners is also provided. In Section 7.4, optimization techniques and objectives of EV charging, including operational, economic, and environmental objectives and advanced technologies such asML algorithms and data-driven based methods in electricity markets participation of EV mechanisms are presented. Research opportunities, challenges, and future directions are provided in Section 7.5, and lastly, Section 7.6 presents the chapter's conclusion. Fig. 7.1 illustrates the overview of this chapter, encompassing *Energy Trading and EVs Market Interactions with R2V Smart Charging*.

7.2 Renewable energy-enabled smart charging approaches with EV integration

Depending on the availability of energy sources, such as wind and solar radiation, the electric power generation from these RESs can be either very high (even more than power demand) or very low (less than the power demand). These RESs have time variants with limitations in control, especially while dealing with power system planning. Using stationary energy storage systems or regulated dispatch loads is a potential way to balance the grid's electricity generation, considering RESs penetration. During low power generation or excess power generation, the stationary energy storage systems could either absorb or deliver electricity. In order to leverage the integration of RESs into the electricity market as dynamic energy storage devices, the EV batteries can then be combined and serve as the energy storage system. Through various charging schemes, the EVs can absorb the excess power generated by the RESs.

EVs are increasingly becoming more and more popular, and PV solar electricity is more likely to be used for grid support and charging, which has demonstrated favorable feasibility. These PV solar systems may also conduct V2G transactions, and effective generation scheduling can save operating costs and improve grid performance. In order to determine the influence of wind energy conversion systems and their viability on the power grid, the synergy between wind energy systems and EVs in different scenarios has been examined.

When a charging device and an EV share a common network with a charging operator, this is referred to as smart charging. Smart charging devices, as opposed to traditional dumb chargers, are connected to the cloud, allowing the charging station's owner to efficiently distribute energy by controlling, monitoring, and restricting the use of their devices. Without smart charging, the transmission between EVs and the grid would not be possible, and EVs could begin to put a strain on the grid and the power delivery system. Different types of charging approaches could be expressed as follows (Barman et al., n.d.).

- Shift charging: Shift charging is a smart strategy that seeks to align EV charging with RE generation timing. In high RE generation periods or off-peak, shift charging utility programs promote EV charging, which is more convenient for the grid. During these preferred periods, the utilities give their customers lower rates, which can compensate for any possible RE premium.
- Charging with excess renewable energy: In this method, to facilitate grid integration, smart charging is coordinated with the surplus energy produced by the RE sources. In this case, loads are driven to times when the system encounters higher RE generation than electric demand by varying the unit rates of electricity inversely with the RE generation.
- On-site renewables: For residential or commercial applications, the on-site renewable approach combines EV charging with on-site renewable energy generation, usually solar PV. Then, a number of strategies can be employed, including integrating battery storage, co-locating the systems to minimize the impact of charging times, or using managed charging to control load.
- Managed charging or controlled charging: Bidirectional managed charging (V2G) or unidirectional managed charging (V1G) are two classifications for this approach, and it initiative develops programs that control when EVs should be charged at home or at work. By adapting to pricing signals and power system requirements, the V1G charging system controls and maximizes charging time, rate, and duration. The utility operator transmits the required signals, which contain data on the grid's current state and the availability of RESs. Unlike the V1G, an EV in the V2G acts as a distributed energy storage device that can provide additional power to the grid when required. By adjusting the

charging rate in response to grid demand, utilities can remotely control EV charging.

Methodological framework for EV integration with R2V smart charging and trading ecosystem is illustrated in (Fig. 7.2). The study is arranged into three layers in this figure. Demand models, charging infrastructure features, aggregator rules and policies, and system data from generation, transmission, and microgrid with EV integration are all specified by inputs and models. With the help of data analytics, configuration and parameter setting, monitoring and control, services, network infrastructure, and a market framework, a control center (execution layer) uses ML algorithms and sophisticated techniques to process these inputs. The optimization core (cost, emissions, reliability, and profit objectives) is also supported. Identity, audit, and automation are provided by security and blockchain features. By producing cost and performance indices, grid-performance metrics, ancillary services, and market results, the framework makes it possible to compare R2V strategies under practical operational limitations. A data-driven approach can simplify the forecasting model by integrating temperature, weather, historical charging demand, and other multi-source datasets.

7.2.1 The overview for assessing V2G integration

The integration of REs, grid stability, and economic benefits is all significantly impacted by the deployment of V2G strategies, including unidirectional and bidirectional charging. By integrating V2G, the power grid can be transformed into a more intelligent and sustainable system. Primary advantages of V2G integration are expressed as follows (Kumar et al., n.d.).

- GridStability: Bidirectional charging offers ancillary services like voltage control and frequency regulation, which improve grid reliability. Blackout risk is decreased by EVs' ability to release energy during periods of high demand.
- RenewableEnergyIntegration: During times of high generation, V2G systems allow excess RE to be stored in EV batteries during the day for solar energy. The grid can then receive stored energy during times when renewable output is low at night.
- Economic Advantages: Consumers and utility companies can benefit economically from integrating V2G technology. Utilities can lower the costs of peak demand, improve the efficiency of power plants through more seamless operation, and possibly lower electricity prices

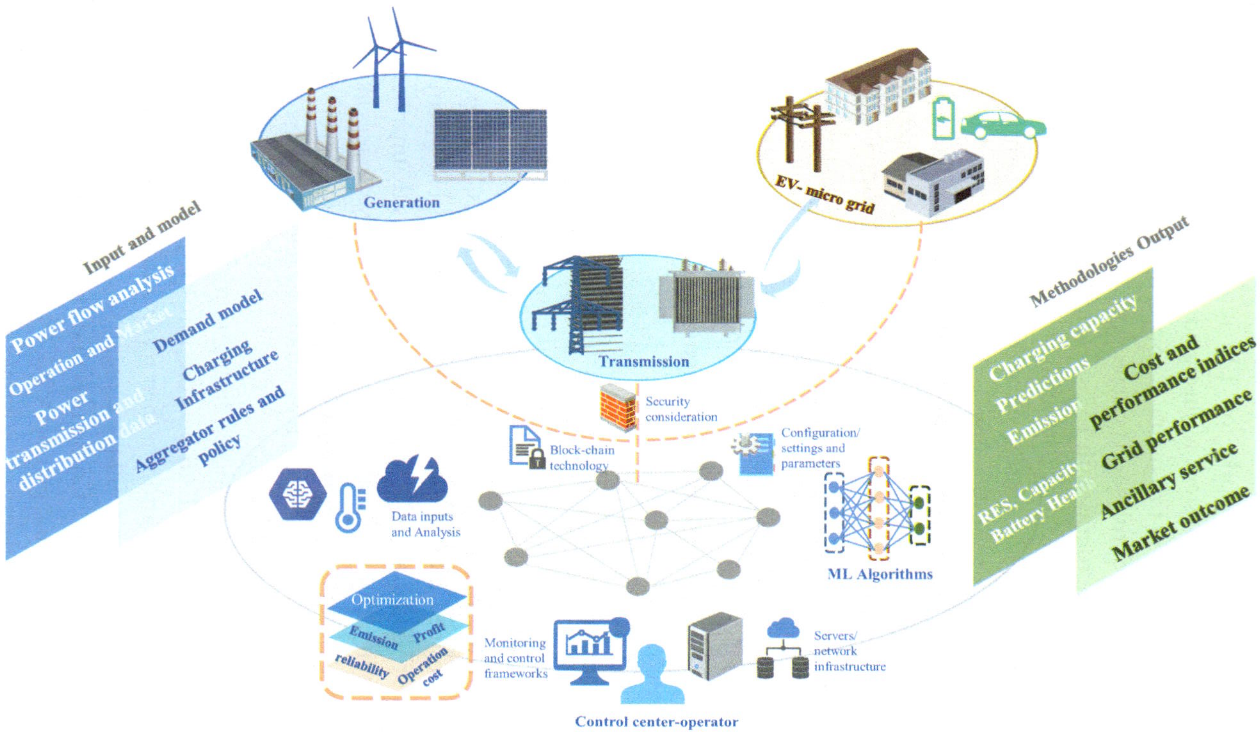

Figure 7.2 Methodological framework for EV integration with R2V smart charging and trading ecosystem.

by integrating more RE. V2G systems save operating costs for utilities by eliminating the need for costly peak power plants. By lowering the costs of peak demand and implementing time-based electricity pricing, which promotes charging during off-peak hours, consumers can save money. V2G users can profit by participating in ancillary service markets or by selling electricity back to the grid during periods of high demand.

- Environmental Impact: V2G technology offers significant advantages in the fight against pollution and climate change.

Table 7.1 summarizes the representative control strategies and principal applications of the five EV charging control modes, including uncontrolled, basic ON/OFF, unidirectional G2V, V1G, bidirectional V2G, and bidirectional V2X (Deb et al., 2022).

7.2.2 Modeling smart charging of EV and V2G

EV electricity consumption, battery state-of-charge dynamics (involving self-discharge and charge/discharge profiles), nodal/hourly energy balance between on-site generation, storage, grid exchange, and EV demand, aggregator economics (profit with prices and battery-wear costs), and parking-lot storage dynamics are all covered in Table 7.2. Through Eqs. (7.1)–(7.8), this table provides a summary of the variables and units is including key formulations for modeling smart charging of EV and V2G.

(Fig. 7.3) shows the overview of the prominent positive and negative effects of EVs when integrated with the power grid. Benefits of EV integration include lowering peak loads, increasing the RESs usage with financial gains, and providing ancillary services. Potential feeder overloading/congestion, power-quality and stability risks, and renewable intermittency are among the drawbacks, which call for forecasting, controls, and cooperation with network operators (Taghizad-Tavana et al., n.d.).

7.2.3 Role of electric vehicle aggregator and market in V2G participation for smart charging ecosystem

In order to allow distributed generation and responsive loads to provide electricity services appropriately and increase the power grid's flexibility, aggregators are for-profit intermediaries that coordinate between the power system and end users. One of the aggregator's responsibilities, for example, is to directly control end-user appliances (Sadeghian et al., n.d.).

Table 7.1 Different EV charging types with control modes, control strategies, and uses.

Type	Description	Control strategy	Uses
Uncontrolled with Time of Use (TOU) tariff	Mutual authentication embedded in a permissioned blockchain.	None; charging follows user plug-in and tariff windows (implicit price response).	Peak shaving through off-peak charging allows for simple long-term capacity planning.
Basic Control	Central or local start/stop scheduling.	Central or local start/stop scheduling; curtail or defer charging; enforce site power caps.	Avoiding overload and managing feeder/transformer congestion.
Unidirectional V1G	Standard EV infrastructure is used for one-way charging, which is less complicated and expensive than V2G. It offers less battery stress and limited grid support.	Smart charger modulates charging rate using real-time price/grid signals and forecasts.	Improved alignment with renewables and ancillary services through controllable load, particularly frequency regulation.
Bidirectional V2G	Two-way power flow between EV and grid; Vehicles can support system requirements by storing energy and releasing it back to the grid when needed.	Two-way power flow; rapid response to grid conditions; requires V2G-capable vehicles and EV supply equipment.	Short-term RES integration, load following, frequency and voltage regulation, ancillary services, and reliability enhancement.
Bidirectional V2X	Self-consumption and resilience are made possible by a bidirectional interface with building/home energy management systems for backup power and behind-the-meter optimization.	Integration with home/building energy-management systems; behind-the-meter optimization and backup.	Resilience (backup power), self-consumption/peak shaving, and microgrid optimization.

Table 7.2 Key formulations for modeling smart charging of EV and V2G.

Title	Equation		Description
Smart charging consumption (Bogdanov and Breyer. n.d.)	$\text{SmartChCons}_{segy,h} = \left(1 - \text{dumpChargeShare}_{segy}\right) \cdot \text{totElCons}_{segy,h}$	(7.1)	SmartChCons: electricity consumption by the smart charged EV in a year; h: hour; $\text{dumpChargeShare}_{segy}$: share of the dump charged EVs in the fleet; totElCons: total electricity consumed by plug-in hybrid electric vehicles and battery electric vehicles
Battery state-of-charge (Bogdanov and Breyer. n.d.)	$\text{batterySoC}_{segy,h} = \left(1 - \text{selfDisch}_{segy}\right) \cdot \text{batterySoCs}_{segy,h-1} + \text{chargeEff}_{segy} \cdot \text{chargeSmart}_{segy,h} \frac{-\text{dischargeSmart}_{segy,h}}{\text{dischargeEff}_{segy}}$	(7.2)	selfDisch: battery self-discharge rate per hour; batterySoC: smart charging EVs fleet battery state-of-charge; chargeSmart: charging profile of smart charging EVs; dischargeSmart: discharging profile of smart charging EVs; dischargeEff: charge and discharge efficiency of the battery
Electricity supply and demand equation; Energy balance between on-site generation, grid exchange, storage and EV charging demand assurance (Bogdanov and Breyer. n.d.)	$\sum_{g}^{gm} \text{elGen}_{g,h} + \sum_{s}^{\text{stor}} \text{storDisch}_{\varsigma_h,h} + \text{dischargeV2G}_{\text{mode},h} + \text{fromGrid}_h = \sum_{s}^{\text{stor}} \text{storCharge}_{\varsigma_h,h} + \text{toGrid}_h + (1 + T\&D\text{Loss}) \cdot \left(\text{elDem}_h + \text{chargeSmart}_{seg,h}\right)$	(7.3)	$\text{elGen}_{g,h}$: electrical generation in the system at node g, hour h (kWh); storCh: storage technologies discharge; toGrid: electricity flow to interregional grid; elDem: electricity demand; T&D Loss: transmission and distribution grid losses within the region; chargeSmart: charging profile of smart charging EVs
State-of-charge of the EV battery of length Δt; necessary energy for the next trip (Shafie-khah et al., 2016)	$SOC_{n,\Delta t} = \frac{k_{m+1}\eta_{km}}{E_{bat}}$	(7.4)	$SOC_{n,\Delta t}$: State of charge (SoC) change over the charging interval Δt (p.u.); η_{km}: per-km energy consumption (kWh/km); E_{bat}: battery capacity (kWh); K: Energy exchanged with the battery
Objective function of EV aggregator profit increase (Taghizad-Tavana et al., n.d.)	$\sum_{t=1}^{T} \left(\varsigma^{ow} P_t^{EV} - \varsigma_t P_t^{EV} - \varsigma^{FT} P_t^{DCA}\right) \times \Delta$	(7.5)	ς^{ow}: contract price between EV owners and the aggregator; ς_t: market/grid price at time t; ς^{FT}: battery wear cost per discharge; P_t^{EV}: net EV power exchange; P_t^{DCA}: aggregate discharging power (kW).
Battery degradation (throughput) cost (Taghizad-Tavana et al., n.d.)	$\varsigma^{FT} = \frac{1000COK}{LCO \times DOD \times EC}$	(7.6)	ς^{FT}: battery throughput degradation cost (€/MWh); COK: battery cost; LCO: cycle life to EoL (cycles); DOD: depth of discharge; EC: battery capacity
Reduced parking energy to moved EVs (Taghizad-Tavana et al., n.d.)	$E_t^{\text{Maxtmum}} = \sum_{U=1}^{U} \sum_{EV=1}^{EV_U} G_U^{\text{Maxtmum}}$	(7.7)	E_t^{Maxtmum}: maximum aggregate energy available from connected EVs; G_U^{Maxtmum}: battery capacity of class-U EV (kWh);
Energy balance of the parking-lot storage (Taghizad-Tavana et al., n.d.)	$E_t = E_{t-1} + E_t^{ARR} + \omega^{CA} p_t^{CA} \times \Delta_t - \left(E_t^{DERP} + \frac{p_t^{DCA} \times \Delta_t}{\omega^{DCA}}\right); \forall t$	(7.8)	E_t: stored energy at time t (kWh); E_{t-1}: previous stored energy; E_t^{ARR}: arrivals energy; E_t^{DERP}: departures energy; ω^{CA}: charge efficiency; ω^{DCA}: discharge efficiency; p_t^{CA}, p_t^{DCA}: charge and discharge power (kW); Δt: time-step duration (h).

Figure 7.3 Overview of the prominent positive and negative effects of EVs when integrated with the power grid.

Data on vehicle characteristics, such as fuel efficiency, vehicle capacity, arrival times, driving mileage, etc., are gathered during the input stage. Data acquisition receives the information, analyzes it, and calculates the energy storage capacity of electric vehicles available at different parking stages for charging and discharging. Vehicle capability, parking availability, charging infrastructure, and other factors are among the G2V and V2G power selection constraints. Through aggregated connected channels, various ancillary services markets are dedicated to the output energy storage capacity (Fig. 7.4). Shows block diagrams of an aggregator scheduling activity with the R2V/V2G ecosystem while taking uncertainties into account. Through bidirectional chargers (G2V–V2G), the figure shows how an aggregator and infrastructure, which includes communication/data storage, charging management, optimization, and service modules, coordinate EVs with upstream sources (power grid, PV, wind). While controlling battery storage and stability, the V2G system provides services to both residential and commercial customers as well as the power system (load balancing, peak shaving, frequency regulation). The optimizer is informed by and shapes operational decisions by key uncertainties, including arrival/departure times, state of charge, user behavior, load variability, electricity prices, and RES intermittency. In order to integrate and manage the fleet of electric vehicles, also known as "virtual power plants," "EV aggregators," and

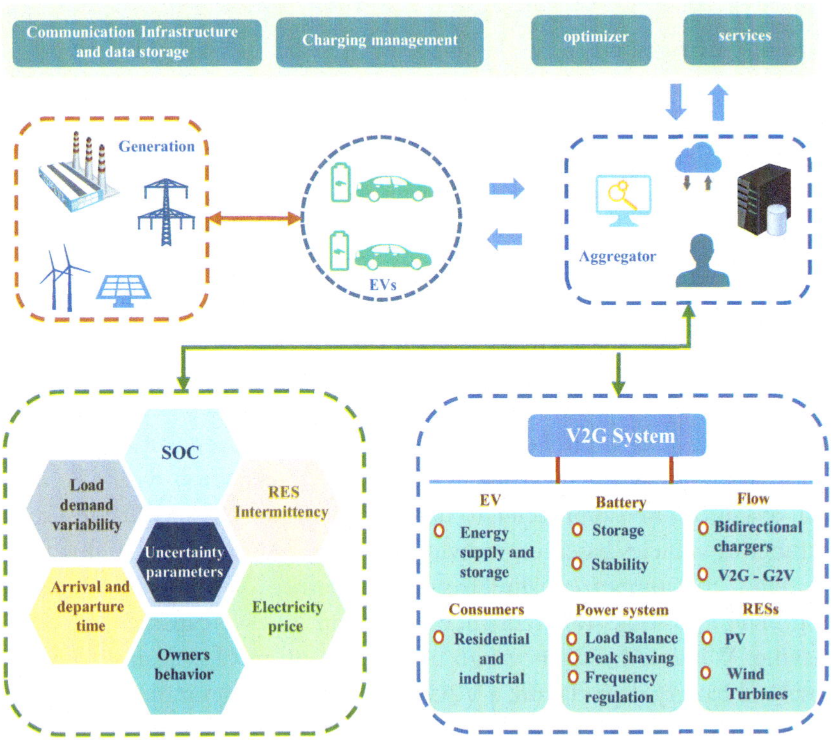

Figure 7.4 An aggregator scheduling activity with the R2V/V2G ecosystem with uncertainties.

"charging services providers," the electric grid requires intelligence to track the vehicles' recharging batteries in order to increase efficiency. An aggregator handles the EV fleet management problems listed below.

- CS backup power: When a plug-in EV needs to recharge its battery to a suitable level for a planned trip, the chosen charging station (CS) needs to have enough energy.
- Minimize the cost of charging: When recharging the battery, the energy sources may use variable charging prices. However, since prices fluctuate according to the grid's load, it is essential to provide enough energy at all affordable prices.
- Taking a lot of cars into account when localizing CS over an extended period of time is known as load handling.

Managing demand, real-time EV charging, and its impact on the smart grid are the objectives of charging scheduling in an energy management system. Typically, CS, EV, Road Side Units (RSUs), and Global Controller

(GC) are the nodes that communicate with one another during EV charging management.

The GC or central aggregator (CA) optimizes each requested EV charging scheduling under the global control charging scheduling mechanism. In order to determine the need for an EV through communication, the CA can obtain the information from the requested EV. Only information pertaining to the EV's electric parameters can be shared with the CA. Every EV makes its own decisions about whether and where to charge under a distributed control scheduling mechanism. In these situations, the requested vehicle shares with the CA its charging goals, including its comfort zone, charging frequency, and charging price. After this, the CA decides which CS is best for the vehicle and distributed control is the only way to achieve this. However, it may schedule and achieve sub-optimal CS if any requested vehicle shares inadequate information with CA. The goal of a distributed approach is to reduce reliance on GC and charge a large number of vehicles based on individual choices (Ahmad et al., n.d.).

(Fig. 7.5) summarizes value streams and market roles for V2G participation. The grid looks for frequency regulation, reserve capacity, and peak shaving. While EV owners weigh user rights and privacy against charging costs and battery degradation, a V2G aggregator manages multiple or single EV participation in peak-valley arbitrage, spot-market trading, demand response, and ancillary-service markets, which is depicted in (Fig. 7.5a). In the market and institutional interactions, which are shown in (Fig. 7.5b), an aggregator connects EVs and charging stations to the load-serving entity and the market to acquire and provide these services; consumers, the Distribution System Operator (DSO), Transmission System Operator TSO, and Generation Company (GENCO) transact through the market operator.

Applications for EV smart charging typically fall into two categories such as tracking and monitoring. Monitoring applications are used for real-time CS and smart grid analysis, supervision, and careful operation. On the other hand, tracking applications are used to generate event notifications in network infrastructure and predict changes in power generation. The following are some notable applications: remote reservation of CS, where a traveling EV uses RSUs to communicate its status and Estimated Time of Arrival (ETA) to a selected CS when a recharge is required; battery-swap services, which offer an automated platform to replace a depleted battery with a fully charged one; and predictive routing and scheduling that minimizes waiting times and avoids congestion at CS.

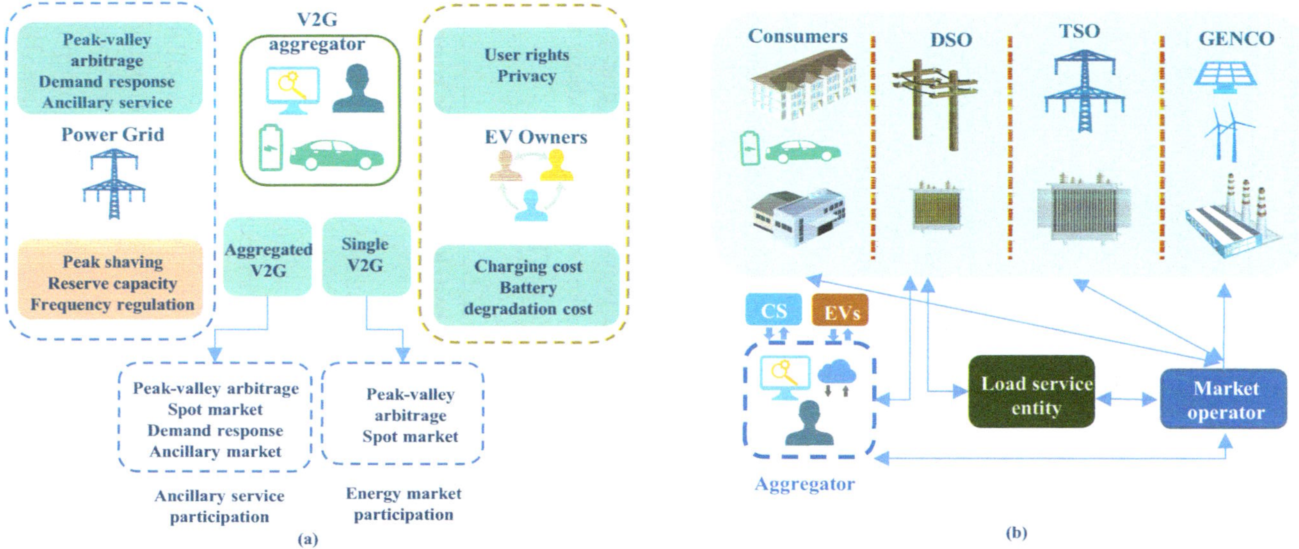

Figure 7.5 Value streams and market roles for V2G involvement: (a) Aggregated V2G and single V2G interactions, and (b): Market and institutional interactions.

7.2.4 Functionality, communication, and safety standards

V2G standards are still in the early stages of development and do not fully address every aspect of the system. These issues are made worse by supply chain constraints, cybersecurity, and insufficient service-oriented control systems. To encourage the safe, effective, and broad use of V2G-enabled EVs, regulators must address upcoming issues. The compatibility and safe operation of electric vehicle charging infrastructure in V2G systems are ensured by establishing communication and safety protocols (Das et al., n.d.). (Table 7.3) summarizes importantstandards for interconnection, safety, functionality, and communication protocols.

In (Kanellos, 2021), a distribution system network-constrained EV scheduling framework based on multi-agents is proposed, where each charging point agent injects active and reactive power into the network via V2G mode. In order to maximize the benefits of aggregators, stakeholders, EVs, and the grid and minimize grid load variation, a V2G trading system based on a hierarchical three-layer blockchain (data, operation, and transaction) is proposed in (Luo et al., n.d.) using the cellular technique. Reference (Aljaidi et al., 2022) suggested an effective algorithm based on Q-learning as well as a charging scheme with the goal of minimizing charging and travel expenses, assigning EVs to the best CS when their batteries need to be recharged, and preventing smart grid overload. The summary of energy trading and market model methods in the literature for R2V Smart Charging is provided in (Table 7.4).

7.3 Energy trading and market network mechanism with EV integration

There are three distinct pricing strategies for EV charging, frequency regulation, and load-leveling in electricity markets: fixed tariff, dynamic tariff, and demand response tariff. The spot price of electricity serves as the basis for EV charging dynamic tariffs. Under fixed tariffs, the wholesale market determines an EV charging lower fixed price. Demand response tariffs give EV owners the chance to use their vehicles to provide ancillary services in return for a reward. The most intricate of these three pricing schemes is the electricity dynamic pricing. This pricing type does not specify the total charging cost in advance because the electricity charging price fluctuates over time and is not known prior to the charging process. By controlling how they use electricity, dynamic pricing schemes are implemented to increase users' flexibility (Sadeghian et al., n.d.).

Table 7.3 Key standards for interconnection, safety, functionality, and communication protocols.

Standard name	Domain (purpose)	Brief Explanation
IEC 62196–1 & 62,196–2	Conductive charging hardware	Defines plugs, sockets, vehicle connectors/inlets for AC-DC interoperability.
IEC 61850-x	Substation communication.	Defines communication infrastructure for V2G systems
ISO/IEC 15118	V2G communication	Establishes communication protocols between EVs and grid.
IEC 61851 series	EV charging systems	Covers conductive charging, requirements, specs, and communication.
IEC 61140	Electric shock protection	Sets protection standards for V2G installations and equipment.
IEC 60529	Enclosure protection (IP Code)	Specifies environmental protection for V2G system components.
IEC 60364–7-722	Low-voltage EV installations	Covers electricity provision for EVs in V2G systems.
ISO 6469–3	EV safety on roads	Ensures safety against electric shock in public-road EVs.
ISO 15118-1-20	charger communication (smart charging-V2G)	Supports bidirectional power transfer and negotiates the Plug and Charge schedule for AC/DC charging.
IEC 61980	Wireless EV charging	Standards for wireless power transfer up to 1000V AC/1500V DC.
SAE J2293	EV energy transfer system	Covers architecture, energy, and communication for AC/DC/inductive systems.
SAE J1772	Conductive charging ratings	Defines equipment ratings for EV charging.
SAE J1773	Inductive charging	Sets standards for inductive charging compatibility in North America.
SAE J2847	Charging communication	Specifies communication requirements for EV infrastructure.
SAE J2931	Digital communication	Defines architecture and protocols for EV-related digital communications.
UL 1741	Inverters/controllers	Standards for grid-tied and standalone power system devices.
UL 1741 SA	Smart inverter grid stability	Defines modern smart inverter functions for DER integration.
IEC 63110	charger backend management (aggregator and DSO)	Interoperable smart charging and V2G are made possible by the protocol for running and managing charging infrastructure, which includes onboarding and configuration, monitoring, and remote control.

Table 7.4 The summary of energy trading and market model methods for ***R2V smart charging ecosystem in the*** literature.

References	Goal/Strategy	Algorithm & method	Results & description
Deb et al. (2022)	Comprehensive review of smart charging systems.	Smart charging system technologies, algorithms, and tactics, with an emphasis on optimization methods.	Identifies technical and financial obstacles in smart charging solutions and highlights best practices
Luo et al. (n.d.)	Trustworthy V2G market trading system.	Three-layer blockchain with smart contracts for data, operations, and transactions in a corporation;	Exhibits a lower peak-valley gap, low-latency settlement, and less than 2-year payback.
Aljaidi et al. (2022)	EV and CS assignment scheme.	Q-learning considering travel, SoC and charging costs	Reduction of overall energy costs; adaptation to changing urban conditions.
Mastoi et al. (2023)	Survey of charging-dispatch strategies with V2G.	More than 200 references are included	Identifies research gaps in cybersecurity, coordination, and standardization.
Geng et al. (n.d.)	Smart charging in transport and power systems.	An elastic price-responsive charging model that optimizes both traffic and power	Under user elasticity, operational costs are reduced and voltage quality is enhanced.
Aoudia et al. (n.d.)	Blockchain-enabled V2G trading in IoT environments.	Blockchain integration with IoT devices in V2G systems	Enhances efficiency, transparency, and trust through the integration of blockchain and IoT
Bharathidasan et al. (2022)	Cybersecurity and EV technologies in energy trading.	Blockchain technology for safe transactions; cybersecurity for EV energy trading.	Emphasizes how critical it is to protect EV energy trading networks.

Johnsen et al. (n.d.)	Market-based incentives for smart charging.	Review of grid-friendly charging techniques and market-based mechanisms for charging during off-peak hours.	Describes the ways in which market-driven incentives can lower energy costs and balance grid load.
Tirunagari et al. (2022)	Smart charging and V2G policy implications.	Technical, policy, and regulatory analysis of V2G and smart EV charging technologies.	Advantages and difficulties of V2G, especially with regard to infrastructure development and regulation.
Sameer et al. (n.d.)	Fair energy trading in blockchain-based smart grids.	Smart grid system powered by blockchain that prioritizes equity and justice in energy trading.	Enhances market efficiency and suggests a transparent and equitable energy trading model.
Gnann et al. (n.d.)	Smart charging's impact on EV market shares.	Using market-penetration models, the effect of smart charging on EV market shares in Germany	By facilitating wider EV adoption and balancing grid load during peak hours, smart charging boosts market share.
Baza et al. (2021)	Privacy-preserving blockchain-based energy trading for EVs.	Blockchain-based EV energy trading solution with privacy-preserving features	Protects user privacy when trading EVs, lowers energy costs, and makes transactions more transparent
Kim et al. (n.d.)	Incentive system for EV energy trading based on auctions.	An auction-based system that encourages charging stations to exchange EV energy	Efficiency, lowers consumer costs, and guarantees equitable energy distribution throughout the network
Chen and Zhang (2019)	Secure electricity trading and incentive contract for EVs.	Secure energy trading model with incentive contracts enabled by blockchain.	The model makes pricing and trading between EV owners and grid operators more transparent and equitable.

(Continued)

Table 7.4 The summary of energy trading and market model methods for ***R2V smart charging ecosystem in the*** literature.—cont'd

References	Goal/Strategy	Algorithm & method	Results & description
Huang et al. (n.d.)	EV charging station market mechanism based on blockchain technology.	Smart-contract clearing with decentralized bids and prices	Enhances transparency, lowers counterparty risk, and makes clearing easier
Rücker et al. (n.d.)	Effects of smart-charging strategies.	Scenario analysis for frequency restoration reserves using grid/market models.	Battery impact from the Frequency Restoration Reserve can be minimized with controlled charging
Sagawa et al. (n.d.)	Bidding agents for PV/EV users in the P2P market.	Agent-based market with automated bidding and learning	When compared to naive bidding, agent strategies improve market efficiency and participant utility.
Aghajan-Eshkevari et al. (n.d.)	Power management and EV routing optimization in coupled power-transport systems.	Combined route and charging optimization while taking distribution limitations into account.	Lowers energy and travel expenses and lessens grid overloads in comparison to decoupled planning.
Aghajan-Eshkevari et al. (2022)	EV services to support the grid.	A book chapter about control/market access and V2G/V2B/V2H services.	Indicates that EV fleets have the potential to provide profitable ancillary services.
Amiri et al. (n.d.)	Use of EVs for power-quality mitigation.	Reactive power and harmonic filtering converter control are reviewed.	When safety and interoperability are guaranteed, EVs can increase PQ

7.3.1 Architecture of energy trading

The smart grid's energy trading mechanism could be illustrated through a four-tier architecture, including the market tier, communication network tier, data acquisition tier, and energy trading tier, which are depicted in (Fig. 7.6). These levels make up the energy trading architecture, which is described in the following (Aggarwal et al., 2021a,b).

- Energy tradingtier: The energy trading layer consists of smart meters, energy nodes (EVs, charging stations, smart buildings), and energy aggregators (local aggregators, advanced metering infrastructure (AMI)). In order to oversee an energy exchange and offer the network communication services, the energy aggregators work as energy brokers.
- Data acquisitiontier: Through intelligent electronic devices, sensors, and monitoring devices, this layer gathers crucial data on energy distribution and consumption from various energy nodes. In a V2G system, sensors such as load, charging, and discharging status, and temperature are used to track the battery health of EVs.

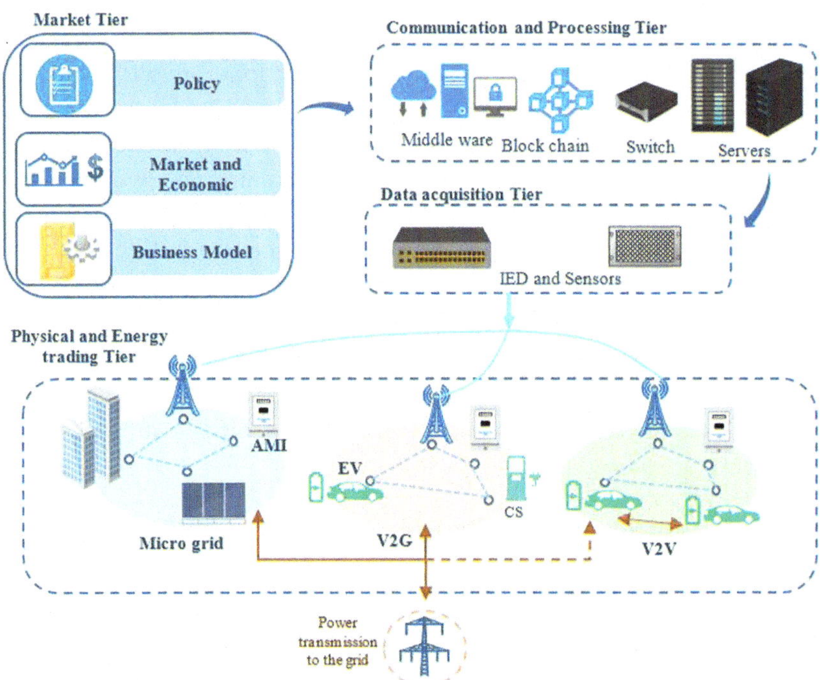

Figure 7.6 The architecture of EV grid integration energy trading in the smart grid.

- Communication network tier: The goal of information and communication technology (ICT) is to facilitate, regulate, coordinate, and oversee the energy exchange between EVs, charging stations, and the electrical grid. In the smart grid, this layer makes it easier for various energy nodes to exchange energy in real time.
- Market tier: The business perspective of energy trading in the smart grid is shown in this layer. The market consists of the retail market and the wholesale market, business models and policies. Energy buyers and sellers, as well as the DSO, play a significant role in the market domain. Bidding, decision-making, energy exchange, and market layer energy settlement are the key procedures. It also takes into account a number of variables, including smart grid energy trading pricing mechanisms, electricity tariffs, and investments.

Across tiers, incentives, compliance, and monetization are provided by policy, market, economic, and business-model blocks in this architecture. Data, identity, and control are managed via the communication and processing tier, which includes middleware, blockchain, switches, and servers. In the physical tier, there is an interface between the microgrid, V2G, V2V, and the power grid. The power system's physical components, such as transformers, feeders, smart meters, electrical loads, distributed energy resources, home chargers, and public charging stations, are all included in the power grid module. The V2V module is essential to determining how energy is allocated and priced in a V2V energy-trading framework. The V2V module serves as a mediator in this energy-trading framework, coordinating financial transactions and information sharing (Xu et al., n.d.). The principal benefits of V2G are as follows (Mwasilu et al., 2014).

- Grid benefits: Peak load reduction, improvement of power system reliability, and help to expand the power grid scale.
- Load aggregator benefits: The cost of services to support the operation of the grid, and Management fees paid by the user.
- User benefit: User Benefits are participants' Compensation costs and electricity cost reduction.

7.3.2 EV energy trading models and networks with P2P and smart contracts

There are several options available in P2P Energy Trading for prosumers who have extra electrical energy, including storing it for later use, transferring it to the electrical grid, or even selling it to other energy users. P2P energy trading is the general term used to describe this direct energy exchange between prosumers and consumers. Unlike consumers, who only use electricity, prosumers also produce and consume electricity. Both prosumers and consumers are motivated by the P2P energy trading of EVs through a block-chain technology framework.

A Self-executing code that operates on the blockchain without the assistance of a middleman or other third party is known as a smart contract. Peer trust is increased in P2P energy trading because the contract is executed transparently and cannot be altered. Code, rules, objects, and data models can be inserted into smart contracts to adapt them to particular needs. Depending on predetermined guidelines and conditions, smart contracts can automate a variety of payments and transactions between the parties. EV owners can view P2P electricity trading as a business model that functions as an online marketplace thanks to the blockchain-oriented charging system.

A centralized P2P energy trading system's primary drawbacks include its heavy reliance on regional laws, the enormous prosumers and consumers, market dynamics, protracted disputes, and the particular objectives of the involved parties. The P2P energy trading network is a decentralized network of energy exchange and transactions between consumers and prosumers that is carried out by individuals. Trades usually take place inside the distribution grid and are governed by regulatory supervision. Figure 7.7 illustrates the centralized and decentralized network P2P energy trading with blockchain technology (Khan et al., n.d.).

7.3.3 Smart grid energy trading incentive model

Theoretical techniques based on incentive economic approaches, with an emphasis on contract theory, game theory, auction theory, bargain theory, and dynamic pricing, are presented in this section (Aggarwal et al., 2021a,b).

Auction Theory: In the energy market, auctions are a way for buyers and sellers to exchange energy, which enhances their utilities by acquiring the goods. There are four different kinds of auction mechanisms, including the first-price sealed bid auction, the ascending-bid auction, the descending-bid auction, and the second-price sealed-bid auction. The ultimate cost of the goods used for trading is the outcome of an auction. Many auctioneers use evaluation criteria to determine the value of the goods for sale in an auction theory.

Price Theory: It is an effective strategy for getting customers to behave in the most cost-effective way possible. The smart grid load fluctuates over time based on the demand for electricity, which is comparable to the price

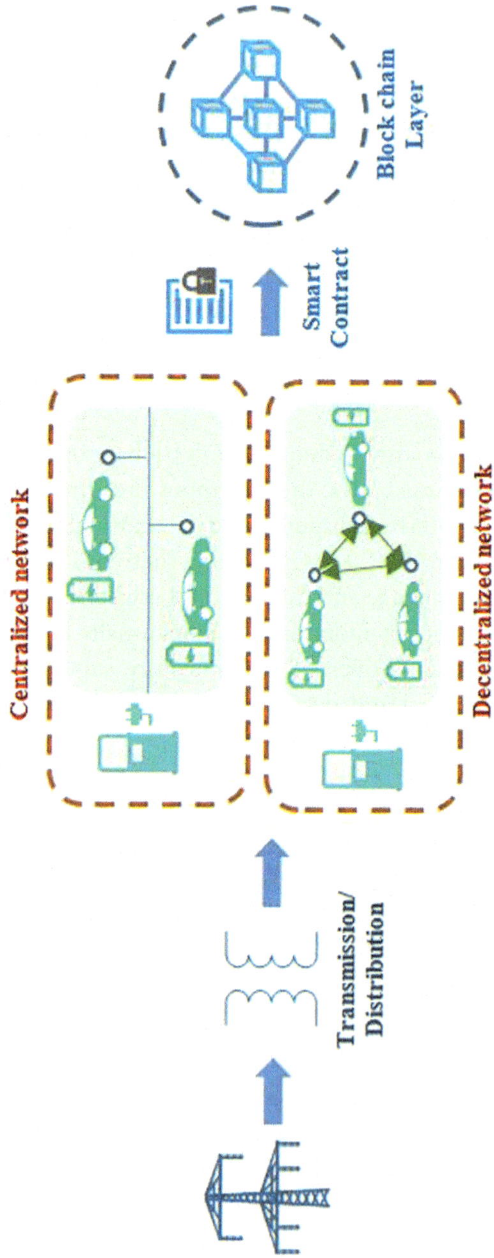

Figure 7.7 Centralized and decentralized network P2P energy trading with blockchain technology.

of electricity at various points in time. Three types of pricing schemes can be distinguished in this order, and they are explained as follows:

- Real-Time Pricing (RTP): Customers who use electricity on an hourly basis are typically charged the hourly rate known as RTP. Customers must be promptly and accurately informed of this pricing, which varies over time based on the state of energy demand. Therefore, it is the most effective kind to raise the energy markets' efficacy and efficiency.
- Time of Use (ToU) Pricing: Unlike RTP, this pricing is set in advance and remains consistent over an extended period of time. It remains constant despite daily fluctuations in the energy market. It can lower overall expenses for both customers and the utility. EVs can modify charging times, lower costs, and fill fewer demand hours by using the TOU price method, which also lowers the number of peak hours, which is a technique for regulating EV load at Charge Stations.
- Inclining Block Rate (IBR) Pricing: This type of pricing is intended for consumers and is determined by the amount of electricity used. At higher energy usage levels, it charges a higher rate per kWh, and at lower usage levels, it charges a lower rate. With the fixed thresholds, these levels are determined by the average amount of electricity used during a given time period.

In the following, Eq. (7.9) is the definition of the relationship between the pricing scheme and the demanded loads.

$$L=\beta\cdot P^{el} \tag{7.9}$$

Where β is a constant, P is the price of electricity, el is the price elasticity, and L is the demanded loads. To balance the load through two-way energy flow between EVs and the smart grid, the pricing could be applied, and the quantity and duration of their energy exchange in the V2G system are resolved by this theory. Reference (Morstyn et al., 2020) created a marginal pricing-based strategy to manage and control energy price uncertainty and local energy trading between producers and consumers.

Bargain Theory: It can be described as a process of negotiating during meetings between workers in order to come to a consensus or to enhance wages and working conditions in the markets for electricity. According to the bargaining theory, customers can negotiate energy use for the smart grid's preferred payment. In contrast to auction theory, which aims to maximize the utility function of both auctioneers and bidders, bargain theory focuses on reaching a self-executing and equitable agreement.

Reference (Kim et al., 2020) suggested a two-stage method to deal with the non-convexity of generalized Nash Bargaining for direct energy trading between several micro-grids and the market price clearance is ascertained in the method. In the following Eqs. (7.10–7.11) show a definition of the Nash bargaining problem for negotiations between N players.

$$\max \prod_{n \in N} \left(U_i^c - U_i^d \right) \tag{7.10}$$

$$\left\{ B_n^* \right\} \text{s.t.} \left(U_i^c \geq U_i^d \right), \forall i \in N \tag{7.11}$$

where U_i^c and U_i^d are the utilities of player I gained with and without collaboration, respectively, and B_n^* is the Nash equilibrium solution with the constraint that the utility of the player gained with collaboration is greater than the utility of the player gained without collaboration.

Contract Theory: Contract theory, which incentivizes participants under asymmetric information, can be a workable solution to the issue of asymmetric information in energy trading in energy trading. Given that there are N energy traders and that each trader has a type (a_i, b_i), where $i \in$ N. In this case, the reward for trading twice as much electricity is a_i.

The following criteria for incentive compatibility and individual rationality need to be met in order to design workable contracts.

- Individual Rationality: A contract complies with this constraint if each participant type's utility (U_i) is non-negative, which is as follows in Eq. (7.12).

$$U_i(a_i, b_i) \geq 0, i \in N \tag{7.12}$$

- Incentive Compatibility: This constraint is satisfied when a contract's ith participant achieves the maximum utility U_i possible, as follows in Eq. (7.13).

$$U_i(a_i, b_i) \geq U_j\left(a_j, b_j\right), i, j \in N, i \neq j \tag{7.13}$$

A system for separating energy suppliers for energy trading between an aggregator and electricity suppliers was proposed in (Amin et al., 2019). In order to enable energy suppliers to sell their energy at varying prices, they created an ideal contract-based system. The cost of producing the unit that maximizes the aggregator's overall cost determines their energy prices.

Game Theory: In the local energy market of the smart grid, there are producers (suppliers) and consumers (demanding users). The strategies of the other party may be impacted by changes in one party. Thus, game theory can be used to balance and analyze the energy trading strategies. There are two categories of game theory with respect to players including cooperative games and non-cooperative games. While cooperative games only allow players to play for self-enforcement, non-cooperative games allow players to compete with one another. Accordingly, P2P energy trading between prosumers and consumers can be facilitated by non-cooperative games. Instead, with the aid of a communication network, cooperative games can be used to enhance social welfare in energy trading. The three primary elements of a game theory are the action (Ai), the set of players (N), and the associated utility function (U_i). According to this theory, each player maximizes the U_i by selecting its Ai. In addition to its own actions, a player's utility function is also influenced by the actions of other players. The mixed-strategy profile $s_j = (s_1,.,s_n)$ has an expected utility U_i for player i. The definition of $U_i(s)$ in a normal-form game (N, A, U) is as follows in Eq. (7.14).

$$U_i(s) = \sum_{a \in A} U_i(a) \prod_{j=1}^{n} s_j a_j \tag{7.14}$$

In Fig. 7.8, a variety of trading models solution methods are classified, including mathematical models (MILP, convex optimization, metaheuristics, and stochastic programming); simulation and ML models (Reinforcement Learning (RL), such as Q-learning) for policy learning; and incentive models (auction, price, bargaining, contract, and game theory) (Aggarwal et al., 2021a,b).

Table 7.5 shows the consolidated taxonomy of smart-charging and energy-trading methods along with representative approaches and the primary objectives they address (Tirunagari et al., 2022).

To provide demand response, ancillary services, and energy trading under time-of-use or dynamic prices and via charging stations, a load aggregator coordinates customer resources (EVs), equipment, and communication/control systems while accounting for market incentives. Intermediaries such as financial services, energy-management platforms, and electricity sales agents enable transactions and risk management. Interactions with DSOs/TSOs and renewable sources deliver system and user-level benefits, including peak-load shifting, improved grid reliability, and lower billing costs (Mwasilu et al., 2014). (Fig. 7.9) shows the ecosystem of EV smart charging and V2G trading design schematic.

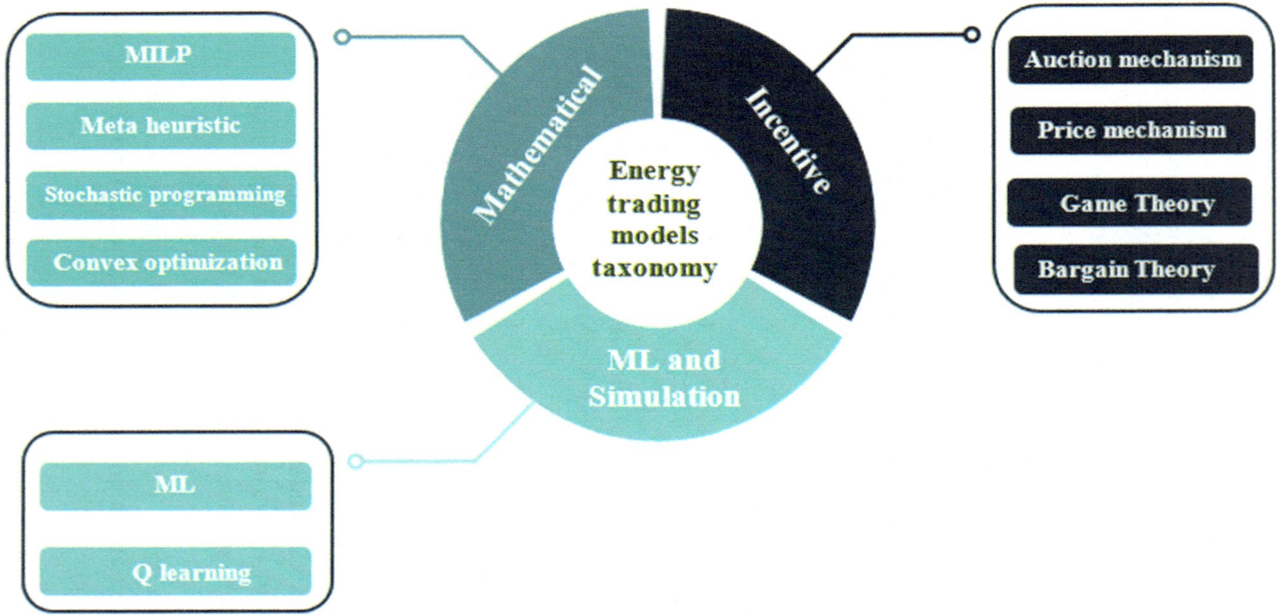

Figure 7.8 Energy trading models solution methods in the smart grid.

Table 7.5 Taxonomy of smart-charging and energy-trading methods.

Smart charging algorithm	Representative approaches & key References (Ref.)	Primary objectives
Probabilistic and statistical scheduling	Bayesian-Network (BN) real-time charging scheduler (Ren et al., 2020);	To optimize the percentage of fully charged EVs during assigned parking hours in the face of uncertainty, SoC statistics are utilized. Optimize real-time schedules and reduce CS operating costs.
Mathematical programming, game theory methods and optimization	MILP Nash-bargaining P2P trading (Zhong et al., 2021a,b); Robust two-stage day-ahead scheduling (Benders) (Deng et al., n.d.); Min-Cut/Max-Flow smart-contract settlement (Wang et al., 2021);	In order to reduce electricity costs, battery aging, peak load, and voltage violations throughout the network, charging should be designed as constrained optimization (power-grid, station, and vehicle limits). It achieves fair, robust energy allocation;
Meta-Heuristic and evolutionary search	NSGA-II multi-objective GA (Houbbadi et al., 2019); Ant Colony Optimization (ACO) station-selection (Moghaddam et al., 2018); GA-ISS for single/massive EV charging (Mao et al., n.d.);	Search-based approaches investigate various charging scenarios, modes, and solutions without listing them. Reduce cost, battery wear, and queueing.
Predictive control and price responsive algorithms	Expected Load Flattening Model Predictive Control with traffic patterns (Tang and Zhang, 2017);	Leverage short-term forecasts (arrival, parking time, price) to update charging power; maximum load-flattening while balancing grid demand and minimizing user cost.

(Continued)

Table 7.5 Taxonomy of smart-charging and energy-trading methods.—cont'd

Smart charging algorithm	Representative approaches & key References (Ref.)	Primary objectives
ML method, forecasting, and reinforcement learning-based smart charging	Bi-LSTM demand forecasting for P2P market (Jamil et al., 2021); Internet of vehicle (IoV) LSTM hybrid demand model (Kumar et al., n.d.); Markov-chain based EV load synthesis (Li et al., 2022); Q-learning load forecasting for EV stations (Dabbaghjamanesh et al., 2021); Deep-Q-Network scheduling for heterogeneous fleets (Hao et al., n.d.); Switched-Integral RL (SIRL) for V2G frequency control (Song et al., n.d.);	Forecast error reduction, increased scheduling precision, decreased over procurement, and a smooth net-load profile. Minimize charging cost, flatten load peaks; grid frequency regulation.
Blockchain-enabled energy trading	DV2G DAG ledger for fee-less micro-transactions (Hassija et al., 2020); Consortium blockchain and contract-theory incentives (Zhou et al., 2020); Hyper ledger Fabric smart-contract P2P market (Jamil et al., 2021);	Tamper-proof settlement; privacy and auditability; trading without per-transaction fees at both high and low latency.
Hybrid AI and blockchain solutions	ANN prediction and federated learning on blockchain ledger (Wang et al., 2020); AI-driven smart-contract adaptive pricing (conceptual extension of (Wang et al., 2020)).	Keep raw data local and combine auditable, decentralized settlement with high-accuracy forecasts.

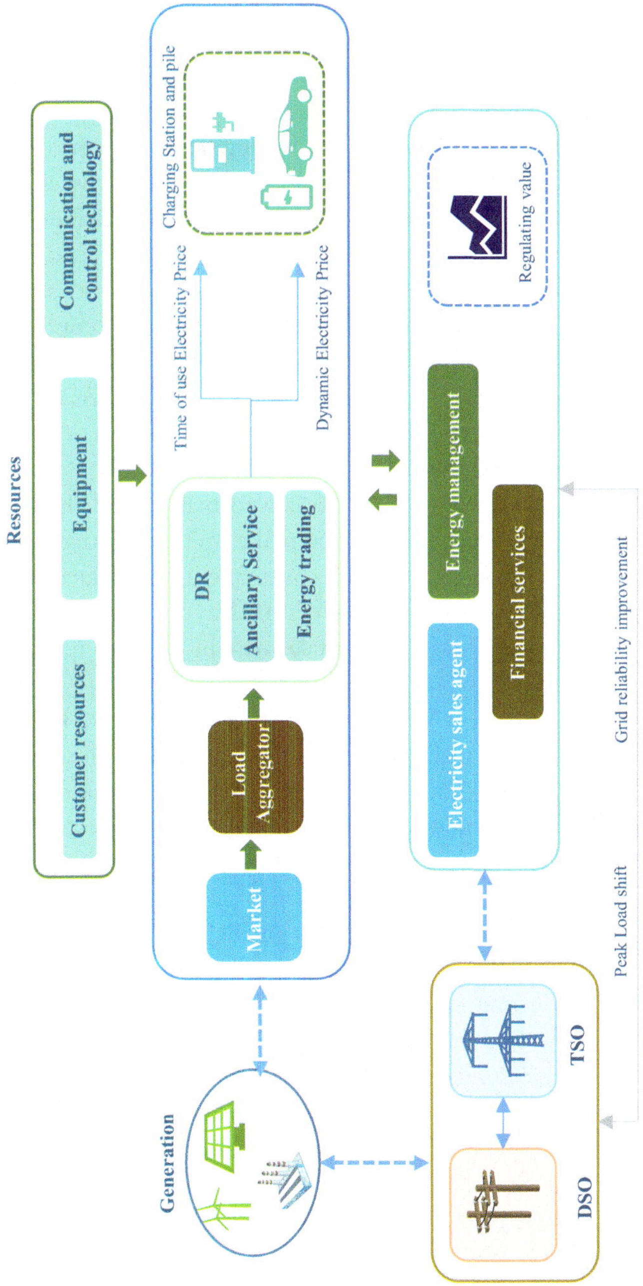

Figure 7.9 Ecosystem of EV smart charging and V2G trading design schematic.

7.3.4 Ancillary services offered by electric vehicles to the power system

Ancillary services are defined as a range of supportive services that could increase the stability, efficiency, and reliability of electricity transportation and could help in voltage regulation and decrease power quality, power loss, and peak load in the power grid. These services support power imbalances between the supply and demand sides, maintain the intended power flow and associated direction of electricity within transmission or distribution grids, and aid in system recovery following an incident or failure. The dispatching center first establishes the EVs' response mechanism and control strategy after identifying the ancillary services that EVs must offer. Second, the grid layer's control strategy is chosen to produce the appropriate control signal based on the kind of ancillary services. Lastly, the load aggregator receives it for implementation and control (Sadeghian et al., n.d.). As a crucial component of the electricity market, the spot market offers a way to buy and sell electricity immediately. The day-ahead market and the real-time market are the two main components that make up the spot market, which operates on supply-demand balance the principle and can facilitate more effective optimization of electrical resources. By buying or selling electricity back to the grid, V2G technology can take part in the spot market (Wan et al., n.d.).

Different services such as Voltage Regulation, Frequency Regulation, Spinning Reserve, power quality improvement, Black Start, loss reduction, peak shaving, and congestion management could be considered in this group. Some of the important ancillary services that EVs can offer are also shown in (Fig. 7.10). In the following, prominent services are explained (Wan et al., n.d.):

- Frequency Regulation: One of the key ancillary services markets is frequency regulation, which also happened to be the first market to use energy storage for commercial purposes. V2G technology is capable of providing comparable services. Consequently, V2G aggregators play a crucial role in the V2G technology participation business plan in the frequency regulation market. A numerical simulation method using time series data was proposed in (Jie et al., 2024) to better observe the contribution of EV clusters to the power system's frequency regulation with the goal of addressing the issue of a large number of RESs being connected to the grid.

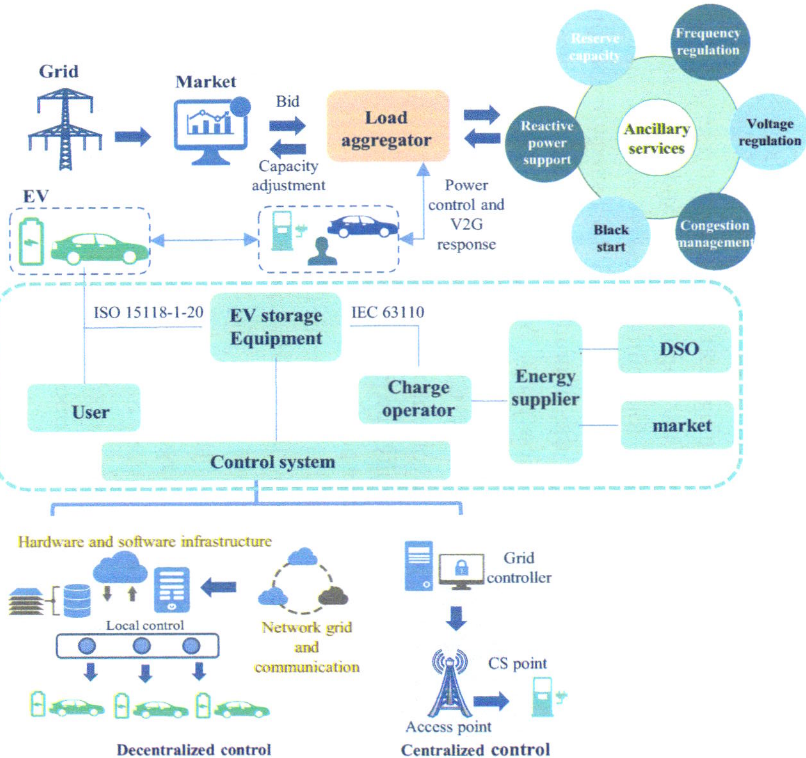

Figure 7.10 Control architecture for EV smart charging and V2G, considering ancillary services.

- Voltage Regulation: The distribution network voltage stability is greatly impacted by the large number of RESs that are currently connected to the power grid. Reactive power can be released from the EV battery via the charging piles' power electronic device. EVs are limited to supporting voltage within a distribution network due to the limitation caused by reactive power. This makes it difficult for V2G technology to enter the voltage regulation market. Reference (Liu and Liang, 2019) examined the role of V2G technology in the market for voltage regulation and suggested a method based on discounted stochastic multiplayer games to analyze the competition between V2G aggregators and determine the best course of action for each of them.
- Spinning capacity: The spinning reserve market is a reserve capacity-based market mechanism. Fast response times and high emergency backup power are essential for spinning reserve sources, and V2G technology satisfies these requirements. The V2G aggregators can profit from their participation in the spinning reserve market and use that money to fund their business plans. A two-layer optimal scheduling method was developed in (Li et al., 2021) based on the community-integrated energy system of EV CSs. The numerical findings demonstrated that EVs' active involvement in spinning reserve service could greatly boost the community-integrated energy system's economy.
- Black Start: In order to start power for other vital pieces of equipment or power plants in the event of a blackout, the black start power sources must be able to start on their own. EV batteries can be used as black start power sources and as energy storage devices that can release power without an external power source. As black start power sources, EV clusters can be made profitable by V2G aggregators. It is important to remember that black start emergency situations are extremely uncommon and that there are more social advantages to using V2G technology as a black start power source than there are financial ones (Wan et al., n.d.).

Mechanisms for trading electric power that take into account the requirements of various V2G stakeholders, considering the ancillary services are provided in Table 7.6 (Du et al., n.d.).

There are several ways that V2G technology can take part in the electricity market. The most popular method involves V2G aggregators indirectly participating in the electricity market, which could lead to their receiving a share of the profits. Individual-V2G and industrial-V2G are the names of the two additional direct participation modes in the electricity market that are available with V2G technology. The most popular mode is

Table 7.6 Trading Mechanisms considering requirements of various V2G stakeholders, considering ancillary services.

References	Market type	Objectives/Strategy	Pricing method
Luo et al. (n.d.)	Spot market (Day-ahead markets)	The market costs of aggregators minimization	Intelligent contracts in a hierarchical blockchain-based power trading model.
Gan et al. (2024)	Ancillary services (Peak shaving)	Load imbalance minimization The power grid profit maximization	Stochastic planning and deterministic optimization of EVs' (or aggregators') response to grid peaking demand.
Zheng et al. (n.d.)	Ancillary services (Frequency regulation)	Maximizing the profits of EV users, aggregators, and the V2G system	A deterministic model that takes demand response and frequency regulation into account.
Lotfi et al. (2024)	Ancillary services (Spinning reserves)	The operating profit of aggregators maximization	Game theory based models for cooperative and independent decision-making.
Zhang et al. (2021)	Spot market (Intraday markets)	The operational cost based on real-time pricing minimization	Decentralized power trading model using smart contracts and blockchain technology.

aggregated-V2G, in which EVs are combined via V2G aggregators to create a VPP or take part in the station-level energy market. Individual-V2G can only participate in the spot market and peak-to-valley arbitrage because ancillary services need a specific capacity to do so (Wan et al., n.d.).

7.3.5 EV charging and discharging control structures in the smart grid

Using the following three complementary control topologies, which are selected to balance system-wide optimality, scalability, resilience, and privacy, EV charging and discharging can be indicated (Aghajan-Eshkevari et al., n.d.).

- Centralized control structure: In a centralized control structure, EV owners relinquish control over EV charging/discharging, and aggregators usually oversee and regulate EV charging and discharging behavior directly. Aggregators first compile each EV's data and charging needs under this scheme. The aggregator determines the charge or discharge rate of EVs at each time based on a specific purpose and the network conditions. To maximize the EV's charging or discharging rate, the aggregator lastly transmits a control signal to the controller that is installed on each EV's charger.
- Decentralized control structure: EV owners decide whether to charge/discharge the battery in decentralized control, as opposed to centralized control, based on their own goals, which are typically to reduce charging expenses. Thus, with the aid of pricing strategies, the aggregator or system operator can indirectly regulate the charging/discharging of EVs behavior and move EV charging from peak to off-peak hours with appropriate financial incentives.
- Hierarchical control structure: Both centralized and decentralized control are characteristics of the hierarchical control framework, which has advantages over both in terms of computational load and communication network requirements. The structure of hierarchical control typically has two layers. All EV aggregators are scheduled by a central controller, like a DSO, in the upper layer. Each aggregator in the lower layer is in charge of managing multiple EVs and planning when each one should be charged and discharged.

In Fig. 7.10, a load aggregator mediates the grid market interface by bidding, capacity adjustment, offering ancillary services, and providing EV assets with power control V2G set points. While IEC 63110 controls charger-backend interactions between the charge operator, energy supplier, DSO, and market, which are regulated by a central control system, ISO 15118–20 regulates EV charger communication within the EV domain. Depending on operational and reliability requirements, deployments can follow either decentralized control or centralized control.

After examining blockchain technology and fixed contracts (Wang et al., 2019), enhanced the durability and efficiency of energy trades by utilizing edge computing and smart contracts. The authors used the backward induction method and the Stackelberg game to solve the computational problem. Two auction mechanisms that guarantee economic properties were proposed by the authors in (Zhong et al., 2017) for a two-layered V2G architecture. To effectively manage energy trading in this context, some researchers have applied pricing theory. Reference (Wu et al., 2015) suggested a pricing theory-based smart microgrid model for regional energy traders. Energy trading under this model benefits both producers and consumers. They have employed two-layered optimization algorithms, where a bottom-layer optimization explains how producers and consumers trade energy based on the price that is announced. A top-layer optimization, on the other hand, explains how local energy traders profit from the advantages of energy providers and consumers. In the same way (Morstyn et al., 2020), created a marginal pricing-based strategy to manage and control energy price uncertainty and local energy trading between producers and consumers. Theenergy trading and market model methods for *R2V smart chargingin the* literature are provided in (Table 7.7).

7.4 Optimization techniques and advanced methods in R2V smart charging ecosystem

The efficiency of interactions between EVs and the grid has greatly increased in recent years due to the growing demand for the application of ML in the V2G field. V2G greatly enhances the grid's capacity for frequency control by dynamically modifying the EV charging and discharging procedures, providing strong support for the power system's sustainability. Optimalscheduling and energystorage scheduling improve the system's energy efficiency in addition to guaranteeing system stability. In V2G systems, energy storage scheduling is primarily used to manage grid load fluctuations and optimize the bidirectional energy flow between EVs and the grid. Electric vehicle aggregators can compete with other market participants by using carefully crafted bidding strategies, which optimize their energy

Table 7.7 The summary of energy trading and market model methods for ***R2V smart charging ecosystemin the***literature.

References	Goal/Strategy	Algorithm & method	Results & description
Morstyn et al. (2020)	P2P trading combined with locational marginal price and probabilistic distribution.	Optimal power flow with distribution locational marginal price settlement that is chance-constrained.	Cuts expected congestion cost by 23%; enables 20% more renewable uptake.
Aggarwal et al. (2021b)	Lightweight blockchain authentication in V2G trading.	Blockchain with permissions that incorporate mutual authentication.	Reduces computational load by 47% preserving privacy.
Li and Hu (2020)	Optimizing EV charge and discharge trading through iterative two-layer optimization.	On the consortium blockchain, real-time allocation and day-ahead dispatch are implemented.	System cost decrease: 11.4%; net-load variance decrease: 18%.
Liang et al. (2022)	Centralized network; reliable blockchain technology for reliable EV-grid commerce.	Smart contracts based on blockchain; performance and security evaluation.	Balances demand and supply in the V2G mode, secures transactions, and is resilient to common attacks.
Al-Obaidi et al. (n.d.)	Centralized network; bidirectional smart charging with various pricing plans that take user preferences into account.	Schedule-based optimization that combines grid services, P2P exchange, and user utility with real-time energy management for EVs with Res.	Enhances aggregator revenue and user satisfaction in comparison to unidirectional or non-trading baselines within grid boundaries.
Saha et al. (2021)	Centralized network; Secure the IoE–EV interface using blockchain.	Access control and lightweight cryptography are features of the blockchain security architecture.	Effectively cuts down on energy waste while guaranteeing confidentiality, integrity, and non-repudiation.
Zaman et al. (2022)	P2P trading platform on a multilayer, semi-permitted blockchain; decentralized network.	Smart contracts and permissioning in a layered design (data, control, and settlement).	Prototype indicates higher throughput and governance flexibility than monolithic designs.
Moniruzzaman et al. (n.d.)	Cooperative game theory based on blockchain technology and decentralized networks.	A cooperative game for fair P2P trading.	Provides improved privacy and scalability, lowers negotiation overhead, and generates stable, transparent payoff allocations.

buying and selling decisions. Pricing strategies are essential for the implementation of V2G because they provide real-time data on electricity prices, allowing for precise control over when electric vehicles should be charged during times when demand for electricity is low and when power should be returned to the grid during peak hours. This effectively optimizes the allocation of grid resources and enhances overall grid operation efficiency.

7.4.1 Optimization objectives of EV charging and discharging in the smart grid

The optimization goals for scheduling EV charging and discharging are provided in this section. It frames how operational objectives, environmental, and economic goals could be evaluated quantifiably. The goals of demand-side management are to minimize peak demand and peak-to-average energy ratios, maximize the use of local energy generation, and minimize customer discomfort. Smart charging can maximize fully charged EVs, minimize energy losses, and increase the grid load factor, in addition to the other advantages it offers power grids. Additionally, smart charging helps to accomplish socio-environmental goals such as lowering CO2 emissions and financial goals like maximizing benefits for aggregators and minimizing generation costs (Bems and Anvari-Moghaddam, n.d.). In Fig. 7.11a, the overview of operational, economic, and environmental objectives of EV charging and discharging is illustrated. EV smart charging procedure objective function based on minimization and maximization classification is depicted in Fig. 7.11b.

Table 7.8. EV services can be divided into three categories: active, reactive, and RESs integration power support. Frequency control services, load variance reduction, valley filling, peak shaving, loss reduction, and voltage regulation are all examples of active power support. By adding reactive power, EVs can also lower losses and control the voltage in the distribution network. Additionally, EVs can facilitate the widespread integration of RESs, like solar and wind, into the power system. In the following, each of these services will be classified into two main taxonomies: operational objectives and economic and environmental objectives of EV charging and discharging. In Table 7.9, operational objectives, including services, description, and their applicability example from the literature are provided. Likewise, economic and environmental objectives are provided in Table 7.10 (Aghajan-Eshkevari et al., n.d.).

Coordinated charging is an energy management strategy that optimizes the power and timing of charging and discharging modes. EV battery

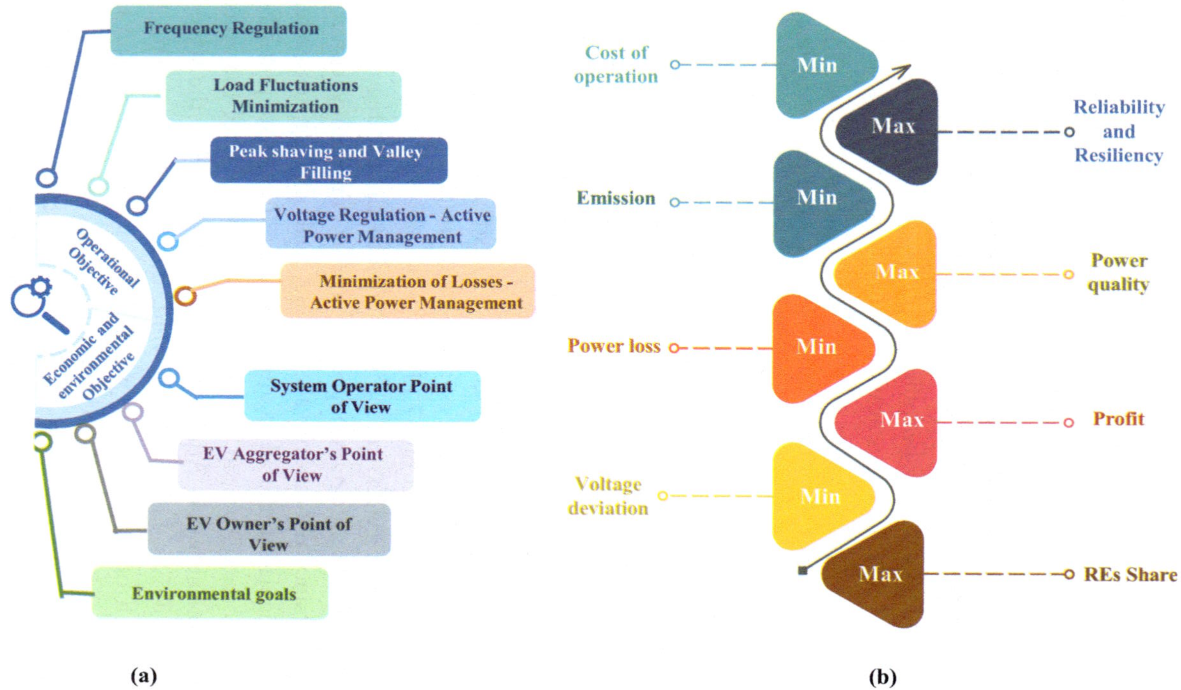

Figure 7.11 (a): The overview of operational, economic, and environmental objectives of EV charging and discharging, (b): EV smart charging procedure objective function based on minimization and maximization classification.

Table 7.8 The summary of blockchain methods in R2V smart chargingenergy trading from the literature.

References	Goal/Strategy	Algorithm & method	Results & description
Wang et al. (2021)	Distributed energy trading management.	min-cost flow scheduling with blockchain smart contracts	A 8% cost reduction is demonstrated in actual data and auditable energy flows
Jamil et al. (2021)	P2P energy crowd sourcing with prediction.	Bi-LSTM forecasting, hyperledger fabric, and smart contracts	Approximately 500 TPS throughput, with a latency of less than 2s, enhances scheduling with a low RMSE.
Hassija et al. (2020)	Enable scalable V2G micro-transactions.	DV2G ledger and game theory-based pricing mechanism	Compared to Proof of Work, it has a low confirmation latency and a throughput about 30 times higher
Zhou et al. (2020)	Secure and incentive V2G trading.	Blockchain and contract theory incentives; Stackelberg game	Enhancement of social welfare and reduction of trading delay.
Wang et al. (2020)	ML-based EV power management platform.	Blockchain ledger load prediction using ANN and federated learning	Reduces power fluctuation; prediction with $R^2 = 0.938$
Krishnamoorthi et al. (2022)	Safe blockchain energy transactions framework.	Blockchain-based architecture; smart contract; threat analysis	Maps security vulnerabilities and offers conceptual design principles.
Khorasany et al. (n.d.)	Location-aware P2P energy market.	Anonymous Proof-of-Location;	Reducing line losses by 15% and trading costs by 12% while maintaining privacy.

Table 7.9 Operational objectives of EV charging and discharging.

Service	Description	Application in the literature
Frequency Regulation	Because EV batteries respond quickly, they are utilized to deliver this service. Stability and economic concerns are the primary challenges.	In order to provide a frequency regulation the use of EV batteries was investigated in (Das et al., n.d.).
Minimizing Load Fluctuations	When load fluctuations are minimized, the system load rises during non-peak times (charging EVs) and falls during peak times (discharging EVs using a V2G capability), improving the power system's overall efficiency and helping to flatten the system load curve. The objective function for reducing load fluctuations is provided in Eq. (7.15): $\min \sum_{t=1}^{T} \left(P_{c.t} \times N_{c.t} + P_{\mathrm{Loadt}} - P_{d.t} \times N_{d.t} - \overline{P}\right)^2 \quad (7.15)$ $P_{c.t}$: EV charging power; $P_{d.t}$: EV discharging power; $N_{c.t}$: the number of EVs in charging mode; $N_{d.t}$: charging mode number of EVs; P_{Loadt}: baseload at time t; $\overline{P}$: load during a day without EV	The particle swarm optimization algorithm is used to address two objectives in a multi-objective optimization problem that is presented in (Wang et al., n.d.): scheduling system operator cost minimization and load variance minimization.
Peak shaving and valley filling	While valley filling involves adding demand to the grid during off-peak hours, peak shaving involves lowering the grid's peak load by managing the load. Peak shaving EV charging and discharging techniques minimize the need for grid investments to expand equipment capacity. Peak shaving and valley filling objective functions are provided in Eq. (7.16) and Eq. (7.17), respectively. $\min \sum_{t=1}^{T} \left(P_{\mathrm{peak.t}} - P_{\mathrm{target.t}}\right) \quad (7.16)$ $\min \sum_{t \in T_{\mathrm{off-pak}}} \left(P_{\mathrm{target.t}} - P_{\mathrm{base}}\right) \quad (7.17)$ $P_{\mathrm{peak.t}}$: The peak load; $P_{\mathrm{target.t}}$: target load at time t; P_{base}: grid's base load.	An EV real-time smart charging algorithm installed at commercial and industrial locations is presented in (Zhang et al., 2018) to decrease Peak demand.

Voltage regulation - active power management	During times of peak load, uncoordinated EV charging within the grid may result in a drop in voltage and an increase in load. Controlling EVs' active or reactive power can lower the incidence of these issues. In this regard, Eq. (7.18) and Eq. (7.19) could be indicated $\min \sum_{i=1}^{N_{bus}} (V_r - V_i)^2 \quad (7.18)$ $\min \sum_{t=1}^{T} \sum_{i=1}^{N_{bus}} \left\| \frac{V_{t,i}^-}{V_{t,i}^+} \right\| \quad (7.19)$ V_i: voltage of bus i, ; V_i: the reference voltage considered (1 p.u.), ; N_{bus}: the number of grid buses; $V_{t.i}^-$ and $V_{t.i}^+$ are the negative and positive sequence voltages at moment t and bus i, respectively.	In (Islam et al., 2020), a multi-objective optimization approach that takes into account power quality concerns like voltage imbalance, energy loss, and bus voltage issues is presented for the coordination of EV and distributed generation.
Minimization of losses - active power management	A higher peak load and disorganized EV charging will also result in more network losses. The network's power losses and operating expenses can be decreased by controlling the active and reactive power of EVs. Eq. (7.20) provides the objective function associated with power loss minimization. $\min \sum_{t=1}^{T} \sum_{l=1}^{N_{lne}} R_l \times I_{l.t}^2 \quad (7.20)$ N_{lne}: the number of network lines; R_l: the resistance of the lth line; $I_{l,t}$: the current of the lth line at time t	In (Azad-Farsani et al., n. d.), the local marginal price in fleets and at DG-connected buses is used to coordinate EVs and DGs in day-ahead electricity markets. The goal is to reduce the costs associated with the distribution grid, including those related to EVs, reliability, and losses.

Table 7.10 Economic and environmental objectives of EV charging and discharging.

Service	Description	Application in the literature
System operator point of view	The system operator can integrate more REs into the grid, minimize the cost of starting, stopping, and fueling generators, and cut down on power purchases numbers from the upstream network during peak hours with the help of optimal EV management. The objective function for reducing the grid's operating costs in the presence of thermal units, renewable energy sources, and electric vehicles are present is stated in Eq. (7.21) (Li et al., 2022): $$\min \sum_{t=1}^{T} P_t \times \pi_t + \sum_{t=1}^{T} \sum_{k=1}^{K} \left(C_{k.t}^{DG} + SUC_k \times u_{k.t}^{on} + SDC_k \times u_{k.t}^{off} \right) + \sum_{t=1}^{T} \sum_{i=1}^{N_i} P_{i.t}^{Dch} \times C_{i.t}^{Dch} \quad (7.21)$$ P_t: power purchased from the grid; π_t: the price of electricity at time t; $C_{k.t}^{DG}$: fuel costs; SUC_k: start-up cost; and SDC_k: start-down cost of DG k; N_i: the total number of EVs; $P_{i.t}^{Dch}$: the discharge power of EV i at time t; $C_{i.t}^{Dch}$: the discharge price of EV i at time t; $u_{k.t:\ on}^{on}$states of DG k and $u_{k.t}^{off}$: off states of DG k respectively.	An ideal active and reactive power exchange between the power grid and EVs is suggested in (Shahkamrani et al., n.d.), which boosts the advantages for both grid operators and EV owners at the same time.
EV aggregator's point of view	To provide the energy that EVs require, aggregators can make money by implementing the right strategies and engaging in a variety of markets, including the energy and ancillary services markets. Additional battery deterioration happens when the EV's battery is discharged into the power grid. To satisfy EV owners, the aggregators must give them a cut of their earnings.	In (Tookanlou et al., n.d.), a comprehensive day-ahead scheduling framework is developed for electric vehicle operation, including types of stakeholders: electric vehicles, retailers, and charging stations.

EV owner's point of view	By charging their EV batteries during times when energy prices are low and discharging them during times when energy prices are high, EV owners can lower their charging expenses and even make money. Eq. (7.22) provides the objective function for reducing the EV owner's charging expenses. $\min \sum_{t=1}^{T} \left[\left(P_t^{\text{dharge}} \times C_t^{\text{charge}} \right) - \left(P_t^{\text{discharge}} \times C_t^{\text{discharge}} \right) + C_{\deg}(E_{\text{trans}}) \right] \quad (7.22)$ P_t^{dharge}: the charging rates of the EV battery at time *t*; $P_t^{\text{discharge}}$: discharging rates of the EV battery at time *t*; C_t^{charge}: the charge and prices of the EV power; $C_t^{\text{discharge}}$: discharge prices of the EV power; $C_{\deg}(E_{\text{trans}})$: the battery's destruction cost as a function of the energy	In order to reduce EV owners' operating expenses while keeping feeder voltages within predetermined bounds, a centralized charging and discharging control method is suggested in (Nimalsiri et al., n.d.).
Environmental goals	In certain situations, charging and discharging strategies for EVs have been designed with the objective of minimizing greenhouse gas emissions, such as CO_2. The power system's structure, the weather, and the EV's economic power are some of the variables that affect greenhouse gas emissions from EVs. The objective function for minimizing CO_2 emissions, which includes the CO_2 emissions rate in the main system (Em^{grid}) system at time interval, is shown in Eq. (7.23). $Em^{\text{grid}} = \sum_{t=1}^{T} E_{CO_2}^{\text{grid.t}} \times P_{\text{grid}}(t) \quad (7.23)$	In order to improve the environmental performance and economy of parking, a time-of-use strategy in the demand response program is suggested in (Jannati and Nazarpour, 2019), along with a multi-objective model for smart EV parking based on PV panels.

specifications, charging station capacity, and technical limitations of the electrical grid are constraints that need to be considered. In addition to the main objective of EVs connecting to the grid, which is to obtain enough charge for the next trip, other objectives can be met to optimize the EV charging/discharging process, such as reducing voltage imbalances, minimizing losses, load flattening, frequency regulation, minimizing the charging cost for the EV owner, reducing greenhouse gas emissions, etc. (Fig. 7.12) shows the overview of objective functions in charging management coordination.

7.4.2 Block-chain based energy trading

Blockchain technology is a distributed ledger technology (DLT) that securely and irrevocably records P2P energy trading data over a network. It is made up of a series of blocks that are connected to digital currencies and contain transaction data. Because the DLT ensures that no third parties participate in the transaction process, blockchain technology is more dependable and secure for P2P energy trading (Khan et al., n.d.). Blockchain guarantees tamper-proof payment transactions, immutability, security, and privacy in smart grid energy trading. It permits distributed public information storage, verification, and exchange. Without the intervention of a central authority, it offers user anonymity, verifiable historical events, and protection against information manipulation and alteration. Compared to a traditional platform, a distributed platform offers greater identity privacy and authentication for energy transactions (Aggarwal et al., 2021a,b). Because of its decentralized architecture, blockchain does not require a central command authority to resolve business disputes between energy trading participants, including power grid stations, energy charging stations, and EVs. P2P technology enables direct communication between electric vehicle owners in order to conduct business. Adversarial and dishonest drivers in the electric vehicle network are unable to alter, remove, or update the blockchain's existing records because the transactions and data it stores are unchangeable. The local drivers can also be encouraged to trade V2G energy through incentives such as energy tokens. Blockchain technology can help automate and secure the use of digital signatures, smart contracts, and efficient consensus algorithms in V2G energy trading. It can guarantee that the owner of an electric vehicle who sells energy to the smart grid receives cryptocurrency-based incentives right away in their wallet. Additionally, a smart grid can use smart contracts to aggregate the energy

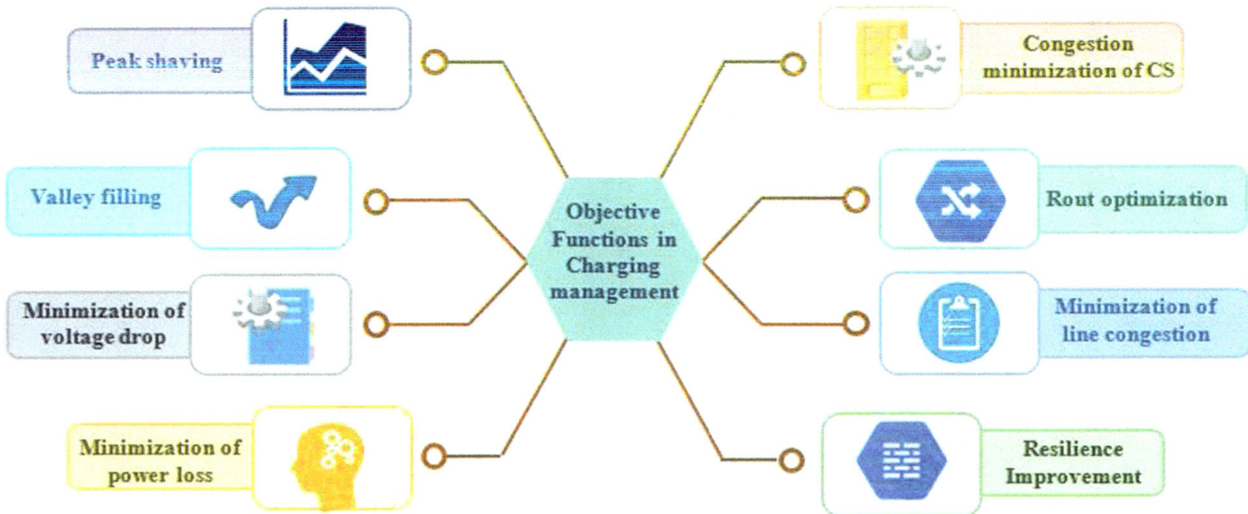

Figure 7.12 The overview of objective functions in charging management coordination.

demands of multiple electric vehicle owners by using smart contracts (Al-Saif et al., 2021). To guarantee the system's reliability, they are able to efficiently monitor grid stability and optimize resource allocation. On the other hand, decentralized P2P trading networks operate by using distributed ledger technology, like blockchain, in which a network of users validates transactions without the involvement of a central authority. Decentralized network implementation and maintenance still present technical challenges and are seeking a reliable architecture utilizing blockchain and state-of-the-art technologies (Khan et al., n.d.).

7.4.3 Blockchain technology requirement

Key requirements for blockchain technology are outlined in this section to protect data and maintain trust in decentralized environments, including data security and privacy protection, and privacy objectives.

Since any data privacy breach can result in identity or information theft, data privacy assurance is one of EV communication networks critical requirements. The centralized data storage makes the system vulnerable to attacks since it gives system members access to the energy trading parties' credentials. Since there is no risk of data hacking, alteration, or security breach, blockchain is regarded as an appropriate technology for energy trading in electric vehicles. Participants in the framework for electric vehicle energy trading need to share and process energy trading-related data in a completely safe and dependable manner. Device spoofing, data manipulation, and fraudulent user authentication are frequently the outcomes of the centralized systems used to manage energy-trading data. For participants in EV energy trading to operate safely, dependably, and with confidence, the stakeholders in these applications demand that energy trading-related data be extremely accurate, transparent, and trustworthy (Al-Saif et al., 2021). Considering security and Privacy objectives, anonymity, authentication, integrity, and transparency can be described as follows (Iqbal et al., 2021):

- Anonymity: Each electric vehicle uses a distinct public key to communicate with the aggregator instead of its true identity, preventing malicious users from learning the identity of the EV. In order to prevent linking attacks, an electric vehicle can also change its public key after every round by using the updated private key that the aggregator sends, with the goal of determining the actual identity of the EV.
- Authentication: The mutual authentication technique employed in this suggested approach enables the aggregator and EVs to authenticate one

another, ensuring that only authorized users are able to access the network.

- Integrity: After every round is finished, the transaction is saved in a block in the blockchain network, each of which has both its own hash and the hash of the block before it. Consequently, an attacker is unable to change or modify any data within the block of any network node.
- Transparency: Every network node can see the data in the blockchain technology with a private key, a software developer or service provider on the network can view and track the transaction data within the block without changing it. This makes it simple to identify and track down any malicious data alteration.

To guarantee the overall security and stability of the V2G system, various nodes must be carefully evaluated and deployed for cybersecurity defense. One of the main targets of cybersecurity defense is endpoint devices, like charging stations and electric cars. These devices must be protected from malware infections and physical assaults. Device authentication, data encryption, frequent security updates, and intrusion detection systems are examples of specific security measures (Xie et al., n.d.).

The block diagram architecture of a blockchain-based P2P energy trading system, including energy trading transactions from energy generation to consumers, that allows energy prosumers to sell energy to customers, is shown in (Fig. 7.13). This Figure shows a blockchain-based energy trading with Authority-managed identities and policies, where EVs,

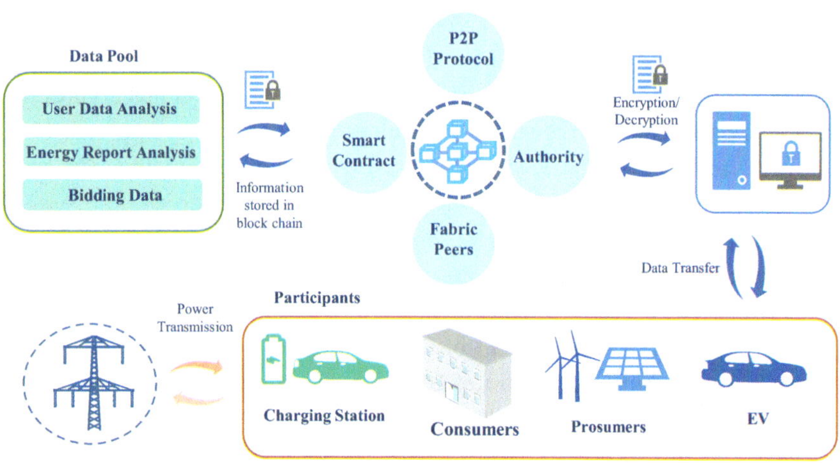

Figure 7.13 The block diagram architecture of a blockchain-based P2P energy trading system from energy generation to consumers.

prosumers, and other participants sell energy over P2P protocols enforced by smart contracts on a permissioned blockchain (Fabric peers). User data analysis, energy reports, and bidding data are produced at the data center and construct a data pool that is used to store information in blockchain. While the power transmission provides energy and the blockchain layer records and audits peer-to-peer transactions, a data center compiles user data for analysis and energy reports, with the ability to be integrated with blockchain-based methods.

7.4.4 Blockchain-enabled process flow, consensus mechanism, and energy auctions

In the smart contract process flow, the smart contract as an electronic agreement between various parties on the blockchain, allows the consumer to offer the charging station any extra energy that is available. The smart contract verified the user after receiving the request from the smart charging station operator. Following validation, the user receives a confirmation message and asks the producer to send energy. The EV owner will receive information after the charging process is finished, and the system will then create payment requests. Lastly, the operator of the charging station will ask the manufacturer to pay. According to the agreement, the smart contract will take the money from the EV owner's electronic wallet after receiving the charging amount from the producer. They have the ability to sell electricity to charging stations in addition to charging it themselves. The Smart contracts allow consumers to sell extra energy to charging stations. It can be used to store all billing and charging information on the block-chain where anyone can subsequently confirm it. Using smart contracts, consumers can start selling, and the person who requires more energy at a set rate can get in touch with the consumer (Khan and Byun, n.d.). After transferring energy to the consumer, the reward management smart contract sends cryptocurrency tokens into the registered prosumers' wallets. By calling sell, buy, confirm, and confirm-receive functions, the order management smart contract helps energy traders carry out their business operations. The system could buy energy from a variety of renewable energy sources, including hydropower plants, solar energy, and wind energy (Al-Saif et al., 2021).

Using a consensus mechanism, every transaction is verified before it is recorded in the chain in the blockchain. The process of reaching a consensus regarding a transaction's correctness guarantees that no malicious transactions can be added to the blockchain. Additionally, consensus aids in establishing dependability in a large network with numerous nodes. The process by which nodes in the network establish a clear transaction sequence and decide which transaction blocks need to adhere to the ledger is known as consensus. According to the guarantee and consensus policy, all transactions in the proposed block must be verified by consensus. The arrangement, accuracy, and implementation results must also be agreed. A set of transactions in the block is validated by a layer of smart contracts through the consensus mechanism. Three distinct types of nodes or peers, including endorsement nodes, ordering nodes, and committer peers, are the foundation of consensus in a permissioned blockchain. A transaction is started by the client and sent to the endorsement nodes. These nodes sign the endorsed transactions in addition to simulating and carrying out the transaction. Following receipt of the endorsed transaction, the client forwards it to the ordering node for read-set and endorsement verification. They forward it to the committer peer if they believe it to be accurate. The transition is finally committed by the committer peer (Khan and Byun, n.d.). One strategy that owners of electric vehicles can use to increase their profits is auctioning, which allows them to sell any leftover energy at competitive prices. Electric vehicles typically use the services of centralized brokers to buy and sell excess energy. By automating the auction process for energy sales and purchases from the most suitable clients, blockchain technology can eliminate the need for intermediaries. Consequently, it aids in reducing the transaction costs associated with energy trading. Blockchain guarantees a high degree of trust between the parties involved in energy auctions.

In Table 7.8, some relevant references of blockchain methods in *R2V smart chargingenergy trading and optimization from the* literature are summarized. Reference (Jamil et al., 2021) suggested a block-chain-based energy model that uses a Hyperledger Fabric network to connect prosumers and consumers and aggregate data for real-time load monitoring. In order to manage energy distribution and make the best decisions, they have also employed data analytics to uncover hidden patterns and information. A minimum cut maximum flow theory for scheduling distributed energy sources was put forth by (Wang et al., 2021). The information and administration of power energy trading have been recorded using blockchain technology. According to their simulation results, the suggested system uses less energy and is more cost-effective than the ones that are currently in use.

7.4.5 ML-based method application in R2V smart charging ecosystem

Due to AI's suitability for nonlinear and complex problems such as optimizing charging schedules and predicting EV charging demand, AI models are more frequently used in the literature than probabilistic models, and most researchers prefer them. Furthermore, AI models like Reinforcement Learning (RL) techniques and neural networks can learn by interacting with their environment and extracting nonlinear relationships from historical data, respectively (Chen and Komla, n.d.).

To improve EV charging scheduling performance, prediction results pertaining to EV charging are typically fed into optimization models. Both regression and classification problems can be solved with a vast variety of ML techniques. Support vector machines (SVM), artificial neural networks (ANNs), k-nearest neighbor (KNN), convolutional neural networks (CNN, RNN, deep convolutional neural networks (DCNN), gated recurrent units (GRUs), and long short-term memory (LSTM) are frequently employed machine learning techniques for forecasting tasks (Chen and Komla, n.d.).

Deep neural networks, including CNN, RNN, GRU, and LSTM, are highly effective and well-known because of their robust computing capabilities and abundance of datasets. Consequently, GRU is also widely used to solve long sequence sample problems, including load demand, PV power generation, and spot electricity price prediction. In order to lower the operating EV cost of charging stations, GRU in (Jin et al., n.d.) predicts the PV power production and load demand future trend. A GRU-based based is suggested in (Lu et al., 2019) to predict spot electricity prices, which are then used to plan charging stations and travel routes. Simulation results show that the route planning algorithm can accurately and efficiently give EV users the best charging routes.

LSTM has been widely used to address issues with lengthy and intricate sequence samples (Van Kriekinge et al., 2021). suggests using an LSTM model to forecast load demand. EV charging and discharging are then scheduled by schedulers using the predicted load demand.

Combining the benefits of various mechanisms to create hybrid models should be able to improve forecasting performance. A combination of an LSTM-based and CNN model is suggested in (Shipman et al., n.d.) based on data of the previous 24h to forecast the aggregated 48 vehicles' total available State of Charge (SoC) for V2G services over the next 24 h.

CNN and a deep belief network (DBN) are combined to create the CNN-DBN forecasting model, which was proposed by (Zhong et al., 2021a,b) as a renewable energy output forecasting and load demand model. The predicted outcomes are then utilized to plan when EVs will be charged in order to reduce the distribution network's operating expenses.

Typically, the Markov decision process (MDP) incorporates RL-based EV charging scheduling. A forecasting model using artificial neural networks (ANN) was proposed in (Dang et al., 2020) to predict the price of electricity for the following day. The charging scheduling model based on Q-learning is then trained using the predicted results. To provide EVs with real-time charging and discharging schedules (Mhaisen et al., 2020), suggested a Q-learning model. According to the simulation results, the suggested model can save roughly 68% more money than the uncontrolled one. However, the curse of dimensionality is one of the drawbacks of Q-learning algorithms.

The frequency of price updates is among the primary characteristics that set RTP apart from ToU, as RTP usually has a higher updating frequency than ToU. Power system operators and EV owners can both gain from RTP if it can accurately depict the conditions of the power system in real time. In order to create price signals that alter charging and discharging behaviors for the benefit of power grids, operators of power systems can utilize real-time power grid conditions, including electricity price, renewable energy generation, and load demand.

A DQN technique has been employed in (Maestre et al., 2018) to offer dynamic pricing models that balance equality and fairness with revenue maximization. In an effort to alleviate peak demand (Moghaddam et al., 2020), suggested an online RL approach to station real-time pricing. The simulation results show that the RTP based on online RL can successfully lower peak demand and boost charging stations' earnings. Large volumes of text data are used to train machine learning models known as Large Language Models (LLMs) for tasks involving natural language processing and V2G systems can be used to analyze human behavior and tackle complicated, large-scale problems. Although it needs a large amount of labeled data, using LLMs and Concerns have also been raised regarding the computational capacity's ability to handle large-scale problems in real time. Reinforcement Learning can produce vast amounts of data with reward values by interacting with the environment. By utilizing LLMs to improve RL, deployment costs can be decreased by taking advantage of LLMs' capacity to record spatiotemporal information while training on pre-existing models

(Xie et al., n.d.). In (Dabbaghjamanesh et al., 2021), a technique based on reinforcement learning, known as the Q-learning technique, which is based on recurrent neural networks and artificial neural networks, has been used to predict the load on EVs. The findings demonstrated that the Q-learning method outperforms the ANN and the recurrent neural network in terms of forecast error.

The summary of advanced methods in *R2V Smart ChargingEnergy Trading and optimization in the* literature is provided in (Table 7.11).

7.5 Research opportunities, challenges, and future directions

There are significant research opportunities when it comes to optimizing energy flows while EVs are integrated into energy markets through R2V smart charging ecosystem. Nevertheless, these advantages necessitate overcoming technological and financial obstacles, underscoring the necessity of multidisciplinary, forward-thinking research to direct future advancements. In this section, technical and financial challenges are listed below.

7.5.1 Technical challenges and concerns

Ensuring system efficiency and dependability requires addressing technical issues, the majority of which are listed below.

- EVs may increase power losses because the additional load would be distributed in the grid.
- Large power requirements for EV charging infrastructure can put stress on the electrical grid and cause instability; therefore, grid capacity and stability are among the challenges that need to be addressed (Bems and Anvari-Moghaddam, n.d.).
- A massive number of EVs while charging simultaneously may cause electrical line congestion and problems with voltage regulation, especially during peak hours. To reduce the strain on distribution lines and prevent instability and voltage drops, CSs must effectively manage EV charging (Bems and Anvari-Moghaddam, n.d.).
- The rise in EVs on the road and the lack of adequate charging infrastructure, which causes charging spots congestion in city centers and necessitates an increase in charging infrastructure, are other challenges of the Scalability concept. Additionally, the current state of blockchain

Table 7.11 The summary of advanced ML methods in ***R2V smart chargingenergy trading and optimization in the***literature.

Reference	Goal/strategy	Algorithm/method	Results and description
Zhong et al. (2021a,b)	Cooperative P2P trading in active distribution networks.	MILP-based game theory formulation	IEEE-33-bus simulations show that integrated Volt-VAR control produces a social welfare gain of more than 10% and a fair profit distribution.
Deng et al. (n.d.)	Day-ahead scheduling considering the effect of wind random output.	A potential-game equilibrium using a two-stage robust optimization method	This approach yields less conservative, individually optimal schedules that enhance expected cost and renewable utilization compared to stochastic and robust baselines
Kumar et al. (n.d.)	Electricity-demand forecasting in V2G.	IoV-enabled data pipeline and hybrid LSTM	By lowering MAE, MSE, and MAPE, the suggested IoV framework improves short-term demand forecasting for aggregator scheduling.
Li et al. (2022)	EV-induced alleviation of peak phenomena.	Stochastic Markov chain model	Using Monte-Carlo load synthesis, unmanaged charging lifts peak by 8.2%
Dabbaghjamanesh et al. (2021)	EV-CS load Forecasting with benchmarks.	Q-learning with RNN/ANN comparison	With the lowest RMSE and MAE in smart, uncoordinated, and coordinated charging scenarios, Q-learning outperforms traditional AI
Hao et al. (n.d.)	V2G-oriented charging management.	Deep-Q-network using 5 M km real travel data	The suggested DQN method lets individual EVs save more than 98% of the charging cost
Song et al. (n.d.)	frequency control with asymmetric V2G capacity.	Switched-integral reinforcement-learning (SIRL)	This technique reduces control costs by approximately 86% compared to baselines and restricts frequency deviation to ±0.2 Hz (±0.4–0.5 Hz for PID/MPC/ADHDP)
Shateri et al. (2020)	Joint privacy cost management.	Deep-Double-Q-learning (DDQL)	Compared to classical Q-learning, DDQL reduces information leakage while maintaining a comparable energy cost

technology is not able to host billions of electric vehicles within a smart city with real-time data processing requirements. Future technologies like fog and edge computing could be solutions for this concern (Al-Saif et al., 2021).

- Other issues include the stochastic behavior of EVs and the RE generation intermittency. Deterministic planning for EV charging and discharging is difficult due to unpredictable arrival times, SoC, and user preferences. Nonetheless, human mobility patterns can be incorporated as an additional validation technique to enhance the reliability of charging management, and EVs can function as flexible energy storage devices (Bems and Anvari-Moghaddam, n.d.).
- Effective communication is essential for facilitating smooth data exchange and communication amongst various EV stakeholders.
- Battery degradation results in increased resistance, which shortens battery life. The most affecting factors on increasing resistance are low temperatures and high SoC levels. Battery degradation is caused by scenarios of continuous charge and discharge in V2G and contributes to the electrical grid supply (Bems and Anvari-Moghaddam, n.d.).
- Network and data security are major concerns; during V2G communication, personal information, EV information, CS information and identity, transactions, and banking information must all be protected. The smart charging service's logging and monitoring, the firmware update process, and the compatibility of different charging types are the primary security issues (Barman et al., n.d.). Billing and tariff information, SoC level, and other data are included in an ecosystem that attackers could gain access to. A variety of objectives are pursued by cyberattacks on smart charging infrastructure, including tampering or falsifying charging data to prevent billing loss, stopping EV power supplies, and encrypting charging data to reveal the charging account and associated location. The ability to swiftly spot anomalous behavior, create strong defenses, and preserve learning effectiveness in erratic settings are some of the cyber concerns in V2G systems (Sadeghian et al., n.d.).
- The stochastic nature of the RE limits the availability of charging access based on RESs. Government policymakers must collaborate with the private sector and researchers to adopt strategies for appropriate location planning, installation of additional RE sources, and infrastructure development in order to address the availability and policy issues.

- It is undeniable that RE-based charging access is limited due to the stochastic nature of the RE. To address the issue of availability and policies, Government policymakers must collaborate with researchers and the private sector to adopt strategies for installing more RE sources, developing infrastructure, and properly planning their locations (Barman et al., n.d.).
- The current distribution networks' inability to handle bidirectional power flow is another obstacle, which restricts the mechanism and service capabilities (Sadeghian et al., n.d.).
- Large-scale electric vehicle discharges in a short time have the potential to cause significant grid fluctuations, increasing the peak-valley difference of grid load and compromising grid stability and power quality. The grid's strain during peak hours and managing grid fluctuations may be made worse by EVs' disorganized charging behavior. The uncertainty of predictions and the incompleteness of data could be addressed by machine learning algorithms (Bems and Anvari-Moghaddam, n.d.).

7.5.2 Economic challenges

The unpredictability of market electricity prices presents further challenges for system cost optimization. The design and capacity constraints of charging infrastructure and market incentives are among the factors that impact cost optimization in the process of cost optimization systems (Xie et al., n.d.). The prominent economic challenges in this regard are listed below.

- Ensuring that current CSs are profitable in providing charging facilities while taking into account technical factors, their expenses when interacting with the electricity market, and the high cost of CS installation and maintenance (Bems and Anvari-Moghaddam, n.d.).
- EV and charging costs, as well as optimizing their advantages in the electricity market by engaging in V2G mode. Expenses for building new charging stations while taking into account the profits from already-existing ones and effective CS pricing schemes (Bems and Anvari-Moghaddam, n.d.).
- The advantages of using aggregators to provide service for assigned EV charging and discharging in a scheduling system based on aggregators (Bems and Anvari-Moghaddam, n.d.).

- The goal of the optimization challenge is a maximize social welfare and economic benefits while involving the framework operator and counterparties (Bems and Anvari-Moghaddam, n.d.).
- Multiple buyers and sellers are involved in V2V energy trading. A rule that determines how a buyer and seller can be matched and carry out a transaction should be created in order to address the matching challenge (Sadeghian et al., n.d.).
- The pricing challenge is figuring out a clearing price to finish the transaction after buyers and sellers have been successfully matched. In addition to considering whether it is profitable for both parties involved in the transaction, this price must also take into account the market's supply and demand relationship and factors like fairness and transparency (Sadeghian et al., n.d.).

The overview of challenges and future directions in *energy trading mechanisms, EVmarket interactions for R2V smart charging* are illustrated in (Fig. 7.14).

7.5.3 Blockchain challenges in EV energy trading

Blockchain technology presents viable options for decentralized, transparent, and safe energy transactions involving electric vehicles. However, in R2V ecosystems, research needs to overcome challenges like scalability, awareness, traceability, and irreversibility, which are primarily mentioned in the following (Al-Saif et al., 2021).

- Energy traceability improvement: The blockchain traceability feature is designed to manage the asset records while the participating organizations process them. Blockchain would increase the efficiency of the energy market and help the energy grid station operators to better regulate, utilize, and manage electric vehicle energy demand (Aghajan-Eshkevari et al., n.d.).
- Reputation-aware energy trading: Blockchain technology can be used to detect, validate, and resolve frauds that result in an unfair distribution of energy resources to electric cars. When dishonest or self-centered drivers come across energy trading fraud, a smart contract can penalize them.
- Irreversibility in Blockchain platforms: One of the challenges facing public blockchain platforms is protecting data privacy because transactions and data are transparent and open on these platforms, and it is necessary to make sure that data stored on the ledger cannot be altered,

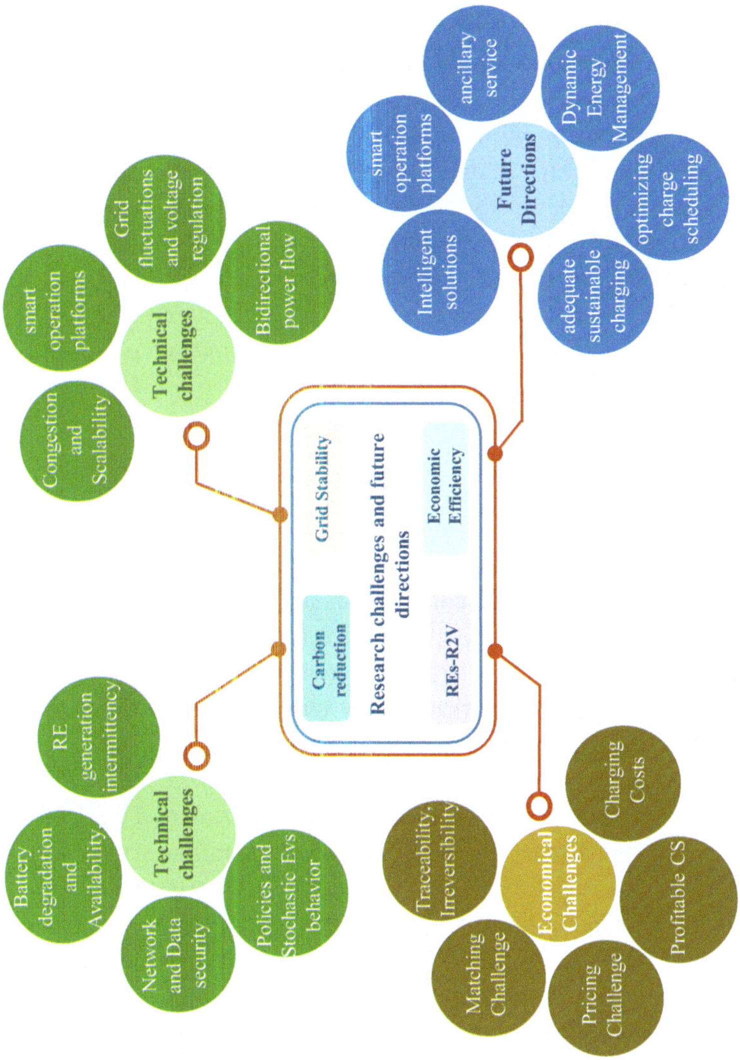

Figure 7.14 The overview of challenges and future directions.

removed, or compromised. Blockchain platforms now face additional security challenges because of computers' growing computational capacity due to quantum computing.

7.5.4 Research direction and future trends

Despite the growing popularity of R2V energy trading, there are still significant research gaps and areas that require improvement. These include security challenges, optimization models, V2G operation, and efficient charging scheduling. Future studies must fill these gaps in order to fully utilize EVs in dynamic electricity markets, which are mainly listed in the following (Bems and Anvari-Moghaddam, n.d.; Khan et al., n.d.).

- Developing charging stations' optimal smart operation platforms for effective simultaneous charging and discharging.
- Optimizing charge scheduling within each CS and improving coordination between charging stations.
- Creating an adequate, sustainable charging infrastructure to meet the EV market's rapid expansion.
- Installing reservation systems for EV charging and incorporating RESs into EV charging systems.
- EVs in public transportation renovation of existing infrastructure, including public parking lots, commercial, and residential buildings.
- Allowing EVs to take part in energy markets through demand response, capacity markets for peak load management and backup power, frequency regulation services that modify charging rates, and spinning reserve services for grid outages.
- In order to meet the minimum market entry requirement, a new structure or standard should be created for individual customers. To improve its adaptability, the authorities must thus enforce a unified standardized framework.
- Developing intelligent and connected EVs is in line with new technologies such as intelligent EV charging, decentralized energy trading, IoT, and ML applications, which have shown efficiency gains in manufacturing industries (Bems and Anvari-Moghaddam, n.d.).
- A significant financial gain from developing a blockchain-supported P2P electricity trading system with an energy-efficient consensus algorithm could occur.
- Optimized energy utilization will be made possible by the establishment of new smart contracts for P2P energy trading, which includes a creative strategic approach to handling environmental compliance requirements.

- Energy systems can become more effective and adaptable to changing demands by leveraging blockchain technology and considering dynamic energy management. This will improve overall efficiency and facilitate the smooth integration of renewable energy sources into the grid.

7.5.5 Future trends in R2V integration in railway stations as high-power energy hubs

Railway stations, particularly those with Park & Ride (P&R) facilities, are emerging as strategic locations for integrating R2V smart charging, transforming them into multifunctional energy hubs (Brenna et al., 2020). These hubs can efficiently combine EV charging infrastructure, renewable energy sources (RESs), energy storage systems (ESSs), and demand-side management to optimize energy flows while supporting mobility and grid services. By leveraging the high turnover of parked EVs and predictable train schedules, railway stations can serve as ideal nodes for both energy exchange and storage in urban and suburban areas (Kaleybar et al., 2025).

One significant trend is the co-optimization of EV charging and renewable energy utilization. Photovoltaic installations on station roofs or nearby areas can directly feed P&R EVs, while excess renewable generation can be stored in local ESSs for later use or injected back into the grid. Smart charging algorithms, combined with real-time data from trains and EVs, can schedule charging and discharging events to align with train arrival/departure patterns and grid demand fluctuations, maximizing both economic and environmental benefits. Additionally, EVs can provide ancillary services to the grid, such as frequency regulation and spinning reserve, leveraging their presence at P&R facilities.

Another emerging trend is the integration of blockchain-enabled P2P energy trading within railway stations. EV owners can trade surplus energy with other vehicles or station loads through secure, transparent platforms. This system supports dynamic pricing and incentive mechanisms, encouraging EV users to participate in energy markets and promoting renewable energy consumption. Combined with IoT-enabled smart meters and digital twins of station energy systems, operators can monitor, forecast, and optimize energy flows while maintaining high reliability and safety standards.

Furthermore, intermodal transportation synergy offers opportunities for enhanced energy efficiency. P&R areas can be designed to coordinate EV charging with public transport schedules, reducing idle times, and integrating demand response programs with train regenerative braking systems (Çiçek et al., n.d.). This enables energy recovered from trains to be stored

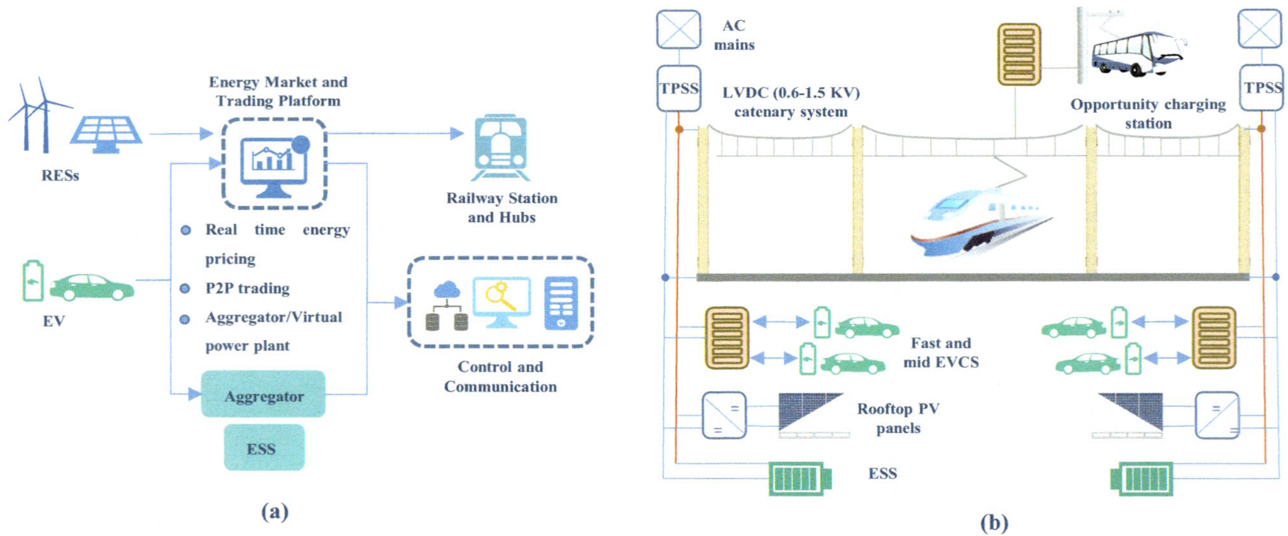

Figure 7.15 (a): Energy trading mechanisms for R2V smart charging integrated in P&R railway hubs, (b): The topological concept of R2V smart charging integrated in P&R area close to the railway line.

Table 7.12 Summary of main challenges and potential solutions, future research direction.

Challenges and solutions		Future direction
Main challenges	**Potential solutions**	
Capacity, stability, RE intermittency	Time-ahead capacity, demand forecasting improvement	Advanced AI intelligent methods, and probabilistic forecasts
Congestion	Charging control algorithms	Multi-agent congestion control
Voltage regulation	Voltage control devices and real-time SoC estimation	Co-optimize V2G, V2X integration with Volt/VAR
Effective communication	Advanced EV communication protocols, edge computing	Regulatory and Policy Research for low-latency standard, interoperability IOT and digital twins
Battery degradation	Adaptive charging policies, battery health monitoring and AI-powered optimization algorithms	Advanced scheduling mechanism and health forecasting
Data security	Blockchain integration, secure multi-party computation	Secure market clearing, anomaly identification
Availability and Policy Issues	Standardized regulations	Regulatory software for dynamic tariffs
Grid fluctuations	Frequency regulation and demand response programs with R2V and an incentive mechanism	EV frequency response
Economic challenges and energy trading	Efficient pricing mechanisms and cost minimization models, market-responsive incentives	Incentive compatible design
Charging costs	Incentive-based and real-time pricing mechanisms, off-peak charging policies,	Personalized tariffs, consumer behavior analysis
Matching challenge	Better matching with AI-driven energy algorithms	Learning based EV charging matching and reservation design

(Continued)

Table 7.12 Summary of main challenges and potential solutions, future research direction.—cont'd

Challenges and solutions		Future direction
Main challenges	**Potential solutions**	
Pricing challenge	Dynamic pricing models, transparent price negotiation	Demand and elasticity estimation for dynamic prices
Optimization challenge	Multi-objective optimization, DR algorithms	Multi objective scheduling, Dynamic energy management
Improve Energy Traceability	Blockchain for transparent transaction records	End-to-end renewable certificate traceability
Reputation-Aware Energy Trading	Stakeholder rating systems, smart contracts for trust assurance	Market framework development
Irreversibility in Blockchain Platforms	Use of reversible smart contract templates, blockchain tools	Upgradable contracts

locally and used for EV charging or sold to the grid, creating a closed-loop energy ecosystem that benefits both mobility and energy management.

Finally, policy and regulatory frameworks are expected to support the standardization and incentivization of EV integration in railway stations. Unified protocols for communication, dynamic tariffs, and energy market participation will allow operators to efficiently manage large-scale EV fleets, ESSs, and RESs while ensuring economic feasibility. In the coming years, railway stations, especially P&R hubs, are likely to evolve into smart energy hubs, bridging transportation and energy sectors while contributing significantly to sustainable urban development (Davoodi et al., 2025).

Figure 7.15(a) shows energy trading mechanisms for R2V smart charging integrated in P&R Railway hubs, which illustrates the interactions of the energy market and trading platform, railway station and hubs, EVs, aggregator and control center. The topological concept of R2V smart charging integrated in P&R area close to the railway lineis depicted in Fig. 7.15(b).

The summary of main challenges and potential solutions, alongside future research direction, is provided in Table 7.12.

7.6 Conclusion

Digital platforms and aggregators allow prosumers and EVs to trade energy with each other on a P2P basis, optimize charging to smooth out peaks, and bid EV flexibility into the electricity market. Smart contracts and blockchain hold promise as facilitators of safe, transparent energy transactions in the systems. The performance of R2V systems is further improved by advanced ML and optimization methods, which minimize costs while enabling EV charging in conjunction with variable renewable generation. Grid and energy costs can also be reduced with effective smart charging.

This review maps incentive smart charging approaches with mathematical modeling support and market mechanisms, and ancillary services. This chapter clarifies the significance of market models, blockchain-based energy trading, and smart contracts foratransparent, efficient, and secure trading mechanism. Furthermore, optimization objectives of EV charging and advanced ML-based methods in electricity market participation of EV mechanisms were also provided. Based on the insights presented in the chapter and the current research landscape for the interactions of EV market interactions with R2V Smart charging ecosystem and energy-trading mechanisms, notable points arise.

- Appropriate results from smart charging/scheduling that takes into account battery health could take place using shifting charging off-peak and more RESs integration with R2V value yields by EV charging and discharging co-optimization with network constraints. Blockchain can automate P2P trades and add auditability with privacy preservation, and the upcoming step could be scaled pilots that compare learning-plus optimization methods for cost, reliability, and emissions.
- Given the nascent nature of these technologies, additional study is essential for the development and enhancement of energy trading, EVs in the electricity market, and R2V smart charging, improving optimization efficiency and better applicability in R2V integration.
- There are still open issues, especially with developing equitable incentive programs, cybersecurity vulnerability, and modeling uncertainty in REs and driving patterns, and operational obstacles have emerged that need further research to be addressed.

R2V smart charging is a crucial enabling technology for a low-carbon, renewable energy system as EV usage keeps rising. The future of *EV market interactions* and *R2V smart charging,* ML-based scheduling, and sophisticated market platforms seems promising, particularly in light of the recommendations presented in recent research.

References

Aggarwal S, Kumar N, Tanwar S, Alazab M. A survey on energy trading in the smart grid: taxonomy, research challenges and solutions. IEEE Access 2021a;9:116231–53. https://doi.org/10.1109/access.2021.3104354.

Aggarwal S, Kumar N, Gope P. An efficient blockchain-based authentication scheme for energy-trading in V2G networks. IEEE Trans Ind Inf 2021b;17(10):6971–80. https://doi.org/10.1109/tii.2020.3030949.

Aghajan-Eshkevari S, Ameli MT, Azad S. Electric vehicle services to support the power grid. In: Electric vehicle integration via smart charging: technology, standards, implementation, and applications. Cham: Springer International Publishing; 2022. p. 129–48.

Aghajan-Eshkevari S., Ameli M.T., Azad S. Optimal routing and power management of electric vehicles in coupled power distribution and transportation systems. Appl Energy 341: 121126.

Aghajan-Eshkevari, S., Azad S., Nazari-Heris M., Ameli MT., Asadi S.. Charging and discharging of electric vehicles in power systems: an updated and detailed review of methods, control structures, objectives, and optimization methodologies. Sustainability; 14(4): 2137.

Ahmad A., et al. Electric vehicle charging modes, technologies and applications of smart charging. Energies; 15(24): 9471.

Aljaidi M, et al. A reinforcement learning-based assignment scheme for evs to charging stations. In: 2022 IEEE 95th vehicular technology conference:(VTC2022-Spring). IEEE; 2022.

Al-Obaidi A., et al. Bidirectional smart charging of electric vehicles considering user preferences, peer to peer energy trade, and provision of grid ancillary services. Int J Electr Power Energy Syst; 124: 106353.

Al-Saif N, Ahmad RW, Salah K, Yaqoob I, Raja J, Omar M. Blockchain for electric vehicles energy trading: requirements, opportunities, and challenges. IEEE Acc 2021;9: 156947–61. https://doi.org/10.1109/access.2021.3130095.

Ameli MT, Jalilpoor K, Azad S, Daneshvar M, Sepasian MS, Mohammadi-Ivatloo B, et al. Future modern distribution networks resilience: from Passive operation to strategic active Paradigms future modern distribution networks resilience: from Passive operation to strategic active Paradigms. Iran: Elsevier; 2024. p. 1–415. https://doi.org/10.1016/C2022-0-02390-4.

Amin U, Hossain MJ, Fernandez E, Mahmud K, Tiezheng G. A contract-based trading model for electricity suppliers in smart grids. In: 2019 20th International Conference on intelligent system application to power systems, ISAP 2019. Australia: Institute of Electrical and Electronics Engineers Inc; 2019. https://doi.org/10.1109/ISAP48318.2019.9065981.

Amiri, MM, Aghajan-Eshkevari S, Ali Rahimi M, Ali S. A review on utilization of electric vehicles for mitigating the power quality issues in power systems. Power Qual-New Insights.

Aoudia M, et al. Toward better blockchain-enabled energy trading between electric vehicles and smart grids in Internet of Things environments: a survey. Front Energy Res; 12: 1393084.

Azad-Farsani E, Abedini S, Sardou, I.G. Optimal coordination of plug-in hybrid electric vehicles: a stochastic market-based approach. J Clean Prod; 321: 128990.

Barman P, et al. Renewable energy integration with electric vehicle technology: a review of the existing smart charging approaches." Renew Sustain Energy Rev; 183: 113518.

Baza M, Ahmed S, Mahmoud MMEA, Bakiras S, Alasmary W, Mohamed A, et al. Privacy-preserving blockchain-based energy trading schemes for electric vehicles. IEEE Trans Veh Technol 2021;70(9):9369–84. https://doi.org/10.1109/tvt.2021.3098188.

Bems, Anvari-Moghaddam A. A review of the latest trends in technical and economic aspects of EV charging management. Energies; 16(9): 3669.

Bharathidasan M, Indragandhi V, Suresh V, Jasiński M, Leonowicz Z. A review on electric vehicle: technologies, energy trading, and cyber security. Energy Rep 2022;8:9662–85. https://doi.org/10.1016/j.egyr.2022.07.145.

Bogdanov D, Breyer C. Role of smart charging of electric vehicles and vehicle-to-grid in integrated renewables-based energy systems on country level. Energy 301: 131635.

Brenna M, Foiadelli F, Jafari Kaleybar H. The Evolution of Railway Power Supply Systems toward Smart Microgrids: the concept of the energy hub and integration of distributed energy resources. IEEE Electrif Mag 2020;8(1):12–23. https://doi.org/10.1109/mele.2019.2962886.

Chen X, Zhang X. Secure electricity trading and incentive contract model for electric vehicle based on energy blockchain. IEEE Access 2019;7:178763–78. https://doi.org/10.1109/access.2019.2958122.

Chen Q, Komla AF. Application of artificial intelligence for EV charging and discharging scheduling and dynamic pricing: a review. Energies; 16(1): 146.

Çiçek A, et al. Integrated rail system and EV parking lot operation with regenerative braking energy, energy storage system and PV availability. IEEE Trans Smart Grid; 13(4): 3049-3058.

Dabbaghjamanesh M, Moeini A, Kavousi-Fard A. Reinforcement learning-based load forecasting of electric vehicle charging station using Q -learning technique. IEEE Trans Ind Inf 2021;17(6):4229–37. https://doi.org/10.1109/tii.2020.2990397.

Dang Q, Wu D, Boulet B. EV charging management with ANN-based electricity price forecasting. In: 2020 IEEE transportation electrification Conference and Expo, ITEC 2020. Institute of Electrical and Electronics Engineers Inc; 2020. https://doi.org/10.1109/ITEC48692.2020.9161659.

Das HS. et al. Electric vehicles standards, charging infrastructure, and impact on grid integration: a technological review." Renew Sustain Energy Rev 120: 109618.

Das R. et al. Multi-objective techno-economic-environmental optimisation of electric vehicle for energy services. Appl Energy; 257: 113965. .

Davoodi M, Jafari Kaleybar H, Brenna M, Zaninelli D, Mishra AK. Sustainable electric railway system integrated with distributed energy resources: optimal operation and smart energy management system. Int Trans Electr Energy Syst 2025;2025(1). https://doi.org/10.1155/etep/6626245.

Deb S, Pihlatie M, Al-Saadi M. Smart charging: a comprehensive review. IEEE Access 2022; 10:134690–703. https://doi.org/10.1109/access.2022.3227630.

Deng H., et al. Distributionally robust day-ahead scheduling for power-traffic network under a potential game framework. Int J Electr Power Energy Syst; 147: 108851.

Du P. et al. Enhancing green mobility through vehicle-to-grid technology: potential, technological barriers, and policy implications. Energy Environ Sci.

Gan W, Wen J, Yan M, Zhou Y, Yao W. Enhancing resilience with electric vehicles charging Redispatching and vehicle-to-grid in traffic-electric networks. IEEE Trans Ind Appl 2024;60(1):953–65. https://doi.org/10.1109/tia.2023.3272870.

Geng L. et al. Smart charging management system for electric vehicles in coupled transportation and power distribution systems. Energy; 189: 116275.

Gnann T., et al. The value of smart charging at home and its impact on EV market shares–A German case study. Appl Energy; 380: 124997.

Gopinathan N, Karthikeyan Shanmugam P. Energy anxiety in decentralized electricity markets: a critical review on EV models. Energies; 15.14: 5230.

Hao X. et al. A V2G-oriented reinforcement learning framework and empirical study for heterogeneous electric vehicle charging management. Sustain Cities Soc; 89: 104345.

Hassija V, Chamola V, Garg S, Krishna DNG, Kaddoum G, Jayakody DNK. A blockchain-based framework for Lightweight data sharing and energy trading in V2G network. IEEE Trans Veh Technol 2020;69(6):5799–812. https://doi.org/10.1109/tvt.2020.2967052.

Houbbadi A, et al. Optimal charging strategy to minimize electricity cost and prolong battery life of electric bus fleet. In: 2019 IEEE vehicle power and Propulsion Conference (VPPC). IEEE; 2019.

Huang Z. et al. A novel power market mechanism based on blockchain for electric vehicle charging stations. Electronics; 10(3): 307.

Iqbal A, Rajasekaran AS, Sai Nikhil G, Azees M. A secure and decentralized blockchain based EV energy trading model using smart contract in V2G network. IEEE Access 2021;9:75761–77. https://doi.org/10.1109/access.2021.3081506.

Islam MDR, Lu H, Hossain J, Islam MDR, Li L. Multiobjective optimization technique for mitigating Unbalance and improving voltage considering higher penetration of electric vehicles and distributed generation. IEEE Syst J 2020;14(3):3676–86. https://doi.org/10.1109/jsyst.2020.2967752.

Jamil F, Iqbal N, Imran SA, Kim D. Peer-to-peer energy trading mechanism based on blockchain and machine learning for sustainable electrical power supply in smart grid. IEEE Access 2021;9:39193–217. https://doi.org/10.1109/access.2021.3060457.

Jannati J, Nazarpour D. Optimal performance of electric vehicles parking lot considering environmental issue. J Clean Prod 2019;206:1073–88. https://doi.org/10.1016/j.jclepro.2018.09.222.

Jie B, Baba J, Kumada A. Contribution to V2G system frequency regulation by charging/discharging control of aggregated EV group. IEEE Trans Ind Appl 2024;60(1):1129–40. https://doi.org/10.1109/tia.2023.3292814.

Jin H. et al. Development of charging/discharging scheduling algorithm for economical and energy-efficient operation of multi-EV charging station." Appl Sci; 12(9): 4786.

Johnsen D, et al. Review on smart charging of electric vehicles via market-based incentives, grid-friendly and grid-compatible measures. World Electr Vehic J; 14(1): 25.

Kaleybar HJ, Brenna M, Foiadelli F, Zaninelli D. Electric railway system integrated with EV charging station realizing train-to-vehicle technology. IEEE Trans Veh Technol 2025; 74(2):2572–86. https://doi.org/10.1109/tvt.2023.3294674.

Kanellos FD. Optimal scheduling and real-time operation of distribution networks with high penetration of plug-in electric vehicles. IEEE Syst J 2021;15(3):3938–47. https://doi.org/10.1109/jsyst.2020.3006002.

Khan PW, Byun YC. Blockchain-based peer-to-peer energy trading and charging payment system for electric vehicles. Sustainability; 13(14): 7962.

Khan S, Amin U, Ahmed AS. P2P energy trading of EVs using blockchain technology in centralized and decentralized networks: a review. Energies; 17.9: 2135.

Khorasany M, et al. Lightweight blockchain framework for location-aware peer-to-peer energy trading. Int J Electr Power Energy Syst; 127: 106610.

Kim H, Lee J, Bahrami S, Wong VWS. Direct energy trading of microgrids in distribution energy market. IEEE Trans Power Syst 2020;35(1):639–51. https://doi.org/10.1109/tpwrs.2019.2926305.

Kim OTT, et al. Distributed auction-based incentive mechanism for energy trading between electric vehicles and mobile charging stations. IEEE Access; 10: 56331-56347.

Krishnamoorthi S, Rajasekar V, Balusamy B. Blockchain for energy transactions blockchain-based systems for the modern energy grid. India: Elsevier; 2022. p. 51–69. https://doi.org/10.1016/B978-0-323-91850-3.00013-5.

Kumar N, Kumar Sood S, Saini M. Internet of vehicles (IoV) based framework for electricity demand forecasting in V2G. Energy; 297: 131199.

Kumar P. et al. A comprehensive review of vehicle-to-grid integration in electric vehicles: powering the future. Energy Convers Manag X: 100864.

Li B, Chen M, Kammen DM, Kang W, Qian X, Zhang L. Electric vehicle's impacts on China's electricity load profiles based on driving patterns and demographics. Energy Rep 2022;8:26–35. https://doi.org/10.1016/j.egyr.2021.11.003.

Li Y, Han M, Yang Z, Li G. Coordinating flexible demand response and renewable uncertainties for scheduling of community integrated energy systems with an electric vehicle charging station: a Bi-level approach. IEEE Trans Sustain Energy 2021;12(4):2321–31. https://doi.org/10.1109/tste.2021.3090463.

Li Y, Hu B. An iterative two-layer optimization charging and discharging trading scheme for electric vehicle using consortium blockchain. IEEE Trans Smart Grid 2020;11(3):2627–37. https://doi.org/10.1109/tsg.2019.2958971.

Liang Y, Wang Z, Ben Abdallah A. V2GNet: robust blockchain-based energy trading method and implementation in vehicle-to-grid network. IEEE Acc 2022;10:131442–55. https://doi.org/10.1109/access.2022.3229432.

Liu Y, Liang H. A discounted stochastic multiplayer game approach for vehicle-to-grid voltage regulation. IEEE Trans Veh Technol 2019;68(10):9647–59. https://doi.org/10.1109/tvt.2019.2936786.

Lotfi S, Sedighizadeh M, Abbasi R, Hosseinian SH. Vehicle-to-grid bidding for regulation and spinning reserve markets: a robust optimal coordinated charging approach. Energy Rep 2024;11:925–36. https://doi.org/10.1016/j.egyr.2023.12.044.

Lu Y, Gu J, Xie D, Li Y. Integrated route planning algorithm based on spot price and classified travel objectives for EV users. IEEE Access 2019;7:122238–50. https://doi.org/10.1109/access.2019.2937910.

Luo Q. et al. A hierarchical blockchain architecture based V2G market trading system. Appl Energy; 307: 118167.

Maestre R, et al. Reinforcement learning for fair dynamic pricing. In: Proceedings of SAI intelligent systems conference. Cham: Springer International Publishing; 2018.

Mao T, Zhang X, Zhou B. Intelligent energy management algorithms for EV-charging scheduling with consideration of multiple EV charging modes. Energies; 12(2): 265.

Mastoi MS, Zhuang S, Munir HM, Haris M, Hassan M, Alqarni M, et al. A study of charging-dispatch strategies and vehicle-to-grid technologies for electric vehicles in distribution networks. Energy Rep 2023;9:1777–806. https://doi.org/10.1016/j.egyr.2022.12.139.

Mhaisen N, Fetais N, Massoud A. Real-time scheduling for electric vehicles charging/discharging using reinforcement learning. In: 2020 IEEE international conference on informatics, IoT, and enabling technologies, ICIoT 2020. Qatar: Institute of Electrical and Electronics Engineers Inc; 2020. https://doi.org/10.1109/ICIoT48696.2020.9089471.

Moghaddam V, Yazdani A, Wang H, Parlevliet D, Shahnia F. An online reinforcement learning approach for dynamic pricing of electric vehicle charging stations. IEEE Acc 2020;8:130305–13. https://doi.org/10.1109/access.2020.3009419.

Moghaddam Z, Ahmad I, Habibi D, Phung QV. Smart charging strategy for electric vehicle charging stations. IEEE Trans Transp Electrif 2018;4(1):76–88. https://doi.org/10.1109/tte.2017.2753403.

Moniruzzaman MD, Yassine A. Benlamri R. Blockchain and cooperative game theory for peer-to-peer energy trading in smart grids. Int J Electr Power Energy Syst; 151: 109111.

Morstyn T, Teytelboym A, Hepburn C, McCulloch MD. Integrating P2P energy trading with probabilistic distribution locational marginal pricing. IEEE Trans Smart Grid 2020;11(4):3095–106. https://doi.org/10.1109/tsg.2019.2963238.

Mwasilu F, Justo JJ, Kim E-K, Do TD, Jung JW. Electric vehicles and smart grid interaction: a review on vehicle to grid and renewable energy sources integration. Renew Sustain Energy Rev 2014;34:501–16. https://doi.org/10.1016/j.rser.2014.03.031.

Nimalsiri NI. et al. Coordinated charging and discharging control of electric vehicles to manage supply voltages in distribution networks: assessing the customer benefit. Appl Energy; 291: 116857.

Paeizi A, Ameli MT, Azad S. Distribution systems operation and planning: a literature review on smart resilience enhancement strategies. Future Modern Distribution Networks Resilience: 111-139.

Ren H, et al. Bayesian network based real-time charging scheduling of electric vehicles. In: 2020 International Conference on smart grids and energy systems (SGES). IEEE; 2020.

Rücker F. et al. Evaluation of the effects of smart charging strategies and frequency restoration reserves market participation of an electric vehicle. Energies; 13(12): 3112.

Sameer Q, Sameer, Sameer, et al. Fair energy trading in blockchain-Inspired smart grid: technological barriers and future trends in the age of electric vehicles. World Electr Vehic J; 15(11): 487.

Sadeghian O. et al. A comprehensive review on electric vehicles smart charging: solutions, strategies, technologies, and challenges. J Energy Storage; 54: 105241.

Sagawa D. et al. Bidding agents for PV and electric vehicle-owning users in the electricity P2P trading market. Energies; 14(24): 8309.
Saha R, Kumar G, Geetha G, Hoon-Kim T-, Alazab M, Thomas R, et al. The blockchain solution for the security of internet of energy and electric vehicle interface. IEEE Trans Veh Technol 2021;70(8):7495–508. https://doi.org/10.1109/tvt.2021.3094907.
Shafie-khah M, Neyestani N, Damavandi MY, Gil FAS, Catalão JPS. Economic and technical aspects of plug-in electric vehicles in electricity markets. Renew Sustain Energy Rev 2016;53:1168–77. https://doi.org/10.1016/j.rser.2015.09.079.
Shahkamrani A. et al. A framework for day-ahead optimal charging scheduling of electric vehicles providing route mapping: Kowloon case study. J Clean Prod; 307: 127297.
Shateri M, et al. Privacy-cost management in smart meters using deep reinforcement learning. In: 2020 IEEE PES innovative smart grid technologies Europe (ISGT-Europe). IEEE; 2020.
Shipman R. et al. We got the power: predicting available capacity for vehicle-to-grid services using a deep recurrent neural network. Energy; 221: 119813.
Song X. et al. Cooperative grid frequency control under asymmetric V2G capacity via switched integral reinforcement learning. Int J Electr Power Energy Syst 155: 109679.
Taghizad-Tavana K. et al. A comprehensive review of electric vehicles in energy systems: integration with renewable energy sources, charging levels, different types, and standards. Energies 16.2: 630.
Tang W, Zhang YJ. A model predictive control approach for low-Complexity electric vehicle charging scheduling: optimality and scalability. IEEE Trans Power Syst 2017; 32(2):1050–63. https://doi.org/10.1109/tpwrs.2016.2585202.
Tirunagari S, Gu M, Meegahapola L. Reaping the benefits of smart electric vehicle charging and vehicle-to-grid technologies: regulatory, policy and technical aspects. IEEE Acc 2022;10:114657–72. https://doi.org/10.1109/access.2022.3217525.
Tookanlou MB, Ali Pourmousavi Kani S, Marzband, M. A comprehensive day-ahead scheduling strategy for electric vehicles operation. Int J Electr Power Energy Syst; 131: 106912.
Van Kriekinge G, De Cauwer C, Sapountzoglou N, Coosemans T, Messagie M. Peak shaving and cost minimization using model predictive control for uni- and bi-directional charging of electric vehicles. Energy Rep 2021;7:8760–71. https://doi.org/10.1016/j.egyr.2021.11.207.
Wan, M. et al. Feasibility and challenges for vehicle-to-grid in electricity market: a review. Energies; 17(3): 679.
Wang H, Ma S, Guo C, Wu Y, Dai H-N, Wu Di. Blockchain-based power energy trading management. ACM Trans Internet Technol 2021;21(2):1–16. https://doi.org/10.1145/3409771.
Wang H, Wang Q, He D, Qi L, Liu Z. BBARS: blockchain-based anonymous rewarding scheme for V2G networks. IEEE Internet Things J 2019;6(2):3676–87. https://doi.org/10.1109/jiot.2018.2890213.
Wang Z, Ogbodo M, Huang H, Qiu C, Hisada M, Ben Abdallah A. AEBIS: AI-enabled blockchain-based electric vehicle integration system for power management in smart grid platform. IEEE Access 2020;8:226409–21. https://doi.org/10.1109/access.2020.3044612.
Wang N, et al. A multi-energy scheduling strategy for orderly charging and discharging of electric vehicles based on multi-objective particle swarm optimization. Sustain Energy Technol Assessments; 44: 101037.
Wu Y, et al. Optimal management of local energy trading in future smart microgrid via pricing. In: 2015 IEEE Conference on computer communications Workshops (INFOCOM WKSHPS). IEEE; 2015.

Xie H. et al. Reinforcement learning for vehicle-to-grid: a review. Adv Appl Energy: 100214.

Xu Y, Ali AP, Balta-Ozkan N. Vehicle-to-vehicle energy trading framework: a systematic literature review. Sustainability; 16.12: 5020.

Zaman I, Hasan Md M, He M, Giesselmann MG. Design of a peer-to-peer energy trading platform using multilayered semi-permissioned blockchain. Int J Commun Netw Syst Sci 2022;15(07):94–110. https://doi.org/10.4236/ijcns.2022.157008.

Zhang G, Tim Tan S, Gary Wang G. Real-time smart charging of electric vehicles for demand charge reduction at non-residential sites. IEEE Trans Smart Grid 2018;9(5): 4027–37. https://doi.org/10.1109/tsg.2016.2647620.

Zhang Y, et al. Profit sharing mechanism for incentivizing vehicle-to-grid in regional energy internet. In: 2021 IEEE sustainable power and energy Conference (iSPEC). IEEE; 2021.

Zheng Y, et al. Modeling the temporal and economic feasibility of electric vehicles providing vehicle-to-grid services in the electricity market under different charging scenarios. J Energy Storage; 68: 107579.

Zhong J, Xiong X, Jiang Y-Z. An orderly EV charging scheduling method based on deep learning in cloud-edge collaborative environment. Adv Civ Eng 2021;2021(1). https://doi.org/10.1155/2021/6690610.

Zhong W, Xie S, Xie K, Yang Q, Xie L. Cooperative P2P energy trading in active distribution networks: an MILP-based Nash bargaining solution. IEEE Trans Smart Grid 2021;12(2):1264–76. https://doi.org/10.1109/TSG.2020.3031013.

Zhong W, Xie K, Liu Y, Yang C, Xie S. Efficient auction mechanisms for two-layer vehicle-to-grid energy trading in smart grid. In: IEEE International Conference on communications. China: Institute of Electrical and Electronics Engineers Inc; 2017. https://doi.org/10.1109/ICC.2017.7997118.

Zhou Z, Wang B, Dong M, Ota K. Secure and efficient vehicle-to-grid energy trading in cyber physical systems: integration of blockchain and edge computing. IEEE Trans Syst Man Cybern: Syst 2020;50(1):43–57. https://doi.org/10.1109/tsmc.2019.2896323.

CHAPTER EIGHT

The application of fuel cell vehicles in hydrogen-based renewable systems

Sina Ahmadi[1], Mehrdad Setayesh Nazar[1], Morteza Nazari-Heris[2]
[1]Faculty of Electrical Engineering Shahid Beheshti University, Tehran, Iran
[2]Department of Electrical Engineering, East Carolina University, Greenville, NC, United States

Abbreviations

FC Fuel Cell
FCV Fuel Cell Vehicle
FCVS Fuel Cell Vehicle charge Station
RESS Residential Energy Supply System
HSS Hydrogen Storage System
ESS Electrical Storage System
MG Microgrid
P2H Power-to-Hydrogen
H2P Hydrogen-to-Power
V2G Vehicle-to-Grid
PV Photovoltaic
WT Wind Turbine
MU Monetary Unit

Indices

t Index of time
sc Index of scenario
u Index of storage
T Traffic network nodes

Parameters

C_{cost} Time and distance costs
D_R/T_R Distance/time spent on a possible trip
θ_1/θ_2 Weight related to the importance of time/distance spent
L_{ab} Distance between two nodes a and b
S_{ini}^{SOC} State of charge at the beginning of the possible trip
$\pi_{sc,t}^{Elec}\Big/\pi_{sc,t}^{Hyd}$ Electricity/hydrogen market price
$P_{sc,t}^{Load}$ Electrical load

Renewable-to-Vehicle Smart Charging
ISBN: 978-0-443-36684-0
https://doi.org/10.1016/B978-0-443-36684-0.00008-4

Variables

$P_{sc,t}^{FCV}$ Proposed power of FCV
$P_{sc,t}^{FCVS}$ Accepted power from FCVS
$P_{sc,t}^{UP}/H_{sc,t}^{UP}$ Electricity/hydrogen exchanged with the upstream network
$P_{sc,t}^{PV}/H_{sc,t}^{WT}$ PV/WT output power
$P_{sc,t}^{FC}$ FC generation power
$H_{sc,t}^{EL}$ FC generation hydrogen
$P_{sc,t}^{dch,ESS}/P_{sc,t}^{ch,ESS}$ ESS charge/discharge power
$H_{sc,t}^{dch,ESS}/H_{sc,t}^{ch,ESS}$ ESS charge/discharge hydrogen
$x_{sc,t}^{u}$ Binary variable of the charging or discharging status of the u
$H_{sc,t}^{FC}$ Hydrogen consumption in FC
$E_{sc,t}^{EL}$ Electricity consumption in electrolyzer
T_{ab} Time interval between two traffic network nodes
V_{ab} Average speed between two traffic network nodes
$E_{c,ab}$ Energy consumption between two traffic network nodes
x_{ab} Binary variable of path selection status

8.1 Introduction

One of the predominant challenges confronting the contemporary century pertains to the mitigation of carbon dioxide emissions originating from transportation, electricity power plants, thermal, and industrial (Egeland-Eriksen et al., 2021; Aghajan-Eshkevari et al., 2023). Integrating renewable energy sources (RESs) into the electrical power system would decrease reliance on carbon dioxide (Arévalo et al., 2024; Aghajan-Eshkevari et al., 2022a,b) and commonly referred to as green energy sources or alternative energy sources, into the network infrastructure is a sustainable and effective approach (Manousakis et al., 2023; Aghajan-Eshkevari et al., 2022b). Hydrogen has the potential to serve as a fuel source in various vehicles, particularly in the context of extensive heavy transportation modalities, including trucks, maritime vessels, and aircraft, in addition to facilitating the generation and storage of electrical energy while concurrently mitigating environmental pollutants (Singh et al., 2015; Dincer and Aydin, 2023). Hydrogen can serve as an additive to natural gas for heating and replacing fossil fuels in industrial processes (Qazi, 2022). One important application of hydrogen is its use alongside renewable systems, referred to as hydrogen-based renewable systems. Hydrogen-based renewable systems are technologies that utilize hydrogen as an energy source and typically include the production, storage, and consumption of hydrogen, as well as its conversion to other energy carriers, such as electricity and heating (Rolo et al., 2024). These systems can act as part of the transition toward clean energy and reducing greenhouse gas emissions. A hydrogen-based renewable power plant designed for energy storage typically consists of the following components: a renewable energy source, such as wind or solar power; an electrolyzer, which produces hydrogen using electricity; hydrogen storage, which may involve compressed metal hydride or liquid hydrogen; and an energy conversion system, usually in the form of a fuel cell (FC) (Valverde et al., 2016). There is also the potential to integrate fuel cell vehicles (FCVs) into these systems. FCVs will be essential in the shift toward hydrogen-based energy systems (Gupta and Rahaman, 2024).

When combined with an appropriate energy source, FCs have the highest potential efficiency and the lowest possible emissions among all vehicular energy sources. As a result, extensive efforts are underway to develop FCVs (Lin et al., 2018). Proton exchange membrane fuel cells have advanced considerably for use in FCVs and other applications in recent decades (Zhao and Burke, 2009). FCVs operate fundamentally based on a chemical reaction. In these vehicles, hydrogen gas is stored in a specialized tank. When the vehicle is started, hydrogen enters an electrode and combines with oxygen from the air. This combination generates electricity, which is then sent to the motor to propel the vehicle (Nonobe, 2017). The only byproduct of this process is water vapor, which is emitted from the vehicle's exhaust. For this reason, these vehicles are considered clean and environmentally friendly. Fig. 8.1 illustrates a complex hydrogen-based renewable system created through extensive hybridization.

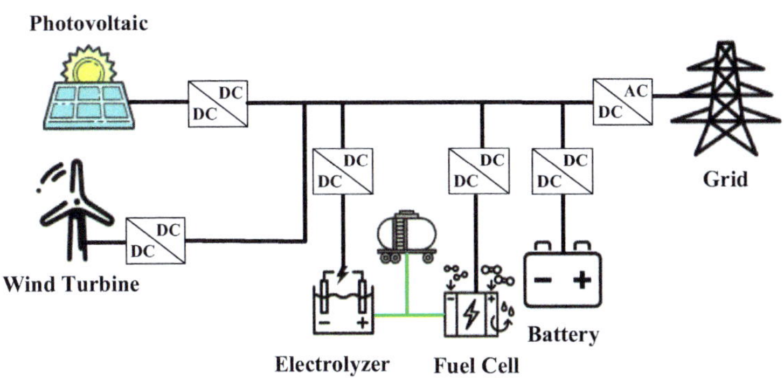

Figure 8.1 An example of a hydrogen-based renewable system.

As hydrogen energy develops, storage is a crucial topic for widespread adoption in hydrogen-based transportation and electrical systems. Currently, the most common methods for hydrogen storage include compressed hydrogen storage (Hydrogen is stored under high pressure in high-strength tanks. It's relatively straightforward and suitable for fuel cell vehicles, but requires robust safety measures due to high pressure.), cryogenic liquid storage (Hydrogen is cooled to cryogenic temperatures to liquefy it, allowing higher density storage. However, it requires significant insulation to minimize boil-off and energy costs.), and metal hydride storage (Hydrogen is absorbed into or reacted with metal alloys to form hydrides. This method offers safety benefits and higher volumetric density but can be heavy and costly, with slower hydrogen release.) (Li et al., 2019).

After assessing the pros and cons of each hydrogen storage method, compressed gas storage has gained popularity due to its cost-effectiveness, technical simplicity, high reliability, and satisfactory efficiency. The storage tank is a key technology for utilizing this technology for hydrogen storage. The material requirements for hydrogen storage tanks include high safety, reliability, cost-effectiveness, and non-reactivity with the stored hydrogen. This method is also considered the most mature and common method for hydrogen storage in FCVs (Xiao et al., 2017).

The advancement of hydrogen as an energy carrier in FC equipment has progressed significantly in recent years (Zhang et al., 2024). One of the outstanding capabilities of this equipment is the continuous generation of energy with low or zero greenhouse gas emissions, which clearly highlights the importance of its use in energy production. The method of energy generation in this equipment provides an alternative to the combustion of fossil fuels, which can, more effectively and with less emissions, convert fossil or renewable fuels into electricity (Brouwer, 2010). FC technology offers notable advantages such as high efficiency, zero pollution, and rapid refueling, which can meet energy demands in various fields, including power systems for transportation, aerospace, and marine applications. An electrolyzer is an electrochemical cell that divides water into hydrogen and oxygen using electricity. This equipment is essential for producing green hydrogen, generated from renewable electricity, and plays a crucial role in decarbonizing various industries and sectors (Mayen et al., 2024). The electrolyzer uses an electric current to split water into hydrogen and oxygen within an electrolysis cell. The current is transferred to the electrodes, where hydrogen is produced at the cathode and oxygen is released at the anode. This process occurs through chemical reactions in which electrons are transferred from the electrodes to the water. By using renewable energy sources, this hydrogen can be produced sustainably. Ultimately, hydrogen gas is stored as a clean energy source (Chisholm et al., 2022).

Hydrogen storage systems (HSS) are an effective method for long-term electricity storage (Fan et al., 2022). While HSSs have high costs and low energy efficiency, they can effectively facilitate the integration of renewable energy sources into the power grid. Therefore, when renewable energy generation is abundant and electric vehicles are unavailable for charging, the surplus energy is stored as hydrogen. The stored hydrogen is then used to supply FCVs. As a result, the energy required for FCVs is provided through renewable sources, achieving the goal of green transportation. In this regard, various studies have been conducted, and several of these studies will be examined further.

Several studies have explored developing and implementing HSSs in various applications. Reference (Egeland-Eriksen et al., 2021). reviews 15 projects in which HSS has been developed and tested. Additionally, several studies on power electronics, control systems, and energy management strategies for HSS were reviewed. The findings from the projects and articles reviewed in this paper suggest that HSSs are technically viable across various contexts, including large-scale industrial facilities, substantial commercial and public buildings, residential homes, and even small-scale mobile systems.

Khan et al. (2023) presents an advanced electrical-to-hydrogen (P2H) system that uses hydrogen as a clean energy storage carrier, which has a wide range of applications. To balance electricity production and consumption, a sophisticated mathematical solution has been developed. This solution regulates the charging and discharging patterns of both the battery and the electrolyzer, ensuring a coordinated power balance. The results demonstrate that the proposed energy management strategy not only maintains power balance in the microgrid (MG) but also significantly lowers operating costs. This research clearly shows the economic advantages of using hydrogen as an energy storage solution and sets the stage for a cleaner and more sustainable energy future.

Shahbazbegian et al. (2023) introduces a new framework to enhance the flexibility of MGs based on the concept of P2H and the capability of MGs to operate independently. For this purpose, a model for the flexible operation of developing MGs is introduced, where compressed hydrogen produced by power systems can be utilized for electricity generation through FC or sold to other industries. This study concludes that the integration of distributed power systems with hydrogen reduces emissions by 20% and cost savings of 30%.

One challenge of renewable energy systems is generating excess electrical capacity, leading to a lack of storage capability. To address this issue, Youssef et al. (2023) proposes two different solutions. The second solution involves an HSS designed to maintain large amounts of energy produced by solar systems. The findings from the scenario involving HSS technology suggest that MG is a viable new solution for rural areas rich in renewable resources. This technology has the potential for wider application as an effective option for promoting clean energy and reducing carbon emissions.

Hou et al. (2023) discusses the energy management of an integrated solar-hydrogen MG, which incorporates zero-emission vehicles, renewable energy resources, electrolyzers, bidirectional charging stations, and hydrogen refueling stations equipped with hydrogen HSS. Vehicle-to-grid (V2G) charging stations can help decrease the variability of renewable electricity by discharging energy from vehicle batteries back to the grid. Additionally, the electrolyzer can utilize excess solar production to balance supply and demand through P2H transactions. The findings show that daily operating costs can be reduced by up to 27.5%. Furthermore, integrating the P2H process with HSS demonstrates improved performance in supporting peak shaving and reducing costs, compared to relying solely on V2G or demand response strategies that use dynamic pricing.

Hwang (2013) examines hydrogen-based energy production using FCs and their application in hybrid vehicles. This study analyzes the use of FCs in hybrid vehicles. With effective control strategies, FCs can be a better option for vehicle applications than batteries. FCs are expected to have a major impact on the transportation sector in the coming years. The price of FCs is anticipated to decrease as these devices are produced in larger quantities and commercialized. It is expected that in the coming decades, transportation, power plants, and generator systems based on FCs will become prominent.

Manoharan et al. (2019) discusses the difficulties associated with the integration and optimization of hybrid renewable energy systems, especially those that incorporate hydrogen energy technology. It also explores strategies for incorporating hydrogen energy technology into hybrid energy systems, particularly emphasizing electricity generation through hydrogen FCs. A four-dimensional multi-objective meta-heuristic function is introduced, balancing technical, economic, environmental, and socio-political aspects to meet the design objectives for energy systems. The advancement of hydrogen energy technology within the global energy system has notably lagged behind that of electric energy-based technologies. This is evident in the limited number of optimization case studies that incorporate hydrogen.

Robledo et al. (2018) has shown that parked FCV can provide balancing electricity for all-electric residential buildings equipped with building-integrated photovoltaics (PVs), helping to achieve net-zero energy goals. This research indicated that the availability of vehicles, considering European parking patterns, complements periods of insufficient solar energy, primarily during nighttime and winter months. In this context, a case study was conducted at the Green Village testing site in the Netherlands. The performance of this MG was evaluated over 2 weeks in winter, during which the FCV delivered 3 and 10 kW in the first and second weeks of the experiment, respectively. The outputs demonstrated that the independence of the MG was greatly enhanced through the use of FCV-to-grid, achieving almost 50% independence in both scenarios.

Schrotenboer et al. (2022) is the inaugural study to investigate optimal strategies for managing integrated energy systems that merge renewable energy generation with HSS. The research utilizes real-world cases involving hydrogen gas. By applying the Markov decision process theory, the researchers established optimal policies for daily decision-making regarding energy storage as hydrogen, energy purchases or sales in the electricity market, and the quantity of hydrogen to be sold for gas use. The findings indicate that with the adoption of HSS and competitive prices in the hydrogen market, operational revenues could rise by as much as 51%.

Tatti et al. (2024) examines the viability and advantages of incorporating HSS into off-grid power systems. The study assessed the energy output of a green hydrogen system designed for seasonal storage on a small off-grid island in southeastern Sardinia, Italy. A mathematical model was created using MATLAB, and a parametric analysis was performed to evaluate the annual performance of a microgrid powered by renewable energy sources, taking into account various sizes of HSS and different levels of integration between the hydrogen and battery sectors. The addition of HSS to the existing microgrid led to a notable enhancement in system performance.

Fragiacomo and Piraino (2021) discusses the applications of V2G technology in evaluating combined generation with hydrogen FC-based trams. The scenario featuring the FC battery and supercapacitor connected from the rail to the building has been numerically modeled, simulated, and validated through a real case study. The results were encouraging for both operational modes: functioning as a standard tram during the day and providing electrical power along with thermal energy at night.

Shinoda et al. (2016) presents a mathematical model of an MG that incorporates FCVs, on-site hydrogen stations, PV, and a wind turbine (WT). By solving a mixed-integer linear programming (MILP) problem, the model achieves optimal scheduling for hydrogen production, hydrogen refueling of FCVs, and power supply from FCVs, all while minimizing the power imported from the main grid. The results also indicate that the stability of the MG can be enhanced by increasing the hydrogen production capacity and effectively managing hydrogen demand and V2G participation rates.

Chen et al. (2021) investigates the optimal performance of integrated power and hydrogen networks, focusing specifically on FCEVs and P2G as essential coupling points. The goal is to maximize daily profits through a probabilistic model. P2G converts solar and wind energy into hydrogen, effectively serving as a stable fueling station for FCVs. The hydrogen produced can be stored for later conversion back into electricity and methane, further boosting economic advantages. The study's key findings indicate that coupling FCEVs with P2G can enhance daily profits by up to 25.32%, while also reducing power fluctuations from wind and solar sources by 36.4%.

Table 8.1 provides a summary of the objectives and conclusions drawn from recent studies in this field. As indicated, the results of these studies present a promising future for HSS, FCVs, P2H, and hydrogen-to-power (H2P). It was also observed in most studies that operational costs decrease with the use of HSS or FCVs, leading to beneficial effects for operators. On the other hand, challenges in the integration and optimization of hydrogen-based renewable systems were identified in the studies, primarily due to the lag in electric energy-based technologies.

Table 8.2 shows a comparison between recent studies. The described papers do not consider the traffic patterns for FCVs while simultaneously converting electricity to hydrogen and hydrogen back to electricity. This chapter discusses the impact of FCVs on renewable hydrogen-based systems and subsequently presents a comprehensive model to illustrate the effects of these technologies.

In summary, the key innovations presented in this chapter include:

- Introduction of a new stochastic operational model that considers electrical and hydrogen markets, and the energy contributions from FCVS.
- Considering traffic patterns for hydrogen vehicles.
- Considering the simultaneous interaction of the residential energy supply system (RESS) and FCVS in electricity and hydrogen.

Table 8.1 Objectives and results of the studies.

Ref	Goal	Result
Egeland-Eriksen et al. (2021)	Review of 15 projects involving a hydrogen storage system.	Hydrogen storage systems are technically feasible in many different situations, including small-scale mobile system installations.
Khan et al. (2023)	Providing an advanced hydrogen power framework to achieve a balance of electricity production and consumption.	1- Reducing operating costs. 2- The economic feasibility of using hydrogen storage system as a clean energy.
Shahbazbegian et al. (2023)	Presenting a new framework for improving the flexibility of microgrids based on the power-to-hydrogen concept.	Integration of distributed power systems with hydrogen reduces emissions by 20% and cost savings of 30%.
Youssef et al. (2023)	Solving the challenge of high production of renewable resources.	Designed a hydrogen storage system to maintain large amounts of energy produced by photovoltaic.
Hou et al. (2023)	Energy management of an integrated solar-hydrogen microgrid.	1- Reduced operating costs. 2- The power-to-gas process with hydrogen storage provides better performance compared to using vehicle-to-grid.
Hwang (2013)	Investigating hydrogen-based energy production using fuel cells and their potential use in hybrid vehicles.	Fuel cells with adequate control strategies can become a more advantageous choice for vehicle applications than batteries.
Robledo et al. (2018)	Evaluation of a system including building-integrated photovoltaic, a residential building, and a fuel cell vehicle.	Greatly increasing microgrid autonomy using a fuel cell vehicle-to-grid.
Schrotenboer et al. (2022)	Investigating optimal strategies for the operation of integrated energy systems consisting of renewable energy sources and a hydrogen storage system.	The possibility of operating income of up to 51% with the introduction of hydrogen storage systems.

(*Continued*)

Table 8.1 Objectives and results of the studies.—cont'd

Ref	Goal	Result
Tatti et al. (2024)	Feasibility and benefits of integrating hydrogen storage systems into off-grid power systems.	By integrating the hydrogen storage system into the existing microgrid, a significant improvement in system performance was achieved.
Fragiacomo and Piraino (2021)	Vehicle-to-grid applications in evaluating cogeneration with fuel cell-based trams.	Ability to provide electrical and thermal energy by standard trams around the clock.
Shinoda et al. (2016)	The presentation presents a mathematical model of a microgrid that includes a fuel cell vehicle, on-site hydrogen stations, photovoltaics, and a wind turbine.	The sustainability of the microgrid can be enhanced by increasing hydrogen generation capacity and controlling hydrogen demand and V2G participation rates.
Chen et al. (2021)	Optimal performance of integrated power and hydrogen networks.	The integration of fuel cell vehicles and power-to-hydrogen increases daily profits by 25.32%.

Table 8.2 Comparison of the proposed framework with other research.

Ref	Mayen et al. (2024)	Mayen et al. (2024)	Fan et al. (2022)	Khan et al. (2023)	Hou et al. (2023)	Hwang (2013)	Schrotenboer et al. (2022)	Tatti et al. (2024)	Fragiacomo and Piraino (2021)	This chapter
Hydrogen market	×	×	×	×	×	✓	×	×	✓	✓
Electricity market	✓	✓	×	✓	✓	✓	×	✓	✓	✓
FCVS	✓	×	×	✓	✓	×	×	×	×	✓
Traffic pattern	×	×	×	×	×	×	×	×	×	✓
P2H	✓	✓	✓	✓	×	✓	×	✓	×	✓
H2P	✓	✓	✓	✓	✓	✓	✓	✓	×	✓
HES	✓	✓	✓	✓	×	✓	✓	✓	✓	✓
EES	✓	×	×	×	×	×	×	×	✓	✓
Uncertainty	✓	×	✓	✓	×	✓	×	✓	×	✓

8.2 Problem modeling and formulation

This study proposes a two-stage energy management algorithm, as shown in Fig. 8.2, for the economically efficient energy supply of a residential building for the day ahead. The first stage consists of two levels: in the first level, the traffic pattern for the day ahead is forecasted, which will output the required travel time between two nodes in the transportation network in hours. In the second level, the proposed electrical energy for each FCV is specified by the fuel cell vehicle charging station (FCVS) operator. In this context, the proposed hydrogen amount required for the FCVS, which entails uncertainty, will be forecasted. The aggregated proposed electric amounts from the FCs and the suggested hydrogen amounts are fed as inputs into the second stage of the proposed algorithm. In the second stage, by executing the optimization problem, the amount of energy produced by the equipment, the energy exchanged with upstream networks, and the accepted amounts from the FCVS will be determined. Finally, the amounts accepted by the FCVS operator will be sent. The

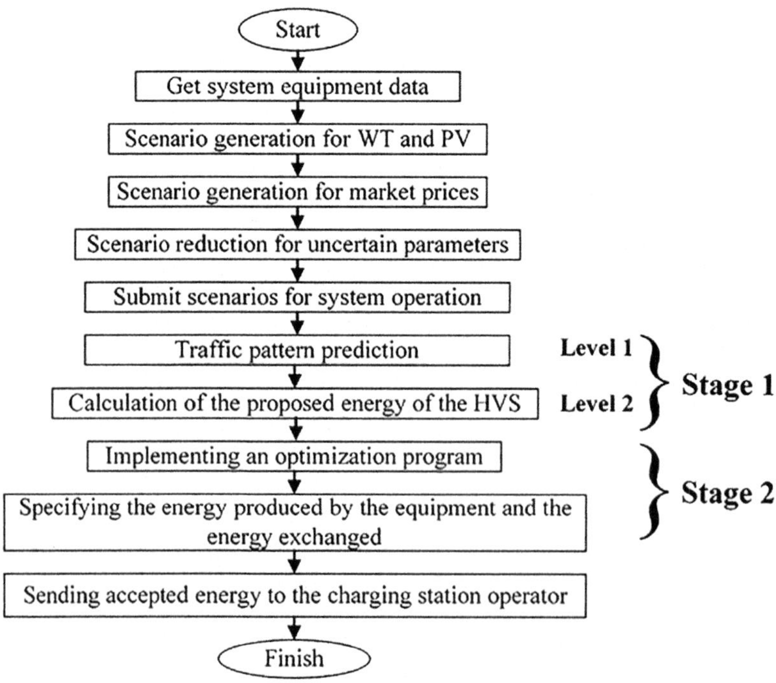

Figure 8.2 Framework of the proposed method.

proposed model is a MILP model, which is solved using GAMS software in conjunction with the CPLEX solver.

The proposed linear model aims to provide an economically efficient energy supply for a residential building energy system equipped with various electrical energy generation facilities for the day ahead. The operator's objective in this optimization problem is to develop an economical energy production and exchange schedule for the day ahead. The presented system features a bidirectional connection with both the upstream electric and hydrogen networks, allowing for energy buying and selling. Another capability of this system is the P2H and H2P, which complicates the decision-making process for optimal operation. Additionally, the system can exchange energy with FCVs through FCVSs.

Fig. 8.3 illustrates the structure of a hydrogen-based renewable system for supplying electrical energy to a residential building. The proposed system consists of a WT, PV, FC, electrolyzer, electric storage system (ESS), and an HSS. The connection to the upstream electric network is facilitated through a transformer, while a pipeline is tapped into the hydrogen network. Additionally, the FCVS serves as an interface between FCVs and the residential energy supply system. The P2H and H2P are enabled through the electrolyzer and FC, respectively. Furthermore, an energy management system is utilized within the proposed structure.

Considering the uncertain nature of parameters such as electrical load, wind speed, solar radiation, and market prices of electricity and hydrogen, a scenario-based approach is used to model these uncertain parameters. The

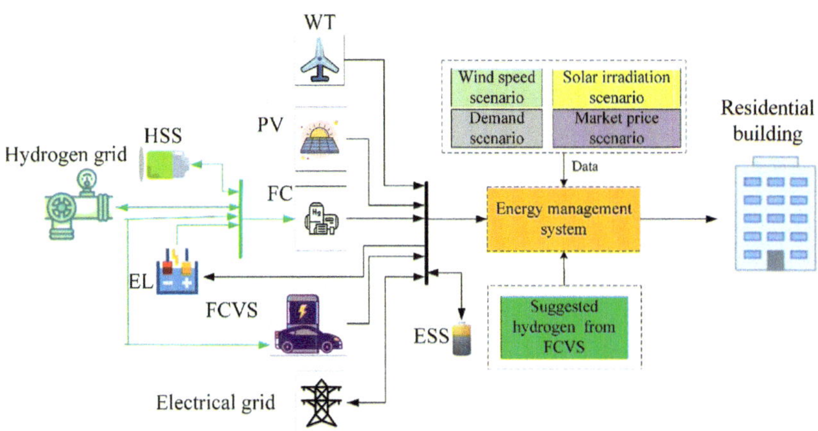

Figure 8.3 Schematic of the presented system.

optimization problems consist of various scenarios with high complexity, computational volume, and time constraints. By reducing the number of scenarios, both computational load and simulation time are decreased. To generate random data and subsequently reduce the scenarios, a data-generating function must first be established. Each uncertain parameter is modeled with a probability distribution function.

In this study, wind speed is modeled using a Weibull distribution function. This function is presented in Eq. (8.1). The shape and scale parameters of this probability distribution function are represented by λ and k, respectively (Yarmohammadi and Abdi, 2024).

$$f(x_t;\lambda,k)=\begin{cases}\frac{k}{\lambda}\left(\frac{x_t}{\lambda}\right)^{k-1}e^{-\left(\frac{x_t}{\lambda}\right)^k}, x_t\geq 0\\ 0, x_t<0\end{cases} \tag{8.1}$$

Solar radiation is modeled with a beta distribution function, which is expressed in Eq. (8.2). This function is characterized by two parameters that determine its shape (α and β), and B is a normalization factor (Yarmohammadi and Abdi, 2024).

$$f(x_t;\alpha,\beta)=\frac{1}{B(\alpha,\beta)}x^{\alpha-1}(1-x_t)^{\beta-1}, \alpha=\frac{\mu_{x_t}\beta}{1-\mu_{x_t}}\&$$
$$\beta=\left(1-\mu_{x_t}\right)\left(\frac{\mu_{x_t}\left(1-\mu_{x_t}\right)}{\sigma_{x_t}^2}-1\right) \tag{8.2}$$

Finally, market prices and electrical load models are represented using a normal distribution function that is expressed in Eq. (8.3). σ_x^2 denote the mean, and μ indicates the standard deviation of the available historical data (Ahmadi and Nazar, 2025).

$$f\left(x;\mu_x,\sigma_x^2\right)=\frac{1}{\sqrt{2\pi\sigma_x^2}}\exp\left[-\frac{\left(x-\mu_x\right)^2}{2\sigma_x^2}\right] \tag{8.3}$$

The roulette wheel mechanism is employed to create a set of scenarios whereby a random number between 0 and 1 is generated for each uncertain parameter. This process continues until a sufficient number of scenarios are produced, and eventually, scenarios with low probability and similar probabilities are eliminated (Azizivahed et al., 2024).

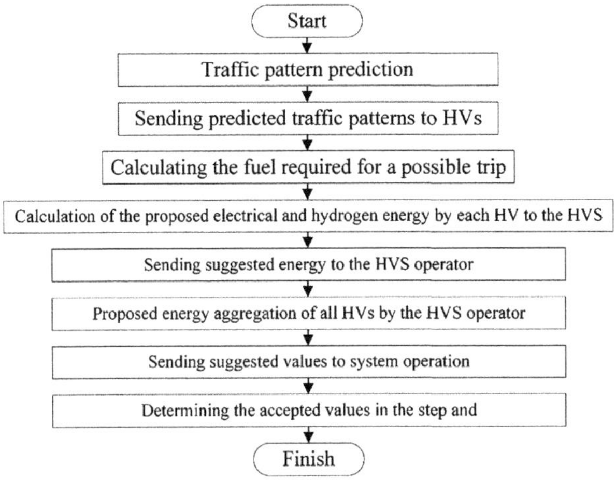

Figure 8.4 Flowchart of the first stage of the proposed optimization model.

FCVs can exchange energy with the residential energy supply system through the connected charging station. These devices function as mobile storage units for the system. The charging station operator acts as an aggregator, initially forecasting an hourly traffic pattern for the day ahead, which is then communicated to each FCV. The energy consumption for the potential trip is calculated accordingly. Subsequently, the proposed energy amount for each FCV is computed based on the objective function and sent to the charging station operator. The operator receives and aggregates all proposed values from the FCVs and ultimately forwards the consolidated information to the energy supply system operator. Fig. 8.4 illustrates the steps involved in modeling the electrical behavior of FCVs.

8.2.1 Modeling of the first stage

In this research, a model for traffic patterns is presented, designed to optimize traffic conditions and make efficient use of available resources. This model includes an objective function and various constraints. The primary goal of this model is to minimize the total costs associated with distance and travel time ($C_{\cos\ t}$) in the traffic network. Eq. (8.4) represents the objective function of the traffic pattern, where the first term accounts for distance costs, and the second term accounts for time costs. In this equation, the time taken for a potential trip and the distance traveled for a potential trip are denoted by T_R and D_R, respectively. θ_1 and θ_2 represent the weights corresponding to the importance of distance and time, respectively. These coefficients may differ based on the preferences of vehicle owners.

$$\min C_{\cos\text{ t}} = \theta_1 D_R + \theta_2 T_R \tag{8.4}$$

The distance between the origin and destination of a potential trip is calculated using Eq. (8.5). In this equation, x_{ab} represents the binary variable indicating the selected route for the potential trip, as well as L_{ab} is the distance of each path. a and b denote the nodes of the traffic network.

$$D_R = \sum_{ab \in T} x_{ab} L_{ab} \tag{8.5}$$

The travel time cost is also expressed in Eq. (8.6). By allowing the objective function to depend on both time and distance, decision-makers can adjust the coefficients to achieve an optimal distribution between time and distance, thereby meeting traffic management goals and reducing costs. Here, T_{ab} represents the time interval between two nodes in the transportation network.

$$T_R = \sum_{ab \in T} x_{ab} T_{ab} \tag{8.6}$$

The time required to traverse between two points is dependent on the length of the route and the speed of the vehicle, as expressed in Eq. (8.7). Here, V_{ab} represents the average speed of the FCV.

$$T_{ab} = \frac{L_{ab}}{V_{ab}} \tag{8.7}$$

Eq. (8.8) expresses the amount of fuel consumption, a crucial parameter for FCV owners' decision-making. This Equation calculates the energy consumption between two nodes ($E_{c,ab}$) and relates it to the selected routes, speed, and distance (Ahmadi and Nazar, 2025).

$$E_{c,ab} = \left(\left(\frac{1.359}{V_{ab}} \right) - 0.003 V_{ab} + 2.981 \times 10^{-5} V_{ab}^2 + 0.218 \right) L_{ab} \tag{8.8}$$

The energy consumption for a potential trip is subject to limitations, as expressed in Eq. (8.9). These constraints are defined to prevent excessive energy consumption and to ensure the safety of the journey. In this Equation, $S_{\text{ini}}^{\text{SOC}}$ represents the hydrogen tank charge status at the beginning of the trip, which is expressed as a percentage. E_{car} indicates the capacity of hydrogen stored in the vehicle's tanks.

$$\sum_{ab\in R} x_{ab} E_{c,ab} \leq S_{\text{ini}}^{\text{SOC}} E_{\text{car}} \tag{8.9}$$

After forecasting the traffic pattern and calculating the fuel for a potential trip, the proposed electrical and hydrogen energy amount for each FCV is determined based on the objective function expressed in Eq. (8.10). According to this equation, the goal of each FCV owner is to minimize costs. In this objective function, the first term represents the hydrogen charging cost for each FCV, while the second and third terms reflect the profit obtained from selling hydrogen and electricity through the charging station to the residential energy supply system. In this equation, $\pi_{sc,t}^{\text{Hyd}}$ and $\pi_{sc,t}^{\text{Elec}}$ represent the market prices of hydrogen and electricity, respectively, while $P_{sc,t}^{\text{FCV}}$ denotes the electrical power proposed by each FCV. $H_{sc.t}^{ch,FCV}$ and $H_{sc,t}^{FCV}$ are FCV's hydrogen charge and sold to the RESS, respectively.

$$M = \sum_{sc} \sum_{t} \left(\pi_{sc,t}^{\text{Hyd}} \cdot H_{sc,t}^{ch,FCV} - \pi_{sc,t}^{\text{Hyd}} \cdot H_{sc,t}^{FCV} - \pi_{sc,t}^{\text{Elec}} \cdot P_{sc,t}^{FCV} \right) \tag{8.10}$$

The state of hydrogen stored in each FCV hydrogen storage device can be expressed by Eq. (8.11). This equation expresses the state of charge instantaneously. In this equation, $\text{SOC}_{sc,t}^{FCV}$ shows the state of charge of FCV's hydrogen storage device. $\eta_{ch,FCV}$ and $\eta_{dch.FCV}$ are, respectively, the charging and discharging efficiencies of the FCV's hydrogen storage system. $H_{sc,t}^{dch.FCV}$ is FCV's hydrogen discharge.

$$\text{SOC}_{sc,t}^{FCV} = \text{SOC}_{sc,t-1}^{FCV} + \left(H_{sc,t}^{ch,FCV} \eta_{ch,FCV} - \frac{H_{sc,t}^{dch,FCV}}{\eta_{dch,FCV}} \right) \Delta t \tag{8.11}$$

Eqs. (8.12) and (8.13) represent the charging and discharging constraints of the hydrogen storage system in FCV. These equations indicate that the hydrogen charging and discharging must be within the minimum and maximum nominal capacity limits. The maximum charging and discharging capacities are respectively determined by $H_{\max}^{ch,FCV}$ and $H_{\max}^{dch,FCV}$. The binary variable $x_{sc,t}^{FCV}$ indicates the charging state of the hydrogen storage system. As shown by the equations, this storage system cannot charge and discharge simultaneously Eq (8.14) (states that the discharge hydrogen of each FCV is divided into two parts: hydrogen that is discharged directly ($H_{sc,t}^{FCV}$) and hydrogen converted to electricity ($H_{sc,t}^{FCV2P}$).

$$0 \leq H_{sc,t}^{ch,FCV} \leq x_{sc,t}^{FCV}.H_{\max}^{ch,FCV} \tag{8.12}$$

$$0 \leq H_{sc,t}^{dch,FCV} \leq \left(1 - x_{sc,t}^{FCV}\right).H_{\max}^{dch,FCV} \tag{8.13}$$

$$H_{sc,t}^{dch.FCV} = H_{sc,t}^{FCV} + H_{sc,t}^{FCV2P} \tag{8.14}$$

The amount of hydrogen stored in each vehicle is limited to be between its minimum ($\text{SOC}_{\min}^{FCV}$) and maximum ($\text{SOC}_{\max}^{FCV}$) nominal capacities, as shown by Eq. (8.15). Eq. (8.16) states that the stored energy at the beginning of the operation period must equal the stored energy at its end.

$$\text{SOC}_{\min}^{FCV} \leq \text{SOC}_{sc,t}^{FCV} \leq \text{SOC}_{\max}^{FCV} \tag{8.15}$$

$$\text{SOC}_{sc,t0}^{FCVS} = \text{SOC}_{sc,t24}^{FCVS} \tag{8.16}$$

The electrical energy produced by each FCV is defined by Eq. (8.17). In this equation $\eta_{FCV,H2P}$ represents the H-to-P conversion efficiency for each FCV.

$$P_{sc,t}^{FCV} = \eta_{FCV,H2P} \cdot H_{sc,t}^{FCV2P} \tag{8.17}$$

8.2.2 Modeling of the second stage

The residential energy supply system operator aims to minimize operating costs while determining the equipment status and the amount of energy exchanged with external factors. For this purpose, the operator provides an optimal combination of resource output power and energy exchanged with upstream networks. The objective function to minimize operating costs over 24 h is presented in Eq. (8.18). In this Equation, the first term represents the cost of energy exchanged with the upstream network, calculated by multiplying the market price of electricity ($\pi_{sc,t}^{\text{Elec}}$) by the amount of electrical energy exchanged ($P_{sc,t}^{\text{UP}}$). The second term accounts for the cost of hydrogen exchanged with the upstream network, calculated by multiplying the market price of hydrogen ($\pi_{sc,t}^{\text{Hyd}}$) by the hydrogen energy exchanged ($H_{sc,t}^{\text{UP}}$). The third term indicates the cost of electricity purchased from the FCVS, where $P_{sc,t}^{FCVS}$ represents the accepted energy from the proposed values. The fourth term represents the cost of hydrogen accepted from the proposed values, calculated by multiplying the price by the accepted quantities. The fifth expression shows the profit from selling hydrogen to FCVS. The sixth term is the operating cost of the PV system, dependent on the amount of energy produced ($P_{sc,t}^{PV}$) and the operating cost factor

(K_{OM}^{PV}). The seventh term represents the operating cost of the WT, calculated by multiplying the produced power ($P_{sc,t}^{WT}$) by the operating cost factor (K_{OM}^{WT}). The eighth term denotes the operating cost of the FC, which is dependent on the produced power ($P_{sc,t}^{FC}$) and the operating cost factor (K_{OM}^{FC}). The ninth term represents the operating cost of the electrolyzer, where $H_{sc,t}^{EL}$ is the power produced by this equipment and K_{OM}^{EL} is the operating cost factor. The 10th term indicates the operating cost of the ESS based on the charge and discharge energy ($P_{sc,t}^{ESS}$) and the operating cost factor (K_{OM}^{ESS}). Finally, the 11th term reflects the operating cost of hydrogen storage, calculated by multiplying charge and discharge energy ($H_{sc,t}^{HSS}$) by the operating cost factor (K_{OM}^{HSS}).

$$\min: \sum_{sc}\sum_{t}\Big(\pi_{sc,t}^{\text{Elec}}.P_{sc,t}^{Up} + \pi_{sc,t}^{\text{Hyd}}.H_{sc,t}^{Up} + \pi_{sc,t}^{\text{Elec}}.P_{sc,t}^{FCVS} + \pi_{sc,t}^{\text{Hyd}}.H_{sc,t}^{FCVS}$$
$$-\pi_{sc,t}^{\text{Hyd}}.H_{sc,t}^{ch,FCVS} + P_{sc,t}^{PV}\cdot K_{OM}^{PV} + P_{sc,t}^{WT}\cdot K_{OM}^{WT} + P_{sc,t}^{FC}\cdot K_{OM}^{FC} + H_{sc,t}^{EL}\cdot K_{OM}^{EL}$$
$$+ P_{sc,t}^{ESS}\cdot K_{OM}^{ESS} + H_{sc,t}^{HSS}\cdot K_{OM}^{HSS}\Big) \tag{8.18}$$

Establishing electrical power balance in any system means achieving equilibrium between the produced and injected power of that system. Maintaining power balance in the electrical network contributes to the stability of the system. This constraint prevents fluctuations in frequency and voltage. The electrical power balance is depicted in Eq. (8.19). In this Equation, $P_{sc,t}^{ch,ESS}$ and $P_{sc,t}^{dch,ESS}$ represent the charge and discharge energy of the ESS, respectively, while $P_{sc,t}^{Load}$ denotes the load on the system.

The balance of hydrogen energy in a system is of great significance as it ensures that the input and output energy quantities are in equilibrium for the system to operate stably. This balance contributes to optimizing system efficiency and reducing energy losses. Furthermore, maintaining energy balance leads to lower costs and mitigates negative environmental impacts. This balance is expressed in Eq. (8.20). This Equation indicates the energy balance between hydrogen purchased from the upstream network, the hydrogen produced by the electrolyzer ($H_{sc,t}^{EL}$), hydrogen accepted from FCVS, the hydrogen discharged from the HSS ($H_{sc,t}^{dch,HSS}$), the hydrogen demand of the system ($H_{sc,t}^{Load}$), the hydrogen charged in the HSS ($H_{sc,t}^{HSS}$), and the hydrogen consumed by the FC ($H_{sc,t}^{FC}$).

$$P_{sc,t}^{\mathrm{UP}} + P_{sc,t}^{FCVS} + P_{sc,t}^{WT} + P_{sc,t}^{PV} + P_{sc,t}^{FC} + P_{sc,t}^{dch,ESS} = P_{sc,t}^{Load} + P_{sc,t}^{ch,ESS} + P_{sc,t}^{EL} \tag{8.19}$$

$$H_{sc,t}^{\mathrm{UP}} + H_{sc,t}^{EL} + H_{sc,t}^{FCVS} + H_{sc,t}^{dch,HSS} = H_{sc,t}^{Load} + H_{sc,t}^{ch,HSS} + H_{sc,t}^{FC} \tag{8.20}$$

The charging and discharging processes for ESS and HSS are identical and adhere to a series of constraints, such as capacity and technical limitations. Accordingly, the overall dynamic model for ESS and HSS is expressed in Eq. (8.21) to Eq. (8.25) (Aghajan-Eshkevari et al., 2024). In these equations, the type of energy storage, which includes both electrical and hydrogen, is indicated by u index. Eq. (8.21) pertains to the energy level of the storage, while Eq. (8.22) and Eq. (8.23) relate to the maximum charging and discharging energy, respectively. Eq. (8.24) addresses the constraint on the energy stored, and Eq. (8.25) illustrates the equality of the energy in the storage at the beginning and end of the operation. Here, $P_{sc,t}^{ch,u}$ represents the energy charged in the storage, $P_{sc,t}^{dch,u}$ denotes the energy discharged from the storage, and $\eta_{ch,u}$ and $\eta_{dch,u}$ indicate the charging and discharging efficiencies, respectively. The maximum charging and discharging energy are represented by $P_{\max}^{ch,u}$ and $P_{\max}^{dch,u}$, while the maximum and minimum energy stored in the storage systems are denoted by $E_{\max}^{u}$ and $E_{\min}^{u}$ as well. The binary variable $x_{sc,t}^{u}$ indicates the charging and discharging status of the storage system (Bostan et al., 2020).

$$E_{sc,t}^{u} = E_{sc,t-1}^{u} + \left(P_{sc,t}^{ch,u} \eta_{ch,u} - \frac{P_{sc,t}^{dch,u}}{\eta_{dch,u}} \right) \Delta t \tag{8.21}$$

$$0 \leq P_{sc,t}^{ch,u} \leq x_{sc,t}^{u} . P_{\max}^{ch,u} \tag{8.22}$$

$$0 \leq P_{sc,t}^{dch,u} \leq \left(1 - x_{sc,t}^{u} \right) . P_{\max}^{dch,u} \tag{8.23}$$

$$E_{\min}^{u} \leq E_{sc,t}^{u} \leq E_{\max}^{u} \tag{8.24}$$

$$E_{sc,t0}^{u} = E_{sc,t24}^{u} \tag{8.25}$$

The FC converts the incoming hydrogen into electrical energy to supply electric loads. The electricity produced by the FC is expressed in Eq. (8.26). This device converts the incoming hydrogen into electricity based on a conversion efficiency factor η_{H2E}^{FC}. Additionally, this device has a production capacity limitation, which is described in Eq. (8.27). The output of this device must be between the minimum ($P_{\min}^{FC}$) and maximum ($P_{\max}^{FC}$) production levels (Mohammadi et al., 2014).

$$P_{sc.t}^{FC} = \eta_{H2E}^{FC} \cdot H_{sc.t}^{FC} \tag{8.26}$$

$$P_{\min}^{FC} \leq P_{sc.t}^{FC} \leq P_{\max}^{FC} \tag{8.27}$$

Eq. (8.28) represents the model of the electrolyzer, which decomposes water using electricity to produce hydrogen. In this process, electrical energy is converted into chemical energy. The conversion efficiency factor (η_{E2H}^{EL}) serves as a measure of the electrolyzer's performance, indicating how much electrical energy is transformed into hydrogen. This factor depends on the type of electrolyzer, as well as the operational temperature and pressure. Optimizing this factor can enhance productivity and reduce hydrogen production costs. Additionally, Eq. (8.29) illustrates the limitation on the hydrogen production of this device, stating that the hydrogen output must be between its maximum ($H_{\max}^{EL}$) and minimum ($H_{\min}^{EL}$) values (Mohammadi et al., 2014).

$$H_{sc.t}^{EL} = \eta_{E2H}^{EL} \cdot P_{sc.t}^{EL} \tag{8.28}$$

$$H_{\min}^{EL} \leq H_{sc,t}^{EL} \leq H_{\max}^{EL} \tag{8.29}$$

Considering the installed capacity of the PV system and the WT at the system, these renewable sources have maximum production capacities, which impose limitations on the output power of each of these renewable resources. Eqs. (8.30) and (8.31) express these constraints, where the installed capacity of the PV system is equal to $P_{\max}^{PV}$, and the installed capacity of the WT is equal to $P_{\max}^{WT}$.

$$0 \leq P_{sc,t}^{PV} \leq P_{\max}^{PV} \tag{8.30}$$

$$0 \leq P_{sc,t}^{WT} \leq P_{\max}^{WT} \tag{8.31}$$

Considering the capacity of the branching electrical line ($P_{\max}^{\mathrm{UP}}$), Eq. (8.32) represents a limitation on the exchange of electrical power with the upstream network. This limitation also applies to the hydrogen network's branching pipeline, as indicated in Eq. (8.33), where $H_{\max}^{\mathrm{UP}}$ represents the capacity of the branching pipeline (Ahmadi et al., 2025).

$$-P_{\max}^{UP} \leq P_{sc,t}^{UP} \leq P_{\max}^{UP} \tag{8.32}$$

$$-H_{\max}^{UP} \leq H_{sc,t}^{UP} \leq H_{\max}^{UP} \tag{8.33}$$

8.3 Numerical analysis

8.3.1 Input data

Electrical power exchanged between the energy supply system of the residential house and the upstream network passes through the transformer between them; therefore, power losses during the exchange are neglected, assuming a transformer efficiency of 1. The capacity of the electrical branch from the upstream network is considered to be 80 kW. After generating scenarios to reduce computational load, 10 reduction scenarios remain for each uncertain parameter, and the averages of these scenarios are illustrated in Fig. 8.5. Information regarding the scenario generation and reduction methods is provided in (Azizivahed et al., 2024). Additionally, the system equipment details can be found in Table 8.3.

8.3.2 Simulation results

In the first level of the optimization problem, the proposed values of hydrogen and electricity for each FCV are calculated, and finally, they are aggregated by the FCVS operator. Fig. 8.6 shows the proposed hydrogen values for FCVS over the operational period. These values represent the aggregate of all vehicles belonging to FCVS, where the proposed amounts depend on two factors: the traffic pattern and the hydrogen market price. These values include recommendations for buying and selling hydrogen.

In the following section, the numerical results obtained from the simulation are examined. The operator must determine how much energy each of the production units should generate, the behavior of the storage systems, the amount of exchanged energy with the main network, and the accepted electricity and hydrogen from the FCVS. Fig. 8.7 depicts the proposed electrical energy for the FCVS. These results are derived from aggregating the proposed electrical energy for each FCV by the operator of the charging station during the first stage of the proposed algorithm. Throughout the operational period, the energy proposal from the charging station to the operator of the residential energy supply system is equal to 161 kWh. These

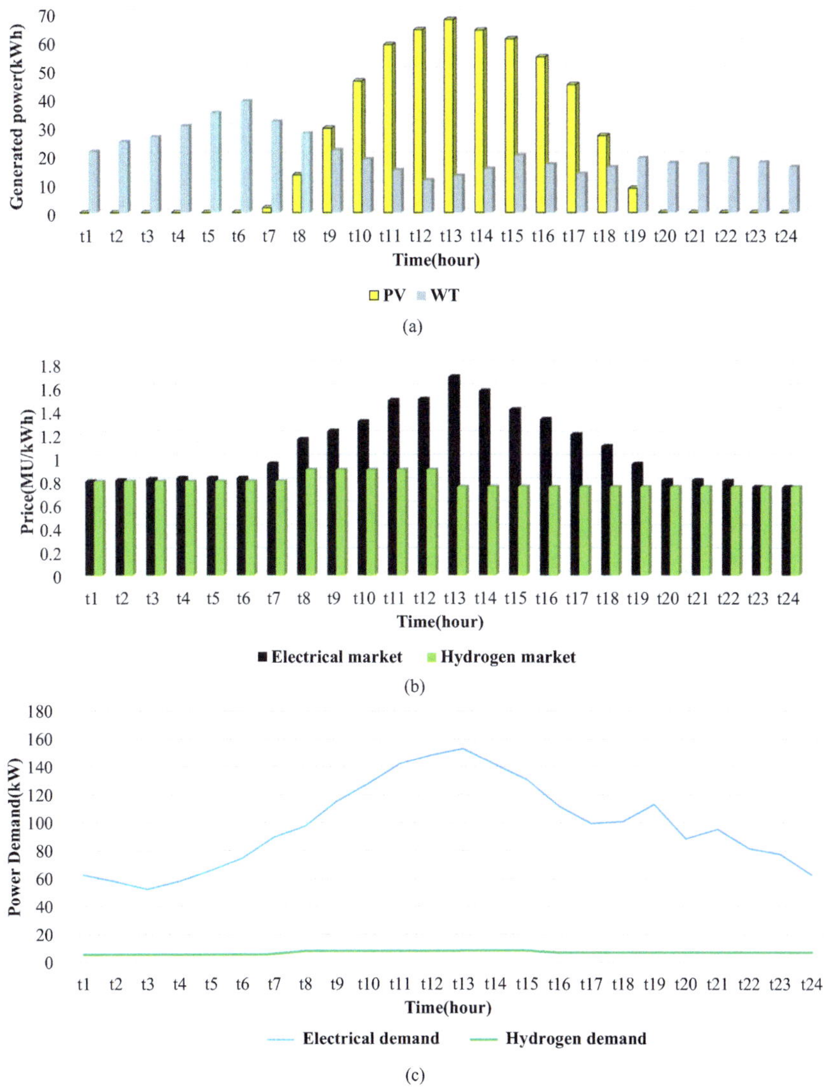

Figure 8.5 Average of scenarios with uncertainty for (a) PV and WT generation (b) electrical and hydrogen market price (c) electrical and hydrogen demand.

proposed values depend on the market price and traffic pattern. Therefore, FCV owners tend to suggest higher values during times of high electricity prices to maximize their profits.

Fig. 8.8 illustrates the electrical energy accepted from the FCVS. These values were calculated in the second stage of the proposed algorithm,

Table 8.3 System equipment data (Bostan et al., 2020).

Equipment	Data
Fuel cell	$\eta_{H2E}^{FC} = 0.9, P_{\max}^{FC} = 80, P_{\min}^{FC} = 10$
Electrolyzer	$\eta_{E2H}^{EL} = 0.85, H_{\max}^{EL} = 100, H_{\min}^{EL} = 20$
Photovoltaic	$P_{\max}^{PV} = 300$
Wind turbine	$P_{\max}^{WT} = 300$
Electrical energy storage	$\text{SOCSOC}_{\max}^{EES} = 200, P_{\max}^{ch} = 50, P_{\max}^{dch} = 50 \quad \eta_{dch}^{EES} = 0.96, \eta_{ch}^{EES} = 0.96$
Hydrogen energy storage	$\text{SOC SOC}_{\max}^{HES} = 200, H_{\max}^{ch} = 50, H_{\max}^{dch} = 50, \eta_{dch}^{HES} = 1, \eta_{ch}^{HES} = 1$
Grid	$P_{\max}^{\text{UP}} = 80, H_{\max}^{\text{UP}} = 150$

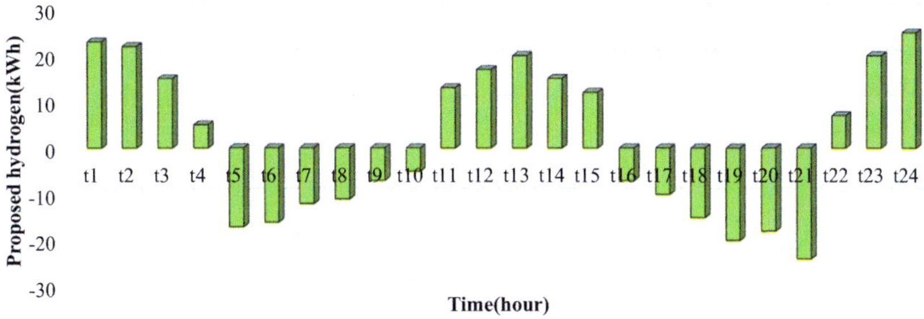

Figure 8.6 Proposed hydrogen to the residential building electrical energy supply system by the FCVS operator.

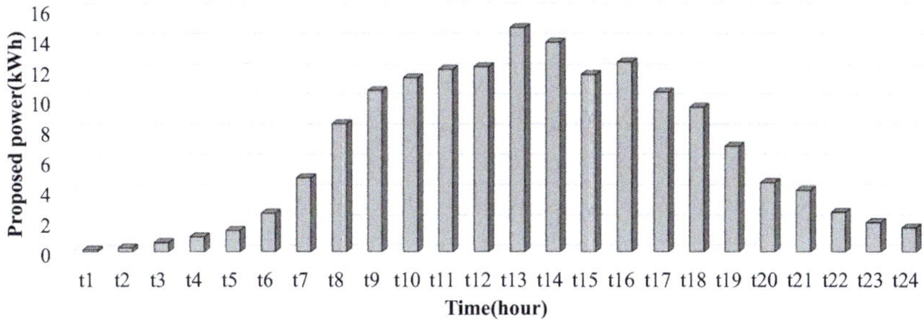

Figure 8.7 Proposed electricity to the residential building electrical energy supply system by the FCVS operator.

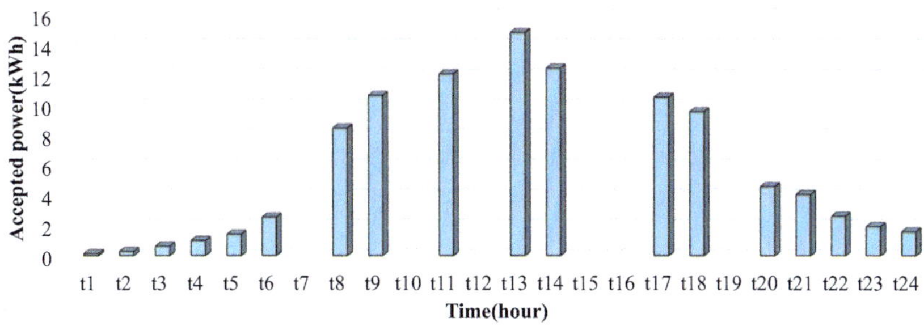

Figure 8.8 Electrical energy accepted from FCVS.

employing an economic approach from the perspective of the residential electricity supply system operator. During the operational period, 100 kWh of electrical energy was accepted from the FCVS, which is 62.5% of the proposed values. The basis for accepting the proposed electrical energy is rooted in economic considerations, determining the accepted amounts based on the reduction of operational costs. The cost of the

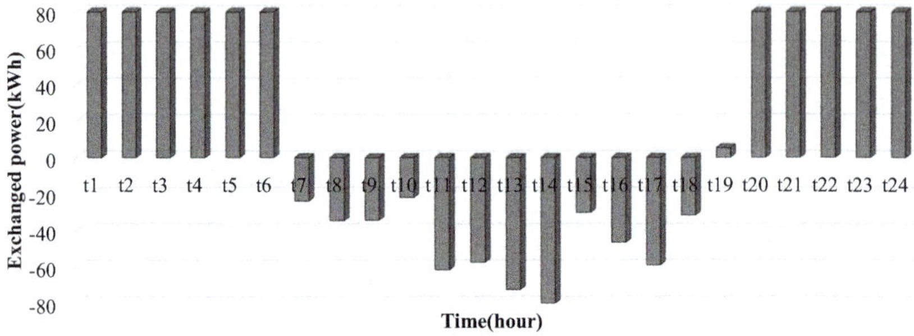

Figure 8.9 Electricity exchanged with the upstream network.

electricity purchased from the FCVS amounts to 125.5 MU over the operational period. The higher the accepted electrical energy, the more it contributes to reducing pollutant emissions.

Fig. 8.9 shows the electrical energy exchanged with the upstream network. The amount of exchanged power depends on market prices. For example, during the hours from 01:00 to 06:00, when market prices are low, it is more economical for the system to buy electrical energy from the main grid. In contrast, during the hours from 07:00 to 18:00, when market prices are high, the operator decides to produce energy internally to meet system loads and sell excess energy to the upstream network for profit. Throughout the operational period, the energy purchased from the upstream network and the energy sold back to it amount to 885 kWh and 553 kWh, respectively. The cost of purchasing electrical energy from the upstream network is 712 MU, while the profit from selling electrical energy is 770 MU, resulting in a net profit of 58 MU for the operator.

Fig. 8.10 shows the electricity generated by the FC. Due to the conversion efficiency of hydrogen gas to electricity, the performance of this device

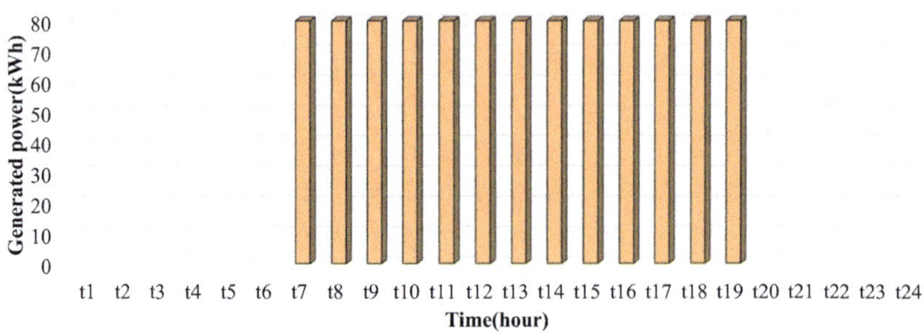

Figure 8.10 Electricity generated by FC.

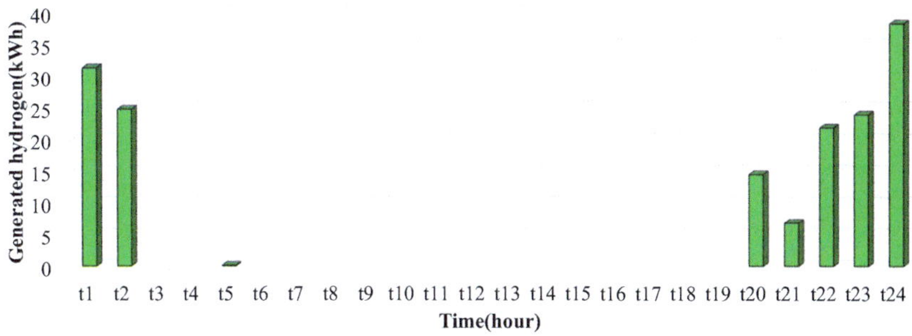

Figure 8.11 Hydrogen generated by electrolyzer.

depends on the market prices of electricity and hydrogen. During the hours from 07:00 to 19:00, when the hydrogen market price is lower, it is economical to purchase hydrogen and convert it to electricity. Although the hydrogen price is also lower at 20:00, the conversion efficiency makes it uneconomical at that time. Over 24 h, this device generates 1040 kWh of electrical energy at its maximum capacity. During the hours when this device is operational, it acts as a seller of electrical energy to the upstream network.

Fig. 8.11 illustrates the hydrogen produced by the electrolyzer. The electrolyzer establishes the connection between electrical energy and hydrogen by converting the system's electricity into hydrogen. This device decides whether to commit to an operation based on a comparison of market prices for electricity and hydrogen, as well as the system's hydrogen needs. For example, at 02:00, when the electricity market price is slightly higher than the hydrogen market price, the electrolyzer commits to operation and injects hydrogen into the system, which is an economical action. Throughout the operational period, the hydrogen produced by this device amounts to 148 kWh, with a financial value of 113.9 MU.

Fig. 8.12 illustrates the hydrogen exchanged with the upstream network. The price of hydrogen is categorized into three levels: 0.75, 0.80, and 0.90 MU. Given that the price of electricity is higher, converting hydrogen to electricity becomes more economical to meet energy needs. For example, during the period from 05:00 to 18:00, when the system acts as a seller of electrical energy to the upstream network, hydrogen is purchased from the upstream network to be converted into electricity, and the generated electricity is subsequently sold back to the upstream network for profit. Overall, the energy exchange with the upstream network incurs costs, amounting to 803 MU.

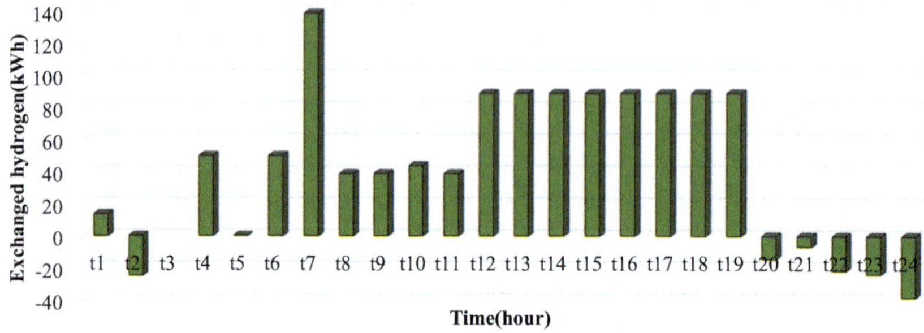

Figure 8.12 Hydrogen exchanged with the upstream network.

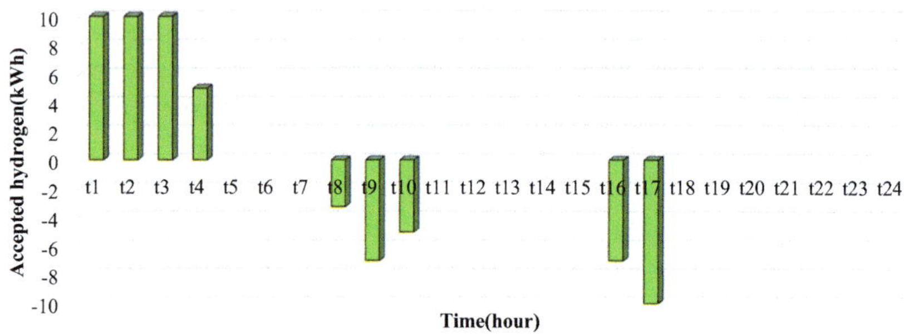

Figure 8.13 Hydrogen energy accepted from FCVS.

Fig. 8.13 shows the accepted amounts of proposed hydrogen. Negative values indicate sales by the residential energy supply system operator, while positive values indicate purchases. Over 24 h, the accepted amounts for hydrogen purchases and sales are 35 kWh and 32 kWh, respectively. The cost of purchases in this 24-hour transaction amounts to 28 MU, while the cost of sales is 26 MU. Due to the limited capacity of this charging station located within the system, the exchanged amounts are relatively low. Consequently, this energy exchange contributes to green fuel production.

Fig. 8.14 depicts the charging and discharging behavior of the ESS and HSS. These storage systems charge during periods of low price and demand, and discharge during periods of high price and demand. The costs associated with charging the ESS and HSS are 168 and 156 MU, respectively. Conversely, during discharge, these storage systems have values of 293 MU and 175 for ESS and HSS, respectively. Thus, the profit derived from using the ESS and HSS amounts to 125 and 20 MU, respectively. The higher profit from the ESS is attributed to the significant price difference during the charging and discharging periods.

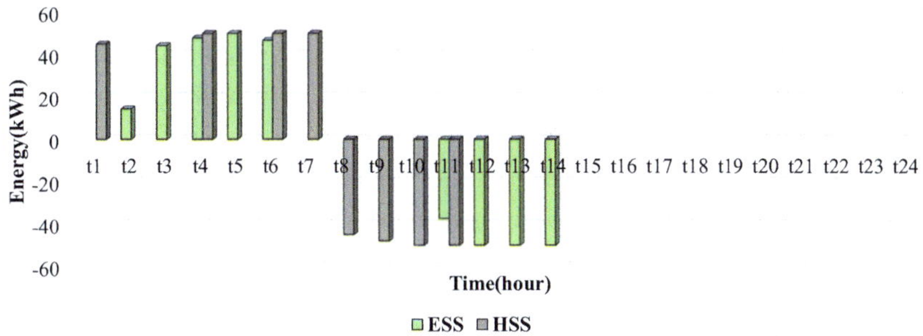

Figure 8.14 Charging and discharging behavior of the ESS and HSS.

The operator of the residential energy supply system, aiming to minimize operating costs, determined the commitment status of the equipment, the energy exchanged with the upstream network, and the energy accepted from the FCVS in such a way that the operating cost of the system was calculated to be 859 MU over a 24-hour period. A significant portion of the costs is attributed to the purchase of hydrogen from the upstream network, amounting to 803 MU. The cost of energy purchased from the FCVS was recorded as 125MU. Furthermore, the exchange of electrical energy resulted in a profit of 58 MU. Other costs related to the system pertain to the operational expenses of the equipment.

In another case study, it was assumed that the FCVS was not included in the proposed system structure. In this case, the operating cost of the system was calculated to be 904 MU, reflecting a 2.5% cost increase. It is observed that the use of FCVs not only reduces pollutants but also leads to a decrease in economic costs.

Table 8.4 presents the operational costs for case studies 1 and 2. In this table, negative values indicate profit, while positive values denote costs. As shown, since there is no connection with FCVS in Case Study 2, the energy exchange cost is zero. In contrast, in case study 1, the costs for purchasing

Table 8.4 Costs in case study 1 and 2.

Case study	Case study 1 (MU)	Case study 2 (MU)
Electricity	−58	112
Hydrogen	803	781
Electricity of FCVS	125	0
Hydrogen of FCVS	2	0
Other costs	10	11
Sum	882	904

electricity and hydrogen from FCVS are 125 and 2 MU, respectively. The exchanged electrical energy with the upstream grid in case study 1 resulted in a profit, whereas in case study 2 it incurred costs. This difference is due to the connection with FCV, which enhances the system's capabilities. Hydrogen exchange in both case studies results in costs for the operator. This is due to the lower price of hydrogen compared to electricity, making its conversion into electricity ultimately more economical. The cost of hydrogen purchased in case study 1 is higher than in case study 2, because more hydrogen was bought and converted into electricity. As mentioned, electricity exchange led to a profit. All other system costs are related to operational expenses.

8.4 Technical challenges of employing FCVs

In this section, the technical challenges and system problems in FCVs are presented. There are various technical challenges in employing FCVs, as illustrated in Fig. 8.15 (İnci et al., 2021).

In the current vehicle industry, fuel cells must be as durable and resilient as traditional engines. Environmental conditions—including startup and shutdown cycles, freezing temperatures, oxygen pressure levels, humidity, and temperature—are critical factors that influence the lifespan of fuel cells (İnci et al., 2021). These challenges necessitate the development of advanced materials and engineering strategies to enhance the durability and reliability of fuel cells under varying environmental conditions.

The integration of FCVs is similar to that of conventional vehicles. With current FC technology, it's feasible to incorporate a high-capacity drivetrain into passenger cars. There are no significant technical barriers to

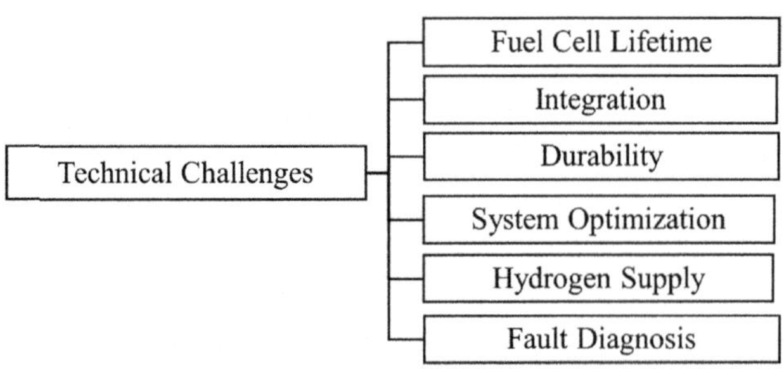

Figure 8.15 Technical challenges of employing FCVs.

implementing a fuel cell system within a vehicle. Thanks to the scalability of fuel cell systems during manufacturing, they can be adapted to fit different vehicle sizes and designs. This allows a fuel cell system initially developed for one vehicle model to be modified for use in other models. Generally, the optimal placement for hydrogen storage tanks is at the rear of the vehicle. However, a key challenge is that these tanks are limited in the amount of hydrogen they can store, primarily because larger tanks take up more space, reducing available room for other vehicle components (von Helmolt and Eberle, 2007).

FCVs require more extensive maintenance and repair than conventional battery EVs to ensure durability, and the current infrastructure for servicing these vehicles is insufficient. As this technology is relatively new, dedicated maintenance facilities are essential, unlike battery EVs that benefit from well-established service networks. Additionally, unbalanced flow distribution within the fuel cell stacks can lead to improper operation of the cells, causing increased uncertainty and reduced efficiency. Frequent maintenance needs and repair downtime can further affect vehicle availability and performance (İnci et al., 2021).

System optimization is a significant challenge in FCVs because it directly impacts performance, efficiency, and reliability. Achieving optimal operation involves managing complex interactions between various components such as the FC stack, HES, and auxiliary systems, as well as balancing power demands in real-time.

Hydrogen supply is a key challenge for fuel cell vehicles, involving the production, storage, and distribution of hydrogen fuel. Developing efficient and sustainable methods, like green electrolysis, is necessary for environmental reasons. Safe storage and expanding refueling infrastructure also pose significant technical hurdles. Overcoming these issues is crucial for making hydrogen-powered vehicles more practical and widely accessible.

In FCVs, faults can arise in various components such as the FC system, storage system, power converters, measurement instruments, and electric motor. These issues may include FC overloading, overcharging of storage units, short circuits or open circuits in the motor stator windings, faulty switches in power converter parts, overheating of components, and more. Some of these faults can impair the vehicle's performance, cause vibrations or noise, or even damage the vehicle's dynamic systems (Zhang et al., 2017).

8.5 Conclusion

In this chapter, a two-stage optimization algorithm is introduced to examine the impact of fuel cell vehicles on hydrogen-based renewable energy systems. In the first stage, the traffic pattern for the day ahead is predicted, and then the proposed values for each fuel cell vehicle are calculated and aggregated with the operator. These results are then entered as proposed values into the second stage. In the second stage, by executing the optimization problem, it will be determined how much of the proposed values from the first stage are accepted, and an optimal combination of production from the equipment and energy exchanged with the upstream network will be provided. This algorithm was tested on a residential energy supply system consisting of various energy production and conversion equipment. Two case studies have been examined: the first case involves a fuel cell vehicle connection, while in the second scenario, the connection to fuel cell vehicles is absent. The operating costs in the first and second scenarios are 882 and 904 MU, respectively, resulting in a profit margin of 2.5% from the interaction with the charging station.

References

Aghajan-Eshkevari S, Azad S, Nazari-Heris M, Ameli MT, Asadi S. Charging and discharging of electric vehicles in power systems: an updated and detailed review of methods, control structures, objectives, and optimization methodologies. Sustainability 2022a; 14(4):2137. https://doi.org/10.3390/su14042137.

Aghajan-Eshkevari S, Ameli MT, Azad S. Electric vehicle services to support the power grid. In: Electric vehicle integration via smart charging: technology, standards, implementation, and applications. Springer; 2022b. p. 129–48.

Aghajan-Eshkevari S, Ameli MT, Azad S. Optimal routing and power management of electric vehicles in coupled power distribution and transportation systems. 03062619 Appl Energy 2023;341. https://doi.org/10.1016/j.apenergy.2023.121126.

Aghajan-Eshkevari S, Ameli MT, Azad S, Pashiri NT. Resilience enhancement of distribution networks by optimal scheduling of hydrogen systems future modern distribution networks resilience: from passive operation to strategic active paradigms. Iran: Elsevier; 2024. p. 353–74. https://doi.org/10.1016/B978-0-443-16086-8.00003-8.

Ahmadi S, Nazar MS. Stochastic day-ahead economic operation scheduling of multi-carrier microgrids considering energy markets and demand response. 2025/01. In: 2025 12th Iranian conference on renewable energies and distributed generation, ICREDG 2025. Iran: Institute of Electrical and Electronics Engineers Inc; 2025 1. https://doi.org/10.1109/ICREDG66184.2025.10966103.

Ahmadi S, Amiri MM, Nazar MS, Ameli MT. Day-ahead optimal scheduling for a energy hub operation considering markets for electricity, heat, cooling and gas. In: The 14th international conference on smart grids; 2025. https://sgc2024.khatam.ac.ir/en/.

Arévalo P, Cano A, Jurado F. Large-scale integration of renewable energies by 2050 through demand prediction with ANFIS, Ecuador case study. Energy 2024;286:129446. https://doi.org/10.1016/j.energy.2023.129446.

Azizivahed A, Gholami K, Arefi A, Li L, Arif MT, Haque ME. Stochastic scheduling of energy sharing in reconfigurable multi-microgrid systems in the presence of vehicle-to-grid technology. Elec Power Syst Res 2024;231. https://doi.org/10.1016/j.epsr.2024.110285.

Bostan A, Nazar MS, Shafie-khah M, Catalão JPS. Optimal scheduling of distribution systems considering multiple downward energy hubs and demand response programs. Energy 2020;190. https://doi.org/10.1016/j.energy.2019.116349.

Brouwer J. On the role of fuel cells and hydrogen in a more sustainable and renewable energy future. Curr Appl Phys 2010;10(2):S9. https://doi.org/10.1016/j.cap.2009.11.002.

Chen H, Song J, Zhao J. Synergies between power and hydrogen carriers using fuel-cell hybrid electrical vehicle and power-to-gas storage as new coupling points. Energy Convers Manag 2021;246:114670. https://doi.org/10.1016/j.enconman.2021.114670.

Chisholm G, Zhao T, Cronin L. Hydrogen from water electrolysis storing energy: with special reference to renewable energy sources. United Kingdom: Elsevier; 2022. p. 559–91. https://doi.org/10.1016/B978-0-12-824510-1.00015-5.

Dincer I, Aydin MI. New paradigms in sustainable energy systems with hydrogen. Energy Convers Manag 2023;283:116950. https://doi.org/10.1016/j.enconman.2023.116950.

Egeland-Eriksen T, Hajizadeh A, Sartori S. Hydrogen-based systems for integration of renewable energy in power systems: achievements and perspectives. Int J Hydrogen Energy 2021;46(63):31963–83. https://doi.org/10.1016/j.ijhydene.2021.06.218.

Fan G, Liu Z, Liu X, Shi Y, Wu Di, Guo J, et al. Two-layer collaborative optimization for a renewable energy system combining electricity storage, hydrogen storage, and heat storage. Energy 2022;259:125047. https://doi.org/10.1016/j.energy.2022.125047.

Fragiacomo P, Piraino F. Vehicle-to-grid application with hydrogen-based tram. Energy Convers Manag 2021;250:114915. https://doi.org/10.1016/j.enconman.2021.114915.

Gupta T, Rahaman. A critical review on hydrogen based fuel cell technology and applications. Chem Record (New York, N.Y.) 2024;24.

Hou L, Dong J, Herrera OE, Mérida W. Energy management for solar-hydrogen microgrids with vehicle-to-grid and power-to-gas transactions. Int J Hydrogen Energy 2023;48(5):2013–29. https://doi.org/10.1016/j.ijhydene.2022.09.238.

Hwang JJ. Sustainability study of hydrogen pathways for fuel cell vehicle applications. Renew Sustain Energy Rev 2013;19:220–9. https://doi.org/10.1016/j.rser.2012.11.033.

İnci M, Büyük M, Demir MH, İlbey G. A review and research on fuel cell electric vehicles: topologies, power electronic converters, energy management methods, technical challenges, marketing and future aspects. Renew Sustain Energy Rev 2021;137. https://doi.org/10.1016/j.rser.2020.110648.

Khan MA, Bayati N, Ebel T. Techno-economic analysis and predictive operation of a power-to-hydrogen for renewable microgrids. Energy Convers Manag 2023;298:117762. https://doi.org/10.1016/j.enconman.2023.117762.

Li M, Bai Y, Zhang C, Song Y, Jiang S, Grouset D, et al. Review on the research of hydrogen storage system fast refueling in fuel cell vehicle. Int J Hydrogen Energy 2019;44(21):10677–93. https://doi.org/10.1016/j.ijhydene.2019.02.208.

Lin T, Wu Y, He XY, Zhang SJ, Hao JM. Well-to-Wheels fossil energy consumption and CO 2 emissions of hydrogen fuel cell vehicles in China. Huanjing Kexue/Environ Sci 2018;39(8):3946–53. https://doi.org/10.13227/j.hjkx.201712113.

Manoharan Y, Ehsan Hosseini S, Butler B, Alzhahrani H, Senior BTF, Ashuri T, et al. Hydrogen fuel cell vehicles; current status and future prospect. Appl Sci 2019;9(11):2296. https://doi.org/10.3390/app9112296.

Manousakis NM, Karagiannopoulos PS, Tsekouras GJ, Kanellos FD. Integration of renewable energy and electric vehicles in power systems: a review. Processes 2023;11(5). https://doi.org/10.3390/pr11051544.

Mayen EV, Ndiwulu GW, Kluyskens V, De Jaeger E. Non-linear control of PEM electrolyzers in DC microgrids. In: 2024 IEEE International Conference on environment and electrical engineering and 2024 IEEE industrial and commercial power systems Europe (EEEIC/I&CPS Europe). IEEE; 2024. p. 1–6.

Mohammadi S, Soleymani S, Mozafari B. Scenario-based stochastic operation management of MicroGrid including wind, photovoltaic, micro-turbine, fuel cell and energy storage devices. Int J Electr Power Energy Syst 2014;54:525–35. https://doi.org/10.1016/j.ijepes.2013.08.004.

Nonobe Y. Development of the fuel cell vehicle mirai. IEEJ Trans Electr Electron Eng 2017;12(1):5–9.

Qazi UY. Future of hydrogen as an alternative fuel for next-generation industrial applications; challenges and expected opportunities. Energies 2022;15(13). https://doi.org/10.3390/en15134741.

Robledo CB, Oldenbroek V, Abbruzzese F, van Wijk AJM. Integrating a hydrogen fuel cell electric vehicle with vehicle-to-grid technology, photovoltaic power and a residential building. Appl Energy 2018;215:615–29. https://doi.org/10.1016/j.apenergy.2018.02.038.

Rolo I, Costa VAF, Brito FP. Hydrogen-based energy systems: current technology development status, opportunities and challenges. Energies 2024;17(1). https://doi.org/10.3390/en17010180.

Schrotenboer AH, Veenstra AAT, uit het Broek MAJ, Ursavas E. A Green Hydrogen Energy System: optimal control strategies for integrated hydrogen storage and power generation with wind energy. Renew Sustain Energy Rev 2022;168. https://doi.org/10.1016/j.rser.2022.112744.

Shahbazbegian V, Shafie-khah M, Laaksonen H, Strbac G, Ameli H. Resilience-oriented operation of microgrids in the presence of power-to-hydrogen systems. Appl Energy 2023;348:121429. https://doi.org/10.1016/j.apenergy.2023.121429.

Shinoda K, Lee EP, Nakano M, Lukszo Z. Optimization model for a microgrid with fuel cell vehicles. In: 2016 IEEE 13th international conference on networking, sensing, and control (ICNSC). IEEE; 2016. p. 1–6.

Singh S, Jain S, Ps V, Tiwari AK, Nouni MR, Pandey JK, et al. Hydrogen: a sustainable fuel for future of the transport sector. Renew Sustain Energy Rev 2015;51:623–33. https://doi.org/10.1016/j.rser.2015.06.040.

Tatti R, Petrollese M, Lucariello M, Serra F, Cau G. Hydrogen storage integrated in off-grid power systems: a case study. Int J Hydrogen Energy 2024;79:164–76. https://doi.org/10.1016/j.ijhydene.2024.06.308.

Valverde L, Pino FJ, Guerra J, Rosa F. Definition, analysis and experimental investigation of operation modes in hydrogen-renewable-based power plants incorporating hybrid energy storage. Energy Convers Manag 2016;113:290–311. https://doi.org/10.1016/j.enconman.2016.01.036.

von Helmolt R, Eberle U. Fuel cell vehicles: status 2007. J Power Sources 2007;165(2):833–43.

Xiao J, Bénard P, Chahine R. Estimation of final hydrogen temperature from refueling parameters. Int J Hydrogen Energy 2017;42(11):7521–8. https://doi.org/10.1016/j.ijhydene.2016.05.213.

Yarmohammadi H, Abdi H. Optimal operation of multi-carrier energy systems considering demand response: a hybrid scenario-based/IGDT uncertainty method. Elec Power Syst Res 2024;235:110877. https://doi.org/10.1016/j.epsr.2024.110877.

Youssef AR, Mallah M, Ali A, Shaaban MF, Mohamed EEM. Enhancement of microgrid frequency stability based on the combined power-to-hydrogen-to-power technology under high penetration renewable units. Energies 2023;16(8). https://doi.org/10.3390/en16083377.

Zhang J, Yao H, Rizzoni G. Fault diagnosis for electric drive systems of electrified vehicles based on structural analysis. IEEE Trans Veh Technol 2017;66(2):1027–39. https://doi.org/10.1109/TVT.2016.2556691.

Zhang L, Jia C, Bai F, Wang W, An S, Zhao K, et al. A comprehensive review of the promising clean energy carrier: hydrogen production, transportation, storage, and utilization (HPTSU) technologies. Fuel 2024;355:129455. https://doi.org/10.1016/j.fuel.2023.129455.

Zhao H, Burke AF. Optimization of fuel cell system operating conditions for fuel cell vehicles. J Power Sources 2009;186(2):408–16. https://doi.org/10.1016/j.jpowsour.2008.10.032.

CHAPTER NINE

Reliability-aware reinforcement learning for renewable-to-vehicle smart charging

Amirreza Shafieinejad and Alireza Sobbouhi
Department of Electrical Engineering, Shahid Beheshti University, Tehran, Iran

9.1 Introduction

9.1.1 Motivation and background

The transition toward renewable-dominated energy systems introduces both opportunities and challenges for modern power grids. Intermittent sources such as wind and solar are central to decarbonization efforts but create significant uncertainty in balancing supply and demand. At the same time, the rapid electrification of transportation through electric vehicles (EVs) introduces large, highly variable loads to the distribution network. Smart charging and Renewable-to-Vehicle (R2V) systems have therefore emerged as key enablers for aligning EV charging with renewable energy availability and feeder constraints (Aghajan-Eshkevari et al., 2022). Yet these opportunities bring new concerns regarding adequacy, reliability, and long-term system resilience.

Classical reliability theory, as formulated by Billinton and Allan, defines reliability as the ability of the power system to provide electricity of acceptable quality and continuity (Billinton and Allan, 1992). Reliability is typically decomposed into adequacy, the sufficient capacity to meet expected demand, and security, the ability to withstand and recover from disturbances (Billinton and Allan, 1992; Karki et al., 2014). Fig. 9.1 illustrates this decomposition and the corresponding reliability indices. Traditional indices such as the System Average Interruption Duration Index (SAIDI), System Average Interruption Frequency Index (SAIFI), Energy Not Supplied (ENS), and Loss of Load Probability (LOLP) have long been used to quantify these aspects (Billinton and Allan, 1992). However, such indices are insufficient for emerging R2V networks, where both user-centric outcomes and resource efficiency must be evaluated simultaneously. To address this gap, new indices have been proposed, notably the Charging Reliability

Renewable-to-Vehicle Smart Charging
ISBN: 978-0-443-36684-0
https://doi.org/10.1016/B978-0-443-36684-0.00006-0

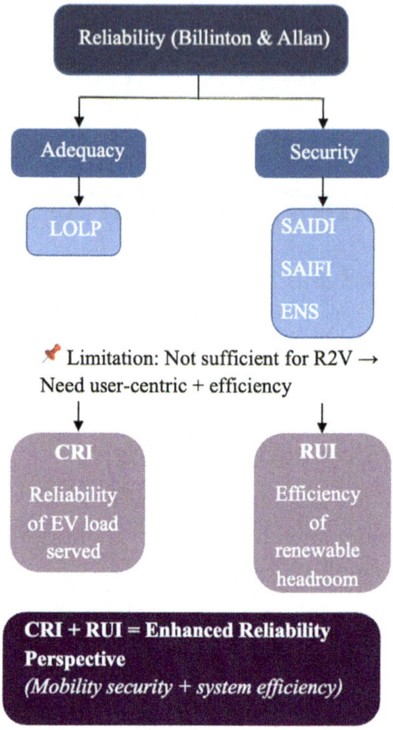

Figure 9.1 Evolution of reliability indices in power systems.

Index (CRI), which measures the fraction of EV demand reliably served, and the Renewable Utilization Index (RUI), which captures how effectively renewable headroom is used (Shaaban et al., 2015). Together, CRI and RUI provide complementary perspectives on mobility security and system efficiency Fig. 9.1.

Assessing these indices is complicated by the stochastic nature of EV charging, which is driven by user behavior, mobility patterns, electricity prices, and renewable variability. Monte Carlo (MC) simulation has become a cornerstone for capturing such uncertainty, but purely probabilistic methods can be computationally intensive and limited in their ability to represent adaptive, real-time strategies (Soares et al., 2010). This motivates the use of reinforcement learning (RL), which provides a principled framework for sequential decision-making under uncertainty (Pei et al., 2024). By learning adaptive charging schedules, RL agents can minimize ENS, reduce LOLP, improve CRI, and maximize RUI. Among RL algorithms, modern actor–critic approaches, such as Proximal Policy

Optimization (PPO), have attracted attention for their stability and scalability, making them particularly suitable for reliability-aware EV smart charging (Abdullah et al., 2021).

9.1.2 Relevant literature

The reliability of power systems under the dual influence of renewable energy integration and large-scale adoption of EVs has been a central topic of research for over 2 decades. Classical methods relied heavily on probabilistic models and MC simulation, whereas recent advances in artificial intelligence (AI), particularly RL, have introduced adaptive strategies for managing uncertainty. This subsection reviews relevant contributions, categorized into traditional reliability studies, the impacts of EV integration, emerging indices for R2V networks, RL approaches, and hybrid simulation–control frameworks.

Pioneering works, such as those by Billinton and Allan, laid the foundation for reliability engineering, introducing indices such as SAIDI, System Average SAIFI, ENS, and LOLP. These indices have been extensively applied to quantify adequacy and security in generation and distribution systems. More recently, studies such as (Zhang et al., 2024; Park and Moon, 2022) applied MC simulation to quantify the impacts of EV loads on reliability indices, explicitly accounting for stochasticity in demand and outages. Similarly, reference (Zhang et al., 2018) advanced MC-based methods to capture renewable intermittency alongside EV variability. These works underscore the robustness of probabilistic methods but also highlight their computational burden when scaling to high-resolution datasets or adaptive strategies. The rise of EV adoption has shifted attention toward distribution-level impacts. Stochastic arrival times, variable dwell durations, and heterogeneity in charging demands complicate system planning. Research in (Wang et al., 2016) developed stochastic EV load models that integrate user behavior and mobility patterns, which were later used for reliability assessment. Comparative analyses, such as (Archana and Rajeev, 2021), have demonstrated how unmanaged plug-and-charge behavior amplifies peak demand and increases LOLP, while controlled charging mitigates these risks. Further (Shahbazi et al., 2023), explored smart charging strategies designed explicitly for reliability enhancement, contrasting unmanaged and rule-based scheduling. Taken together, these studies demonstrate that unmanaged charging consistently degrades reliability metrics, motivating the development of adaptive scheduling policies.

While SAIDI, SAIFI, ENS, and LOLP remain important, they fall short of capturing the dual objectives of user satisfaction and renewable utilization in R2V networks. To address this gap, this chapter introduces two new indices, the Charging Reliability Index (CRI) and the Renewable Utilization Index (RUI), to reflect both user satisfaction and renewable resource efficiency in R2V systems. While (Mhaisen et al., 2020) emphasized the limitations of traditional reliability metrics in renewable-integrated grids, the proposed CRI quantifies the portion of EV demand served within reliability thresholds, directly linking to mobility security. RUI evaluates how effectively renewable headroom is exploited, reflecting economic and environmental efficiency. Together, these indices extend the reliability paradigm from conventional adequacy toward a joint evaluation of service quality and resource sustainability.

RL has gained traction as a promising method for sequential decision-making in uncertain environments. Early applications in energy management, such as (Hashemi-Dezaki et al., 2015), highlighted the suitability of RL for dynamic load control, contrasting value-based, policy-based, and actor–critic methods. More recent work, including (Archana and Rajeev, 2020; Shahbazi et al., 2023), demonstrated the application of deep RL (DRL) to EV scheduling and demand response. Actor–critic algorithms, particularly PPO, have emerged as preferred choices due to their stability and scalability in high-dimensional control problems. These methods enable an agent to adaptively modulate EV charging, thereby minimizing ENS, reducing LOLP, and enhancing CRI and RUI under uncertainty. Importantly, RL frameworks address the limitations of purely probabilistic MC approaches by incorporating feedback and adaptation over time.

Several studies have sought to combine probabilistic reliability assessment with adaptive control. For instance (Zhang et al., 2018; Wang et al., 2016), integrated stochastic EV demand models with simulation-based evaluation, while (Shahbazi et al., 2023; Archana and Rajeev, 2020) extended this approach by embedding RL into the decision-making loop. This hybrid perspective, MC for uncertainty quantification, RL for adaptive scheduling, bridges classical reliability assessment with modern control. The synergy is evident: MC captures variability across runs and seasons, while RL enables policies that respond to observed conditions in real time. Such integration is especially relevant for R2V systems, where both renewable intermittency and EV charging patterns must be jointly addressed.

The reviewed literature highlights a clear progression: from classical indices and probabilistic simulations (Zhang et al., 2018, 2024; Park and Moon, 2022), through EV-specific stochastic modeling (Wang et al., 2016; Archana and Rajeev, 2021), toward new indices tailored for R2V systems (Mhaisen et al., 2020; Diaz-Londono and Li 2025; Wang et al., 2018), and finally to RL approaches (Hashemi-Dezaki et al., 2015; Archana and Rajeev, 2020; Shahbazi et al., 2023). However, gaps remain. There is a lack of integrated frameworks that combine emerging indices such as CRI and RUI with RL-based scheduling. Moreover, these policies are rarely evaluated under realistic market and operational conditions using data from large-scale operators like CAISO. This motivates the present chapter, which synthesizes probabilistic reliability methods, R2V-specific indices, and RL into a coherent methodology. The chapter further contributes by comparing RL-based scheduling with unmanaged and rule-based baselines, providing actionable insights for grid operators and policymakers responsible for managing EV integration under renewable energy uncertainty. A summary of the key studies, their methods, and contributions is provided in Table 9.1 to offer a clearer comparison across the literature.

9.1.3 Contributions

The chapter begins with an overview of traditional reliability indices in power systems, before introducing emerging metrics specifically designed for R2V networks, such as the CRI and the RUI. It then turns to the foundations of RL, clarifying the differences between value-based, policy-based, and actor–critic methods, with particular emphasis on PPO as the most suitable algorithm for reliability-aware control. Building on this background, the methodology for smart EV charging is described, which combines RL with MC simulations to capture the stochastic nature of renewable energy generation and EV demand. The case study focuses on California's Independent System Operator (CAISO) data and compares RL–based scheduling against unmanaged and rule-based baselines. Performance is assessed using metrics such as ENS, LOLP, CRI, RUI, and the Peak-to-Average Ratio (PAR), with seasonal variations also examined. The discussion highlights the broader implications of AI-driven reliability assessment for grid operators, policymakers, and system planners, particularly in managing EV integration under renewable uncertainty. The chapter concludes by summarizing key insights and outlining directions for future research. By combining classical indices, emerging R2V-specific metrics, and advanced AI methodologies, this chapter aims to provide a comprehensive framework

Table 9.1 Summary of the key studies.

Study/Reference	Main focus	Methodology	Key contribution
Billinton & Allan (1992)	Foundational reliability metrics	Probabilistic modeling	Introduced SAIDI, SAIFI, ENS, LOLP
Zhang et al. (2024); Park and Moon (2022)	Impact of EVs on traditional indices	Monte Carlo Simulation (MC)	Quantified reliability with EV loads
Zhang et al. (2018)	Renewable + EV variability	Probabilistic Simulation	Modeled renewable intermittency with EV demand
Wang et al. (2016)	Stochastic EV behavior modeling	EV behavior modeling	Simulated realistic EV load patterns
Archana and Rajeev (2021)	Controlled versus uncontrolled EV charging	Comparative Analysis	Demonstrated how smart charging enhances reliability
Shahbazi et al. (2023)	Smart charging for reliability	Rule-based simulation	Showed rule-based scheduling reduces ENS, LOLP
Mhaisen et al. (2020)	Limits of traditional metrics	Theoretical analysis	Highlighted inadequacy of SAIDI/SAIFI for R2V systems
Hashemi-Dezaki et al. (2015)	RL for energy control	RL (value/policy-based)	Established RL feasibility for load management
Archana and Rajeev (2020)	Deep RL for EV scheduling	DRL (Actor–Critic)	Applied PPO to dynamic EV scheduling
[This chapter]	Unified R2V reliability framework	Hybrid (MC + RL), Novel Indices	Introduced CRI & RUI, evaluated PPO under real data

for understanding, evaluating, and enhancing power system reliability in the age of renewable energy and electric mobility.

Fig. 9.2 illustrates the integration of renewable energy sources, EV smart charging, and artificial intelligence/machine learning (AI/ML) techniques to support reliability-aware decision-making in R2V systems. Renewable generation (solar and wind) feeds into the grid and EV charging infrastructure, while AI/ML models analyze system conditions to optimize scheduling and resource allocation. The resulting decisions balance user charging reliability with efficient utilization of grid resources, ensuring both operational stability and long-term sustainability.

9.1.4 Organization

The remainder of this chapter is organized as follows. Section 9.2 introduces traditional reliability indices such as System Average Interruption Duration Index (SAIDI), System Average Interruption Frequency Index (SAIFI), Energy Not Supplied (ENS), and Loss of Load Probability (LOLP), alongside emerging metrics tailored for Renewable-to-Vehicle (R2V) systems, including the Charging Reliability Index (CRI) and Renewable Utilization Index (RUI), with their mathematical formulations. Section 9.3 outlines the foundations of reinforcement learning, distinguishing value-based, policy-based, and actor–critic methods, with a focus on Proximal Policy Optimization (PPO) as the chosen algorithm. Section 9.4 describes the methodology for smart EV charging, integrating Monte Carlo simulations with RL, and presents a case study using California ISO (CAISO) data to compare unmanaged, rule-based, and RL-based scheduling policies, assessing performance with metrics and seasonal variations. Section 9.5 presents the results and analysis from these simulations across various renewable (RENx), outage (OUT), and EV penetration (EVx) levels. Finally, Section 9.6 concludes with a discussion of key findings, the advantages of RL-based scheduling for balancing efficiency and robustness, and future research directions, including the potential of Agentic AI systems.

9.2 Overview of reliability indices in power systems

In this section, the traditional reliability metrics, along with tailored indices R2V networks (CRI and RUI), are discussed in full detail.

9.2.1 Traditional metrics: SAIDI, SAIFI, LOLP, ENS

Reliability assessment is fundamental to power system operation, ensuring electricity is consistently delivered to consumers with minimal interruptions. Traditional reliability metrics provide a quantitative assessment of system performance, enabling utilities and policymakers to design robust, resilient power networks. Among the most widely used metrics are the SAIDI, SAIFI, LOLP, and ENS. These indices provide insight into various aspects of reliability, ranging from service continuity at the consumer level to system-wide performance and risk assessment (Kovalev and Lebedeva, 2019).

9.2.1.1 System average interruption duration index (SAIDI)

SAIDI is a key indicator used to measure the average duration of power interruptions experienced by consumers over a given period, typically expressed in minutes or hours per year (Karki et al., 2014), which is calculated as follows:

$$\mathrm{SAIDI} = \frac{\sum (N_i \times T_i)}{N_{\mathrm{total}}} \tag{9.1}$$

where.

- N_i = Number of customers affected by outage i,
- T_i = Duration of outage i,
- N_{total} = Total number of customers served.

A lower SAIDI value indicates a more reliable power system, suggesting that outages are infrequent and short-lived. Utilities use SAIDI to assess the effectiveness of their maintenance strategies, outage response protocols, and infrastructure investments.

9.2.1.2 System average interruption frequency index (SAIFI)

SAIFI measures the average number of outages a customer experiences over a specific period, reflecting the frequency of service interruptions (Karki et al., 2014). It is calculated as:

$$\mathrm{SAIFI} = \frac{\sum N_i}{N_{\mathrm{total}}} \tag{9.2}$$

where.

- N_i = Number of customers affected by outage i,
- N_{total} = Total number of customers served.

SAIFI is a crucial indicator for evaluating system reliability from a customer perspective. A lower SAIFI value signifies fewer service interruptions, implying a higher level of reliability. Utilities leverage SAIFI data to identify weaknesses in the grid and implement targeted infrastructure upgrades.

9.2.1.3 Loss of load probability (LOLP)

LOLP is a probabilistic measure that quantifies the likelihood of system demand exceeding available generation capacity, potentially leading to blackouts (Karki et al., 2014). It is mathematically represented as:

$$\mathrm{LOLP} = P(D > G) \tag{9.3}$$

where.

- P is the probability function of the power shortage

- D = System demand,
- G = Available generation capacity.

A higher LOLP indicates an increased risk of supply shortfalls, necessitating robust contingency planning, enhanced demand response mechanisms, and investment in grid flexibility. Given the growing reliance on intermittent renewable energy sources, LOLP analysis has become critical in ensuring system adequacy, particularly in energy systems with high solar and wind power penetration.

9.2.1.4 Energy not supplied (ENS)

ENS is a reliability index quantifying the total amount of unserved energy due to system interruptions (Karki et al., 2014). It is defined as:

$$\mathrm{ENS} = \sum_{i=0}^{Nn} \left(P_{\mathrm{lost,i}} \times T_{\mathrm{i}} T\right)\left(P_{\mathrm{lost}} \times T\right) \tag{9.4}$$

where.

- $P_{\mathrm{lost,i}} P_{\mathrm{lost}}$ = Power lost during the i-th outage (MW),
- T_{i} = Duration of the i-th outage (hours).
- N = is the total number of outage events.

ENS is particularly useful for assessing the economic and operational impact of outages. High ENS values indicate substantial losses in energy supply, often translating to financial repercussions for industries, businesses, and consumers. Policymakers use ENS to prioritize grid reinforcement, redundancy planning, and resilience enhancement investments.

Algorithm1 Proximal Policy Optimization(PPO) (Rio et al., 2024).

1: Initialize policy parameters θ
2: Initialize value function parameters ϕ
3: **For** each iteration **do**
4: Collect a set of trajectories D by running policy π_θ in the environment
5: **For** each trajectory **do**
6: **For** each time step t **do**
7: Compute advantage estimate $\widehat{A}_t$
8: **end for**
9: **end for**
10: Update the policy by maximizing the PPO-Clip objective:
$\mathscr{L}^{\mathrm{CLIP}}(\theta) = \hat{\mathbb{E}}_t\left[\min\left(\frac{\pi_\theta(a_t|s_t)}{\pi_{\theta_{\mathrm{old}}}(a_t|s_t)}\hat{A}_t, \mathrm{clip}\left(\frac{\pi_\theta(a_t|s_t)}{\pi_{\theta_{\mathrm{old}}}(a_t|s_t)}, 1-\epsilon, 1+\epsilon\right)\hat{A}_t\right)\right]$
11: Update the value function by minimizing the loss:
$\mathscr{L}^{\mathrm{VF}}(\phi) = \hat{\mathbb{E}}_t\left[\left(V_\phi(s_t) - R_t\right)^2\right]$
12: Update the policy parameters: $\theta \leftarrow \theta + \alpha \nabla_\theta \mathscr{L}^{\mathrm{CLIP}}(\theta)$
13: Update the value function parameters: $\phi \leftarrow \phi - \alpha_\phi \nabla_\phi \mathscr{L}^{\mathrm{VF}}(\phi)$
14: **end for**

Table 9.2 Comparative analysis of traditional reliability metrics.

Metric	Definition	Unit	Interpretation
SAIDI	Average outage duration per customer	Minutes/ Year	Lower values indicate fewer service interruptions
SAIFI	Average number of outages per customer	Outages/ Year	Lower values imply fewer disruptions
LOLP	Probability of power supply failure	Probability (0–1)	Lower values suggest a more stable grid
ENS	Total unserved energy due to outages	MWh	Lower values indicate less energy loss

Table 9.2 provides a comparative overview of the most widely used reliability metrics in classical power system analysis. SAIDI and SAIFI focus on the customer experience of outages, capturing average duration and frequency, respectively. These indices are particularly relevant for distribution utilities, as they reflect service continuity from the end-user perspective. In contrast, LOLP and ENS are system-oriented indices. LOLP quantifies the probability that available generation capacity is insufficient to meet demand, while ENS measures the total energy curtailed during outages, highlighting both economic and operational consequences.

Together, these metrics have been instrumental in guiding reinforcement, redundancy planning, and investment in conventional grids. However, they remain limited in the context of R2V networks. Specifically, while they measure outage frequency, duration, and unserved energy, they do not capture two critical dimensions introduced by large-scale EV integration: (i) mobility reliability, i.e., the extent to which EV charging demand is met on time, and (ii) renewable utilization efficiency, i.e., how effectively variable renewable headroom is absorbed by flexible loads. This gap motivates the introduction of emerging indices such as the CRI and the RUI, which extend the classical framework by linking reliability not only to supply adequacy but also to user satisfaction and sustainability outcomes (Xue et al., 2021).

Problem 9.1. A city-wide EV smart charging network consists of 200 public charging stations integrated with renewable energy sources (solar and wind). The network's reliability is assessed under different failure scenarios using the SAIDI, SAIFI, LOLP, and ENS. The following scenarios represent common failure modes affecting EV charging infrastructure reliability Table 9.3.

- Total number of chargers in the network = 200

Table 9.3 Scenario data.

Scenario	Affected chargers	Outage duration (hours)	Critical notes
Peak hour overload	30	1.5	High demand causes transformer overloading and localized failures.
Renewable energy fluctuation	40	2.0	Low solar/wind availability reduces available charging power.
Cyberattack on charging Network	20	4.0	A cyberattack disables charger communication, preventing operation.
Equipment failure at Major Hub	50	6.0	Transformer failure disrupts an entire cluster of charging stations.
Grid-wide power outage	200	3.0	A blackout causes all charging stations to shut down.

- Power demand per charger = 50 kW
- Total operational days in a year = 365

Solution.

1. SAIDI

$$\mathbf{SAIDI} = \frac{\sum (\text{Affected Chargers} \times \text{Outage Duration}(\mathbf{mins}))}{\text{Total Chargers}}$$

2. SAIFI

$$\text{SAIFI} = \frac{\text{Affected Chargers}}{\text{Total Chargers}}$$

3. LOLP

$$\text{LOLP} = \frac{\text{OutageDuration}(\text{days})}{\text{Total Days}}$$

4. ENS

$$\text{ENS} = (\text{PowerDemandperCharger} \times \text{AffectedChargers}) \times \text{OutageDuration}(\text{hours})$$

Table 9.4 consolidates the reliability indices for all five identified failure scenarios affecting EV charging infrastructure over a one-year period. Rather than treating each scenario independently, the values are summed to provide total annual metrics, as is customary in reliability engineering. The aggregated SAIDI of 331.5 minutes and SAIFI of 1.75 outages per

Table 9.4 Combined reliability indices (SAIDI, SAIFI, LOLP, ENS) for EV charging failures over 1 year, with total values reflecting system-wide impact across all scenarios.

Scenario	SAIDI (min)	SAIFI (outages/charger)	LOLP (%)	ENS (MWh)
Peak hour overload	13.5	0.15	0.017	2250
Renewable energy fluctuation	24.0	0.20	0.023	4000
Cyberattack on charging Network	24.0	0.10	0.046	4000
Equipment failure at Major Hub	90.0	0.30	0.057	15,000
Grid-wide power outage	180.0	1.00	0.034	30,000
Total annual value	331.5	1.75	0.177	55,250

charger reflect the cumulative duration and frequency of service interruptions. The LOLP, expressed as a percentage, indicates the probability that the system will fail to meet demand due to these disruptions. Meanwhile, the total ENS of 55,250 MWh quantifies the magnitude of unserved energy, highlighting the severe energy loss potential of widespread outages. These comprehensive metrics offer a more realistic and policy-relevant understanding of the system's performance, particularly in the context of increasing electrification, distributed energy sources, and demand variability in R2V networks.

9.2.2 Emerging indices tailored for R2V networks: CRI, RUI

9.2.2.1 Charging reliability index (CRI)

The CRI is an emerging reliability metric specifically designed for R2V networks. While traditional indices such as SAIDI or LOLP focus on supply interruptions at the grid level, CRI quantifies the probability that EV charging demand is successfully met within the required time window. It answers the critical question: "How reliably can an EV connect to the grid and obtain the energy it needs for mobility?". Formally, CRI is defined as:

$$\mathrm{CRI} = \frac{\text{Total EV energy successfully delivered within deadline}}{\text{Total EV energy requested}} \tag{9.5}$$

This ratio ranges between 0 and 1 , where 1 represents perfect charging reliability. Unlike aggregate energy indices such as ENS, CRI incorporates both probabilistic availability of supply and temporal urgency of EV charging requests.

Example: Suppose 10 EVs each request 40 kWh during a peak evening period (total = 400kWh). Due to renewable fluctuations and feeder limits, only 340 kWh is delivered within the charging deadlines. Then:

$$\mathrm{CRI} = \frac{340}{400} = 0.85$$

This means 85% of the requested charging demand was reliably satisfied. Even though the aggregate shortfall is only 60 kWh (small relative to total feeder load), the reliability experienced by EV users is significantly impacted.

By tracking CRI, operators can directly evaluate user-facing reliability under scenarios of renewable intermittency and high EV penetration. As a proposed criterion, in R2V applications, a CRI above 0.9 is often desirable, while values below 0.8 indicate severe degradation in user satisfaction and potential mobility risks.

9.2.2.2 Resource utilization index (RUI)

The RUI is another metric introduced to capture the efficiency of renewable and grid resources in serving EV loads. Unlike CRI, which reflects service quality from the user perspective, RUI measures how well available feeder headroom and renewable output are actually exploited by the charging scheduler. Formally:

$$\mathrm{RUI} = \frac{\text{Energy served to EVs}}{\text{Available headroom capacity over the charging horizon}} \tag{9.6}$$

Here, "available headroom" refers to the capacity margin left after serving base grid load, which could theoretically be allocated to EVs. RUI thus indicates the alignment between EV charging schedules and available capacity.

Problem 9.2. Imagine a feeder with 1,000kWh headroom during the evening, due to a mix of renewables and network capacity. EVs request 600 kWh, but only 400 kWh is served due to poor scheduling (e.g., unmanaged plug-and-charge at the same hour). Then:

$$\mathrm{RUI} = \frac{400}{1000} = 0.40$$

This implies that 60% of the feeder's headroom went unused, representing inefficiency. If a smarter scheduling policy had coordinated EVs better, the RUI could approach 0.6 or 0.7, ensuring both better renewable utilization and higher CRI.

While CRI and RUI each provide valuable insights into R2V network performance, they capture different aspects of system behavior. CRI reflects the reliability of EV charging service and the degree to which user demand is consistently satisfied, whereas RUI highlights how efficiently available resources, such as renewables, grid headroom, and storage, are utilized. Evaluating only one of these indices can give an incomplete or even misleading picture of system performance. For instance, a network might report a very high CRI by over-provisioning grid capacity or frequently curtailing renewable generation, but such practices would lower RUI and waste valuable resources. Conversely, maximizing RUI alone could lead to highly efficient use of generation and grid headroom, but at the expense of unmet charging demand and reduced user confidence. Considering the two indices together, therefore, enables a balanced assessment: CRI emphasizes user-centric reliability and mobility assurance, while RUI emphasizes system-level efficiency and sustainability.

9.2.3 Applications and case studies

Traditional reliability indices such as SAIDI, SAIFI, LOLP, and ENS have been extensively applied in real-world power systems. For instance, SAIDI and SAIFI remain benchmarks for distribution network performance, with utilities striving to keep outage durations and frequencies within acceptable limits through innovations like automated fault detection and self-healing grids. Similarly, LOLP has guided generation planning during critical events such as the 2000–2001 California Energy Crisis, where elevated LOLP values indicated the rising risk of rolling blackouts due to insufficient supply (Joskow, 2001). ENS, on the other hand, has been a valuable tool for assessing the severity of large-scale disruptions, exemplified by the Texas Winter Storm of 2021, when ENS exceeded 1,500 GWh and underscored the vulnerability of power systems to climate-driven shocks (Wu et al., 2021). While these indices remain integral to system planning and resilience assessment, the increasing penetration of renewables, EV charging, and digitalization creates operational challenges that they cannot fully capture. This gap has prompted the development of emerging indices tailored for R2V networks, such as the CRI and RUI, which better reflect service quality and flexibility in modern, electrified grids.

9.3 Fundamentals of reinforcement learning for smart charging control

9.3.1 Introduction

RL is a branch of machine learning concerned with how agents learn to make sequential decisions through interaction with an environment. Unlike supervised learning, which depends on labeled data, RL agents improve their behavior through trial and error, guided by feedback in the form of rewards and penalties. This framework is particularly well-suited for problems that require real-time decision-making under uncertainty, with the goal of optimizing long-term performance rather than short-term gains (Abdullah et al., 2021).

In the context of EV smart charging, RL provides a natural solution for managing the variability of both demand and renewable energy supply. Charging requests arrive stochastically, feeder headroom fluctuates with system load, and renewable availability is inherently intermittent. Traditional deterministic or rule-based schedulers often struggle under these conditions, whereas an RL agent can adapt dynamically, learning strategies that balance competing objectives, such as minimizing ENS, maximizing CRI, and reducing LOLP.

By framing EV charging as a sequential decision-making problem, the RL agent learns how to allocate charging power hour by hour in a way that anticipates future conditions. This adaptive capability makes RL a powerful tool for reliability-aware scheduling in R2V systems, where ensuring both user satisfaction and grid stability is crucial.

9.3.2 Core RL concepts

In the context of smart EV charging, RL is implemented using well-defined states, actions, and rewards that capture the dynamics of the grid and user behavior. The state space typically includes parameters like battery SOC, current energy price, grid headroom, and solar availability. The action space may consist of discrete or continuous charging decisions, such as charging at a reduced rate, deferring charging, or stopping altogether.

The agent receives rewards based on the impact of its actions, such as minimizing the expected ENS, avoiding grid overloads (which reduces LOLP), and maximizing cost efficiency. Over time, the agent uses this feedback to learn a policy that balances competing objectives (e.g., user needs vs. grid constraints).

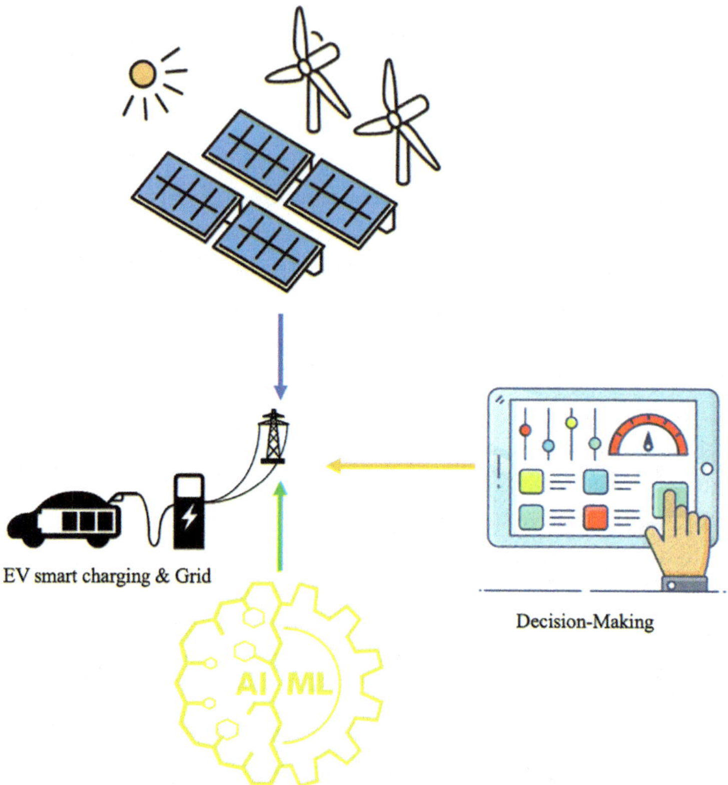

Figure 9.2 graphical abstract of the integration of EV and AL/ML techniques.

Common RL algorithms like Proximal Policy PPO or Deep Q-Networks (DQN) are employed to handle the complexity and uncertainty of real-world environments (Abdullah et al., 2021). This abstraction is illustrated in Fig. 9.3, where the agent learns to make real-time decisions that improve system-wide reliability and user satisfaction.

This process closely resembles how humans and animals learn. A child learning to ride a bicycle, for example, experiments with balance, experiences falls, and gradually refines technique until stability is achieved. Similarly, the agent–environment relationship can be described as a feedback loop: at each time step, the agent observes the current state, selects an action, and the environment responds by producing a new state along with a reward. The agent then updates its internal strategy, or policy, to improve future outcomes. A useful analogy is playing a video game: the screen represents the state, pressing a button is the action, the score is the reward, and

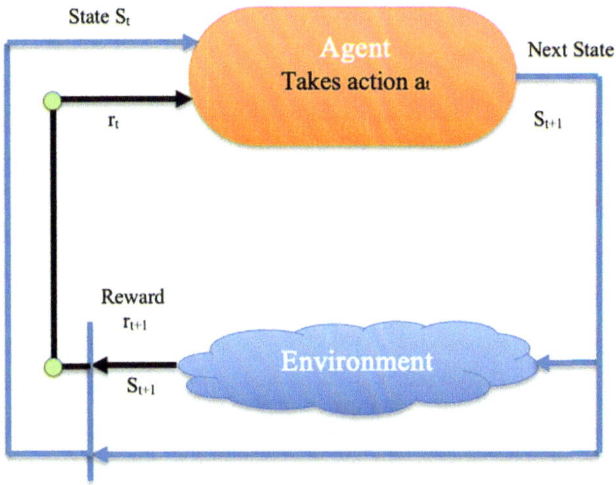

Figure 9.3 Illustration of the reinforcement learning cycle.

through repeated play, the player learns which sequences of actions lead to winning (Mnih et al., 2015).

Formally, RL problems are modeled as Markov Decision Processes (MDPs), defined by the tuple (Bilgin, 2020):

$$\mathcal{M} = (\mathcal{S}, \mathcal{A}, P, R, \gamma) \tag{9.7}$$

where $\mathcal{S}$ is the set of possible states, $\mathcal{A}$ is the set of actions, $P(s'|s,a)$ describes the transition probability of moving from state s to state s' after action a, $R(s,a)$ defines the reward function, and $\gamma \in [0,1]$ is the discount factor, which balances short-term versus long-term outcomes. The goal of RL is to find an optimal policy $\pi(a|s)$, a mapping from states to actions, that maximizes the expected cumulative discounted reward (Bilgin, 2020):

$$J(\pi) = \mathbb{E}_{\pi}\left[\sum_{t=0}^{T} \gamma^t r_t\right] \tag{9.8}$$

The central challenge is that the environment is often **stochastic and uncertain**, meaning the agent must learn strategies that perform well on average rather than in any single trajectory.

9.3.3 Core algorithmic paradigms in RL

Different approaches have been developed to solve MDPs, each with unique strengths and limitations. These are broadly classified into value-based, policy-based, and actor–critic methods.

9.3.3.1 *Value-based methods*

A central concept in RL is the value function, which quantifies how good it is for the agent to be in a particular state or to take a particular action. These functions allow the agent to evaluate and compare alternatives, forming the foundation of most RL algorithms (Mnih et al., 2015). For a given policy π-that is, a rule or probability distribution that tells the agent which action to take in each state, the state-value function is defined as (Sanghi, 2021):

$$V^{\pi}(s) = \mathbb{E}_{\pi}\left[\sum_{t=0}^{\infty} \gamma^t r_t \middle| s_0 = s\right] \tag{9.9}$$

Where.

- s is a state, representing the environment's condition (e.g., system load and renewable availability).
- r_t is the reward received at time step t.
- $\gamma \in [0, 1)$ is the discount factor, which determines how much future rewards are worth compared to immediate ones. A value close to 1 means the agent strongly considers future outcomes; a value closer to 0 means it focuses on short-term rewards.
- The expectation $\mathbb{E}_{\pi}[\cdot]$ means the value depends on the randomness of both the environment and the policy's action choices.

Intuitively, $V^{\pi}(s)$ tells us the expected long-term reward the agent will collect if it starts in state s and follows policy π thereafter. Closely related is the action-value function, or Q-function (Sanghi, N. (2021). *Deep Reinforcement Learning with Python. New York, NY, USA: Apress.*, no date):

$$Q^{\pi}(s, a) = \mathbb{E}_{\pi}\left[\sum_{t=0}^{\infty} \gamma^t r_t \middle| s_0 = s, a_0 = a\right] \tag{9.10}$$

In this case.

- a is an action taken in state s at the first step.
- $Q^{\pi}(s, a)$ measures the expected long-term reward from starting in state s, taking action a immediately, and then following policy π afterward.

The two functions are connected: the state-value is just the average of action-values under the policy's action probabilities,

$$V^{\pi}(s) = \mathbb{E}_{a \sim \pi(\cdot|s)}[Q^{\pi}(s, a)] \tag{9.11}$$

Another important concept is the advantage function:

$$A^{\pi}(s, a) = Q^{\pi}(s, a) - V^{\pi}(s) \tag{9.12}$$

This measures how much better (or worse) an action a is compared to the average action prescribed by the policy at state s. If $A^{\pi}(s, a)$ is positive, the action is better than average; if it is negative, it is worse. Advantage functions are widely used in modern RL algorithms such as actor-critic methods because they reduce variance in gradient estimates (Mnih et al., 2015).

Finally, the Bellman equations provide recursive definitions for these value functions. For a given policy π (Halperin, 2022):

$$V^{\pi}(s) = \mathbb{E}_{a\sim\pi}\left[r(s, a) + \gamma\mathbb{E}_{s'\sim P}V^{\pi}(s')\right] \tag{9.13}$$

$$Q^{\pi}(s, a) = \mathbb{E}\left[r(s, a) + \gamma\mathbb{E}_{a'\sim\pi}Q^{\pi}(s', a')\right] \tag{9.14}$$

where $P(s'|s, a)$ is the transition probability of moving to the next state s' given the current state s and action a.

The optimal value functions, denoted $V^{*}(s)$ and $Q^{*}(s, a)$, are defined with respect to the best possible policy. They satisfy the Bellman optimality equations (Halperin, 2022):

$$V^{*}(s) = \max_{a} \mathbb{E}\left[r(s, a) + \gamma V^{*}(s')\right] \tag{9.15}$$

$$Q^{*}(s, a) = \mathbb{E}\left[r(s, a) + \gamma \max_{a'} Q^{*}(s', a')\right] \tag{9.16}$$

These functions represent the gold standard for decision-making: if the agent knew them exactly, it could act optimally by always choosing the action that maximizes $Q^{*}(s, a)$. In practice, algorithms like Q-learning, policy gradients, and actor-critic methods are designed to approximate these functions or use them as guides (Halperin, 2022).

Problem 9.3. PEV Charging with Limited Headroom. Imagine a simplified EV charging system with only two states:

i. s_1: Evening peak demand, where feeder headroom is tight.
ii. s_2: Midday surplus, when solar generation is high.

At each state, the agent can choose one of two actions.

i. a_1: Conserve charging (serve 50% of the request).
ii. a_2: Boost charging (serve 100% of the request).

The reward is defined as the amount of EV demand successfully served, but with a penalty if headroom is exceeded.

Suppose the following.

i. In state s_1 (evening peak):

ii. If the agent conserves (a_1), it serves 50 units → reward $r = 50$.
iii. If it boosts (a_2), it overloads, serving 80 but with a penalty, so $r = 20$.
iv. In state s_2 (midday surplus):
v. Conserve gives $r = 40$.
vi. Boost gives $r = 90$ with no penalty.

Now, the state-value function under some policy π:

- If the policy always conserves, then

$$V^{\pi}(s_1) = 50, \; V^{\pi}(s_2) = 40.$$

- If the policy always boosts, then

$$V^{\pi}(s_1) = 20, \; V^{\pi}(s_2) = 90$$

The action-value function shows this distinction explicitly. For state s_1:

$$Q^{\pi}(s_1, a_1) = 50, \; Q^{\pi}(s_1, a_2) = 20$$

And for the state $\boldsymbol{s}_2$:

$$Q^{\pi}(s_2, a_1) = 40, \; Q^{\pi}(s_2, a_2) = 90$$

Finally, the advantage function tells us how much better one action is than the average. Suppose in s_2 the policy mixes actions (50% conserve, 50% boost). Then the average value is

$$V^{\pi}(s_2) = 0.5(40) + 0.5(90) = 65$$

So, the advantages are:

$$A^{\pi}(s_2, a_1) = 40 - 65 = -25, A^{\pi}(s_2, a_2) = 90 - 65 = +25.$$

This means boosting is clearly better than conserving at midday, while the reverse holds in the evening.

9.3.3.2 Policy-based methods

PPO is a widely adopted RL algorithm that balances policy improvement with stability, making it particularly suitable for complex control tasks such as smart EV charging. The central idea of PPO is to update the policy parameters in a way that encourages beneficial changes while preventing abrupt or destabilizing shifts. This is achieved by constraining the magnitude of policy updates through a clipped surrogate objective (Halperin, 2022).

The algorithm begins by initializing two sets of parameters: $\boldsymbol{\theta}$ for the policy (actor) and $\boldsymbol{\phi}$ for the value function (critic). During each training iteration, the agent interacts with the environment using its current policy π_θ, generating a collection of trajectories $\mathscr{D}$. Each trajectory consists of sequences of states, actions, rewards, and subsequent states. From these rollouts, the algorithm computes the advantage estimates $\widehat{A}_t$, which quantify how much better (or worse) an action performed is compared to the critic's baseline expectation. A positive advantage indicates that an action was more effective than anticipated, while a negative value implies underperformance.

The policy is then updated by maximizing the PPO-Clip objective (Rio et al., 2024):

$$L^{\mathrm{CLIP}}(\theta) = \hat{\mathbb{E}}_t\left[\min\left(r_t(\theta)\hat{A}_t, \mathrm{clip}\left(r_t(\theta), 1-\varepsilon, 1+\varepsilon\right)\hat{A}_t\right)\right] \tag{9.17}$$

where $r_t(\theta) = \frac{\pi_\theta(a_t|s_t)}{\pi_{\theta_{\mathrm{old}}}(a_t|s_t)}$ represents the ratio of the new policy's probability of choosing an action to that of the old policy. The clipping term ensures that $r_t(\theta)$ remains within the interval $[1-\varepsilon, 1+\varepsilon]$, thereby preventing excessively large updates that could destabilize training. Intuitively, this mechanism encourages the policy to adjust in the correct direction indicated by the advantage, but only within a conservative margin.

In parallel, the critic is updated to minimize the error between its predicted state values and the empirical returns observed during rollouts. This is expressed as (Rio et al., 2024):

$$L^{VF}(\phi) = \hat{\mathbb{E}}_t\left[\left(V_\phi(s_t) - R_t\right)^2\right] \tag{9.18}$$

where R_t is the discounted cumulative reward. By iteratively refining the critic, the algorithm improves the accuracy of advantage estimates, which in turn stabilizes the policy gradient updates.

Training proceeds in cycles of collecting experience, estimating advantages, and updating both actor and critic networks. Typically, batches of trajectories are divided into smaller mini-batches, and several epochs of updates are performed before discarding the data and generating fresh rollouts. This ensures efficient use of data while maintaining stability.

Applied to the smart charging problem, the state vector includes features such as time of day, renewable generation share, feeder headroom, and aggregated EV charging demand. The action space is discretized into three charging modes: Conserve (0.8), Normal (1.0), and Boost (1.2), while the reward function integrates penalties for unmet demand (ENS), unsafe boosting under limited headroom, and frequent action switching, alongside rewards for efficient RUI. The objective of PPO in this setting is to learn charging policies that minimize reliability risks, reduce the LOLP, and improve indices such as CRI and RUI, while avoiding extreme peaks reflected in the PAR.

Through this combination of cautious yet adaptive policy updates, PPO provides a robust framework for sequential decision-making in uncertain and dynamic environments. Its stability, scalability, and ability to balance exploration with exploitation make it particularly advantageous for R2V reliability-aware scheduling.

Table 9.5 summarizes the classification of RL methods, highlighting the progression from early value-based approaches to modern, scalable algorithms (Abdullah et al., 2021; Mnih et al., 2015; Bilgin, 2020. Value-based methods form the foundation, where the agent estimates long-term returns for each action and chooses the one with the highest value. While intuitive, these approaches are ill-suited for EV charging control, where decisions involve semi-continuous adjustments and high-dimensional uncertainty. Policy-based methods address this by directly parameterizing the policy, making them suitable for continuous action domains such as charging rate modulation. However, they often suffer from high variance in gradient estimation, which hampers convergence. To overcome these challenges, actor–critic methods integrate both ideas: the critic provides low-variance estimates of action quality, while the actor updates the policy accordingly. Within this family, PPO has become the algorithm of choice in practice, offering stable updates and robust performance under stochastic conditions. For reliability-aware EV smart charging, PPO's balance of adaptability and stability makes it the most suitable approach.

9.3.4 Monte Carlo evaluation of policies

In reinforcement learning, evaluating a policy based on a single training trajectory is rarely sufficient, particularly in domains characterized by high stochasticity, such as EV charging and renewable energy integration. Random variations in EV arrivals, dwell times, and renewable generation can significantly alter outcomes, meaning that a single trajectory may provide a misleading representation of policy performance. To overcome this limitation, policies are commonly assessed using MC rollouts. Under this approach, each candidate policy is executed repeatedly under many independent random seeds. These repeated simulations yield a distribution of outcomes rather than a single deterministic value, thereby capturing the variability introduced by uncertain demand and fluctuating supply (Bilgin, 2020).

Table 9.5 Summarizing the classification of RL methods.

Approach	Core idea	Strengths	Limitations	Examples
Value-based	Learn a value function (e.g., Q(s,a)) and act greedily by selecting the action with the highest value.	Simple, well-studied; effective in small, discrete action spaces.	Struggles with large or continuous action spaces; unstable with function approximation.	Q-learning, SARSA, DQN
Policy-based	Directly optimize the policy $\pi\theta(a\|s)$, parameterized by θ, via gradient ascent on expected returns.	Handles continuous actions; naturally stochastic policies; flexible for real-time control.	High variance in gradient estimates; can be sample-inefficient.	REINFORCE, vanilla policy gradient
Actor–Critic	Combine policy (actor) with value function (critic) to stabilize learning; critic reduces gradient variance.	More stable than pure policy gradients; scalable to complex problems; efficient exploration.	More complex; requires tuning two models (actor + critic).	PPO, A3C, DDPG, SAC

Formally, if $J(\pi)$ denotes the expected return of a policy π, its MC estimate with N rollouts is expressed as (Bilgin, 2020):

$$\hat{J}(\pi) = \frac{1}{N} \sum_{i=1}^{N} \sum_{t=0}^{T} \gamma^t r_t^{(i)} \tag{9.19}$$

where $r_t^{(i)}$ is the reward received at time t during the i-th rollout, $\gamma \in [0, 1]$ is the discount factor balancing short- and long-term objectives, and T is the episode horizon. This formulation captures the empirical average of cumulative discounted rewards across multiple stochastic simulations.

From these rollouts, statistical summaries of key reliability metrics can be obtained, including ENS, CRI, LOLP, and RUI. Importantly, MC evaluation produces full probability distributions of these indices, visualized through boxplots, histograms, or frontier curves, rather than relying solely on averages. This enables more robust comparisons between alternative policies, highlighting not only expected performance but also variability and worst-case scenarios.

The advantage of this probabilistic approach is that it ensures policies perform reliably across a broad range of operating conditions rather than being tuned to a narrow set of typical scenarios. By embedding MC evaluation into the reliability analysis, performance is assessed probabilistically rather than deterministically, providing grid operators with confidence that the learned scheduling policies can withstand real-world uncertainty. The distributions and frontier plots presented later in Section 9.4 (Case Study) are generated using this methodology, offering deeper insight into the trade-offs between service reliability, resource utilization, and charging performance.

9.4 Case study: Reliability-aware EV smart charging with RL approach

9.4.1 Overview

This section presents a fully integrated RL framework, a machine learning technique where agents learn optimal actions through trial and error with feedback, for optimizing the reliability of R2V smart charging networks. The case study leverages real-world CAISO data (California Independent System Operator (CAISO), n.d.), where renewable generation variability and dynamic EV demand jointly influence system reliability. In this context, feeder headroom (the margin between a feeder's rated capacity and the non-EV baseline load) represents the available capacity for EV charging and serves as the key operational constraint.

The RL controller operates through three discrete actions: Conserve (0.8 multiplier, reducing charging demand to protect feeder capacity during stressed hours), Normal (1.0 multiplier, serving EV requests as-is within feeder headroom, equivalent to unmanaged plug-and-charge), and Boost (1.2 multiplier, temporarily increasing charging to absorb surplus renewable energy or recover from backlog). These three modes form the agent's decision space for real-time adaptive scheduling, balancing user demand, renewable variability, and grid reliability. The study evaluates a reliability-aware EV smart-charging controller using CAISO time-series, structured into four implemented blocks:

1. **Data & Inputs** Hourly CAISO load, solar, and wind profiles for 2020 are used as exogenous signals. EV charging demand is synthesized from empirical diurnal and weekday–weekend patterns, scaled to the study feeder capacity.
2. **Monte Carlo Scenario** Uncertainty in EV arrivals, dwell times, and renewable availability is captured through NN stochastic simulations per policy, stratified by season and week. Each run samples EV arrival streams and renewable perturbations, while network limits (headroom) are enforced by feeder constraints.
3. Scheduling Policies (benchmarked side-by-side)
    1. Unmanaged: Plug-and-charge at arrival, constrained only by feeder capacity caps.
    2. LP Baseline: Deterministic day-ahead allocation that minimizes shortfalls under capacity limits.
    3. PPO-RL Controller: A trained policy selecting hourly actions $a_t \in \{\text{Conserve (0.8), Normal (1.0), Boost (1.2)}\}$.

        3.1. Observation vector: Encodes hour-of-day and day-of-week, current EV request, available headroom (load vs. feeder limit), recent renewable share, and running statistics.

        3.2. Reward design: Penalizes ENS and loss-of-load events (LOLP surrogates), while rewarding higher CRI. Reward shaping encourages energy delivery when headroom exists and conservation when grid limits are binding.

        3.3. Episodes: One-week slices. Training proceeds over rolling seasons; evaluation uses held-out weeks.
4. **Reliability Engine (post-simulation metrics)** For each MC run, the following metrics are computed:

(a) **ENS (MWh):** Total unmet EV energy.
(b) **LOLP (0–1):** Fraction of hours with energy shortfalls.
(c) **CRI (0–1):** Fraction of hours where served energy meets reliability targets.
(d) **RUI (0–1):** Normalized utilization of available headroom by EVs.

Aggregating these metrics across MC runs produces the ENS boxplots, CRI/LOLP/RUI distributions, and the ENS–CRI frontier presented in Section 4.

Structuring the case study into data → MC → policies → reliability metrics ensures that the results are not tied to a single deterministic scenario but instead reflect the stochastic nature of EV charging and renewable availability. This approach allows the findings to generalize beyond CAISO to other regional grids, providing operators with a framework to evaluate trade-offs between service quality and grid stability under uncertainty. To contextualize these metrics in practical operation, Table 9.6 provides a decision aid for system operators. Based on threshold values observed in the simulations, it identifies the preferred scheduling policy and operational cues (e.g., when to apply Conserve during peak stress or Boost during midday renewable surplus). This bridges the simulation framework with actionable strategies, ensuring that reliability metrics translate into clear operator guidance.

Note: Numbers reflect the central tendency and spread observed in the figures (MC evaluation). They are intended as decision thresholds/guides; operators can plug in their own fleet size and limits. The relative ordering (RL > LP > Unmanaged) is robust.

Table 9.6 Quick decision aid for operators.

Objective	If current metric is ...	Preferred policy	Operational cue
Cut shortfalls (ENS)	ENS >20 GWh	RL	Enable peak *Conserve* at evening peaks; *Boost* at midday surplus.
Improve reliability (CRI)	CRI <0.6	RL or LP	Activate reliability-based scheduling; enforce feeder headroom limits.
Reduce risk (LOLP)	LOLP >.015	RL	Tighten transformer/feeder constraints; prioritize flexible loads.
Better asset utilization (RUI)	RUI <0.35	RL	Shift charging to PV/wind hours; consider time-varying incentives.

9.4.2 Data sources

The reliability-aware EV smart charging framework relies on multiple datasets to capture both system-level constraints (e.g., feeder capacity, load) and stochastic influences (e.g., weather, renewable generation, and EV behavior). Using diverse and realistic data sources ensures that the case study reflects operational conditions encountered in real distribution systems. Table 9.7 summarizes the datasets employed. The datasets serve complementary roles:

- **CAISO data** provides the baseline for system load and capacity headroom, anchoring the RL controller's decisions to actual grid conditions.
- **NREL solar/wind data** allows simulation of renewable fluctuations, ensuring the RL agent learns how to exploit surplus energy (via *Boost*) or conserve during shortfalls (*Conserve*).
- **NOAA/OpenEI weather data** adds stochasticity, allowing the MC generator to account for conditions like cloud cover or temperature that strongly influence solar output and EV travel demand.
- **NREL eVI demand profiles** serve as the core driver of EV charging requests, ensuring a realistic representation of how vehicles arrive, dwell, and request energy.

By integrating these datasets, the framework bridges system-level constraints (CAISO), supply-side variability (NREL + NOAA), and demand-side behavior (NREL eVI). Together, they enable a realistic reliability evaluation across multiple seasons and MC runs.

CAISO load profiles provide the backbone of system demand. Fig. 9.4 highlights the relationship between net load and renewable injections across 2020. While demand remains clustered around 20–30 GW, renewable

Table 9.7 Data sources used in the case study.

Data source	Description
CAISO	Grid load profiles (hourly, 2020), outage reports, feeder headroom statistics.
NREL	Solar irradiance and wind speed datasets used to derive renewable generation profiles.
NOAA/OpenEI	Weather data (temperature, wind, cloud cover), providing exogenous uncertainty drivers for demand and renewables (National Oceanic and Atmospheric Administration, n.d.).
NREL eVI	Empirical EV charging demand profiles (arrival times, dwell durations, energy needs), scaled to feeder capacity (National Renewable Energy Laboratory, n.d.).

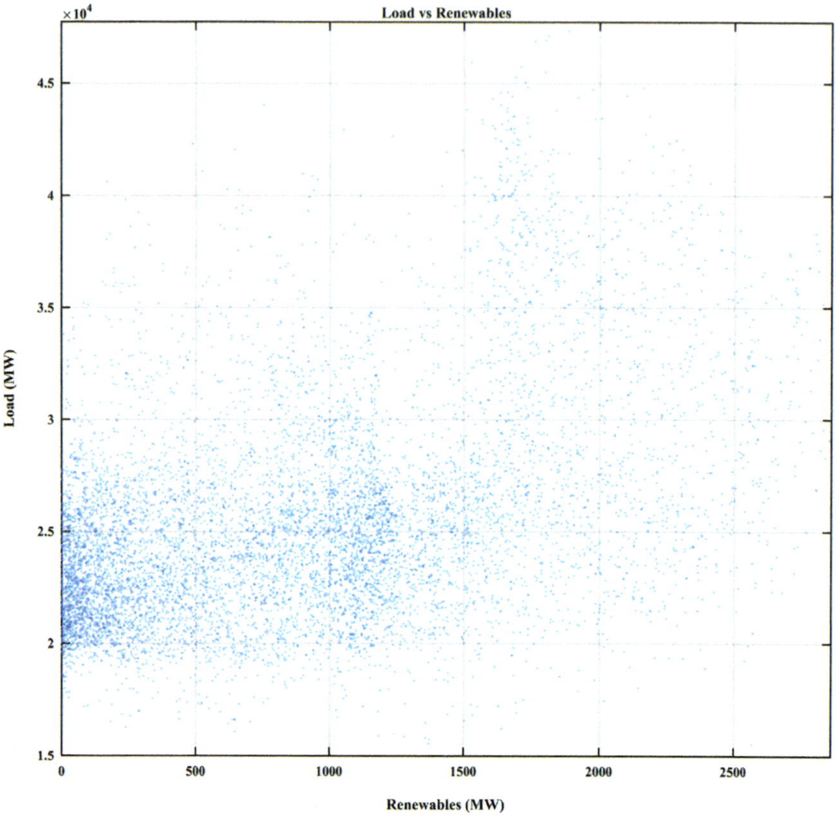

Figure 9.4 Scatter: Load versus renewables.

variability introduces a significant spread, with renewables occasionally displacing conventional load by over 2 GW.

Fig. 9.5 depicts the daily peak headroom available for EV charging, calculated as feeder capacity minus underlying CAISO demand. Peaks occur in spring and early summer, but tighter headroom during late summer and fall signals periods of higher congestion risk for EV integration.

Fig. 9.6 compares solar PV and wind peak generation across the year. PV exhibits strong seasonal cycles with higher summer peaks (>1.5 GW), while wind is more intermittent, fluctuating daily. These differences explain why RL must adapt dynamically to renewable conditions.

Fig. 9.7 presents scatterplots of hourly net load, illustrating clear diurnal variability across the year. The data reveal that demand consistently peaks between 16:00 and 20:00, coinciding with evening EV charging and

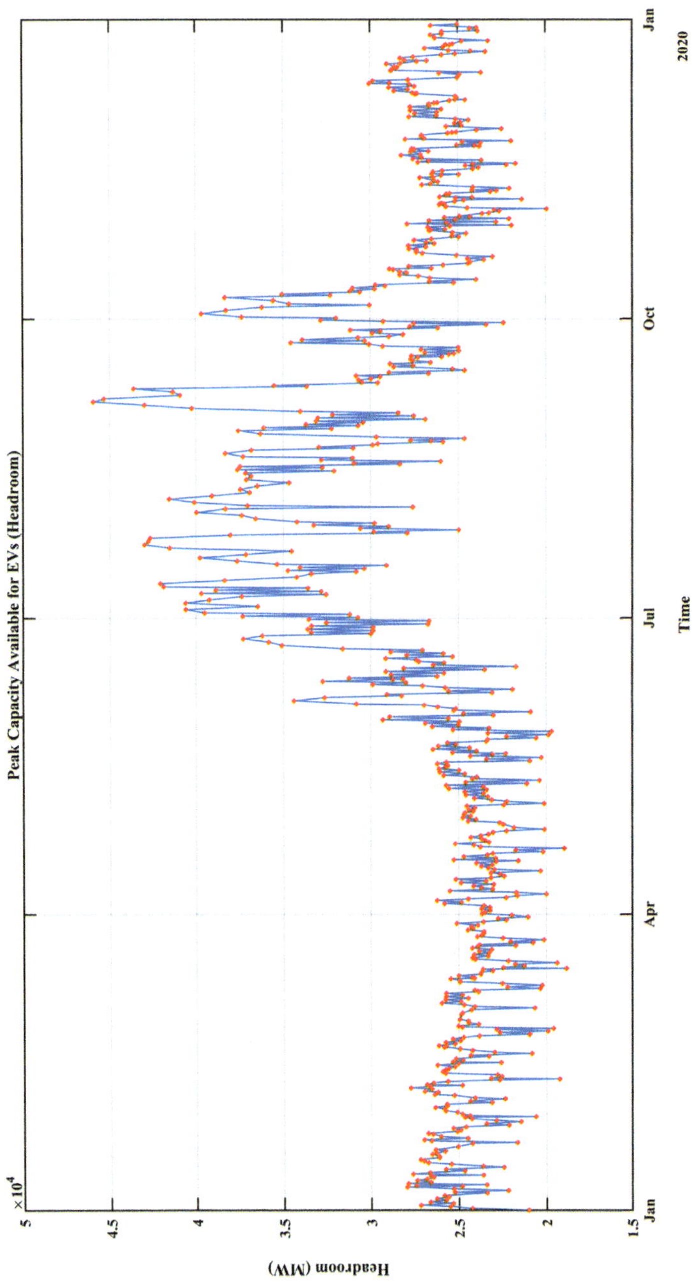

Figure 9.5 Daily peak capacity available for EVs.

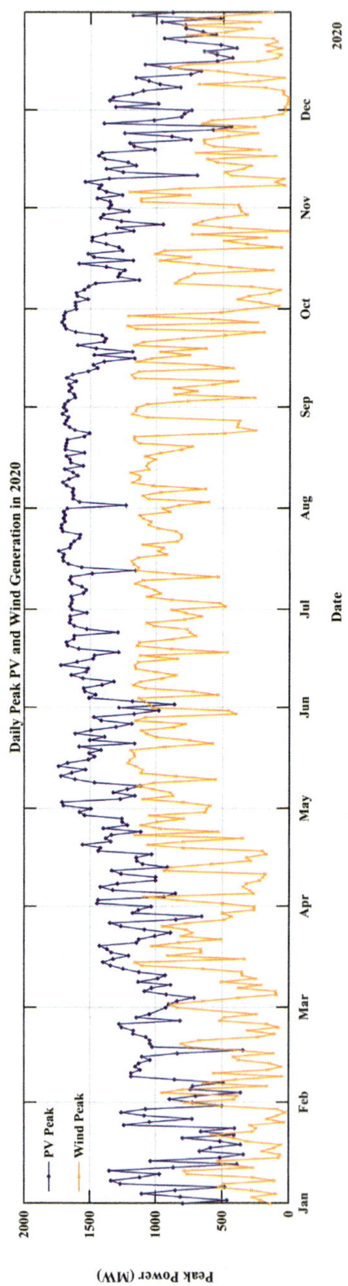

Figure 9.6 Daily Renewable peaks in 2020.

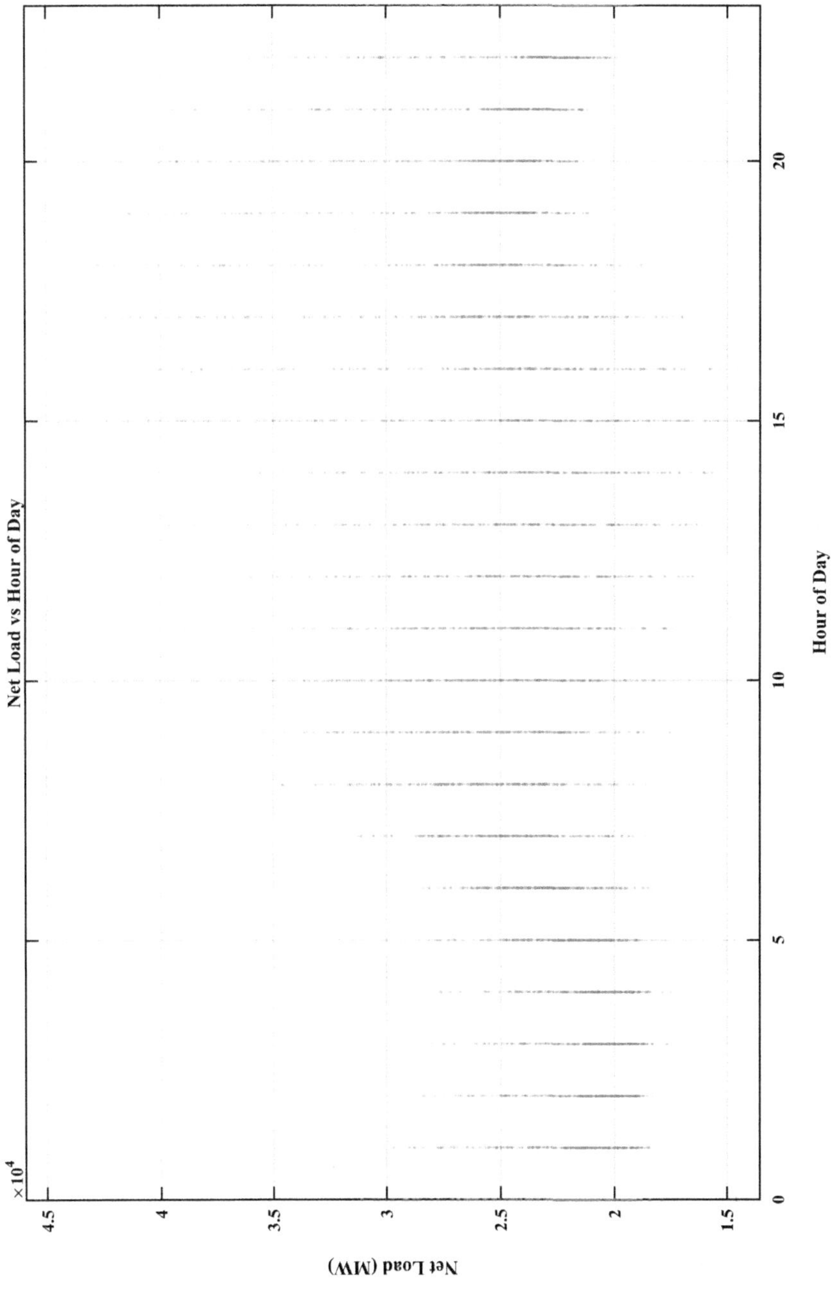

Figure 9.7 Net load versus hour-of-day.

residential consumption. During these hours, the system approaches its maximum load, highlighting the periods of greatest stress on feeder headroom. By contrast, early morning hours maintain lower demand, reflecting reduced household and mobility-related activity. This diurnal cycle underscores the importance of time-sensitive scheduling strategies for effective RL-based EV charging control.

Fig. 9.8 compares average sample daily net load profiles across the four seasons: spring, summer, autumn, and winter. The results indicate that summer demand is most pronounced, with peak values exceeding 34 GW in the evening, driven by the combined effect of cooling loads and EV charging. Autumn profiles lie between summer and winter, reflecting moderate consumption patterns, while spring exhibits comparatively lower peaks. Winter demand remains relatively flat throughout the day, showing less pronounced evening surges due to the absence of cooling loads and more stable consumption patterns. These seasonal variations emphasize the necessity of adaptive policies, as RL agents must learn context-specific behaviors to balance ENS reduction with grid reliability across different operating conditions.

Fig. 9.9 presents the smoothed hourly load profile over the course of 1 year, capturing both daily fluctuations and seasonal stress concentrations. The gray curve represents raw load data, while the red line highlights the smoothed peak trajectory. The results show that system load generally hovers around 25–30 GW, but during July to September, sustained peaks emerge that consistently exceed 35 GW, with extreme cases surpassing 45 GW. These seasonal stress patterns coincide with high summer cooling demand and increased EV charging activity, underscoring the vulnerability of the grid to concurrent peak drivers.

Fig. 9.10 illustrates the load duration curve (LDC), which provides a probabilistic view of system load by plotting demand levels against their exceedance probability. The curve demonstrates that while median load conditions are stable near 25 GW, the tail of the distribution extends above 45 GW, reflecting rare but severe stress events. This steep tail emphasizes the importance of incorporating probabilistic planning and risk-aware scheduling, as system reliability cannot be fully understood by average values alone. The LDC thus reinforces the necessity of robust RL-based controllers capable of adapting to both typical and extreme operating conditions.

Fig. 9.11 illustrates typical weekday–weekend patterns. The top heatmap shows load intensity by hour and day of week, while the bottom

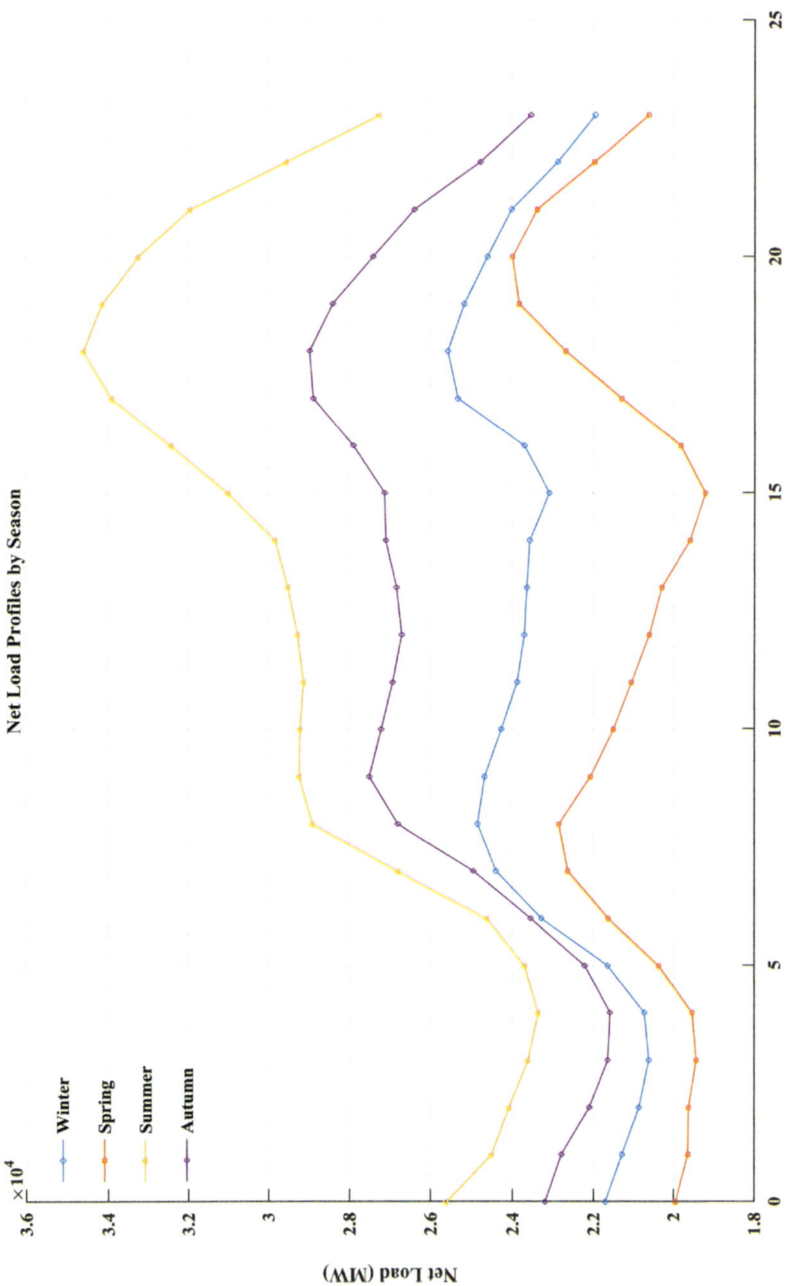

Figure 9.8 Net load profiles by season.

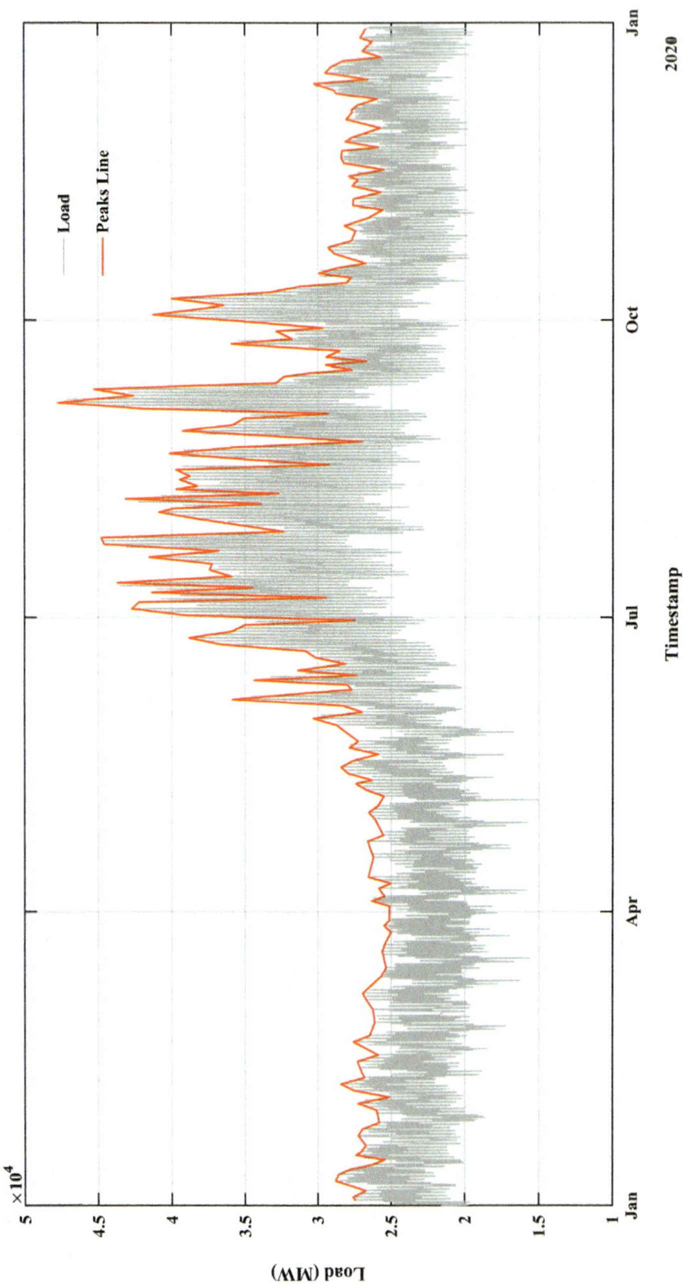

Figure 9.9 Load peaks - smoothed over year.

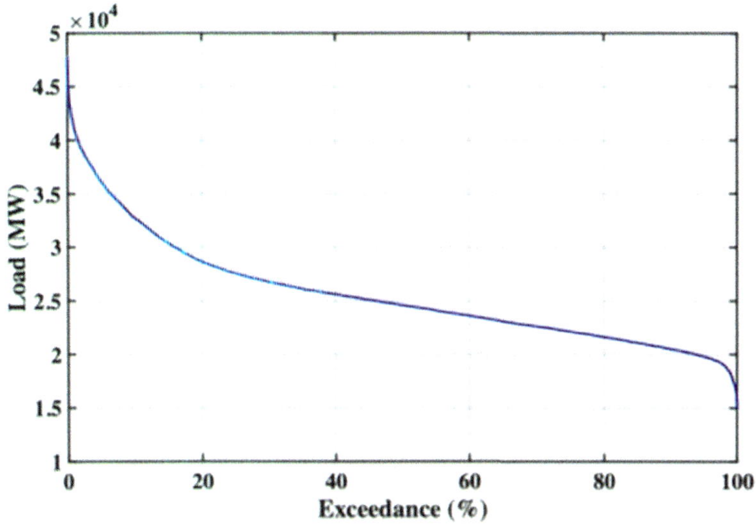

Figure 9.10 Load duration curve.

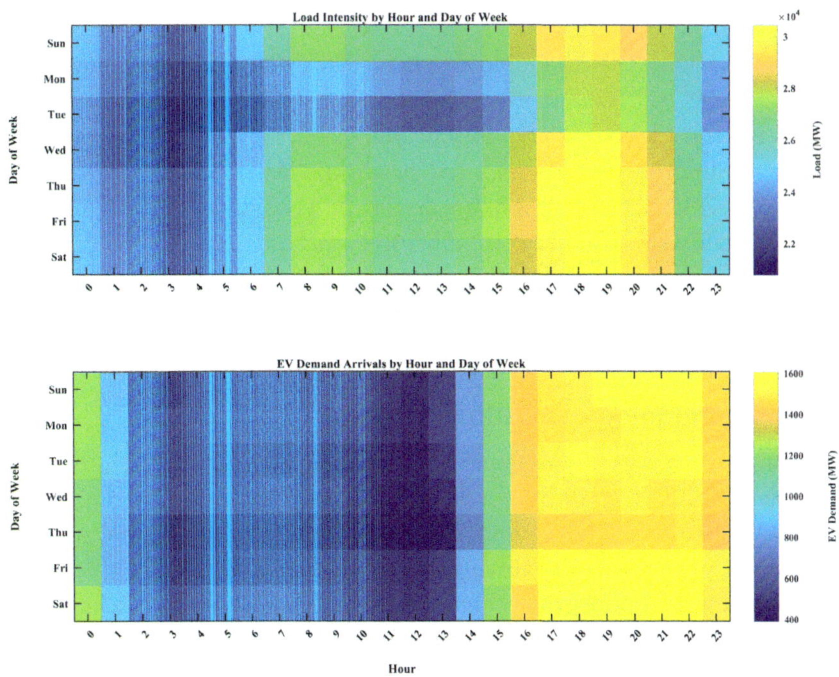

Figure 9.11 Heatmap of net load & EV profiles.

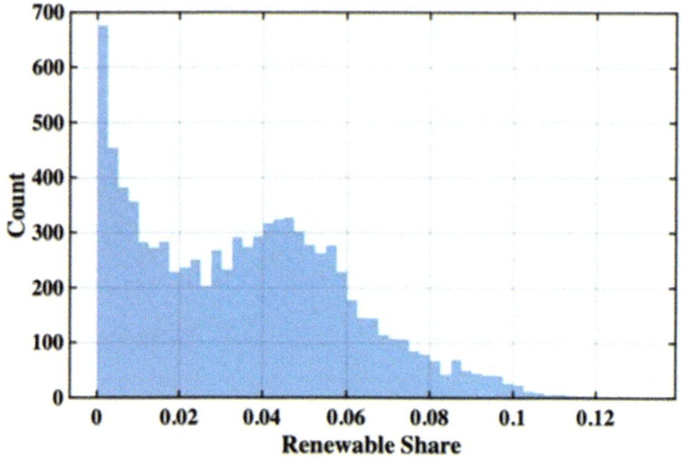

Figure 9.12 Distribution of Renewable share.

heatmap visualizes EV demand arrivals. Both indicate concentrated stress in weekday evenings, aligning with the RL controller's "Conserve" actions during high-load hours and "Boost" actions when renewables are abundant.

Fig. 9.12 shows the distribution of renewable share (renewables/total generation). The distribution is highly skewed toward low values (0%–5%), with a long tail extending to ~0.12. This highlights that renewable penetration is still modest in most hours, but high renewable events do occur and must be considered in scheduling.

By combining CAISO grid and headroom data, NREL's renewable datasets, NOAA/OpenEI's weather drivers, and NREL's eVI charging profiles, the study constructs a comprehensive stochastic environment. These datasets not only ground the RL training in realistic operating conditions but also provide the basis for MC scenario generation used in the reliability evaluation.

9.4.3 Methodology workflow

Building on the general RL framework introduced in Section 9.3, this section develops a case study of reliability-aware EV smart charging using CAISO demand and renewable generation data. The methodology integrates MC simulation with three distinct scheduling policies.

i. **Unmanaged charging** (plug-and-charge) represents the baseline case.
ii. **Linear Programming (LP) scheduling** is a deterministic day-ahead benchmark.
iii. **PPO scheduling** is an adaptive RL-based controller that dynamically allocates charging power based on real-time states.

To ensure robustness, all policies are evaluated under identical stochastic conditions using multiple MC rollouts, thereby capturing the inherent variability of EV demand and renewable generation.

9.4.3.1 Data assembly and pre-processing

The simulation environment is constructed from hourly CAISO data, including system load, solar generation, and wind generation. Three key components are prepared.

(a) Feeder Headroom:

In distribution networks, *feeder headroom* refers to the available margin between the maximum allowable feeder capacity and the baseline non-EV demand already being served. It essentially represents how much additional load can be safely accommodated without exceeding thermal or operational limits of the feeder. Mathematically, it is often expressed as:

$$H_t = C_{\text{feeder}} - L_t^{\text{base}} \tag{9.20}$$

where C_{feeder} is the rated feeder capacity and L_t^{base} is the non-EV demand at time t.

For example, if a feeder is rated at 10 MW and the underlying residential and commercial demand at a given hour is 8 MW, then the feeder headroom is 2 MW. This means that up to 2 MW of additional EV charging can be scheduled without overloading the system. However, if vehicles are also discharging energy back into the grid (e.g., via V2G mechanisms), this affects the headroom calculation: while discharging may temporarily relieve some demand, it also reduces the net available capacity over time and can lead to feeder saturation if not properly managed. If total EV activity (charging and discharging) exceeds this headroom, either curtailment or load shifting becomes necessary.

(b) EV Demand:

EV charging requests are modeled stochastically to capture real-world uncertainty. Arrivals are generated using a nonhomogeneous Poisson process, while dwell durations follow a log-normal distribution conditioned on arrival time. Per-vehicle energy demand is truncated to reflect feasible ranges, with weekday/weekend and seasonal patterns preserved.

(c) Renewable Share

Renewable share is the fraction of the total system load at a given time supplied by renewable generation (e.g., solar and wind). It provides a measure of how much the grid relies on clean, variable sources compared with conventional generation. Mathematically, it can be written as:

$$\rho_t = \frac{P_t^{\mathrm{RES}}}{L_t} \tag{9.21}$$

where P_t^{RES} is the combined renewable output (solar + wind) at time t, and L_t is the total system load at the same time.

For instance, if total demand is 100 MW and renewables produce 30 MW, the renewable share is $\rho_t = 0.3$, meaning 30% of the demand is covered by renewables while the remaining 70% must be supplied by non-renewable sources.

This metric is critical for EV smart charging because high renewable shares indicate abundant low-carbon energy that can be efficiently absorbed by flexible loads like EVs, while low renewable shares signal a heavier reliance on conventional generation and tighter headroom.

9.4.3.2 Monte Carlo scenario

Reliability assessment requires capturing the full spectrum of stochastic variation in EV demand and renewable generation. This is achieved through MC rollouts, where each candidate policy is simulated repeatedly across randomized weekly scenarios. Each rollout consists of 168 hourly steps (1 week) and incorporates:

i. **EV arrivals** are generated as a nonhomogeneous Poisson process.
ii. **Dwell times** sampled from log-normal distributions.
iii. **Per-EV charging requests** are drawn from a truncated normal distribution.
iv. **Renewable fluctuations** are modeled as Gaussian perturbations of CAISO wind and solar profiles.

To ensure comparability, identical random seeds are applied across different scheduling policies, guaranteeing that all methods are exposed to equivalent uncertainty realizations.

9.4.3.3 Feeder and constraints model

At each hour t, the scheduler determines the served EV demand as:

$$S_t = \min\left(m_t \cdot R_t^{\mathrm{EV}}, H_t\right) \tag{9.22}$$

where R_t^{EV} is the aggregate EV charging request, and $m_t \in \{0.8, 1.0, 1.2\}$ is the action multiplier corresponding to Conserve, Normal, and Boost charging modes.

The shortfall (unserved demand) is then:

$$\text{short}_t = \max\left(0, m_t R_t^{EV} - H_t\right) \tag{9.23}$$

and the weekly ENS is:

$$\text{ENS} = \sum_t \text{short}_t \tag{9.24}$$

This formulation ensures that served demand never exceeds feeder headroom while explicitly quantifying unmet energy. ENS thus becomes a primary indicator of reliability for evaluating different policies.

9.4.3.4 Scheduling policies compared

Three scheduling strategies are benchmarked.

i. Unmanaged (Plug-and-Charge)

EVs begin charging immediately at full request ($m_t = 1.0$) until departure, subject only to feeder headroom H_t. This represents the business-as-usual baseline, which often amplifies evening peaks.

ii. Linear Programming (LP) Baseline

A deterministic day-ahead scheduler minimizes total shortfall $\sum_t \text{short}_t$ subject to daily demand fulfillment and hourly headroom constraints. LP reflects an optimized, but inflexible, benchmark without adaptive feedback.

RL-Based PPO Controller.

An RL agent selects $m_t \in \{0.8, 1.0, 1.2\}$ online each hour. This enables adaptive adjustments to demand and renewable variability, balancing service quality with feeder stability.

9.4.3.5 RL environment: Observations, actions, and reward

The RL agent interacts with the environment through a structured set of observations, discrete actions, and a reward function designed to balance service quality, safety, and efficiency.

- State vector (o_t):

  At time step t, the state comprises the following features:

  i. Temporal encoding: Hour-of-day represented using sine and cosine transformations, and day-of-week in one-hot encoding.

  ii. EV charging demand ( ): The aggregate requested charging power from all connected EVs.

  iii. Feeder headroom: The available grid margin between feeder capacity and baseline demand.

iv. Renewable share ( ): The proportion of renewable generation in the supply mix.
v. Previous action ( ): To account for temporal consistency and switching costs.
vi. Rolling averages: 24-hour smoothed histories of demand and feeder headroom to embed short-term trends.

- Action space (a_t): The agent selects a discrete action multiplier from:

$$a_t \in \{0.8, 1.0, 1.2\}$$

corresponding to Conserve, Normal, and Boost charging modes, respectively. The multiplier scales the requested EV charging demand relative to available feeder headroom.

- Reward function (r_t): The agent receives feedback at each step according to:

$$r_t = -\lambda_1 \cdot \text{short}_t - \lambda_2 \cdot 1\left[H_t < \eta C_{\text{feeder}}, a_t = 1.2\right] + \lambda_3 \cdot \text{util}_t - \lambda_4 \cdot \Delta a_t \tag{9.25}$$

where.

i. $\text{short}_t = \max\left(0, a_t R_t^{EV} - H_t\right)$: unmet demand, penalizing energy not served.
ii. $1\left[H_t < \eta C_{\text{feeder}}, a_t = 1.2\right]$: safety penalty applied when the agent attempts to boost charging under tight headroom conditions, with $\eta \in [0.8, 0.9]$.
iii. $\text{util}_t = \frac{S_t}{\max(H_t, \epsilon)}$: renewable utilization reward, incentivizing efficient use of available headroom.
iv. $\Delta a_t = |a_t - a_{t-1}|$: penalty on frequent action switching, promoting operational stability.

The weights $\lambda_1 = 3.0, \lambda_2 = 1.0, \lambda_3 = 0.5$, and $\lambda_4 = 0.04$ are tuned to balance four competing objectives:

i. Reliability (minimizing ENS through λ_1).
ii. Safety (avoiding overloads through λ_2).
iii. Efficiency (maximizing renewable integration through λ_3).
iv. Stability (discouraging oscillatory policies through λ_4).

This formulation ensures that the RL agent learns strategies that minimize unmet demand, respect system constraints, effectively exploit renewable energy, and maintain stable scheduling decisions.

9.4.3.6 PPO training procedure

The PPO controller is designed to balance stability and adaptability in reliability-aware EV smart charging. A relatively deep yet compact policy network with two hidden layers of 64 units each provides sufficient expressive power without incurring excessive computational cost. The discount factor ($\gamma = 0.99$) emphasizes long-term reliability objectives, while the GAE parameter ($\lambda = 0.95$) ensures stable advantage estimation. The clipping parameter ($\varepsilon = 0.2$) constrains policy updates to avoid instability, while entropy regularization (0.01) encourages exploration. The batch size and episodic horizon are selected to capture weekly demand–supply variability, with multiple epochs per iteration enhancing convergence. Curriculum training improves robustness by gradually increasing EV penetration, while early stopping safeguards against overfitting by monitoring trends in the CRI and ENS. A summary of the main hyperparameters used in the PPO training configuration is provided in Table 9.8 for clarity and reproducibility.

9.4.3.7 Evaluation metrics

Each policy is evaluated through MC rollouts across all seasons to capture the variability of EV arrivals, renewable supply, and grid conditions. The following reliability and efficiency metrics by time interval representation are computed.

Table 9.8 PPO controller configuration.

Component	Description
Policy network	Actor–critic PPO with a two-layer MLP (64 units, Tanh)
Discount factor (γ)	0.99
GAE parameter (λGAE)	0.95
Clipping parameter (ε)	0.2
Learning rate	3×10^{-4}
Entropy coefficient	0.01
Batch size	8 episodes × 168 steps
Epochs per iteration	10
Curriculum training	Trained under moderate-to-seasonal EV penetration
Early stopping	With early stopping triggered when CRI plateaus (10-episode moving average) and ENS improvement falls below 2%

i. Energy Not Supplied (ENS, MWh):

ENS quantifies the total unmet charging demand due to feeder or capacity constraints. It reflects the absolute amount of energy that EVs requested but could not be delivered. Lower ENS values indicate that the system is better at fulfilling user charging requirements, thereby enhancing mobility security.

$$\mathrm{ENS} = \sum_t \mathrm{short}_t \tag{9.26}$$

ii. Loss of Load Probability (LOLP: 0–1):

LOLP measures the likelihood that at least some EV demand cannot be served at a given time step. Unlike ENS, which captures magnitude, LOLP captures frequency. A system with high LOLP experiences frequent shortages even if the energy lost per event is small. Lower LOLP indicates a more reliable availability of charging.

$$\mathrm{LOLP} = \frac{1}{T}\sum_t 1[\mathrm{short}_t > 0] \tag{9.27}$$

iii. Charging Reliability Index (CRI, 0–1):

CRI evaluates the fraction of time steps in which EV demand is served at or above a specified threshold. It provides a user-centric reliability perspective by quantifying how consistently drivers receive an adequate share of their requested charging power. For example, with threshold $\tau = 0.9$, CRI indicates the proportion of time that at least 90% of the requested demand was met.

$$\mathrm{CRI}(\tau) = \frac{1}{T}\sum_t 1\left[S_t \geq \tau R_t^{EV}\right], \tau \in [0.9, 1.0] \tag{9.28}$$

iv. Resource Utilization Index (RUI, 0–1):

RUI measures how efficiently available feeder headroom is utilized. A high RUI means that whenever renewable or feeder capacity is available, it is effectively converted into EV charging rather than left idle. By contrast, a low RUI suggests under-utilization of resources, leading to wasted renewable generation or unnecessary curtailment.

$$RUI = \frac{1}{T}\sum_t \frac{S_t}{\max(H_t, \epsilon)}, \text{ capped at1} \tag{9.29}$$

v. Peak-to-Average Ratio (PAR):

PAR captures the degree of load smoothing in charging schedules. It is defined as the ratio between the maximum feeder load and the average load across the evaluation horizon. A lower PAR indicates flatter, more grid-friendly demand profiles that reduce stress on transformers and distribution assets, while higher values suggest spiky and operationally challenging load curves.

$$PAR = \frac{\max_t L_t}{\frac{1}{T}\sum_t L_t} \tag{9.30}$$

9.4.3.8 Uncertainty quantification

i. ENS and CRI distributions via MC runs (boxplots).

ii. ENS-CRI frontier plotted with policy means and 95% ellipses.

iii. Bootstrap confidence intervals (2,000 resamples) for policy comparisons (RL vs. LP, RL vs. Unmanaged).

9.4.3.9 Decision tables (seasonal/weekly)

To translate simulation outcomes into actionable insights for grid operators, decision tables are derived from MC results. These tables summarize the median values of reliability metrics (ENS, LOLP, CRI, RUI, and PAR) for each season and policy, as well as the relative improvements of RL and LP policies against the unmanaged baseline.

Furthermore, heuristic rules of thumb are extracted by linking observed system conditions to the most reliable scheduling actions. For example.

i. If feeder headroom is scarce ($H_t/C_{\text{feeder}} < 0.2$) or renewable share is low ($\rho_t < 0.1$), the Conserve mode ($a_t = 0.8$) is generally preferable to avoid overloads.

ii. If feeder headroom is abundant ($\boldsymbol{H}_t/C_{\text{feeder}} > 0.35$) and the renewable share is high ($\rho_t > 0.25$), the Boost mode ($a_t = 1.2$) becomes advantageous to accelerate charging while exploiting excess renewable supply.

By providing both aggregated seasonal summaries and interpretable operational heuristics, these decision tables serve as a bridge between algorithmic optimization and practical field deployment. They enable operators to make informed, reliability-aware scheduling choices even without direct access to the RL agent.

9.4.4 Results and analysis

This section presents the outcomes of the proposed RL–based EV smart charging framework. The analysis progresses step by step: first examining the training convergence of the PPO agent (Section 9.4.4.1), then evaluating operational performance on representative test weeks (Sections 9.4.4.2–9.4.4.3). Next, system-level performance is quantified using reliability frontiers, ENS distributions, and probabilistic reliability indices (Sections 9.4.4.4–9.4.4.6). Seasonal adaptation of the RL policy is summarized in a decision table (Section 9.4.4.7), before concluding with key insights on robustness and adaptability (Section 9.4.5).

9.4.4.1 RL training convergence

Fig. 9.13 presents the reward trajectory of the PPO agent over 30 training episodes. At the outset, the cumulative reward is low (approximately −120 arbitrary units), reflecting poor initial performance. As training progresses, the reward steadily increases, converging to around −40. This upward trend indicates that the RL agent progressively learns more effective scheduling decisions, balancing charging demand against feeder headroom with greater efficiency.

Fig. 9.14 shows the evolution of normalized ENS per unit EV demand. Early in training, nearly half of the requested charging energy remains unmet, with ENS values near 0.4. As learning advances, the ENS consistently declines, approaching zero by the final episodes. This reduction highlights the agent's increasing ability to allocate charging capacity efficiently, ensuring that a larger fraction of EV demand is met within the feeder's operational limits.

Fig. 9.15 illustrates the progression of the CRI, which captures the proportion of time that EV charging meets predefined reliability thresholds. The CRI begins at roughly 0.4, reflecting frequent service interruptions, but rises steadily to around 0.9 by the end of training. This improvement demonstrates that the RL agent not only reduces shortfalls but also enhances the overall reliability of EV charging service delivery.

Although the plots suggest that convergence occurs within 30 episodes for this case study, it is important to note that stable and generalizable training in larger or more complex scenarios typically requires a greater number of episodes.

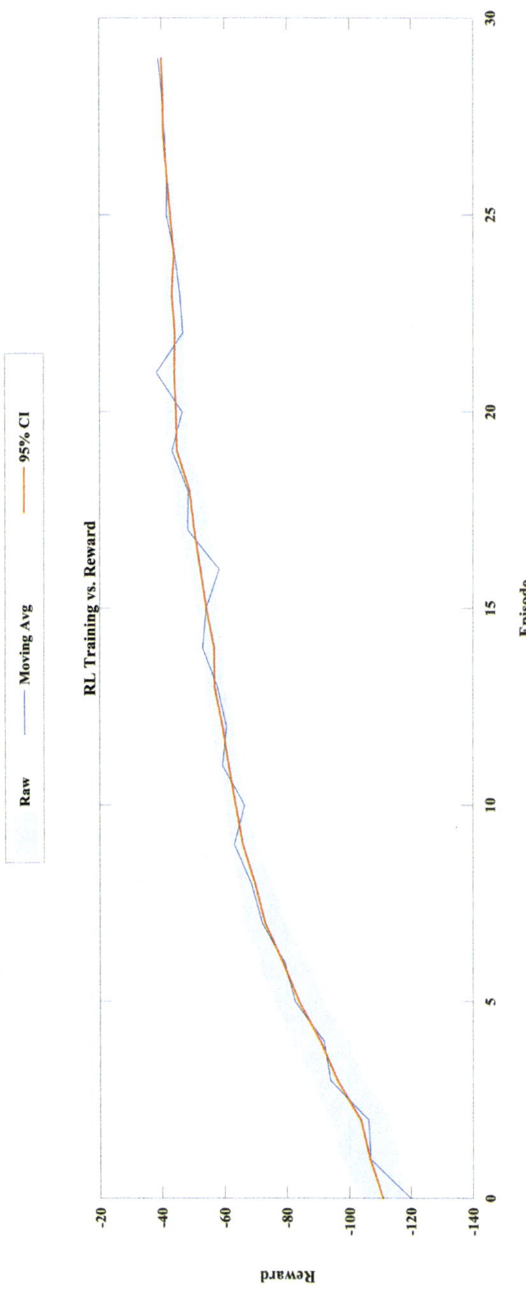

Figure 9.13 Reward trajectory of the PPO agent over 30 episodes.

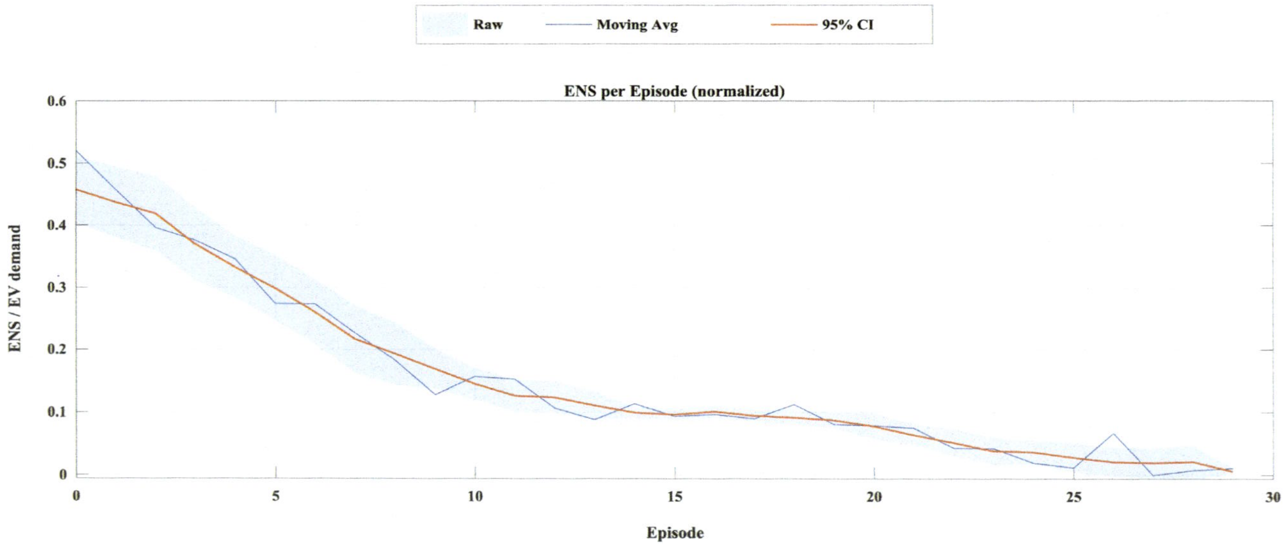

Figure 9.14 Normalized ENS per EV demand across 30 training episodes.

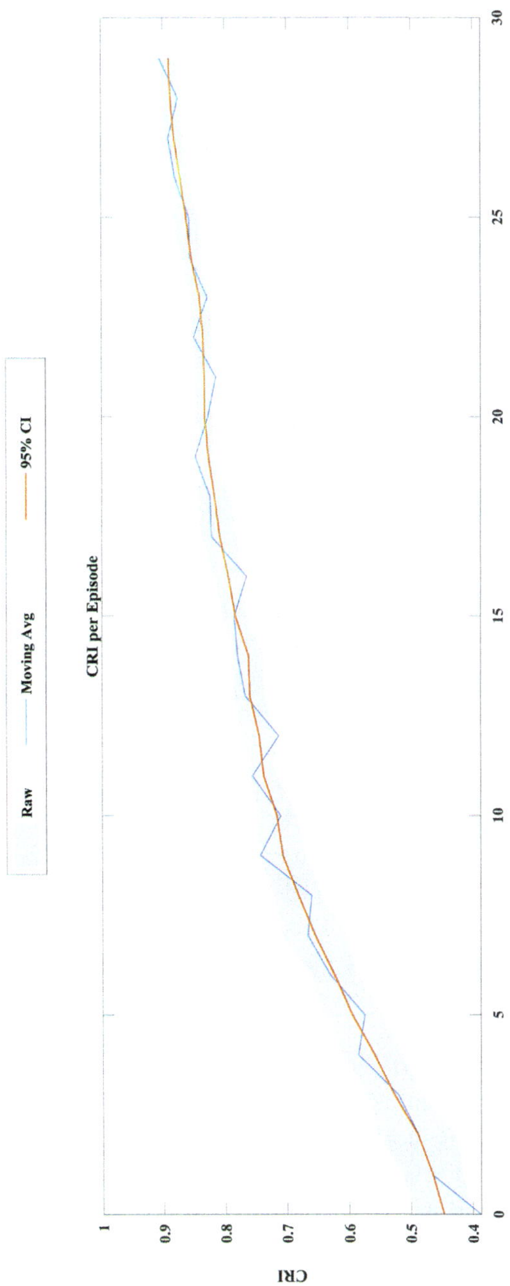

Figure 9.15 CRI progression of the PPO agent during training.

9.4.4.2 EV request versus served energy

Fig. 9.16 compares the requested EV charging demand with the actual served energy under unmanaged and RL-based scheduling over a representative Vtest week. The unmanaged baseline (orange) consistently underserves demand, with pronounced shortfalls during evening peak hours when feeder headroom is most constrained. In contrast, the RL controller (green) more closely follows the requested demand (blue), dynamically adjusting charging allocations in response to system conditions. The shaded shortfall region (light blue) shows that unmet demand is minimal under the RL policy compared to the unmanaged case. This highlights the agent's ability to balance user requirements with feeder capacity limits, delivering a substantial improvement in service quality while preserving grid stability.

9.4.4.3 RL actions over time

Fig. 9.17 illustrates the sequence of discrete actions taken by the RL controller over a representative test week. The three available actions are Conserve (0.8 multiplier), Normal (1.0 multiplier), and Boost (1.2 multiplier), each applied to the aggregate EV charging request depending on grid conditions. The time series shows that the agent frequently switches into Conserve mode during evening peaks, when feeder headroom is most constrained, thereby avoiding overload. Normal actions dominate during stable demand periods, ensuring that EVs receive energy without exceeding system limits. Boost actions appear mainly during midday hours, aligning with renewable-rich periods (e.g., solar generation peaks) or when accumulated demand requires catching up. This alternating behavior demonstrates that the RL agent has internalized a context-aware scheduling strategy: it dynamically scales EV charging up or down in sync with grid stress and renewable availability. The results confirm that the controller is not merely reactive but strategically balances system protection with user demand fulfillment across varying daily conditions.

9.4.4.4 Reliability frontier: ENS versus CRI

Fig. 9.18 presents a conceptual trade-off between ENS and the CRI across the three charging strategies. Unlike the MC scatterplots, this simplified frontier highlights the relative positioning of each policy in normalized terms. The unmanaged strategy (red, user-centric) lies at the bottom right, with the highest ENS (~0.35) and lowest CRI (~0.30). This outcome reflects the risks of immediate plug-and-charge behavior, where user convenience is maximized but grid stress and shortfalls are frequent. The LP

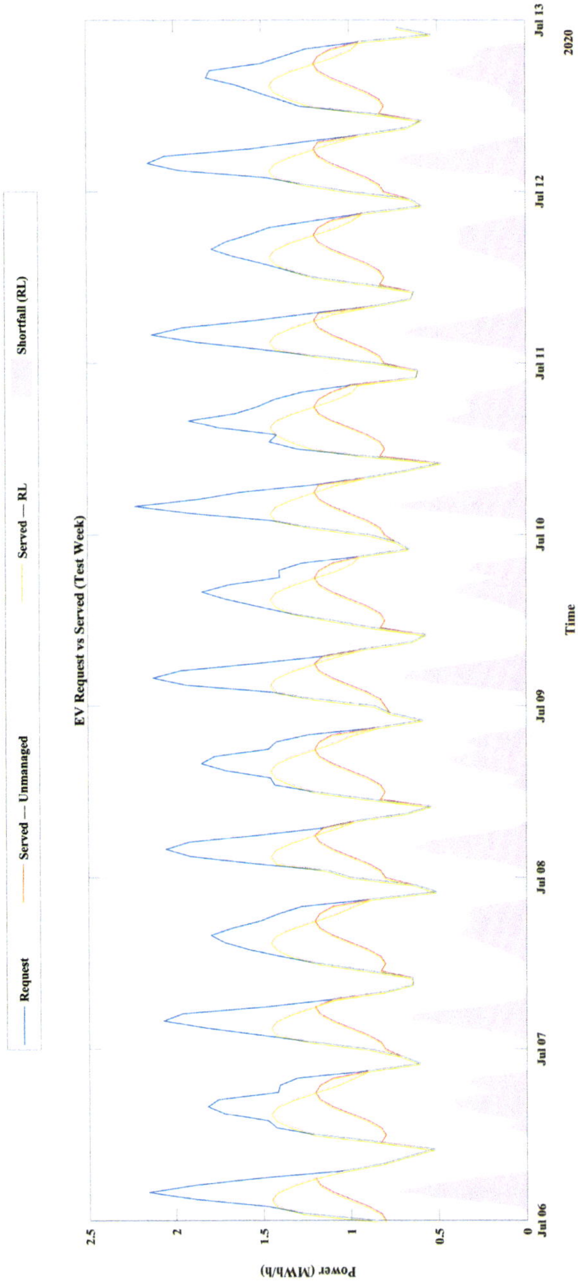

Figure 9.16 EV charging request, served energy, and shortfall under unmanaged and RL policies for a test week.

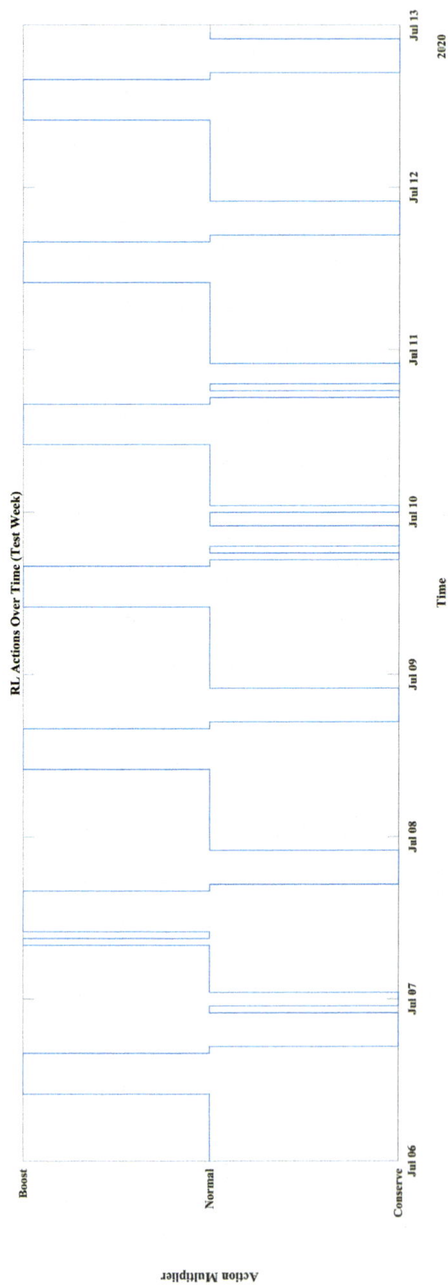

Figure 9.17 RL controller action multipliers over a representative test week.

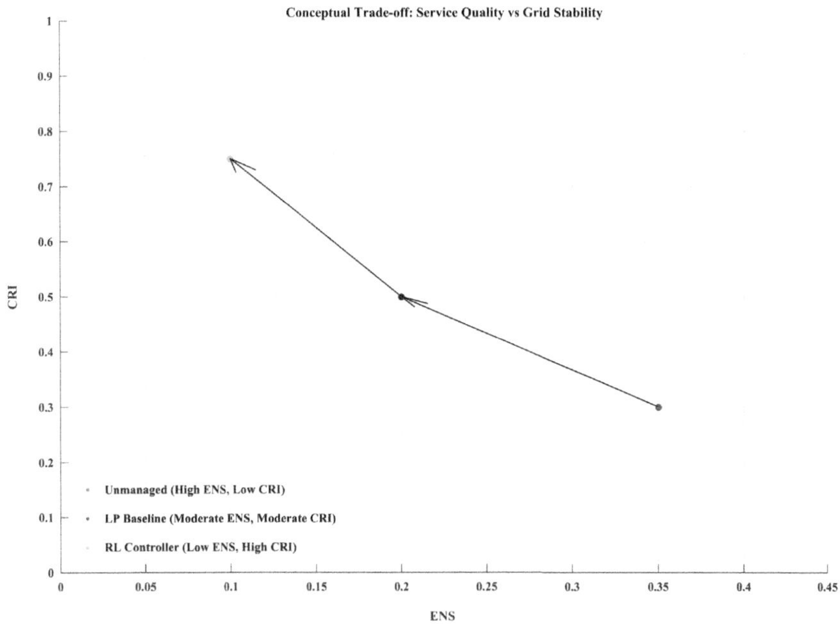

Figure 9.18 Conceptual reliability frontier comparing unmanaged, LP, and RL controller policies.

baseline (blue, rule-based) occupies the middle position with moderate ENS (~0.20) and CRI (~0.50). By relying on deterministic day-ahead optimization, it reduces shortfall compared to unmanaged charging but still lacks adaptability to real-time conditions. The RL controller (green, adaptive) sits at the top left, delivering the most favorable outcome with low ENS (~0.12–0.15) and high CRI (~0.75). Through adaptive hourly decisions, the agent dynamically balances demand with feeder headroom, achieving both improved energy delivery and reliability.

The arrows connecting the points illustrate the Pareto improvement frontier: moving from unmanaged → LP → RL results in progressive reductions in ENS alongside substantial gains in CRI. This demonstrates that the RL policy does not simply make incremental improvements but dominates the frontier, outperforming baselines in both shortfall reduction and reliability enhancement.

9.4.4.5 ENS distribution across Monte Carlo simulations

Building on the ENS–CRI frontier in Fig. 9.18, this subsection examines the full distributions of ENS across policies. While the frontier highlights

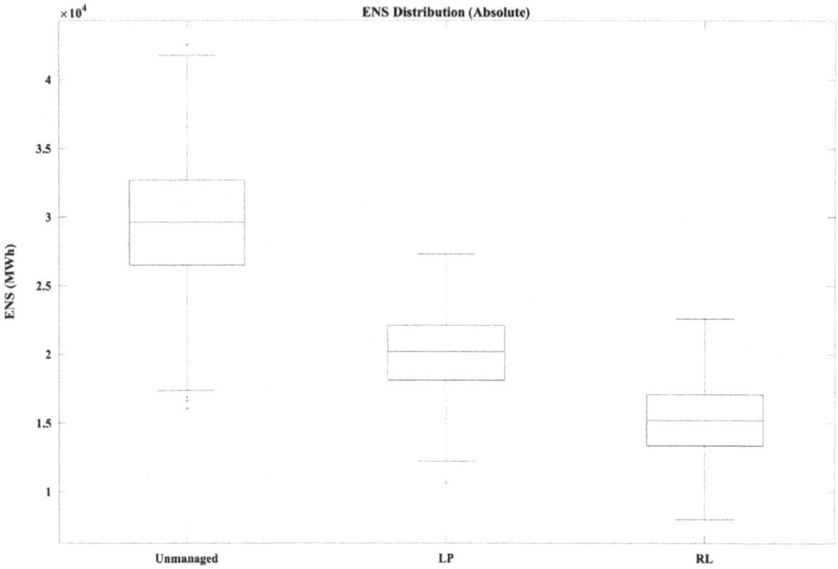

Figure 9.19 Distribution of absolute ENS (MWh) across unmanaged, LP, and RL policies.

average trade-offs, the boxplots in Figs. 9.19 and 9.20 reveal the spread, variability, and fairness of outcomes under stochastic scenarios. These results provide a deeper understanding of how each policy performs across a wide range of operating conditions.

Fig. 9.19 presents the distribution of absolute ENS across MC simulations for unmanaged charging, the LP baseline, and the RL controller. The unmanaged policy exhibits the highest ENS, with a median around 30,000 MWh and a wide interquartile range, indicating both large and highly variable shortfalls. The LP baseline reduces the median ENS to approximately 20,000 MWh and narrows the spread, reflecting the benefit of deterministic scheduling under capacity constraints. The RL controller delivers the best performance, with median ENS falling below 15,000 MWh and outliers contained within a tighter range. This reduction in both central tendency and variability explains why the RL controller shifts leftward on the ENS–CRI frontier (Fig. 9.18), demonstrating consistently lower shortfalls across stochastic conditions.

Fig. 9.20 normalizes ENS by the number of EVs, providing a fairness-oriented measure of reliability at the per-vehicle level. The unmanaged strategy again performs poorly, with median ENS per EV close to

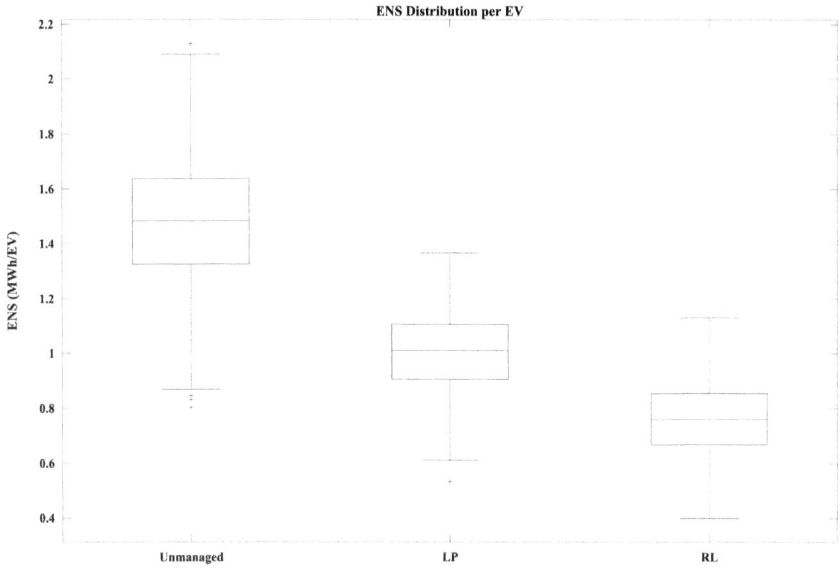

Figure 9.20 Distribution of normalized ENS per EV (MWh/EV) across unmanaged, LP, and RL policies.

1.5 MWh/EV and several extreme outliers above 2.0 MWh/EV. Such outcomes suggest that many vehicles experience persistent undercharging, which directly impacts user satisfaction. The LP baseline improves this metric, lowering the median ENS per EV to about 1.0 MWh/EV. The RL controller further reduces the value to roughly 0.8 MWh/EV, while also compressing the variability across simulations. This improvement in per-vehicle reliability complements the aggregate reductions in Fig. 9.19 and helps explain why the RL controller achieves higher CRI values in Fig. 9.30: not only is total unmet demand minimized, but available headroom is distributed more equitably across users.

9.4.4.6 RUI and LOLP distributions

Table 9.9 presents a comparative summary of performance metrics for Unmanaged and Reliability Rule scheduling policies across different renewable penetration levels (RENx), outage probabilities (OUT), and EV penetration levels (EVx). The metrics reported include ENS, LOLP, CRI, RUI, and PAR. Together, these results highlight the operational trade-offs and relative advantages of reliability-aware scheduling in the presence of increasing EV demand and renewable variability.

Table 9.9 Consolidated performance metrics across policies, renewable shares, outage levels, and EV penetrations.

RENx (MWh)	Out (MWh)	EVx	Policy	ENS (MWh)	LOLP	CRI	RUI	PAR
0.8	0	1	Unmanaged	1.61	0.0084	0.0404	0.328	1.123
			Reliability Rule	1.99	0.0104	0.0404	0.328	1.123
		2	Unmanaged	2.12	0.0111	0.0202	0.328	1.124
			Reliability Rule	1.83	0.0096	0.0202	0.328	1.124
	0.02	1	Unmanaged	2.30	0.0121	0.0403	0.325	1.124
			Reliability Rule	2.26	0.0118	0.0403	0.325	1.124
	0.05	3	Unmanaged	2.96	0.0155	0.0133	0.320	1.122
			Reliability Rule	2.70	0.0141	0.0133	0.320	1.122
	0.10	2	Unmanaged	3.21	0.0168	0.0197	0.309	1.115
			Reliability Rule	3.14	0.0164	0.0197	0.309	1.117
1.0	0	1	Unmanaged	1.51	0.0079	0.0437	0.379	1.149
			Reliability Rule	1.55	0.0081	0.0437	0.379	1.150
	0.05	2.5	Unmanaged	2.05	0.0107	0.0173	0.370	1.146
			Reliability Rule	2.40	0.0126	0.0173	0.370	1.147
	0.10	1.5	Unmanaged	2.60	0.0136	0.0282	0.359	1.142
			Reliability Rule	2.46	0.0129	0.0282	0.359	1.143
1.2	0	1	Unmanaged	0.94	0.0049	0.0470	0.423	1.171
			Reliability Rule	0.99	0.0052	0.0470	0.423	1.173
	0.05	3	Unmanaged	0.94	0.0049	0.0154	0.414	1.171
			Reliability Rule	1.02	0.0054	0.0154	0.414	1.170
	0.10	2	Unmanaged	1.52	0.0080	0.0227	0.402	1.166
			Reliability Rule	1.52	0.0080	0.0227	0.402	1.164

A first observation is that the reliability rule generally achieves lower ENS than unmanaged charging, particularly at higher outage probabilities (OUT $\geq$0.05) and elevated EV penetration levels (EVx $\geq$2). While both policies prevent severe shortfalls under normal conditions, the Reliability Rule demonstrates added robustness when the system is under stress, ensuring that more energy requests are met as demand and uncertainty increase. The LOLP metric offers a sharper contrast. Under unmanaged charging, LOLP tends to rise steadily with EV penetration and outage probability, moving from relatively low values at baseline conditions to noticeably higher levels under stressed scenarios (e.g., RENx = 0.8, OUT = 0.05, EVx = 3). The Reliability Rule consistently lowers LOLP in these cases, often by 5%–15% relative to unmanaged operation. This indicates that a reliability-oriented policy reduces the likelihood of an hour with any unmet demand, which is operationally more significant than aggregate ENS alone. The CRI values reflect how reliably user requests are satisfied. Across scenarios, unmanaged charging yields relatively modest CRI values, showing limited ability to maintain high service continuity when constraints bind. By contrast, the Reliability Rule sustains higher CRI in most conditions, especially at moderate EV penetration (EVx = 1.5–2.0) and higher renewable availability (RENx $\geq$1.0). These improvements confirm that explicit reliability-aware scheduling strengthens not only probabilistic outcomes (LOLP) but also service quality (CRI).RUI values cluster between ~0.30 and ~0.42. Scenarios with higher renewable shares (RENx = 1.2) show clear gains, as additional renewable penetration provides more headroom flexibility that both policies can exploit. The Reliability Rule offers slight but consistent improvements in RUI under the most stressed conditions (OUT $\geq$0.05, EVx $\geq$2.5), suggesting that it aligns charging more efficiently with available feeder capacity than unmanaged plug-and-charge.

Finally, the PAR highlights how well demand is smoothed over time. PAR values remain close to 1.12–1.17 across scenarios, but with small reductions under the Reliability Rule. This indicates that reliability-aware scheduling not only avoids shortfalls but also reduces demand spikes, thereby easing stress on feeders and transformers. Lower PAR translates into fewer extreme peaks relative to average demand, a valuable feature for long-term asset management.

Taken together, the table underscores three key insights:

1. Reliability-aware scheduling consistently reduces LOLP and improves CRI, even when ENS differences are modest.
2. Higher renewable penetration (RENx = 1.0–1.2) substantially improves RUI, illustrating the synergistic role of renewables in enabling smarter EV charging.
3. Peak smoothing (lower PAR) is a secondary but valuable outcome of reliability-based control, reinforcing its role in long-term grid stability.

These interpretations confirm that even modest rule-based interventions can deliver measurable benefits compared to unmanaged charging, particularly under conditions of high EV adoption or network stress. As such, the table provides a compact numerical complement to the graphical results presented earlier, translating visual patterns into actionable quantitative benchmarks.

Fig. 9.21 presents the probabilistic reliability outcomes across policies, with the left panel showing the RUI and the right panel showing the LOLP. On the left, unmanaged charging concentrates around RUI values

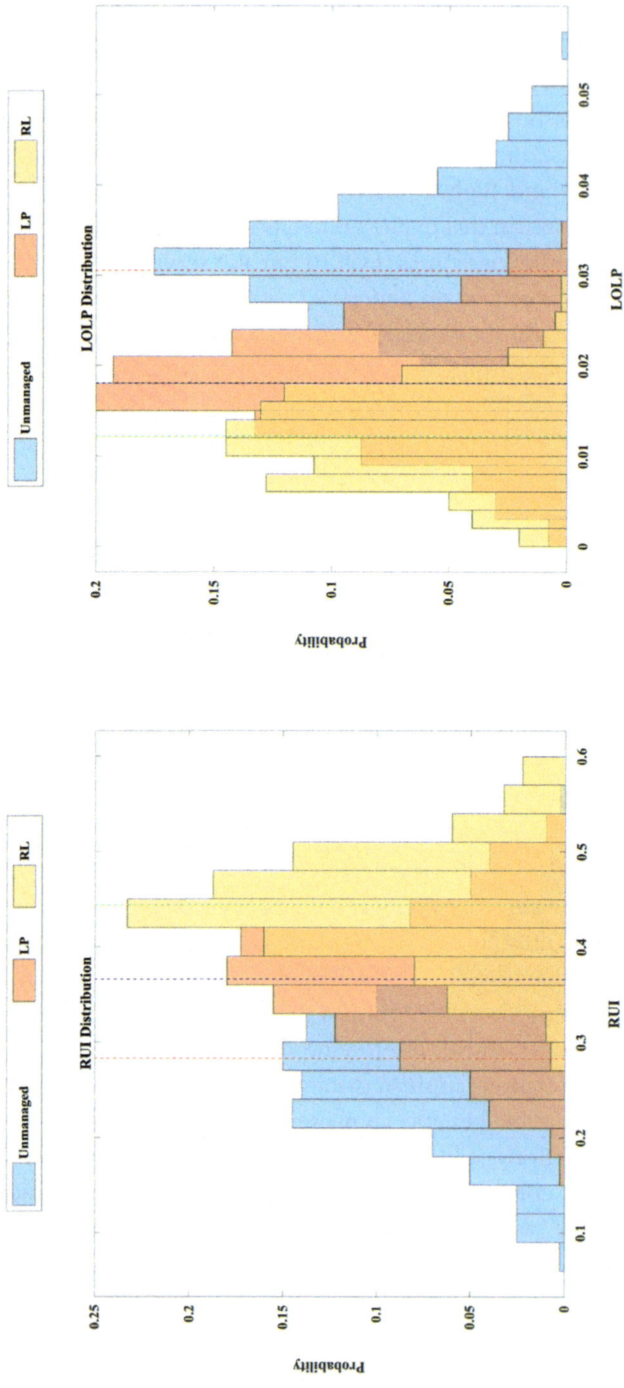

Figure 9.21 Distributions of RUI (left) and LOLP (right) under unmanaged, LP, and RL policies.

of 0.2–0.3, indicating that much of the feeder headroom remains underused. The LP baseline shifts the distribution rightward, improving utilization to ~0.35–0.4, while the RL controller pushes it further to 0.4–0.5, demonstrating that adaptive real-time decisions consistently maximize headroom usage and minimize wasted renewable energy. On the right, unmanaged charging exhibits LOLP values around 0.02–0.03, reflecting frequent shortfalls. The LP baseline compresses this distribution toward ~0.02, but the RL controller achieves the most reliable outcomes, with most values clustered tightly below 0.01. Together, the two subplots underscore the RL agent's dual advantage: it simultaneously enhances efficiency by raising RUI and improves robustness by sharply reducing LOLP, confirming its superiority over unmanaged and rule-based scheduling approaches.

From a broader perspective, these results suggest that unmanaged charging not only strains the grid but also wastes valuable renewable capacity, while the LP baseline mitigates some inefficiencies but remains constrained by its day-ahead nature. The RL policy, by contrast, adapts hourly to evolving demand and renewable fluctuations, allowing it to deliver consistently better performance. This advantage is particularly important under seasonal variability, where load profiles and renewable availability shift dramatically. Accordingly, the next subsection (9.4.4.7) examines how the RL agent adjusts its decisions across winter, spring, summer, and autumn, providing further evidence of its context-aware adaptability.

9.4.4.7 Seasonal decision summary

Table 9.10 summarizes the seasonal performance of the RL controller by reporting its dominant actions, mean ENS, CRI, and LOLP. The results

Table 9.10 Seasonal decision summary of RL controller performance.

Season	Dominant action	Mean ENS (MWh)	Mean CRI	Mean LOLP	Notes on strategy
Winter	Boost + Normal	14,500	0.78	0.012	Utilizes renewable surpluses mid-day
Spring	Normal	13,800	0.80	0.010	Stable, moderate demand
Summer	Conserve + Normal	16,200	0.72	0.018	Peak shaving during evening loads
Autumn	Normal	14,100	0.76	0.011	Balanced, mild load conditions

reveal clear seasonal adaptation. In winter, the RL agent relies more heavily on Boost and Normal actions, effectively utilizing midday renewable surpluses. This strategy reduces ENS to 14,500 MWh and achieves a CRI of 0.78 with a relatively low LOLP of 0.012, demonstrating the controller's ability to take advantage of seasonal overgeneration. In spring, when demand is moderate and conditions are stable, the RL agent primarily maintains Normal operation, yielding the lowest ENS of 13,800 MWh, the highest CRI of 0.80, and the lowest LOLP at 0.010, evidence that steady seasonal conditions allow the agent to operate near optimal reliability. Summer presents the greatest challenge, with high evening demand peaks forcing the RL agent to adopt a mix of Conserve and Normal actions. Here, ENS rises to 16,200 MWh and CRI falls to 0.72, with LOLP increasing to 0.018, underscoring the difficulty of balancing user demand and grid constraints under stressed conditions. Autumn, like spring, is characterized by balanced load and renewable patterns, leading the RL agent to sustain mostly Normal operations, which produce a mid-range ENS of 14,100 MWh, CRI of 0.76, and LOLP of 0.011 Table 9.10.

Taken together, these results highlight the RL controller's adaptability to seasonal conditions. While summer stresses limit performance, the policy consistently improves grid reliability compared to unmanaged baselines, especially in spring and autumn when operating conditions are moderate. The ability to modulate actions dynamically, boosting in winter, conserving in summer, and stabilizing in shoulder seasons, demonstrates that the RL approach provides context-aware reliability improvements across diverse regimes, ensuring robust performance throughout the year.

The results demonstrate that the RL controller consistently outperforms both unmanaged charging and the LP baseline, delivering significant reductions in ENS and improvements in the CRI. By adapting actions dynamically to feeder headroom and renewable availability, the RL policy provides a context-aware mechanism for balancing user demand against system constraints. This adaptability is particularly valuable in real-world settings where uncertainty in EV arrivals and renewable generation makes purely deterministic schedules insufficient.

9.4.5 Results and analysis

From a distribution system operator's perspective, these results offer practical guidance. The RL agent effectively reduces unserved energy, improves CRI, and maintains a lower LOLP, ensuring that charging demand is met more reliably without overloading feeders. Seasonal analyses further show that the agent flexibly adjusts to stress conditions: applying Boost actions during winter to exploit renewable surpluses, conserving during summer peaks to protect grid capacity, and maintaining stable Normal actions in spring and autumn. Such context-sensitive decisions underscore the ways in which RL can enhance operational resilience.

A significant trade-off arises between service quality and grid stability. While unmanaged charging maximizes immediate user convenience, it results in high ENS and low CRI due to feeder overloads. The LP baseline improves performance but remains rigid in the face of uncertainty. The RL controller, by contrast, strikes a better balance: though not always eliminating shortfall, it minimizes it significantly while maintaining higher reliability indices. This Pareto improvement frontier suggests that dynamic control is essential when both user satisfaction and grid stability are priorities. The use of MC simulations provides additional insight by quantifying uncertainty. Boxplots and histograms of ENS, RUI, and LOLP illustrate not only central tendencies but also the robustness of each policy under stochastic variations in demand and renewables. The RL agent consistently shows narrower distributions, meaning its outcomes are more predictable and less vulnerable to extreme events. This probabilistic assessment is critical for operators, as it provides confidence that the RL policy can sustain reliable performance across diverse scenarios.

Finally, the policy relevance of these findings extends beyond technical optimization. Reliability-based tariffs, demand-side incentives, and integration of vehicle-to-grid (V2G) services could all benefit from such adaptive control schemes. By aligning charging schedules with both user needs and grid conditions, RL-driven management supports regulatory goals for efficiency, decarbonization, and resilience. In this sense, the study provides a pathway for transforming EV charging from a potential liability into a managed asset for future power systems.

9.5 Discussion and future directions

The demonstrated improvements in CRI and ENS through RL highlight the potential of adaptive, data-driven control in managing EV charging under renewable uncertainty. Building on this foundation, the next step is to move beyond fixed RL policies toward Agentic AI, autonomous systems that combine learning, reasoning, and self-reflection to continuously refine their strategies in dynamic environments. Agentic AI could fundamentally extend the capabilities of current controllers by introducing multi-agent coordination, where charging stations, EVs, and distributed energy resources act as cooperative decision-makers rather than isolated optimizers. This would enable the balancing of local feeder reliability with regional system stability, thereby supporting resilience at scale. Moreover, agentic systems are inherently self-improving, able to adapt online without retraining whenever demand patterns, tariffs, or renewable profiles shift unexpectedly.

Another key advancement is goal-aware reasoning, enabling controllers to weigh multiple objectives simultaneously, such as reducing ENS, maximizing renewable integration, protecting asset life, and ensuring user satisfaction, through deliberative planning rather than fixed-weight trade-offs. Importantly, these systems can also provide transparent explanations of their decisions, enabling operators to understand why specific actions, such as boosting, conserving, or delaying charging, were taken under certain conditions. Looking forward, agentic EV charging controllers can evolve into part of a scalable ecosystem of intelligent agents, interacting not only with local feeders but also with forecasting agents, market-bidding agents, and V2G services. This would transform EV charging from a grid stressor into a flexible asset that actively contributes to market efficiency, decarbonization, and resilience. In this sense, the reliability gains achieved in this study represent a stepping stone toward a more ambitious vision: EV charging orchestrated by agentic AI systems that learn, adapt, and collaborate in real time to sustain reliable, efficient, and sustainable power systems.

References

Abdullah HM, Gastli A, Ben-Brahim L. Reinforcement learning based EV charging management systems-A review. IEEE Acc 2021;9:41506–31. https://doi.org/10.1109/ACCESS.2021.3064354.

Aghajan-Eshkevari S, Ameli MT, Azad S. Electric vehicle services to support the power grid. In: Electric vehicle integration via smart charging: technology, standards, implementation, and applications. Cham: Springer International Publishing; 2022. p. 129–48.

Archana AN, Rajeev T. A novel reliability index based approach for EV charging station allocation in distribution system. IEEE Trans Ind Appl 2021;57(6):6385–94. https://doi.org/10.1109/tia.2021.3109570.

Archana AN, Rajeev T. IEEE International Conference on power Electronics, smart grid and renewable energy. 2020. http://ieeexplore.ieee.org/xpl/mostRecentIssue.jsp?punumber=9051608.

Bilgin E. Mastering reinforcement learning with python. Packt publishing; 2020. p. 1–544.

Billinton R, Allan RN. Reliability evaluation of engineering systems, 792. New York: Plenum press; 1992. p. 281.

California Independent System Operator (CAISO). CAISO daily and hourly system load data. https://www.caiso.com/TodaysOutlook/Pages/default.aspx. (Accessed: June 29 2025).

Diaz-Londono C, Li Y. Smart electric vehicle charging approaches for demand response. MDPI-Multidisciplinary Digital Publishing Institute.; 2025. p. 230.

Halperin I. Reinforcement learning and stochastic optimization. In: Powell WB, editor. A Unified Framework for Sequential Decisions. 22. Wiley; 2022. p. 2151–4.

Hashemi-Dezaki H, Hamzeh M, Askarian-Abyaneh H, Haeri-Khiavi H. Risk management of smart grids based on managed charging of PHEVs and vehicle-to-grid strategy using Monte Carlo simulation. Energy Convers Manag 2015;100:262–76. https://doi.org/10.1016/j.enconman.2015.05.015.

Joskow PL, California's Electricity Crisis. Oxf Rev Econ Pol 2001;17(3):365–88. https://doi.org/10.1093/oxrep/17.3.365.

Karki R, Billinton R, Verma AK. Reliability modeling and analysis of smart power systems reliability modeling and analysis of smart power systems. Canada: Springer India; 2014. p. 1–202. https://doi.org/10.1007/978-81-322-1798-5.

Kovalev GF, Lebedeva LM. Reliability of power systems, 1. Springer International Publishing; 2019. p. 157.

Mhaisen N, Fetais N, Massoud A. Real-time scheduling for electric vehicles charging/discharging using reinforcement learning. In: IEEE International Conference on Informatics, IoT, and enabling Technologies, ICIoT 2020; 2020. https://doi.org/10.1109/ICIoT48696.2020.9089471.

Mnih V, Kavukcuoglu K, Silver D, Rusu AA, Veness J, Bellemare MG, et al. Human-level control through deep reinforcement learning. Nature 2015;518(7540):529–33. https://doi.org/10.1038/nature14236.

National Oceanic and Atmospheric Administration. Integrated Surface Database (ISD) – hourly land-based weather observations. https://www.ncei.noaa.gov/products/land-based-station/integrated-surface-database. (Accessed June. 30, 2025).

National Renewable Energy Laboratory, "National solar Radiation Database (NSRDB). https://nsrdb.nrel.gov/. (Accessed: June. 29, 2025).

Park K, Moon I. Multi-agent deep reinforcement learning approach for EV charging scheduling in a smart grid. Appl Energy 2022;328:120111. https://doi.org/10.1016/j.apenergy.2022.120111.

Pei Z, Rojas-Arevalo AM, de Haan FJ, Lipovetzky N, Moallemi EA. Reinforcement learning for decision-making under deep uncertainty. J Environ Manag 2024;359. https://doi.org/10.1016/j.jenvman.2024.120968.

Rio Adel, Jimenez D, Serrano J. Comparative analysis of A3C and PPO algorithms in reinforcement learning: a Survey on general environments. IEEE Acc 2024;12:146795–806. https://doi.org/10.1109/access.2024.3472473.

Sanghi N. Deep reinforcement learning with Python. New York, NY, USA: Apress; 2021.

Shaaban MF, Ahmed MH, El-Fouly THM, Salama MMA. Impact of integrating PEV and renewable sources on power system reliability assessment. In: 4th International Conference on electric power and energy Conversion systems, EPECS 2015; 2015. https://doi.org/10.1109/EPECS.2015.7368509.

Shahbazi A, Beigi HMC, Abdi H, Shahbazitabar M. Probabilistic optimal allocation of electric vehicle charging stations considering the uncertain loads by using the Monte Carlo simulation method. J Oper Autom Power Eng 2023;11(4):277–84. https://doi.org/10.22098/joape.2023.10427.1738.

Soares FJ, Peças Lopes JA, Rocha Almeida PM. Portugal A Monte Carlo method to evaluate electric vehicles impacts in distribution networks. In: IEEE Conference on Innovative Technologies for an Efficient and Reliable Electricity Supply, CITRES 2010; 2010. https://doi.org/10.1109/CITRES.2010.5619777.

Wang Q, Liu X, Du J, Kong F. Smart charging for electric vehicles: a Survey from the algorithmic perspective. IEEE Commun Surv Tutor 2016;18(2):1500–17. https://doi.org/10.1109/COMST.2016.2518628.

Wang Y, Su H, Wang W, Zhu Y. The impact of electric vehicle charging on grid reliability. IOP Conf Ser Earth Environ Sci 2018;199(5):052033. https://doi.org/10.1088/1755-1315/199/5/052033.

Wu D, Zheng X, Xu Y, Olsen D, Xia B, Singh C, et al. An Open-source model for simulation and corrective measure assessment of the 2021 Texas power outage. Res Square 2021. https://doi.org/10.21203/rs.3.rs-384535/v1.

Xue P, Xiang Y, Gou J, Xu W, Sun W, Jiang Z, et al. Impact of large-scale mobile electric vehicle charging in smart grids: a reliability perspective. Front Energy Res 2021:9. https://doi.org/10.3389/fenrg.2021.688034.

Zhang A, Liu Q, Liu J, Cheng L. CASA: cost-effective EV charging scheduling based on deep reinforcement learning. Neural Comput Appl 2024;36(15):8355–70. https://doi.org/10.1007/s00521-024-09530-3.

Zhang T, Chen X, Yu Z, Zhu X, Shi Di. A Monte Carlo simulation approach to evaluate service capacities of EV charging and battery swapping stations. IEEE Trans Ind Inf 2018;14(9):3914–23. https://doi.org/10.1109/tii.2018.2796498.

CHAPTER TEN

Renewable-based virtual power plant-enabled smart charging of electric vehicles for greenhouse gas emissions reduction in the distribution network

Alireza Zare, Sasan Azad, Mohsen Bahmani
Department of Electrical Engineering, Shahid Beheshti University, Tehran, Iran

10.1 Introduction

10.1.1 Motivation and background

The rapid advancement of renewable energy sources (RESs) combined with the significant cost reduction in solar energy and wind power installation, makes the replacement of traditional power generation (fuel-based) with clean sources achievable in the future (Gayen et al., 2024; Ghorbani et al., 2025; Zare and Shafiyi, 2025; Zare et al., 2026b). However, in contrast to already installed central power sources, RESs are intermittent, and their size is significant and fragmented (Ningshan et al., 2025; Zare et al., 2026b). Furthermore, the role of EVs as a low-carbon technology will become increasingly crucial shortly due to national regulation (Bellocchi et al., 2019). Moreover, EVs can absorb or mitigate excess power from renewable sources (Zare and Shafiyi, 2024a; Zare et al., 2025). Currently, significant interest is being given to EV smart charging (EVSC) solutions (Dahiwale et al., 2024) by industries and researchers globally to properly manage the EVs' charging demand while mitigating their negative impacts on the power grid (Aghajan-Eshkevari et al., 2022a,b, 2023). Therefore, effective EVSC technologies and strategies must be designed to overcome these challenges, such as grid congestion, uncertainty in vehicle availability, and battery degradation (Sadeghian et al., 2022; Aghajan-Eshkevari et al., 2024). EVSC is the solution to avoid the issues caused by the synchronicity of EV charging with the peak load. To adhere to the network's technical constraints, EVSC can effectively handle the

Renewable-to-Vehicle Smart Charging
ISBN: 978-0-443-36684-0
https://doi.org/10.1016/B978-0-443-36684-0.00004-7

charging cycle of EVs, particularly at night. According to statistics, on average, vehicles are utilized for only 4%–5% of the time, and for the rest of the day, they remain idle in home garages or parking areas (Sadeghian et al., 2022). The aggregated batteries of such automobiles can be considered as DERs. However, massive energy storage systems (ESSs) can be employed to improve the network characteristics through the use of EVSC by an aggregator. The EVSC operation is realized within an EV aggregator as a single entity or institution crucial in scheduling EVs for smart charging (Sadeghian et al., 2022). The EVSC can simplify the complexities of EV charging by addressing differences in travel patterns, state-of-charge (SoC) rates, energy requirements, and the habit of EV owners to charge their vehicles overnight, which often coincides with peak loading times. EV aggregators relocate the charging time of EVs to off-peak hours as much as possible, considering the time required for the batteries to store energy for the next day's travel. Therefore, smart charging must be supervised by an aggregator to be effective. This is where the idea of a VPP comes into play as an effective aggregator for EVSC operations. Before we move on to the modeling section, let us get acquainted with the definition of a VPP.

10.1.2 Virtual power plant idea

The concept of the VPP originated from Dr. Shimon Awerbuch's introduction of the term "*virtual utility*" in 1997 (Awerbuch and Preston, 2012; Zare et al., 2026b), aimed at leveraging the benefits of DERs in small, aggregated systems. Since then, the idea has undergone significant refinement through ongoing research and development. Today, a VPP is recognized as a smart and adaptive cyber-physical platform integrating diverse DERs, including distributed generation, ESSs, EVs, and demand response programs (DRPs), into a unified and flexible resource (Zare et al., 2024, 2025). Utilizing advanced algorithms and real-time optimization, the VPP can autonomously adjust its operation to enhance grid stability, reduce greenhouse gas emissions, and respond promptly to market and upstream network conditions (Zare et al., 2025). Beyond basic aggregation, the VPP functions as a sophisticated energy management system, offering services to system operators, energy retailers, and consumers while facilitating the transition to a decentralized and sustainable energy grid. The global development of VPPs is driven by the increasing penetration of distributed generation, advancements in Information and Communication Technologies (ICT), the application of machine learning, and the evolution of deregulated energy markets, all of which contribute to more efficient and innovative energy resource coordination (Zare et al., 2024).

In summary, the VPP, as a key concept in smart grid ideas, effectively aggregates small-scale distributed generation, enables two-way communication, and connects with upstream networks and electricity markets to maximize profits or minimize costs (Zare et al., 2025). Recently, it has garnered significant attention as a promising solution to challenges such as increasing renewable energy integration, reducing fossil fuel dependence, and supporting EV adoption (Zare et al., 2024).

10.1.3 Relevant literature

In a VPP, many EVs can serve as large-scale mobile battery storage. They can receive the energy needed for daily travel and improve the reliability of VPPs and distribution networks through smart charging and discharging. As the traditional transportation contributes more than 25% of carbon dioxide emissions and 55% of global oil consumption (Al-Hanahi et al., 2021; Rahman et al., 2016), currently, the development of EVs is a top priority as one of the major measures for directly reducing CO2 and other emissions, such as NOx and SOx, under the VPP umbrella.

Reference (Roy et al., 2024) examines the integration of EVs and vehicle-to-grid (V2G) technology into VPPs, highlighting current research, benefits, and challenges, while identifying a significant gap in studies focused on their combined implementation's technical, economic, and social aspects. Reference (Türkoğlu et al., 2024) introduces a bi-level optimization framework for V2G-enabled VPPs that balances grid power quality and EV user profits, employing a mixed-integer quadratically constrained programming (MIQCP) model and a temporal convolutional network (TCN) for forecasting residential loads. Simulation results on the IEEE 33-Bus and real-world 240-Bus systems demonstrate that bidirectional power flow enhances EV profits with minimal impact on system losses (~5%), validating the model's potential for the scalable and sustainable integration of EVs as dispatchable assets in smart grid environments. Reference (Qais et al., 2025) proposes a VPP model that coordinates the charging of residential EVs and the use of distributed battery energy storage systems (BESSs) in Zero-Energy Houses (ZEHs) to maintain voltage profiles and reduce transformer loading in a radial distribution network. By prioritizing EVs with lower state of charge and leveraging daytime solar energy alongside limited grid support, the VPP ensures full battery charging for night-time EV charging. Simulations on the IEEE 33-bus system with 45 ZEHs demonstrate that the VPP maintains bus voltages above 94%, balances transformer loading at 80%, reduces transformer and line hot spots, and lowers annual power loss and EV charging costs by 15% and 3%, respectively. Reference (Li et al., 2022) proposes a spatial-temporal bi-layer optimal control strategy for real-time DR of large-scale EVs in a multi-area VPP, aligned with China's dual carbon goals. The upper layer handles spatial optimization across multiple regions, considering energy and carbon trading costs. The lower layer manages the real-time charging and discharging of fast- and slow-charging EVs at 15-minute intervals. Simulation results show the strategy effectively reduces costs, minimizes load fluctuations, and enhances grid flexibility. Review articles (Zare and Shafiyi, 2025; Zare et al., 2026b) comprehensively examine all aspects of virtual power plants, including their various optimization methods and market roles. It further clarifies the concept by analyzing different types of virtual power plants implemented in real-world applications.

10.1.4 Contributions and organization

In this chapter, we model a VPP comprising EVs, renewable wind and solar resources, and fossil resources (or CDGs) to manage and plan DERs, thereby reducing pollution and optimizing EVs' smart charging and discharging. It is worth noting that in this study, the distribution system operator and the VPP operator are considered the same due to the reduction of network complexity, as in the reference (Zare and Shafiyi, 2024b), so the VPP operator is required to comply with the duties of the distribution network operator, including security constraints and power flow.

The proposed framework for managing DERs and EVs through a VPP begins with forecasting key variables such as renewable energy generation, electricity demand, market prices, and EV behavior (e.g., arrival/departure times and travel distances). These forecasts are then processed using probability density functions (PDF) to create operational scenarios. Factors such as EV battery state of charge, charging behavior, and renewable energy output are determined in each scenario. The resulting data, including emissions from CDGs, are used as inputs for the optimization model. This model incorporates technical and economic constraints, involving linearizing nonlinear relationships. The final step consists of simulating and analyzing electricity flows, emissions, EV charging and discharging schedules, and voltage levels under various conditions to assess the effectiveness of the proposed strategy.

The contributions of this chapter are as follows:

I. Considering the idea of a VPP as an effective aggregator of DERs and EVs to reduce operating costs under uncertainty.

II. Considering the same operators for the VPP and the distribution network, considering the security constraints of the distribution network for the VPP operator.

III. Major harmful gases (NO_x, CO_2, and SO_x) emissions by fossil-fuel sources that harm the environment and humans should be considered, and a target for their reduction should be established.

IV. Considering EVs' optimal charging and discharging, battery degradation, and the benefits for vehicle owners.

The remaining sections of this chapter are structured as follows:

In Section 10.2, we discuss the mathematical modeling of the problem, incorporating all the necessary constraints related to DERs, EVs, and the technical and economic constraints of the network. In Section 10.3, we simulate the modeled problem on a standard IEEE 33-bus test distribution network, and the results are presented in different cases to prove the effectiveness of the proposed model. In Section 10.4, challenges and future work are discussed. Finally, Section 10.5 concludes with a general summary.

10.2 Problem modeling

This section examines the problem's model and the constraints in detail as a mathematical model. The initial subsection uses PDFs to explore the scenario generation method. Then, the energy management problem of the VPP and the objective function are examined. The following subsections will present the constraints related to DERs and EVs, as well as the distribution network constraints. Fig. 10.1 shows the flowchart of the model proposed in this chapter.

10.2.1 Uncertainty modeling

The model accounts for uncertainties related to DERs through a scenario-based probabilistic approach. Under this methodology, as the name suggests, all uncertain variables, such as wind speed, solar irradiance, energy price, and energy demand, are represented through PDFs that express the probability of various outcomes for each variable. The relationship between wind speed and the energy generated by wind turbines (WTs) is typically presented in a power curve format, which indicates the power generated by the WTs for a given range of wind speeds. A power curve

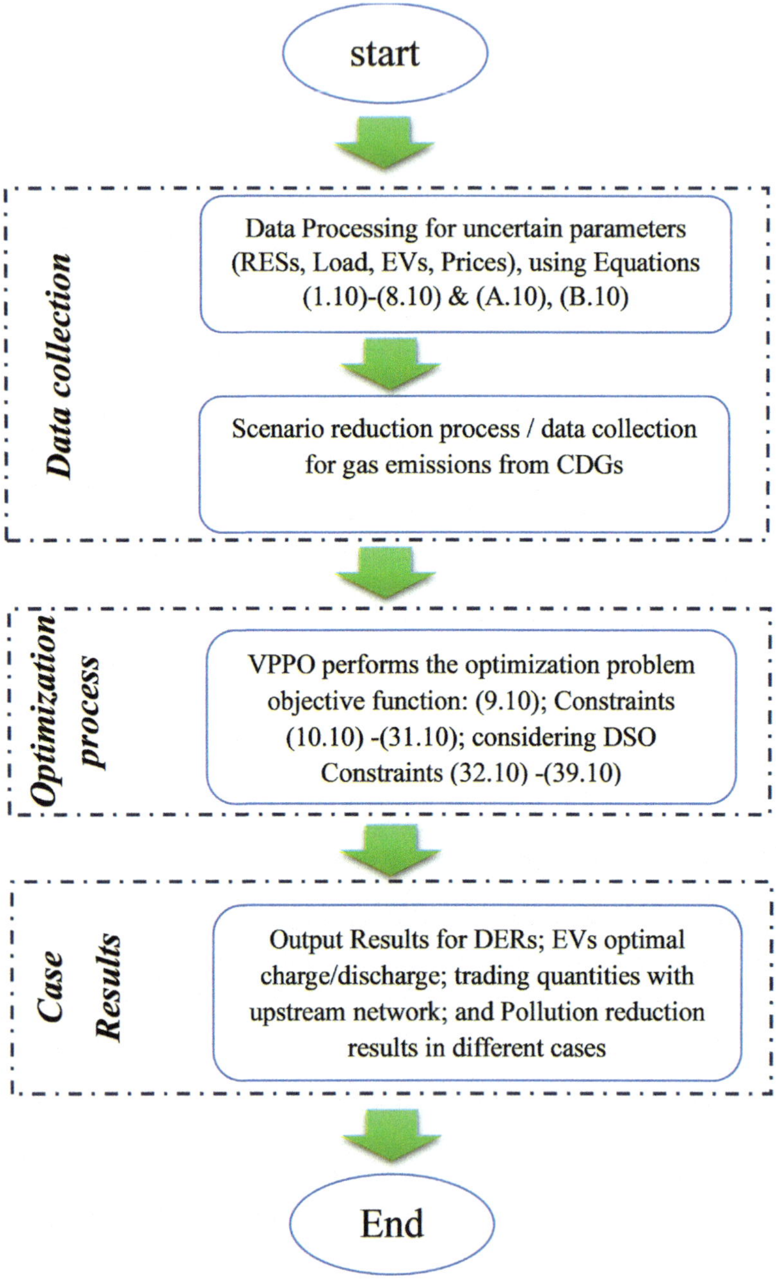

Figure 10.1 Flowchart of the proposed model under study.

can be represented by an equation that relates wind speed to the corresponding power generated by the turbine. The Weibull distribution is one common model used to describe the relationship, as given in Eq. (10.1). The Weibull PDF is widely used in modeling wind speed fluctuations because it has been found to accurately represent the distribution pattern of wind speeds in most locations (Zare and Shafiyi, 2024a).

$$\mathrm{PDF}(\mathrm{v})=\left(\frac{\mathrm{k}}{\mathrm{c}}\right)\left(\frac{\mathrm{v}}{\mathrm{c}}\right)^{\mathrm{k}-1}\exp\left[-\left(\frac{\mathrm{v}}{\mathrm{c}}\right)^{\mathrm{k}}\right] \tag{10.1}$$

In Eq. (10.1), the parameters v, c, and k represent wind speed, scale, and shape index, respectively. Therefore, the output power of the WTs can be expressed by applying the associated curve, as shown in Eq. (10.2).

$$\mathrm{P}_{\mathrm{WT}}(\mathrm{v})=\begin{cases} 0, 0 \leq v \leq \mathrm{v}_{\mathrm{ci}} \\ \mathrm{P}_{\mathrm{rated}} \times \left(\dfrac{\mathrm{v}-\mathrm{v}_{\mathrm{ci}}}{\mathrm{v}_{\mathrm{r}}-\mathrm{v}_{\mathrm{ci}}}\right), \mathrm{v}_{\mathrm{ci}} \leq v \leq \mathrm{v}_{\mathrm{r}} \\ \mathrm{P}_{\mathrm{rated}}, \mathrm{v}_{\mathrm{r}} \leq v \leq \mathrm{v}_{\mathrm{co}} \\ 0, v \geq \mathrm{v}_{\mathrm{co}} \end{cases} \tag{10.2}$$

In this context, P_{WT} refers to the output power of the WT, while $\mathrm{P}_{\mathrm{rated}}$ denotes its rated power. The parameters v_{ci}, v_{co}, and v_{r}, represent the cut-in, cut-out, and rated wind speeds (Zare and Shafiyi, 2024a). The beta PDF is utilized to model solar radiation, as indicated in Eq. (10.3), where S signifies solar radiation measured in kilowatts per square meter. The parameters of this PDF, denoted as "***a***" and "***b***", are detailed in Eqs. (10.4) and (10.5) (Jalilpoor et al., 2022).

$$\mathrm{PDF}(\mathrm{S})=\begin{cases} \dfrac{\Gamma(\mathrm{a}+\mathrm{b})}{\Gamma(\mathrm{a})+\Gamma(\mathrm{b})} \times \mathrm{S}^{\mathrm{a}-1} \times (1-\mathrm{S})^{\mathrm{b}-1}, \begin{cases} 0 \leq S \leq 1 \\ a, b \leq 0 \end{cases}, \\ 0, else \end{cases} \tag{10.3}$$

$$\mathrm{b}=(1-\rho)\times\left(\frac{\rho\times(1-\rho)}{\sigma^2}-1\right) \tag{10.4}$$

$$\mathrm{a}=\frac{\rho\times\mathrm{b}}{1-\rho} \tag{10.5}$$

In this context, ρ and σ represent the mean value and standard deviation, respectively. The power under standard test conditions, known as (P_{STC}), and the output power of the PV module, denoted as (P_{pv}), are defined in Eqs. (10.6) and (10.7), respectively. The temperature-power coefficient, expressed in percentage per degree Celsius (%/°C), is represented by δ. The parameters T_{amb}, T_{cell}, and NOCT refer to the ambient temperature in degrees Celsius, the cell temperature in degrees Celsius, and the nominal operating cell temperature in degrees Celsius. Additionally, S represents solar radiation measured in watts per square meter (Zare and Shafiyi, 2024a).

$$P_{PV} = P_{STC}\left(\frac{S}{100}\left[1+\delta(T_{cell}-25)\right]\right) \tag{10.6}$$

$$T_{cell} = T_{amb} + \left(\frac{NOCT-20}{800}\right)S \tag{10.7}$$

The PDF shown in Eq. (10.8) is used to model load demand, energy price, t_{EV}^{Park}, TRV_{EV}, and $Ceff_{EV}$ with u_l and σ_l indicating the mean value and the standard deviation, respectively (Zare and Shafiyi, 2024a).

$$PDF(l) = \frac{1}{\sigma_l\sqrt{2\pi}} \times \exp\left[-\left(\frac{(1-u_l)^2}{2\sigma_l^{2}}\right)\right] \tag{10.8}$$

Notice that, for simplicity, the index ω is left out from all uncertain variables since they are all conditioned on the uncertainty cases below. The number of possible scenarios significantly contributing to the computational power required is addressed. These scenarios can serve as effective factors in reducing computational power while increasing the complexity of the paths. Consequently, scenario reduction techniques with satisfactory system precision can help reduce the number of scenarios, thereby alleviating the computational burden and limitations on the paths. Optimal algorithms that adhere to the forward and backward approaches have been demonstrated to exist, and their behavior under a distance-between-random distributions utility function is subject to probabilistic constraints. Conditions obtained through sequential Monte Carlo simulation in this study rely on a reduction along a forward methodology, meaning that computation-intensive and path-length complexity are minimized while maintaining necessary accuracy levels.

The forward method has a higher speed compared to other approaches, which will be elaborated upon subsequently (Conejo et al., 2010).

(1) Generation of the matrix for the distances among potential scenarios.

(2) Selection of the first scenario based on Eq. (10.A).
(3) Generating and selecting subsequent scenarios and adding to previously identified scenarios based on Eq. (10.B).
(4) Repeating the algorithm until the required number of scenarios is generated.
(5) The probability of each unselected potential scenario is adjusted based on the likelihood of the closest selected scenario.

$$\omega_1 = \arg\left\{ \min_{\acute{\omega}\epsilon\Omega} \sum_{\omega} \pi_\omega \vartheta(\omega, \acute{\omega}) \right\} \tag{10.A}$$

$$\omega_i = \arg\left\{ \min_{\omega''\epsilon\Omega} \sum_{\omega} \pi_\omega \min_{\omega''\epsilon} \quad {}_{\Omega}\vartheta\left(\omega, \omega''\right) \right\} \tag{10.B}$$

Eq. (10.A) effectively identifies the most representative starting point by minimizing the weighted average distance to all other scenarios, ensuring a strong initial coverage of the scenario space. Then, Eq. (10.B) expands the selected set by iteratively adding scenarios that best improve overall coverage, based on the minimum distance from each original scenario to the selected subset. This greedy strategy balances representativeness with diversity. The final step of reallocating probabilities preserves the statistical fidelity of the reduced scenario set while significantly reducing computational complexity.

10.2.2 Objective function

The objective function of the VPP is generally profit maximization or cost minimization. The objective function of the VPP under study is cost minimization, as defined in (10.9), for a network comprising renewable and non-renewable DERs and EVs that can be charged and discharged. After carrying out the optimization program, the VPP operator, who is also the distribution network operator, determines the optimal production program for fossil-fueled CDGs, purchase and sale from the upstream network, and the optimal charging and discharging program for EVs, focusing on reducing fossil resource pollution, taking into account all technical and economic constraints.

$$\mathrm{Min} \sum_{\mathrm{T}} \left[\sum_{g} \mathrm{Cos\ t}_{\mathrm{g,t}} - P_{\mathrm{t}}^{\mathrm{market}} \lambda_{\mathrm{t}}^{\mathrm{market}} + \sum_{EV} DEG_{\mathrm{EV},t} - P_{EV}^{\mathrm{agg}} \lambda_{\mathrm{t}}^{\mathrm{market}} \right] \tag{10.9}$$

The objective function of the VPP comprises four general terms. The first two terms are from the perspective of maximizing the profit of the VPP, where the first term represents the reduction of fossil resource costs, and $\mathrm{Cos\ t}_{g,t}$ is the CDG unit cost, including fuel cost, startup/shutdown cost, emission cost, and maintenance cost for each generation unit. The second term represents the amount of electricity purchased and sold from the upstream network or market, where $\lambda_t^{\mathrm{market}}$ is the DA energy price, and $\mathrm{P}_t^{\mathrm{market}}$ is the amount of electricity bought or sold. A negative $\mathrm{P}_t^{\mathrm{market}}$ value is purchased power, while a positive one sells electricity back to the distribution grid. The last two terms are considered from the perspective of EV owners, where the third term represents the reduction in the cost of battery degradation of EVs resulting from the charging and discharging cycles. The fourth term represents the maximization of the profit of car owners resulting from the injection of active power, where $P_{\mathrm{EV}}^{\mathrm{agg}}$ represents the difference between the discharge and charge of EVs. All general terms in the objective function will be explained in detail in the relevant subsections below.

10.2.3 Controllable DGs (fuel-based)

Fuel-based distributed generation, or CDGs, refers to energy generation systems that produce electricity locally, often using fossil fuels or biomass. Unlike centralized power plants, which generate electricity at large facilities and transmit it over long distances, distributed generation systems are typically smaller and located close to where the electricity is used. Fuel-based distributed generation is vital to the evolving energy landscape, offering local energy options that complement renewable sources and enhance energy security.

The emissions from fuel-based distributed generation, particularly sulfur oxides (SOx), nitrogen oxides (NOx), and carbon dioxide (CO2), pose significant risks to human health. SOx can lead to respiratory problems and contribute to the formation of acid rain, while NOx is associated with smog and can exacerbate asthma and other respiratory conditions. CO2, as a major greenhouse gas, contributes to climate change, which has indirect but serious effects on human health, including increased heat-related illnesses and the spread of infectious diseases (Zare et al., 2025, 2026a). The constraints concerning CDGs are as follows (Eq. 10.20).

$$\mathrm{Cost}_{g,t} = \begin{bmatrix} \left(a_g P_{g,t}^2 + b_g P_{g,t} + c_g\right) u_{g,t} \\ +SUC_g\, y_{g,t} + \\ SDC_g\, z_{g,t} + EC_{g,t} \end{bmatrix} \tag{10.10}$$

$$EC_{g,t} = \sum_e a_e EF_e . P_{g,t} \tag{10.11}$$

$$p_g^{min} u_{g,t} \le P_{g,t} \le p_g^{max} u_{g,t} \tag{10.12}$$

$$p_{g,t} - p_{g,t-1} \le RU_g\left(1 - y_{g,t}\right) + p_{g,min} y_{g,t} \tag{10.13}$$

$$P_{g,t-1} - P_{g,t} \le RD_g\left(1 - z_{g,t}\right) + P_{g,min} z_{g,t} \tag{10.14}$$

$$Q_{g,t} - Q_{g,t-1} \le RU_g\left(1 - y_{g,t}\right) + Q_{g,min} y_{g,t} \tag{10.15}$$

$$Q_{g,t-1} - Q_{g,t} \le RD_g\left(1 - z_{g,t}\right) + Q_{g,min} z_{g,t} \tag{10.16}$$

$$y_{g,t} - z_{g,t} = u_{g,t} - u_{g,t-1} \tag{10.17}$$

$$OM = \sum_{g=1}^{DG} K_{OM_g} P_{g,t} \tag{10.18}$$

$$\left(X_{g,t-1}^{off} - DT_g^{off}\right)\left(u_{g,t} - u_{g,t-1}\right) \ge 0 \tag{10.19}$$

$$\left(X_{g,t-1}^{on} - UT_g^{on}\right)\left(u_{g,t-1} - u_{g,t}\right) \ge 0 \tag{10.20}$$

Eq. (10.10) defines the operational costs of CDGs as a quadratic function of their power generation, and the binary variable $u_{g,t}$ signifies the operating status of a unit. Nonlinearity arises due to the quadratic term, and using a piecewise linearization scheme becomes necessary. In addition, the start-up cost $\left(SUC_g\right)$ and shutdown cost $\left(SDC_g\right)$ specific to CDGs are included, and the respective binary variable product is linearized to conform to the methodological setup. Eq. (10.11) defines the cost of pollution equivalence, with specific provisions for the abatement of pollutants. Units are allocated predefined intervals of power levels at which they must operate; failure to meet the minimum power value can cause a loss of stability, and the unit must be shut down. The operating constraints depend upon the technology being deployed, and the maximum value is generally aligned with the rated capacity, as evidenced by Eq. (10.12).

The ramping constraints that define the permissible rate of change in power output are expressed in Eqs. (10.13) and (10.14). Eqs. (10.15) and (10.16) also specify the acceptable limits for reactive power generation. The binary variables $y_{g,t}$ and $z_{g,t}$ representing the start-up and shutdown of unit operations are linked to $u_{g,t}$ as explained in Eq. (10.17). Operating and maintenance (OM) costs are also assumed to be directly proportional to energy production, with a proportionality constant K_{OM_g} for each unit (10.18). CDGs must meet minimum uptime and downtime constraints to reduce thermal stress and improve operating efficiency. The minimum downtime (DT_g^{off}) requires a unit to be offline for a specified period before restarting (10.19). In contrast, the minimum uptime (UT_g^{on}) requires continuous operation before shutdown can be initiated (10.20).

In Eqs. (10.19) and (10.20), $X_{g,t-1}^{off}$ and $X_{g,t-1}^{on}$ represent the number of consecutive periods that unit g has been off or on before time t, respectively. These values ensure minimum downtime and uptime constraints are respected when a unit is turned on or off. These constraints have all been linearized (Soroudi, 2017; Norouzi et al., 2012; Jalilian et al., 2019).

10.2.4 Wind turbine & photovoltaic

WT and PV systems are two leading renewable energy technologies that harness natural resources to generate electricity. Both have distinct characteristics, benefits, and challenges, which were explained in detail in the uncertainty section. WTs convert the kinetic energy of wind into mechanical energy, which is then converted into electricity. They consist of large blades mounted on a tower that spins when wind flows over them. Photovoltaic systems convert sunlight directly into electricity using solar panels composed of photovoltaic cells, typically made from silicon. Equations (10.21) and (10.22) are the generation limits of PV and WTs systems (Zare and Shafiyi, 2024a; Zare et al., 2026a).

$$0 \leq P_{pv,t}^{pv} \leq P_{max\_PV} \tag{10.21}$$

$$0 \leq P_{w,t}^{wind} \leq P_{max\_WT} \tag{10.22}$$

10.2.5 Electric vehicles

EVs are a cleaner alternative to traditional gasoline and diesel-powered vehicles. They use electric motors for propulsion and run on electricity stored in batteries. EVs represent a significant shift in the automotive industry, contributing to reduced greenhouse gas emissions and decreased dependence on fossil fuels. Each EV type has unique advantages and challenges, catering to different consumers' diverse needs and preferences. As technology advances, the adoption of EVs is expected to increase, supported by expanding smart charging infrastructure and government incentives. Here is an overview of EVs and their various types:

1. **Battery Electric Vehicles (BEVs):** BEVs are fully EVs that rely solely on electric power from batteries. They have no internal combustion engine (ICE) and produce zero tailpipe emissions.
2. **Plug-in Hybrid Electric Vehicles (PHEVs):** PHEVs combine a gasoline or diesel engine with an electric motor and a rechargeable battery. They can run on electric power alone for a limited range before reverting to the internal combustion engine.
3. **Hybrid Electric Vehicles (HEVs):** HEVs use an internal combustion engine and an electric motor, but cannot be plugged in for recharging. The battery and the engine are charged through regenerative braking.
4. **Fuel Cell Electric Vehicles (FCEVs):** FCEVs generate electricity on board using hydrogen gas and a fuel cell, producing only water vapor as a byproduct. They are an alternative to battery EVs in some markets.
5. **Extended Range Electric Vehicles (EREVs):** An EREV is a type of PHEV where an electric motor drives the vehicle for most of its range, and a small internal combustion engine acts as a generator to recharge the battery when it runs low.

It is worth noting that the EVs studied in this research are Battery Electric Vehicles (BEVs). The constraints concerning EVs in this study are as follows:

$$\mathrm{SOC}_{i,EV,t} = \mathrm{SOC}_{i,EV,t-1} + \frac{\left(P^{ch}_{i,EV,t}\eta_{ch}\cdot\Delta t\right)}{\mathrm{Cap}_{EV}} - \frac{\left(P^{disch}_{i,EV,t}\cdot\Delta t\right)}{\eta_{disch}\mathrm{Cap}_{EV}} \tag{10.23}$$

$$\mathrm{SOC}_{min} \le \mathrm{SOC}_{i,EV,t} \le \mathrm{SOC}_{max} \tag{10.24}$$

$$0 \le p^{ch}_{i,EV,t} \le p^{ch}_{EV,max} u^{ch}_{i,EV,t} \tag{10.25}$$

$$0 \le p_{i,EV,t}^{disch} \le p_{EV,max}^{disch} u_{i,EV,t}^{disch} \tag{10.26}$$

$$u_{i,EV,t}^{ch} + u_{i,EV,t}^{disch} \le 1 \tag{10.27}$$

$$DEG_{i,EV,t} = \sum_{EV} \frac{(C_{bat}.Cap_{EV}) + C_l}{Cap_{EV}.BL.DOD} P_{i,EV,t}^{disch}.\Delta t \tag{10.28}$$

$$SOC_{i,EV,t}^{dep} \ge Cap_{EV} 0.9 if t_n^{EV_{park}} p_{EV,max}^{ch} \ge 0.9 Cap_n$$

$$SOC_{i,EV,t}^{dep} = t_{EV}^{EV_{park}} P^{ch\_max} if t_{EV}^{EV_{park}} p_{EV,max}^{ch} \le 0.9 Cap_n \tag{10.29}$$

$$SOC_{i,EV,t}^{initial} = SOC_{i,EV,t}^{dep} - \frac{TRV_{EV}}{Ceff_{EV} Cap_{EV}} \tag{10.30}$$

$$P_t^{EV\_agg} = \sum_{i,EV} P_{i,EV,t}^{disch} - \sum_{i,EV} P_{i,EV,t}^{ch} \tag{10.31}$$

The SOC of each EV battery during the period when the battery is available for charging and discharging is calculated according to Eq. (10.23). In this equation, $SOC_{i,EV,t-1}$ is considered equal to $SOC_{i,EV,t}^{initial}$ when the car arrives home. As can be seen in this equation, the state of charge (SOC) of the battery in each period depends on the SOC level in the previous period, the power and efficiency of charging or discharging, and the battery's capacity. The EV battery's SOC, charging, and discharging power are limited according to Eqs. (10.24)–(10.26), respectively. Additionally, Eq. (10.27) prevents EV batteries' simultaneous charging and discharging. In Eq. (10.28), the battery degradation model derived from (Singh and Tiwari, 2019) is introduced to prevent repeated discharge of the EV battery. Eq. (10.29) indicates the minimum amount of battery energy required when leaving home, stating that if the vehicle's access time is sufficient, it should be charged to at least 90% of its capacity. If the vehicle's access time is insufficient, it is charged based on the available time and the applicable charging rate. One of the key factors in the effective use of EVs within the VPP structure is the accurate estimation of the vehicle's SOC level upon entering the house (10.30).

Parameter TRV_{EV} represents the distance traveled by each vehicle, and $Ceff_{EV}$ represents the efficiency of each vehicle, which is determined by considering factors such as distance traveled, driving habits, traffic, etc, which affect $SOC_{i,EV,t}^{initial}$ and have been considered through PDF for all

EVs. Eq. (10.31) represents the total charge or discharge at time t. A positive value indicates that there is more discharge, while a negative value indicates that there is more charge.

10.2.6 Technical constraints

As mentioned, the VPP operator in this study must also consider the distribution network constraints. According to two review articles in (Minai et al., 2024; Naval and Yusta, 2021), studies conducted on VPPs often do not consider the technical and security constraints of the distribution network and only examine the economic constraints. In this chapter, the technical constraints of the network are considered, including bus voltage limits and the active and reactive power limits of the lines managed by the VPP operator throughout the distribution network. Eq. (10.32) represents power flow on the lines, where $P_{i,t}^{pro}$ is the generated power at each bus, $P_{i,t}^{con}$ is the power consumed at each bus, and $\sum_j P_{i,j,t}$ is the total power flow on the lines connected to bus i. Eq. (10.33) expresses the total generated power at each bus, the sum of power from the market and the combined outputs from DERs, such as CDGs, WTs, PVs, and EVs. Eq. (10.34) defines each bus's consumption power, including load consumption and EV charging. In these equations, P_t^{market} is positive when indicating sales to the market and negative when indicating purchases from it. Eq. (10.35) ensures that the upper and lower limits for buying and selling from the market are maintained.

$$P_{i,t}^{pro} - P_{i,t}^{con} = \sum_j P_{i,j,t} \tag{10.32}$$

$$P_{i,t}^{pro} = \begin{bmatrix} -P_t^{market} \\ (ifi = 1) \\ +P_{g,t} + P_{pv,t}^{PV} \\ +P_{w,t}^{wind} + P_{EV,t}^{disch} \end{bmatrix} \tag{10.33}$$

$$P_{i,t}^{con} = P_{i,t}^{load} + P_{EV,t}^{charg} \tag{10.34}$$

$$P_t^{market,min} \leq P_t^{market} \leq P_t^{market,max} \tag{10.35}$$

The suggested VPPO model is strengthened by the addition of extra constraints related to the flow of power, bus load voltage levels, and thermal limits related to the power flow of lines, as clarified by Eqs. (10.36) and (10.37) that express the technical aspects of the proposed VPPO. The flow of active and reactive power through the lines is limited by the conductor's thermal capacity, which is inherently related to the square of the current through the lines. This category is characterized by operational limitations represented by a circular area. The limits associated with both active and reactive power flow are simplified through the use of a polygon approximation, according to the formulations detailed in Eqs. (10.38) and (10.39) (Zare and Shafiyi, 2024a, 2024b; Zare et al., 2026a). The decimal outcomes of 0.9239 and 1.3066, achieved for the following equations, result from the linearization method used to convert a circle into an octagon (Jalilian et al., 2019). Additionally, CDGs that are synchronized with the upstream network provide the required reactive power.

$$V_{i,t} - V_{j,t} = 2\left(r_{ij}P_{ij,t} + x_{ij}Q_{ij,t}\right) \tag{10.36}$$

$$V_i^{min} \leq V_{i,t} \leq V_i^{max} \tag{10.37}$$

$$\left|P_{ij,t}\right|, \left|Q_{ij,t}\right| \leq 0.9239 \times S_{ij}^{max} \tag{10.38}$$

$$\left|P_{ij,t} \pm Q_{ij,t}\right| \leq 1.3066 \times S_{ij}^{max} \tag{10.39}$$

10.3 Case study and analysis

As earlier shown, the optimization issues were linearized due to initially nonlinear constraints in the system. The conversion renders the stochastic MILP problem extremely simple to implement while only ensuring the identification of a global optimal solution. The principal advantage of MILP is its computational efficiency, and its use of specialized solvers in modern optimization packages makes it an effective choice for large applications, providing a robust solution to hard optimization problems (Minai et al., 2024; Naval and Yusta, 2021). Linearization follows the procedures outlined in references (Zare and Shafiyi, 2024a; Soroudi, 2017; Jalilian et al., 2019). The effectiveness of the proposed approach was validated on an IEEE 33-bus distribution system with a peak load of 3.715 MW and 2.3 MVAr and a base voltage of 12.66 kV (Jalilpoor et al., 2023). Simulations were conducted in the GAMS environment using the CPLEX solver on a laptop computer with an Intel Core i7-10,750H CPU @ 2.60 GHz and 16 GB RAM. The DA time horizon was set to 1 h. Uncertain

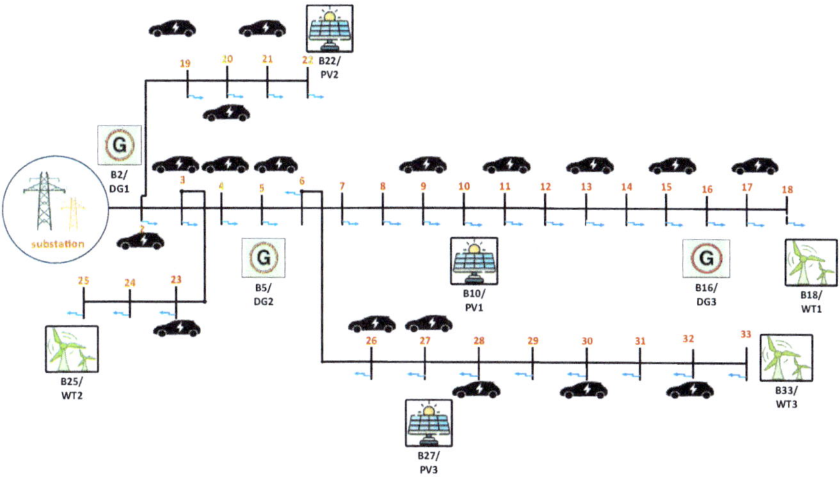

Figure 10.2 The schematic of the modified 33-bus in this study.

parameters were modeled with PDFs, which generated 625 scenarios per parameter, and these were reduced to 10 scenarios. The modified IEEE 33-bus system diagram is shown in Fig. 10.2. As can be seen in the figure, 18 EVs, three 80 kW wind power plants, three 40 kW solar power plants, and three CDGs are considered in the network. CDGs and EVs data are provided in Tables 10.1–10.3, (Zare and Shafiyi, 2024a; Aghajan-Eshkevari et al., 2023; Zare et al., 2026a). It is important to note that the cost of NO_x in Table 10.3 has changed compared to the original references. For the linearization of the quadratic function of CDGs, 10 segments were considered.

In this study, the results will be discussed in three case studies.

- **Case 1: The network contains only distributed wind and solar generation sources, with the remaining electricity demand supplied from upstream.**
- **Case 2: The network includes distributed renewable sources, fossil fuel generators, and EVs; however, there is no pollution control, and EVs are charged without smart management, treated solely as loads by the VPP operator.**
- **Case 3: Similar to Case 2, but EV charging and discharging schedules are optimized by the VPP operator, and CDG schedules are managed considering their environmental impact.**

The electricity price chart over 24 h for a cold region, based on the local final price from the NE-ISO system, is shown in Fig. 10.3 (I. N. England.,

Table 10.1 Data related to CDGs.

	Bus	SUC_g $	SDC_g $	p_g^{max}/Q_g^{max} (kW/kVAr)	p_g^{min}/Q_g^{min} (kW/kVAr)	a_g $/kW^2	b_g $/kW^1	c_g $
DG1	2	2	1	300/400	30/-280	0.001	0.025	0.16
DG2	5	2	1	350/400	35/-280	0.001	0.025	0.16
DG3	16	2	1	500/600	50/-380	0.001	0.03	0.23

Zare A, Shafiyi MA, Nazar MS, Shafie-khah M, Siano P. Modeling the Combined Effect of Demand Response Programs and Virtual Power Plant Facilities in Providing Flexibility.

Table 10.2 Data related to Electric Vehicles (Aghajan-Eshkevari et al., 2023).

EV types	Bus	Capacity (kW)	Battery Life (BL)	Depth of discharge (DOD)	C_{bat} \$/kW	C_l \$	η_c%	η_d%	$p^{charg}_{max,t}/p^{disch}_{max,t}$ kW
Tesla model X	2,9,17,23,30	100	5000	0.8	300	240	0.95	0.95	15
Tesla model 3	3,11,19,26,32	75	5000	0.8	300	240	0.95	0.95	12
Chevrolet Bolt	4,13,20,27	60	5000	0.8	300	240	0.95	0.95	10
Nissan Leaf	5,15,21,28	40	5000	0.8	300	240	0.95	0.95	8

Table 10.3 Coefficients related to the pollution caused by CDGs (Zare and Shafiyi, 2024a; Zare et al., 2026a).

Type of gas released	Externality Costs EF_e $/lb	Emission factors a_e lb/kWh
CO_2	0.014	0.001432
SO_x	0.99	0.000454
NO_x	1.1	0.218

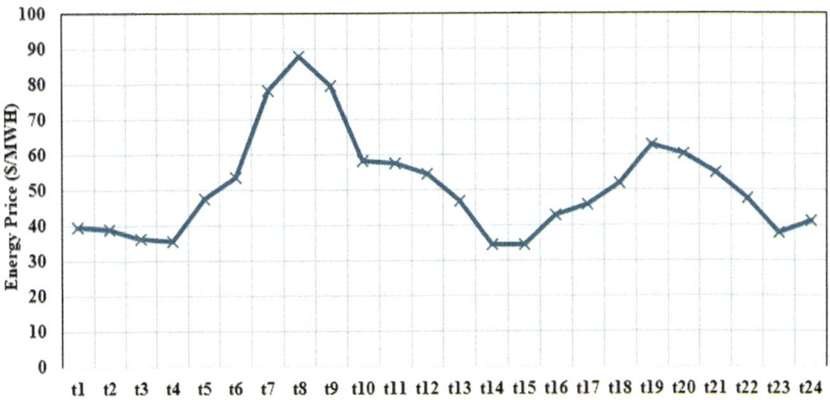

Figure 10.3 The prediction profile of the energy price for the DA energy exchange.

n.d.). Fig. 10.4 shows the predicted values of wind and solar sources (Zare and Shafiyi, 2024a). Fig. 10.5 shows the predicted arrival and departure times for each EV owner.

Fig. 10.6 shows the prediction of the distance traveled by each EV and the vehicle's efficiency based on each driver's behavior.

Fig. 10.7 shows the amount of energy purchased from the upstream or the energy market in all three cases. As can be seen, in case 1, all the electricity required is supplied from upstream, and the energy purchase from upstream is higher in all hours compared to Cases 2 and 3. The second case also has the lowest purchase from upstream, and this is because, in the second case, due to the lack of consideration of environmental issues mentioned in, the production program of CDGs is higher than in case 3. Fig. 10.8 shows the cumulative amounts of CDGs production in Cases 2 and 3. The different amounts of three gases (NO_x, CO_2, and SO_x) emitted from CDG sources in cases 2 and 3 are shown in Figs. 10.9 and 10.10, respectively. It can be seen that the pollution amounts in case 3 are much

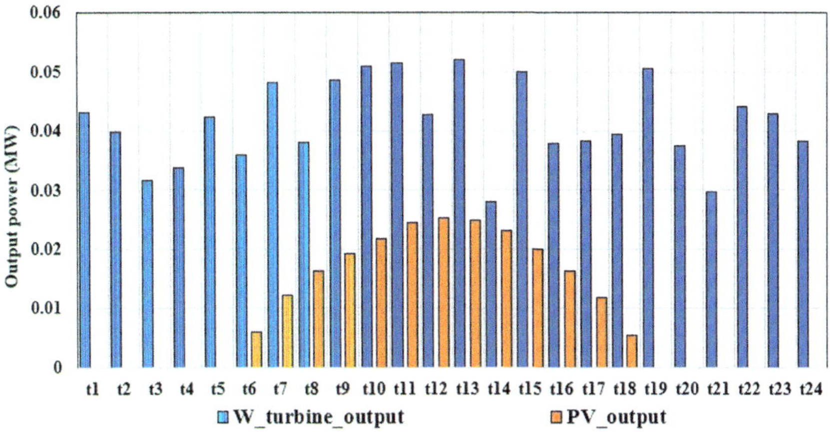

Figure 10.4 The prediction profile of solar and wind energy generation.

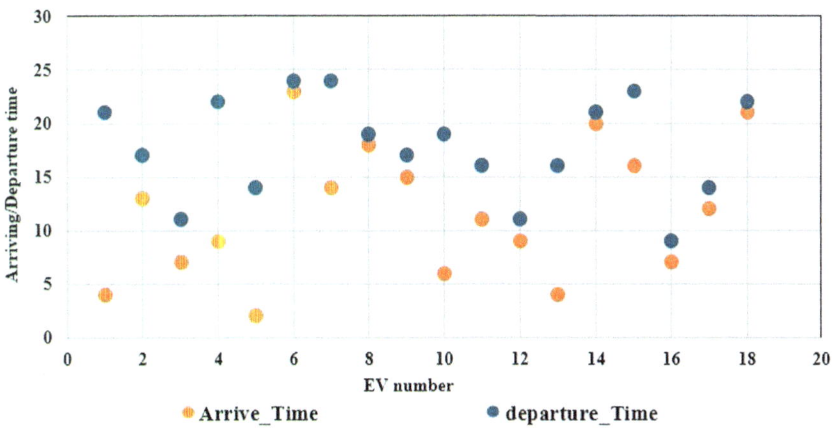

Figure 10.5 The prediction profile of the arrival/departure time of each EV.

lower than in case 2, and this is because in the third case, the VPPO considers environmental issues, and EVs discharging in energy planning.

Figs. 10.11 and 10.12 show the charging and discharging of EVs in two cases, two and 3, respectively. As shown in Fig. 10.11, in the second case, EVs are considered in the distribution system but are not under the supervision of the smart charging of the VPP; they are only charged. However, in the third case, the results differ as the EVs can charge and discharge (as shown in Fig. 10.12). To this end, EVs are supervised by the VPP operator, which can also be the distribution system operator. Aforementioned provides the EV charging and discharging schedule to the vehicle owners,

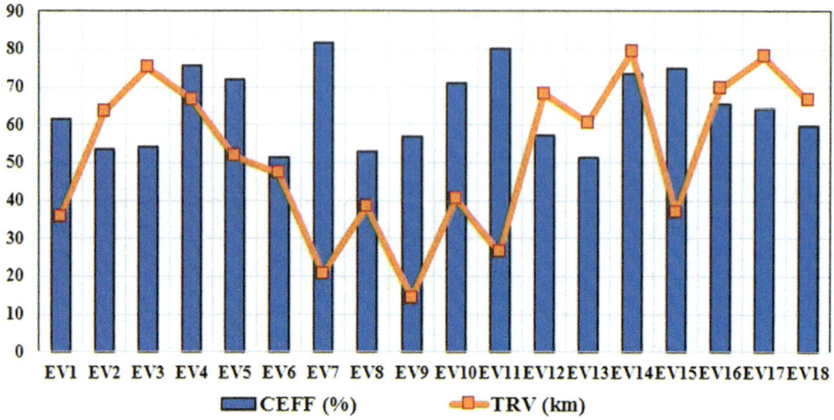

Figure 10.6 The prediction profile of distance traveled and each EV's efficiency.

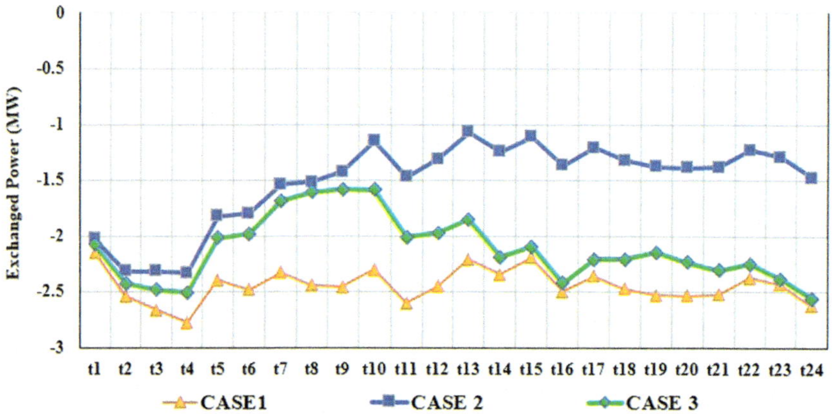

Figure 10.7 Exchanged power between the VPP and the upstream network in 3 cases.

taking into account the constraints of EVs such as the SOC, battery degradation, driver preferences, and technical constraints of the network. In these two figures, the negative power sign means charging EVs, and the positive power sign means discharging the EV battery.

Fig. 10.13 illustrates the average voltage for each case of all 33 buses per hour. Considering that it was previously stated that the VPP operator and the distribution system operator are regarded as the same entity, and taking into account the technical constraints outlined in this study, it is noted that the bus voltage and the average voltage of all buses in each case fall within the permissible range. The smart (case 3) and non-smart (case 2) charging

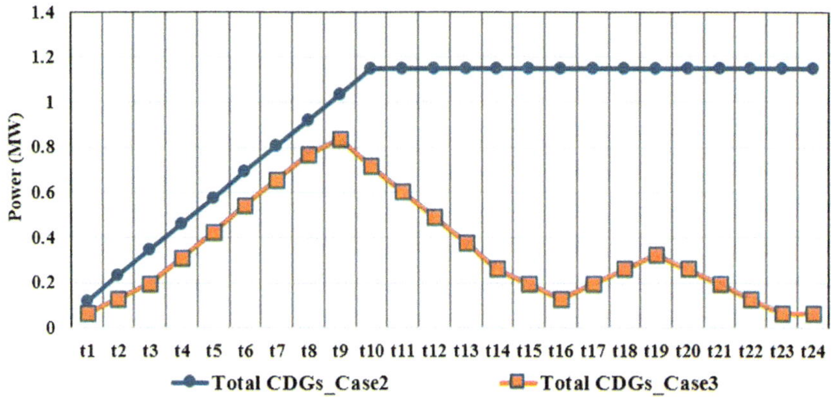

Figure 10.8 Optimal scheduling of CDGs in cases 2 and 3.

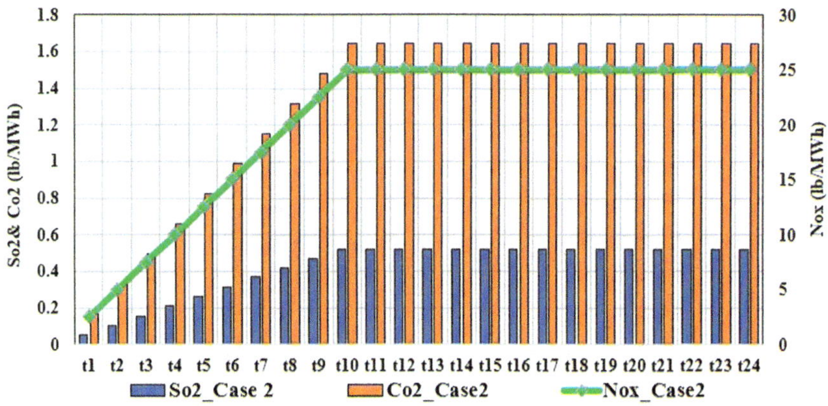

Figure 10.9 NO_x, CO_2, and SO_X emissions in Case 2.

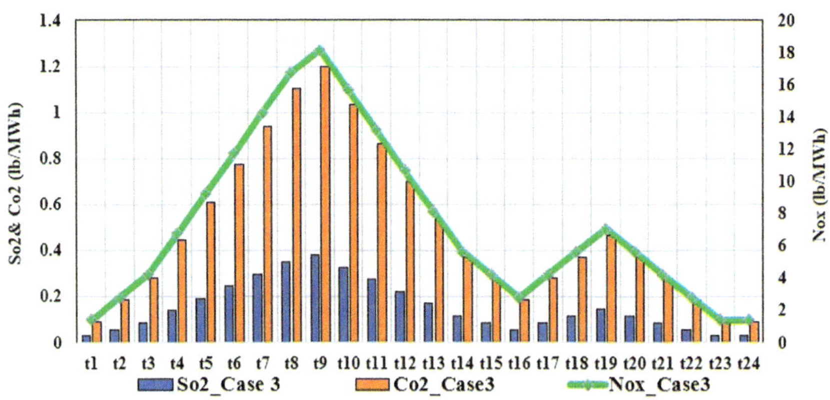

Figure 10.10 NO_x, CO_2, and SO_X emissions in Case 3.

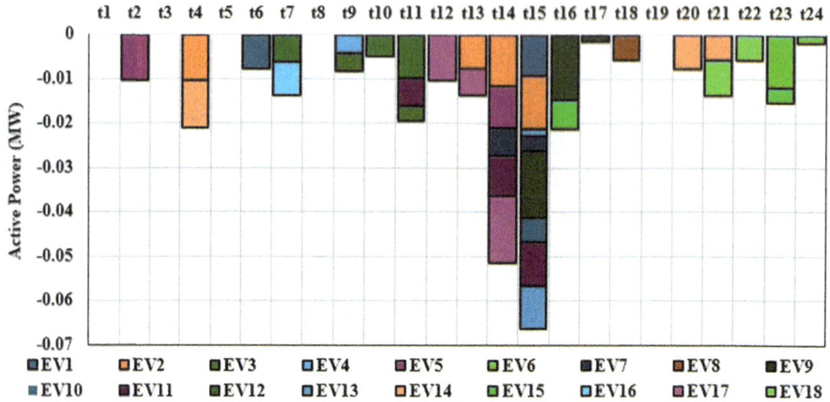

Figure 10.11 EVs charging in Case 2.

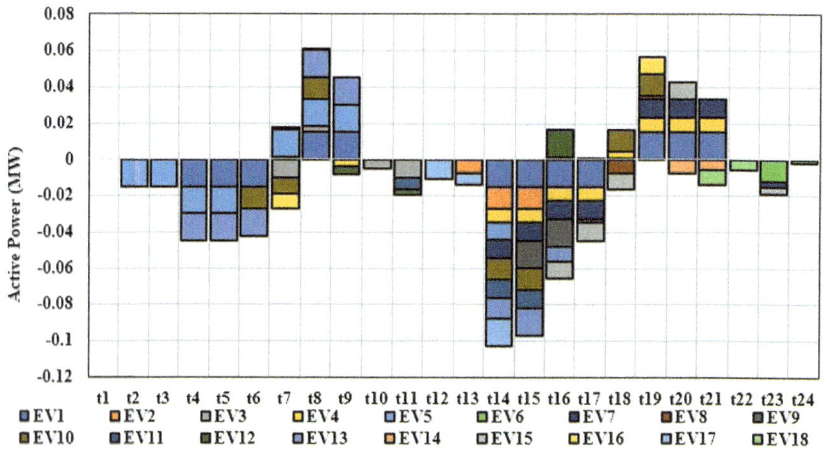

Figure 10.12 EVs' smart charging under VPPO supervision in Case 3.

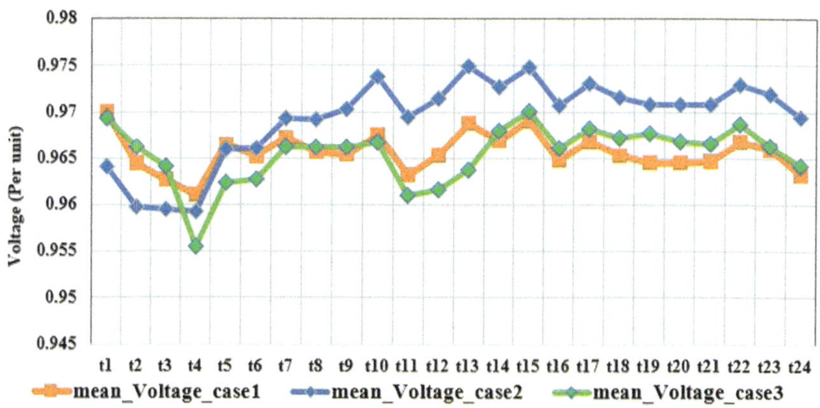

Figure 10.13 Average bus voltage per hour in the three cases.

and discharging of EVs, along with control or lack thereof over greenhouse gas production, as well as transactions with the upstream network, all influence the voltage deviation (i.e., reduction from t3 to t4 in all cases) in the second and third cases compared to the base case. In the third case, due to the VPP operator's effective management of EV smart charging/discharging and pollution control, the voltage deviation from the base case is smaller than that observed in the second case.

10.4 Discussion, challenges, and future work

The VPP modeled in this study optimizes the resources under its supervision and, considering the constraints of the distribution system, schedules DERs and EVs effectively to reduce pollution. The results of this optimization in three cases were explained in the previous section. The VPP planning in this study is based on DA, with pollution reduction and cost minimization as the primary reasons for this choice. Pollution reduction is achieved in DA for future planning, and in this chapter, DA planning was selected. However, other planning methods, such as intra-day and real-time, could be employed to adjust the power values in DA planning, which may cause the pollution values to exceed acceptable levels. This aspect is beyond the scope of the current work and is suggested for future research.

In this study, all EV-related parameters are subject to uncertainty, such as the average distance traveled by each EV. However, it is important to note that future studies and research can address this issue on two levels by considering the transportation network. The first level focuses on optimizing the routing of each vehicle, while the second level deals with power grid-related issues. This type of problem connects the power grid with the transportation network, which is beyond the current study's scope.

As discussed earlier, all nonlinear constraints in this research have been linearized using various methods. Polygonal linearization has also been applied to linearize the distribution system constraints, which presented one of the optimization challenges in this study. This chapter summarizes the linearized formulas and their results to facilitate easier use for future work and researchers. Not all mathematical formulas related to linearization are included, as doing so would overly extend the chapter and detract from its main purpose in the book. However, these linearizations are not essential, since most real-world constraints are nonlinear; they can be addressed similarly, with NLP optimization aiming to bring results closer to reality in the real world. Moreover, heuristic and meta-heuristic methods are often employed, providing greater flexibility and robustness for optimization. The primary reason for linearizing the model and converting it to a MILP problem was to decrease computational complexity and achieve a global optimal result more quickly.

This model considers the DA market, and the VPP is assumed to be a price-taker entity. Meanwhile, the VPP can integrate different types of DERs and participate in various types of energy markets as a price-maker entity, which can lead to better profitability and performance of the VPP in competitive electricity markets. There are also other methods for reducing pollution that are recently being developed, called carbon trading markets, and for future works in VPPs, these markets can be used in conjunction with energy markets for more effectiveness.

Finally, Artificial intelligence (AI) techniques have recently been introduced to forecast uncertain parameters, but further research is needed to refine these methods. AI can enable VPP models to optimize operations by learning from real-world data, improving demand forecasting, and resource management.

10.5 Conclusion

Vehicle-to-grid (V2G) is an appealing technology that enables stationary or parked EVs to serve as decentralized energy resources. The EV can store or release energy during peak times, creating a two-way energy flow between the electric grid and the EV. This functionality enhances the overall capacity for electricity generation while also supporting the electrical network's stability, reliability, environmental friendliness, and efficiency. In a distribution system with a high penetration of EVs, charging them during peak hours may pose risks to the grid, such as voltage drops or surges. In this scenario, with smart charging of EVs managed by the VPP operator, EVs can also discharge energy. This not only helps stabilize the grid and maintain voltage at appropriate levels but also reduces dependence on fossil resources, ultimately benefiting EV owners. This study uses a standard 33-bus distribution network as a test grid, incorporating various renewable resources, CDGs, and EVs across different buses. The simulation results of three case studies were analyzed and compared. The comparison of the second and third cases shows that by implementing the concept of a VPP to control and integrate resources in a distribution system, it is possible to maintain network security constraints and supply electricity to all loads while considering the benefits for EV owners. EVs are discharged, leading to a significant reduction in reliance on fossil resources and emissions of NO_x, CO_2, and SO_x gases. In the second case, the total amount of gases emitted is 531.15,855(lb/MWh), whereas in the third case, with optimized management and planning of the VPP to reduce greenhouse gas emissions, this figure has decreased to 194.10,677(lb/MWh), resulting in a 63.5% reduction in greenhouse gas emissions.

In this study, the VPP—assumed to act as the distribution system operator—effectively integrates diverse resources, optimizes EV charging and discharging while accounting for battery degradation, and ensures benefits for vehicle owners. It also serves as a key market participant, maintaining network security and reducing pollution.

References

Aghajan-Eshkevari S, Ameli MT, Azad S, Pashiri NT. Resilience enhancement of distribution networks by optimal scheduling of hydrogen systems future modern distribution networks resilience: from passive operation to strategic active paradigms. Iran: Elsevier; 2024. p. 353–74. https://doi.org/10.1016/B978-0-443-16086-8.00003-8.

Aghajan-Eshkevari S, Ameli MT, Azad S. Optimal routing and power management of electric vehicles in coupled power distribution and transportation systems. Appl Energy 2023:341. https://doi.org/10.1016/j.apenergy.2023.121126.

Aghajan-Eshkevari S, Azad S, Nazari-Heris M, Ameli MT, Asadi S. Charging and discharging of electric vehicles in power systems: an updated and detailed review of methods, control structures, objectives, and optimization methodologies. Sustainability 2022a; 14(4):2137. https://doi.org/10.3390/su14042137.

Al-Hanahi B, Ahmad I, Habibi D, Masoum MAS. Charging infrastructure for commercial electric vehicles: challenges and future works. IEEE Acc 2021;9:121476–92. https://doi.org/10.1109/access.2021.3108817.

Aghajan-Eshkevari S, Ameli MT, Azad S. Electric vehicle services to support the power grid. In: Electric vehicle integration via smart charging: technology, standards, implementation, and applications. Springer; 2022b. p. 129–48.

Awerbuch S, Preston A. The virtual utility: accounting, technology & competitive aspects of the emerging industry. Springer Science & Business Media; 2012.

Bellocchi S, Klöckner K, Manno M, Noussan M, Vellini M. On the role of electric vehicles towards low-carbon energy systems: Italy and Germany in comparison. Appl Energy 2019;255:113848. https://doi.org/10.1016/j.apenergy.2019.113848.

Conejo AJ, Carrión M, Morales JM. Decision making under uncertainty in electricity markets. Int Series Oper Res Manage Sci 2010;153. https://doi.org/10.1007/978-1-4419-7421-1.

Dahiwale PV, Rather ZH, Mitra I. A comprehensive review of smart charging strategies for electric vehicles and way forward. IEEE Trans Intell Trans Syst 2024;25(9):10462–82. https://doi.org/10.1109/TITS.2024.3365581.

I. N. England. ISO New England. https://www.iso-ne.com/.

Gayen D, Chatterjee R, Roy S. A review on environmental impacts of renewable energy for sustainable development. Int J Environ Sci Technol 2024;21(5):5285–310. https://doi.org/10.1007/s13762-023-05380-z.

Ghorbani Y, Ilankoon IMSK, Dushyantha N, Nwaila GT. Rare earth permanent magnets for the green energy transition: bottlenecks, current developments and cleaner production solutions. Resour Conserv Recycl 2025:212. https://doi.org/10.1016/j.resconrec.2024.107966.

Jalilian A, Taheri B, Safdarian A. Optimal switch placement in power distribution systems with complex configurations. In: Smart Gird Conference, SGC 2019; 2019. https://doi.org/10.1109/SGC49328.2019.9056625.

Jalilpoor K, Nikkhah S, Sepasian MS, Aliabadi MG. Application of precautionary and corrective energy management strategies in improving networked microgrids resilience: a two-stage linear programming. Elec Power Syst Res 2022;204. https://doi.org/10.1016/j.epsr.2021.107704.

Jalilpoor K, Ameli MT, Azad S, Sayadi Z. Resilient energy management incorporating energy storage system and network reconfiguration: a framework of cyber-physical system. IET Gener Trans Distrib 2023;17(8):1734–49. https://doi.org/10.1049/gtd2.12478.

Li J, Peng D, Zhao H, Wang D, Qi E. A 'spatial-temporal' bi-layer optimal control strategy of large-scale electric vehicles (EVs) in a multi-area Virtual Power Plant (VPP) oriented to dual carbon target. IET Renew Power Gener 2022;16(7):1445–61. https://doi.org/10.1049/rpg2.12390.

Minai AF, Ahmad Khan A, Kitmo, Ndiaye MF, Alam T, Khargotra R, Singh T. Evolution and role of virtual power plants: market strategy with integration of renewable based microgrids. Energy Strategy Rev 2024;53:101390. https://doi.org/10.1016/j.esr.2024.101390.

Ningshan H, Dragomir VD, Energy R. Sustainable business models, and decarbonization in the European union: comparative Analysis of corporate sustainability reports. Sustainability 2025;17(8):3646. https://doi.org/10.3390/su17083646.

Naval N, Yusta JM. Virtual power plant models and electricity markets - a review. Renew Sustain Energy Rev 2021;149:111393. https://doi.org/10.1016/j.rser.2021.111393.

Norouzi N, Tavakkoli-Moghaddam R, Ghazanfari M, Alinaghian M, Salamatbakhsh A. A new multi-objective competitive open vehicle routing problem solved by particle swarm optimization. Network Spatial Econ 2012;12(4):609–33. https://doi.org/10.1007/s11067-011-9169-4.

Qais M, Kirli D, Moroshko E, Kiprakis A, Tsaftaris S. A virtual power plant for coordinating batteries and EVs of distributed zero-energy houses considering the distribution system constraints. J Energy Storage 2025;106:114905. https://doi.org/10.1016/j.est.2024.114905.

Rahman I, Vasant PM, Singh BSM, Abdullah-Al-Wadud M, Adnan N. Review of recent trends in optimization techniques for plug-in hybrid, and electric vehicle charging infrastructures. Renew Sustain Energy Rev 2016;58:1039–47. https://doi.org/10.1016/j.rser.2015.12.353.

Roy S, Das DC, Sinha N. Optimizing smart city virtual power plants with V2G integration for improved grid Resilience. In: IEEE International Conference on Interdisciplinary approaches in technology and management for social innovation, IATMSI 2024; 2024. https://doi.org/10.1109/IATMSI60426.2024.10502468.

Sadeghian O, Oshnoei A, Mohammadi-ivatloo B, Vahidinasab V, Anvari-Moghaddam A. A comprehensive review on electric vehicles smart charging: solutions, strategies, technologies, and challenges. J Energy Storage 2022;54:105241. https://doi.org/10.1016/j.est.2022.105241.

Singh J, Tiwari R. Real power loss minimisation of smart grid with electric vehicles using distribution feeder reconfiguration. IET Gener Trans Distrib 2019;13(18):4249–61. https://doi.org/10.1049/iet-gtd.2018.6330.

Soroudi A. Power system optimization modeling in GAMS power system optimization modeling in GAMS. Ireland: Springer International Publishing; 2017. p. 1–295. https://doi.org/10.1007/978-3-319-62350-4.

Türkoğlu AS, Güldorum HC, Sengor I, Çiçek A, Erdinç O, Hayes BP. Maximizing EV profit and grid stability through Virtual Power Plant considering V2G. Energy Rep 2024;11:3509–20. https://doi.org/10.1016/j.egyr.2024.03.013.

Zare A, Shafie-khah M, Siano P, Lazaroiu GC. A systematic review of Virtual Power Plant configurations and their interaction with electricity, carbon, and flexibility markets. Renew Sustain Energy Rev 2026;226. https://doi.org/10.1016/j.rser.2025.116448.

Zare A, Shafiyi MA, Azad S. Virtual power plant: a review on Components, models, types and scheduling. In: Development Engineering Conferences Center Articles Database. 1; 2024.

Zare A, Shafie-Khah M, Shafiyi MA. A review of Microgrid and virtual power plant strategies for effective DERs scheduling. In: 12th Iranian Conference on renewable Energies and distributed generation, ICREDG 2025; 2025. https://doi.org/10.1109/ICREDG66184.2025.10966088.

Zare A, Shafiyi MA. The impact of energy storage system on the flexible operation of the renewable-based virtual power plant. In: 11th Iranian Conference on renewable energy and distribution generation, ICREDG 2024; 2024. https://doi.org/10.1109/ICREDG61679.2024.10607790.

Zare A, Shafiyi MA. Iran virtual power plant as a demand response aggregator. In: 28th International electrical power distribution Conference, EPDC 2024; 2024. https://doi.org/10.1109/EPDC62178.2024.10571710.

Zare A, Shafiyi MA. Virtual power plant models and market participation: a deep dive into optimization and real-world applications. Results Eng 2025;26. https://doi.org/10.1016/j.rineng.2025.105548.

Zare A, Agha Shafiyi M, Setayesh Nazar M, Shafie-khah M, Siano P. Bi-level coordination of demand response and multiple virtual power plants in a distribution network for flexibility assessment. Energy Sci Eng 2026. https://doi.org/10.1002/ese3.70434.

CHAPTER ELEVEN

Smart energy management systems for supporting zero-emission electric vehicles: Challenges and future prospects in renewable energy integration

Nazanin Pourmoradi and Mohammad Setareh
Department of Electrical Engineering, Shahid Beheshti University, Tehran, Iran

Abbreviation Definition

AC Alternating Current
AI Artificial Intelligence
BEVs Battery Electric vehicles
BMS Battery Management System
DC Direct Current
EMS Energy Management System
EVs Electric Vehicles
IoT Internet of Things
ML Machine Learning
PHEVs Plug-In Hybrid Vehicles
RES Renewable energy resources
SAE Society of Automotive Engineers
SEMS Smart Energy Management Systems
V2B Vehicle-to-Building
V2D Vehicle-to-Device
V2G Vehicle-to-Grid
V2H Vehicle-to-Home
V2I Vehicle-to-Infrastructure
V2N Vehicle-to-Network
V2V Vehicle-to-Vehicles
V2X Vehicle-to-Everything
X2V Everything-to-Vehicle
ZEVs Zero Emission Electric Vehicles

Renewable-to-Vehicle Smart Charging
ISBN: 978-0-443-36684-0
https://doi.org/10.1016/B978-0-443-36684-0.00007-2

11.1 Introduction

With the rapid growth of urban populations and the increasing environmental pollution caused by fossil fuels, the need to develop sustainable transportation systems is felt more than ever. EVs play a vital role in the future of energy and the environment as a new and efficient option to reduce greenhouse gas emissions. However, the expansion of these vehicles will only be effective if they can be managed intelligently, especially in terms of energy supply and consumption. In this regard, SEMS play a key role as the main platform for coordination between energy sources, charging infrastructures, and end users (Kumar et al., 2025).

Over the past decade, advancements in RES and smart charging technologies have created new opportunities to enhance the efficiency and sustainability of electric transportation systems (Barman et al., 2023). Combining sustainable energy sources with variable electric vehicle loads poses many technical and operational challenges, including accurately forecasting generation and consumption, managing peak load, and maintaining grid quality and stability (Tan et al., 2016).

One of the emerging concepts in this field is the two-way interactions between vehicles and the power grid, known under titles such as V2G, V2H, V2B, and X2V, as shown in Fig. 11.1. These concepts not only

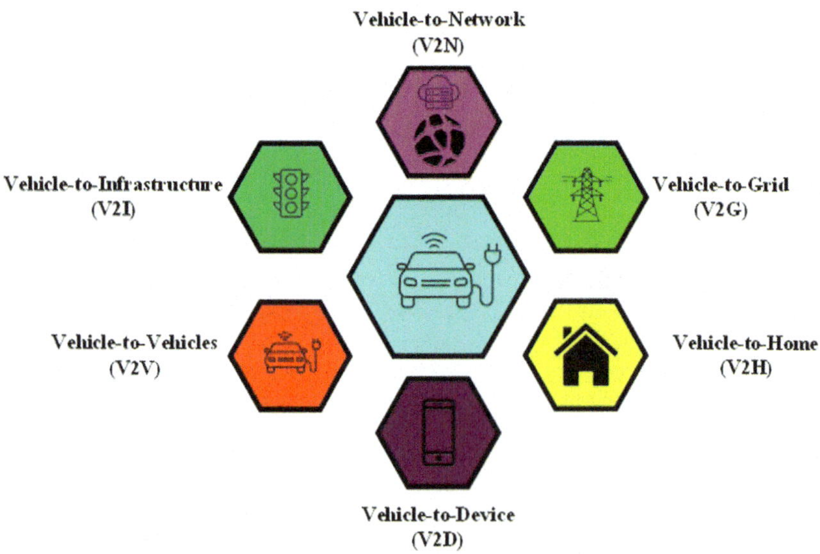

Figure 11.1 Emerging features and capabilities in electric vehicles.

enable the grid to be recharged by electric vehicles but can also play a role in load regulation, reducing energy costs, and improving system reliability (Dong and Guo, no date; Bonfiglio et al., 2025). Consequently, the design and implementation of systems that can manage these interactions in an intelligent, flexible, and sustainable manner has become one of the main research focuses in recent years. On the other hand, to integrate RES with smart charging systems, algorithms and infrastructure need to cope with the inherent fluctuations and uncertainties of these sources. For example, solar and wind energy production depends on weather conditions and varies based on the time of day and season. This inherent unpredictability, along with volatile consumer behavior, complicates the design of energy management systems (Anisie et al., 2023). Therefore, the need for advanced approaches based on machine learning algorithms, predictive optimization, and decentralized technologies such as blockchain is increasingly felt (Shamami et al., 2020).

Despite significant technical advances, serious obstacles remain to the widespread adoption of these systems, from challenges such as standardization and data protection to high initial investment costs and even social acceptance. For this reason, the future path of intelligent energy management systems depends on simultaneous growth in three key areas: technological development, improved policymaking, and improved user awareness and skills (Ghani et al., 2025).

This chapter provides a comprehensive picture of the current status, challenges, and prospects of intelligent energy management systems in supporting zero-emission electric vehicles. First, an overview of the structure and operation of these systems is presented, and then how to integrate RES with smart charging infrastructures is discussed. Next, the main challenges in the implementation and operation of these systems are examined. Finally, future trends and opportunities in the form of new technologies, business models, and supporting policies will be analyzed. The chapter concludes with a summary and suggestions for future research.

11.2 An overview of smart energy management systems

BMS focuses on the health, safety, and optimization of battery performance (monitoring voltage, current, temperature, and charge/discharge cycles) and operates at the equipment (battery) level. In contrast, EMS is a comprehensive system for managing generation, storage, distribution, and consumption at the grid, building, or even city level. It is responsible for optimizing energy consumption, load balancing, and coordinating different resources (including batteries). BMS is a subset of the infrastructure that EMS relies on for macro-level energy management.

BMS and EMS are key components of modern energy structures that, despite their complementary roles, have distinct functions. BMS is dedicated to monitor and control the health, safety, and optimal performance of batteries at the equipment level and performs tasks such as measuring voltage, current, temperature, and managing charge and discharge cycles. Therefore, BMS can be considered a specialized part and subset of EMS, whose function makes sense within the framework of macro energy management (Aruna and Vasan, 2019).

Advances in renewable energy technologies and the increasing penetration of EVs have elevated the importance of SEMS in ensuring efficient energy utilization. By leveraging advanced communication technologies, big data analytics, and intelligent optimization algorithms, SEMS facilitate precise and real-time coordination among energy generation, storage, and consumption across complex and heterogeneous networks (Aruna and Vasan, 2019).

The architecture of the SEMS typically consists of several layers, each with its roles and technologies. In the first layer, data is acquired by sensors and smart meters and subsequently transmitted via secure, high-speed communication networks to processing units located in the middle layers for optimal analysis and management. Finally, the processing layer performs advanced analyses and employs machine learning algorithms to make optimal decisions for efficient energy management (Hu et al., 2024). This layered structure enables fast and flexible responses to variations in energy generation and consumption. An overview of this architecture is shown in Fig. 11.2.

11.2.1 SEMS layered architecture

The architecture of a SEMS is usually designed in a multi-layered manner, with each layer having its specific role and technologies as summarized in Table 11.1. In the first layer, data is collected and monitored from smart sensors and advanced metering equipment. Using Internet of Things (IoT) technologies, this layer can provide accurate and real-time data on energy consumption, battery status, and environmental conditions.

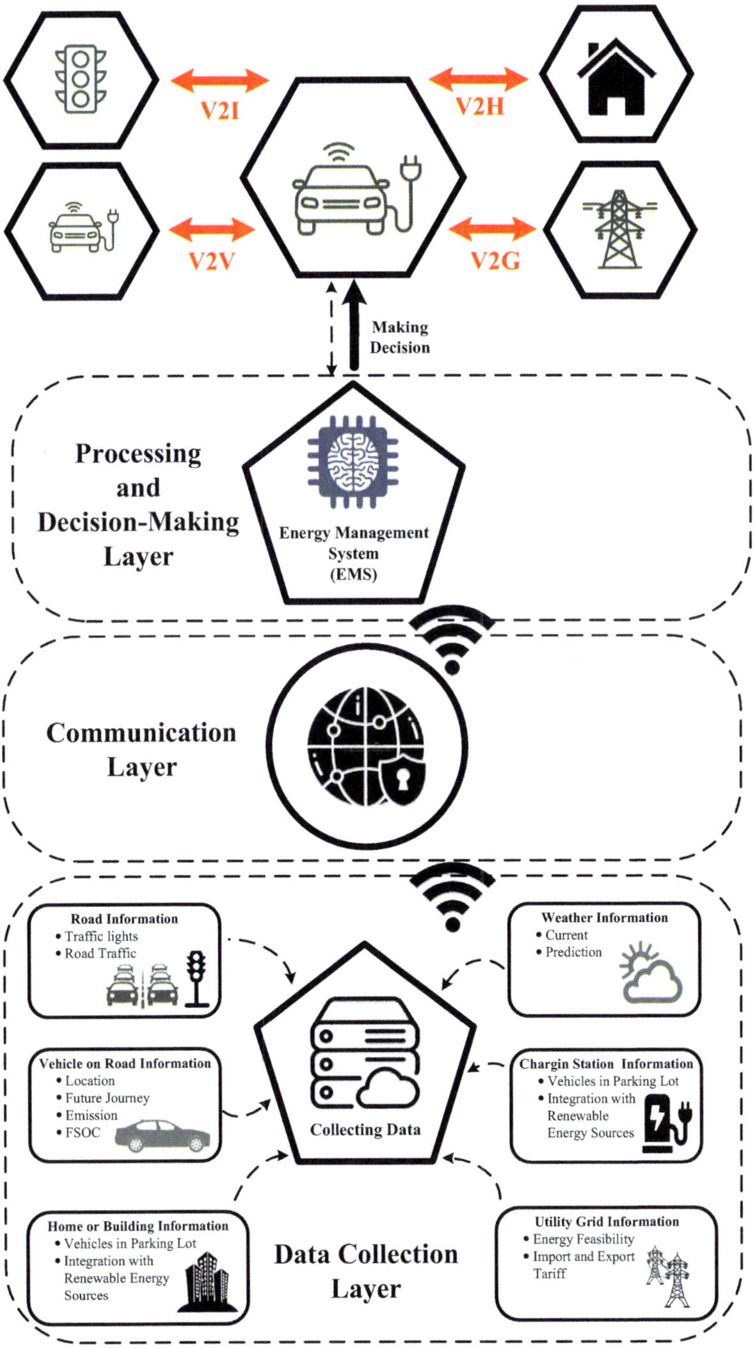

Figure 11.2 Overview of SEMS architecture.

Table 11.1 Layered architecture of SEMS.

SEMS architecture layer	Main function	Technologies used	Advantages
Data Collection Layer	Receiving and monitoring data from sensors and equipment	Smart sensors, IoT, and advanced measuring devices	High accuracy, real-time accessibility, precise and comprehensive data collection
Communication Layer	Transmitting data to processing units	5G networks, secure private networks, encrypted protocols	High speed, low latency, secure, and reliable data transmission
Processing and Decision-Making Layer	Analyzing data and making optimal decisions through intelligent algorithms	Machine learning, deep learning, reinforcement learning, multi-agent systems	Optimization of processes, intelligent decision-making, and adaptability to dynamic conditions

The collected data includes physical parameters such as voltage, current, and temperature, along with ecological and behavioral variables important for future decision-making (Hu et al., 2024; Abir et al., 2021).

In the middle layer, data is transferred to processing units via fast, secure, and stable communication networks such as 5G, proprietary networks, and secure protocols. This layer acts as a bridge between the data collection and processing layers and will ensure the accuracy, security, and speed of information transfer. It is essential to use security protocols and standards such as TLS/SSL and encryption technologies at this stage to protect against cyber threats. 5G networks are known as the ideal communication infrastructure for SEMS due to their capabilities, such as high speed, very low latency, and support for many connected devices (Irshad and Alreshoodi, 2024; Ahmadzadeh et al., 2021).

In the upper layer, or the processing and decision-making layer, data is analyzed using complex machine learning, deep learning, and multi-agent system algorithms, and optimal decisions are made for managing energy consumption, storage, and distribution. In addition to data analysis, this layer is responsible for predicting energy consumption trends, detecting anomalies, and responding to dynamic and unpredictable network conditions. With the help of reinforcement learning algorithms, the system can self-learn and dynamically improve its management to and react in real time (Bilal et al., 2024; Domínguez-Barbero et al., 2020; Ahmed et al., 2020).

11.2.2 Multi-agent systems and their role in SEMS

One of the new and practical approaches in SEMS is the use of multi-agent systems in which several agents operate independently and, by cooperating and exchanging information, pursue the goal of optimizing the performance of the entire system. Each agent represents a specific sector, such as energy production, storage, or consumer. The model improves flexibility, fault tolerance, and stability, which are especially beneficial in large-scale and distributed networks. Furthermore, these systems are robust against cyber-attacks and network disruptions due to their distributed structure (Li et al., 2021).

11.2.3 V2X technologies and their application in SEMS

V2X technologies are an essential aspect of SMES. These technologies enable communication and exchange of energy and information between electric vehicles, the power grid, buildings, and other vehicles. The most important models of this technology include V2G, V2H, and V2B. In this framework, EVs can not only receive energy but also return energy to the grid or building during peak consumption times, while being capable of charging from multiple energy sources such as solar power(X2V). This feature significantly improves grid stability by contributing to load balancing (Brown and Soni, 2019; Colak et al., 2024).

11.2.4 New technologies in improving SEMS

Advances in blockchain technology that enhance the security and transparency of energy transactions are a crucial achievement for SEMS. The use of blockchain increases trust among stakeholders, prevents fraud, and provides an immutable and traceable record of transactions. In addition, artificial intelligence with predictive, automated decision-making capabilities, and edge computing allow data processing at the point closest to the source of production, reducing latency and improving overall system performance (Longo et al., 2018; Sharma et al., 2023).

11.2.5 Challenges facing SEMS

Despite the many benefits of SEMS, these systems face significant challenges. The most important of these challenges include high costs for setup and maintenance, as well as software complexity that requires specialized teams and strong infrastructure. Besides, the lack of international standards creates problems in equipment compatibility and system coordination. Another challenge is maintaining cybersecurity and protecting user privacy, which is of utmost importance given the widespread connection of devices and systems to the Internet (Mao et al., 2019).

SEMS as the brains and coordinators of energy resources, play a vital role in moving toward sustainable energy and reducing environmental impacts. Given the rapid proliferation of RES and EVs, the development and improvement of SEMS through advanced technologies and innovative algorithms have become an inevitable necessity. Investment in this area will not only increase the efficiency and sustainability of energy systems but also help reduce costs and improve the quality of energy services.

11.3 Different methods for charging electric vehicle batteries in the power grid

In general, EVs that depend on grid-based charging can be classified into two categories: 1) battery electric vehicles (BEVs), and 2) plug-in hybrid vehicles (PHEVs). BEVs are powered entirely by energy stored in the battery, while PHEVs can use fossil fuels and electric energy. Therefore, BEVs are equipped with batteries of higher capacity compared to PHEVs (Gong et al., 2020; Momen et al., 2020). In general, EVs battery charging methods are classified into three main categories: 1) conductive charging, 2) inductive charging, and 3) battery swapping, as shown in Fig. 11.3 (Khalid et al., 2019).

11.3.1 Conductive charging

Conductive charging refers to charging EV batteries through a direct physical connection to the power grid. In this method, two types of chargers are used, on-board and off-board chargers. On-board chargers are integrated into EVs and do not require external peripheral equipment to connect to the power grid; hence EVs can be charged using regular power outlets anywhere. However, on-board chargers cannot provide high current due to limitations in power transfer capacity. As a result, charging EVs batteries through this method will be time-consuming. In contrast, external chargers

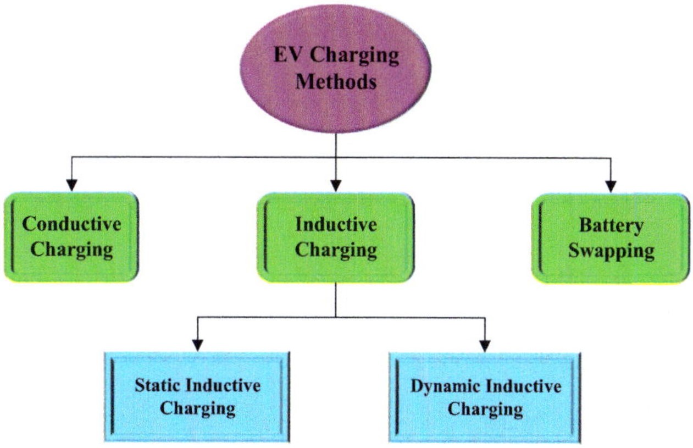

Figure 11.3 Classification of various electric vehicle charging methods.

are not part of the vehicle's internal equipment and are usually installed in public parking lots, highways, and fast-charging stations (Moradewicz, 2019). Since these chargers can transfer higher amounts of power to EVs, the charging time is substantially reduced. However, external chargers cannot be used in all locations due to higher costs, a more complex structure, and limited access (Moradewicz, 2019).

SAE has developed a standard for classifying different charging levels for electric vehicles (Kumar, 2024). This standard defines three different charging levels for both AC and DC charging. A summary of these charging levels is provided in Table 11.2.

11.3.2 Inductive charging

Inductive charging, known as wireless charging, is performed without a direct physical connection between the electric vehicle and the power grid; and energy is transferred by creating an electromagnetic field. The primary advantages of this method are reducing the risk of electric shock and possible injuries, since the energy transfers through an air gap. However, the relatively large air gap and imperfect coil matching reduce the efficiency of the charging process in this mode (Khalid et al., 2019). In general, inductive charging can be performed in two modes: 1) static and 2) dynamic. As shown in Fig. 11.4, in the static mode, EV remain stationary during charging. However, in the dynamic mode, EV can be charged even while moving. Therefore, as shown in Fig. 11.5, installing inductive charging paths in highway pavement can extend the driving range of, thereby reducing the need for larger batteries. In addition, since the battery cost

Table 11.2 *SAE standard for AC and DC charging levels of electric vehicles in the power grid.*

Types of power levels	Location of charger	Typical usage	Interface for energy supply	The expected level of power (KW)
Level 1: convenient Vac: 230 (EU) Vac: 120 (US)	Single-phase On-board	Office or home-based charging	Any convenient outlet	P: 1.4 (12A) P: 1.9 (20A)
Level 2: Main Vac: 400 (EU) Vac: 240 (US)	Single phase/three phase On-board	Publicly & privately-based charging	EV supply equipment	P: 4 (17A) P: 8 (32A) P: 19.2 (80A)
Level 3: Fast Vac: 208-600	Three-phase off-board	Like a filling station, a commercial point	EV supply equipment	P: 50 P: 100
DC Power Level 1: Vdc: 200-450	Off-board	Dedicated charging stations	EV supply equipment	P: 40 (80A)
DC Power Level 2: Vdc: 200-450	Off-board	Dedicated charging stations	EV supply equipment	P: 90 (200A)
DC Power Level 3: Vdc: 200-600	Off-board	Dedicated charging stations	EV supply equipment	P: 240 (400A)

Figure 11.4 View of static induction charging in electric vehicles.

accounts for a large part of the price of EVs, dynamic inductive charging can reduce the initial cost of EVs (Khalid et al., 2019). Finally, dynamic inductive charging can alleviate many common issues of EV users, including limited driving range, long charging time, and a highprice cpmared to conventional internal combustion vehicles. For this reason, this charging method has attracted widespread attention from researchers. However, high investment costs are the most important obstacle to develop dynamic inductive charging technology (Khalid et al., 2019).

11.3.3 Battery replacement

Battery swapping is one of the fastest ways to provide a fully charged battery in EVs. In this process, the vehicle owner replaces the discharged battery with a charged battery at special stations. In addition to significantly reducing charging time for users, this method significantly beneficial for battery swapping stations by optimizing the charging, discharging, and battery swapping processes. Furthermore, by optimally managing the charging and discharging processes of batteries at swapping stations, it is possible to improve the performance and increase the overall efficiency of the power grid (Park and Moon, 2022; Mansour et al., 2025). In Park and Moon (2022), a battery charging schedule at swapping stations was presented, designed to level the voltage profile and free up network capacity. In Mansour et al. (2025), The process of battery swapping for EVs at swapping stations is optimized. This approach effectively reduces the difference between peak and off-peak loads and facilitate the integration of RES into the power grid. On the other hand, the battery replacement method comes with challenges; for example, each battery is from a specific manufacturer and has unique characteristics and compatibility, which makes it difficult

Figure 11.5 Dynamic inductive charging of electric vehicles.

to find a similar battery. Also, the battery must be designed for easy removal from the EV and replacement with a charged battery. In addition, the infrastructure required for this charging method is more complex and costly compared to other methods. The issue of battery ownership also arises; one option is for the car owner to purchase a spare battery as a backup when the main battery is discharged, which increases the costs for the car owner. In another case, the car owner does not have a spare battery, and the charging station owns the battery, so in addition to the charging cost, the car owner must also pay the station owner a rental fee for the battery.

11.4 Integrating renewable energies with smart charging systems

Smart charging systems are a key component in realizing clean and sustainable energy networks. With the increasing share of solar, wind, and other RES in energy production, optimal charging management of EVs has become increasingly important to balance the variable energy supply and charging demand (Dahiwale et al., 2024).

One of the serious obstacles in this area is the constant changes in the rate of renewable energy production, which can upset the balance between production and consumption. To manage this situation, smart charging systems rely on accurate forecasting algorithms to estimate the production rate in advance and set an optimal schedule for charging vehicles based on it. By analyzing weather data, energy consumption patterns, and driver behavior, these algorithms allow for maximum use of clean energy while reducing pressure on the power grid (Zhang et al., 2025).

In this regard, energy storage systems (ESS) play an important role as a key complementary tool. These systems store excess energy generated during off-peak times and return it to the grid when needed, reducing fluctuations and improving power quality. Furthermore, integrating ESS with smart charging infrastructure allows grids to be more resilient to unforeseen disruptions and increase their ability to support critical situations (Islam et al., 2024).

New technologies such as V2G and V2H are essential to this architecture. These technologies allow electric vehicles to return the energy stored in their batteries to the power grid or to use it to meet household needs. This two-way exchange, in addition to helping to balance the grid load, plays an effective role in preventing overloads. In the future, the widespread development of V2G technologies can create a large and distributed network of energy storage devices and play an important role in the long-term sustainability of smart energy grids (Mojumder et al., 2022).

On the other hand, developing advanced management models based on artificial intelligence and machine learning will greatly help improve the forecasting and control processes in smart charging systems. These models can identify consumer behavior patterns and energy production conditions by analyzing a huge amount of historical and real-time data and, based on that, provide optimal solutions for charging and discharging schedules. For example, the use of deep neural networks and reinforcement learning algorithms in forecasting demand and regulating energy flow in charging systems significantly increases energy efficiency and productivity.

In addition, using IoT technology in smart charging infrastructures enables seamless communication between vehicles, charging stations, energy sources, and energy management systems. Using sensors, advanced measurement equipment, and modern communication networks, real-time data on the status of the network, battery charge level, and energy production are collected and analyzed. This information is the basis for decision-making in charging and discharging processes and helps improve the stability and efficiency of the network.

In addition, the implementation of Local Energy Markets has also been proposed as a new approach to integrating renewable energies and smart charging systems. In these markets, consumers and small producers (such as owners of electric cars or home solar systems) can exchange energy with each other and conduct their transactions transparently and securely through smart contracts based on blockchain technology. This mechanism not only optimizes energy consumption at the local level but also creates new economic incentives for users to participate in the energy management process actively.

Finally, attention to infrastructure and policy dimensions also plays a significant role in the success of this integration. Developing appropriate physical infrastructure for fast-charging stations, upgrading the capacity of electricity distribution networks, and developing international standards for equipment interoperability are among the necessary measures to achieve the successful integration of renewable energies and smart charging systems. Incentive policies, such as preferential tariffs for off-peak charging and support for investment in energy storage technologies, can help accelerate the adoption and expansion of these technologies.

To better understand the main components and technologies involved in integrating renewable energy with smart charging systems, Table 11.3

Table 11.3 Key components in integrating renewable energies with smart charging systems.

Technology/Component	Main function	Advantages
Solar and Wind Energy	Production of clean and variable energy	Reduction of greenhouse gas emissions
Energy Storage Systems	Storing and regulating generated energy	Increased grid stability and flexibility
Forecasting Algorithms	Analyzing environmental and consumption data for optimized charging	Optimal use of clean energy and cost reduction
V2G and V2H Technologies	Returning energy from vehicles to the grid or home	Load balancing and improved grid stability

is presented. This table summarizes the different technology layers, their functions, and the benefits of each.

A fundamental requirement for these systems to function properly is the development of communication standards and harmonized protocols so that the different components can work well together. These standards enable data to be exchanged accurately and securely between vehicles, charging stations, and the network, paving the way for the growth and development of smart grids (Hussain et al., 2018).

Overall, integrating renewable energy with smart charging systems is a fundamental step toward reducing carbon emissions and increasing energy efficiency, which is made possible by using modern technologies and smart management.

11.5 Challenges in smart energy management for zero-emission vehicles

Despite the numerous environmental and economic benefits, intelligent energy management in charging and operating systems for Zero-Emission Vehicles (ZEVs) faces complex and multi-layered challenges. These challenges include various dimensions such as technical, economic, security, and political, which need to be examined in an integrated and comprehensive manner to ensure the optimal performance, sustainability, and scalability of these technologies in future energy networks (Subashini and Vijayan, 2023).

One of the most fundamental challenges is the complexity of coordinating the different components of the energy grid. Energy management systems must simultaneously and in real time balance the supply and demand of energy between RES (solar, wind, etc.), ESSs, distribution networks, charging stations, as well as consumers (both domestic and transportation fleets). This coordination is performed in an environment that is highly dynamic and unpredictable (Deb et al., 2020). Sudden changes in weather conditions, energy market price fluctuations, consumer behavior, and network loading require real-time and dynamic decision-making at different operational and management levels.

The inherent fluctuations in RES generation, especially solar and wind power, pose challenges in accurately forecasting energy supply. These fluctuations can lead to serious problems in the optimal scheduling of vehicle charging, especially in scenarios with high charging loads and a major share of RES in the energy supply (Karmaker et al., 2023). The uncertainty in generation reinforces the need to develop intelligent forecasting algorithms, load management strategies, and flexible storage solutions.

In addition to these issues, infrastructure constraints are another major obstacle to the development of smart charging networks. The limited capacity of charging stations, especially in dense urban areas and intercity routes, along with the lack of uniform standards of communication protocols and hardware, creates many technical and operational problems (Kong et al., 2018). This not only hinders the scalability of charging networks but also slows down market development and increases operational and maintenance costs.

Security of smart charging ecosystems has become a serious challenge. Connecting vehicles, charging infrastructure, and energy management systems to the Internet and cloud environments significantly increases the risk of site hijacking (Acharya et al., 2020). Intrusion into control systems, manipulation of consumption, or disruption of the charging process can have serious implications for energy networks and public safety. Therefore, the design and implementation of advanced security solutions such as data encryption, multi-layered authentication, and intrusion detection systems in this area is very necessary.

From an economic perspective, the high initial investment costs of smart charging infrastructure and new energy management technologies represent the most considerable barriers (Burnham et al., 2017). Installing high-power charging stations, developing energy storage systems, implementing advanced communication and analytical systems require significant financial resources. Without government support, financial incentives and targeted strategic policies, the development of these technologies may proceed slowly.

In addition, user adoption of the technology is also a soft but effective barrier. Lack of public awareness of the benefits of zero-emission vehicles, concerns about limited range (range anxiety), long charging times, and easy access to charging stations reduce the adoption of these technologies (Will and Schuller, 2016). Therefore, to accelerate public adoption of smart charging, appropriate educational programs must be held, awareness campaigns must be created, and real incentives such as cheaper electricity tariffs must be provided.

Finally, harmonizing policies and standards at local, national, and international levels is one of the biggest infrastructure challenges (Tirunagari et al., 2022). The lack of complete and up-to-date regulations and standards that keep pace with new technologies can be a major obstacle to the development and implementation of smart charging networks. Developing coherent legal frameworks, international cooperation, and standard protocols for data and energy exchange is a critical requirement in this direction.

To better understand and categorize these challenges, Table 11.4 summarizes the most critical obstacles to intelligent energy management for ZEVs. The table briefly illustrates the nature of each challenge and its potential implications for overall system performance.

This visual structure enables readers to quickly identify key challenges and understand their connections and impacts, which is essential for advancing research and developing practical solutions.

11.6 Future trends and opportunities in smart charging systems

Over the past few years, the rapid growth of ZEVs along with significant advances in renewable energy technologies have made smart charging systems an essential pillar of sustainable energy networks. By utilizing new technologies, these systems enable energy consumption optimization, cost reduction, and network efficiency enhancement, which help accelerate the transition to clean transportation (Bhupathi and Chinta, 2022).

11.6.1 Development of intelligent charging management algorithms

One of the most important future trends in this field is using artificial intelligence and machine learning. By analyzing a tremendous amount of

Table 11.4 Main challenges in integrating renewable energies with smart charging systems and their consequences.

Challenge	Brief description	Consequences/ Importance
System Coordination Complexity	Difficulty in real-time management of various grid components	Reduced efficiency and grid stability
Renewable Energy Fluctuations	Sudden changes in solar and wind energy production	Increased uncertainty and charging scheduling issues
Infrastructure Limitations	Limited station capacity and inconsistent standards	Slower grid development and increased costs
Cybersecurity	Risk of attacks on internet-connected systems	Threats to data security and grid performance
Economic Barriers	High initial investment costs	Need for government support and incentive policies
Technology Adoption	Lack of awareness and user resistance	Reduced efficiency and slower technology deployment
Policy Coordination	Absence of coherent standards and regulations	Legal and technical challenges in development and implementation

consumption and network data, these algorithms determine optimal battery charging schedules to reduce pressure on the power grid and lower energy costs.

Jahan Jui et al. (2024) and Park and Moon (2022). Deep learning algorithms are exceptionally efficient in recognizing complex consumption patterns and changing network conditions and can dynamically make optimal decisions (Mansour et al., 2025).

Recent advances in reinforcement learning have enabled the creation of systems that are self-learning and improve network performance with experience. Using these methods, energy management systems can respond in real time to changes in network conditions and adjust vehicle charging subject to network constraints and user needs (Park and Moon, 2022).

Integrating intelligent algorithms with multi-objective optimization enables the simultaneous optimization of cost, time, and sustainability (Ferraz et al., 2023).

On the other hand, with the advancement of communication technologies such as 5G and 6G, which have high bandwidth and very low latency, suitable infrastructures have been provided for implementing large-scale and fast-response charging management systems (Varga et al., 2025). These technologies significantly improve the possibility of real-time coordination between vehicles, charging stations, and energy grids.

Blockchain technology has also found a special place in the security and transparency of energy transactions. Using blockchain, transactions related to the exchange of energy between vehicles (V2V), grids (V2G), and homes (V2H) can be managed in a secure decentralized manner, that increases user trust in smart systems (Rao et al., 2023).

Alongside these technologies, developing supportive policies and coherent legal frameworks at to facilitate standardization and promote cooperation among different countries will play a key role for faster adoption of these technologies (Brownet al., 2018).

11.6.2 Internet of things and advanced communications in smart charging

IoT technology is fundamental to developing smart charging systems. By connecting sensors, charging devices, and vehicles to a smart network, information is transmitted in real time, and the charging process is coordinated (Dasi et al., 2024). This connectivity gives operators better control over the network, reduced consumption fluctuations, and more precise management of energy flows.

IoT-based systems collect data on charging status, battery capacity, environmental conditions, and consumer demand, and use data analytics to enable more efficient energy management. These capabilities are critical in distributed charging systems and smart energy grids.

11.6.3 Advanced energy storage technologies

Energy storage is a critical part of smart charging systems. It plays an important role in balancing supply and demand in the grid. New technologies such as flow batteries and supercapacitors suit these systems due to their high storage capacity, long life, and high charging and discharging speeds (Dasi et al., 2024).

Flow batteries, due to their separate electrolytes and scalable structure, enable large-scale energy storage and can compensate for fluctuations caused by variable RES. Supercapacitors are ideal for handling sudden demand spikes and load peaks due to their high charge and discharge rates.

Furthermore, the development of distributed energy management systems by combining local energy storage, RES, and consumption management increases the optimal utilization of energy resources and enhances grid flexibility.

11.6.4 Security and trust in V2X interactions

Proliferation of two-way interactions between vehicles and energy networks (V2X) has dramatically increased the importance of data protection and transaction integrity. Meanwhile, blockchain technology, with its decentralized and transparent structure, provides a secure and reliable platform for recording and managing energy transactions and helps maintain system integrity (Will and Schuller, 2016; Ferraz et al., 2023). Smart contracts on blockchain platforms automate and secure the settlement and energy management processes.

In addition to increasing security, these technologies help build trust among consumers and operators while providing a platform for the wider development of smart charging systems.

11.6.5 Supportive policies and legal frameworks

Government support and appropriate policies and regulations are critical factors in developing smart charging systems and facilitating the adoption of EVs. Given the importance of reducing greenhouse gas emissions and dependence on fossil fuels, many countries have implemented strategic policies to develop charging infrastructure, energy management technologies, and support consumers.

11.6.6 Financial incentive and support programs

Common policies in this area include EV purchase subsidies, tax breaks, low-interest loans for installing home and public charging stations, and direct government investments. For example, the European Union and leading countries such as Norway, Germany, and the Netherlands have allocated a significant share of their transportation fleet to EVs through subsidies and other financial incentives (Ali et al., 2022).

In addition to financial support, technical standardization and the development of unified protocols for charging, data exchange, and energy management at national and international levels are key success factors. The lack of harmonized standards can lead to incompatibility between vehicles and charging infrastructure, limiting the growth of smart grids. For this reason, institutions such as International Electrotechnical Commission, International Organization for Standardization, and industry consortia such as CHAdeMO, and Combined Charging System play a pivotal role in developing global standards. For example:

- The European Union is leading the way in creating integrated smart charging networks with the ISO 15118 standard, which defines the vehicle-to-grid (V2G) communication protocol.
- The United States has implemented SAE J1772 standards for AC and DC chargers, and national programs such as the EV everywhere grand challenge to develop charging infrastructure and promote energy management technologies.
- Japan has played a key role in expanding fast charging infrastructure by developing the CHAdeMO standard and focusing on V2G and V2H technologies.
- China has also formed the world's largest electric vehicle market by developing domestic GB/T standards and allocating extensive subsidies to manufacturers and consumers.

All worlwide supportive policies and standards EVs energy management are listed in Table 11.5.

11.6.7 Future vision: integrated and intelligent systems

The future of smart charging systems lies not just in managing the charging process, but in fully integrating with energy grids and transportation infrastructure. In this direction, the use of advanced technologies such as artificial intelligence, IoT, high-capacity energy storage systems, and cybersecurity technologies will pave the way for the creation of intelligent, dynamic, and automated systems (Bhupathi and Chinta, 2022; Rao et al., 2023).

In this vision, charging systems intelligently and adaptively optimize the charging process using real-time data from different resources such as weather conditions, energy consumption patterns, power grid status, and user needs. Advanced machine learning algorithms will enable accurate prediction of supply and demand fluctuations and intelligent management of

Table 11.5 Comparison of supportive policies and standards in leading countries for electric vehicle energy management.

Country/Region	Financial support	Technical standards	Roadmap and key policies
European Union	EV purchase subsidies, tax exemptions, and support for public charging infrastructure	ISO 15118 (V2G), CCS	"Green Deal" program to reduce emissions by 55% by 2030, support for integrated charging networks
United States	Federal tax credit up to $7,500, grants for charger installation	SAE J1772, ISO 15118	"EV Everywhere" program, investment in federal charging infrastructure
Japan	Subsidies for EV purchase and V2H installation, support for storage technologies	CHAdeMO, ISO 15118	Focus on V2G technologies and smart home energy grids
China	Extensive subsidies for manufacturers and buyers, tax exemptions	GB/T, CCS (transitioning to international standards)	"New Energy Vehicles (NEV)" policy, target of 40% EV market share by 2030
Norway	Full tax and fee exemptions for EVs, extensive charging infrastructure	CCS, ISO 15118	100% EV sales target by 2025, world's largest EV market share
Germany	Government subsidies for EV purchase and home charger installation, support for public chargers	CCS, ISO 15118	"National Platform for Electric Mobility (NPE)" roadmap to enhance infrastructure
Netherlands	Tax exemptions, discounts on public charging costs	CCS, ISO 15118	Strategy for urban charging infrastructure development and V2G market expansion

network load, thereby increasing the stability and efficiency of the entire system.

Ultimately, developing innovative security solutions to protect data and prevent cyber threats will be a key pillar in the success of these systems. Combining these technologies will pave the way for sustainable, flexible, and low-carbon energy grids. All future trends and opportunities in smart charging systems that are mentioned in the papers are summarized in Table 11.6.

11.7 Conclusion and suggestions

This chapter comprehensively reviews SEMS that support ZEVs and advanced vehicle-to-grid (V2V and V2X) interactions. First, basic concepts and existing frameworks are introduced. Then, focusing on integrating RES, the challenges and opportunities in this field are discussed. Besides, new technologies such as artificial intelligence, IoT, advanced storage, and blockchain, which play an essential role in improving the efficiency and security of these systems, are examined in detail. Finally, future trends and promising prospects for developing smart charging systems are thoroughly discussed.

Among the outstanding strengths of this chapter is the combination of theoretical and practical discussions concerning up-to-date sources, which provide an accurate picture of the current situation and the path forward. Furthermore, the analysis of the challenges, along with the presentation of technological and policy solutions, has provided the reader with multi-dimensional and realistic perspectives. The use of new technologies and the emphasis on interdisciplinary cooperation are other advantages of this analysis.

The key achievements of this chapter include:

⇒ Comprehensive understanding of the structure and function of intelligent energy management systems for electric vehicles.

⇒ Identify technical, security, and infrastructure challenges in integrating renewable energy and smart charging management.

⇒ Introducing new technologies that effectively improve the efficiency and security of systems, including artificial intelligence and blockchain.

⇒ Analyze future trends and opportunities in developing smart charging networks and increasing energy sustainability.

Table 11.6 Future trends and opportunities in smart charging systems.

References	Advantages	Description	Trend/Opportunity
Karmaker et al. (2023), Kong et al. (2018) and Jahan Jui et al. (2024)	Cost reduction, optimized energy consumption, and fast response to grid conditions	Use of AI and ML algorithms for charging prediction and optimization	Intelligent AI and ML Algorithms
Acharya et al. (2020); Park and Moon (2022)	Improved grid coordination, optimal load management, and increased flexibility	Connecting devices and EVs to the grid for real-time data exchange	IoT
Burnham et al. (2017); Mansour et al. (2025)	Increased storage capacity, reduced fluctuations, enhanced grid stability	Flow batteries and supercapacitors for energy storage and grid balancing	Advanced storage technologies
Will and Schuller (2016); Ferraz et al. (2023)	Increased trust, fraud prevention, transaction transparency	Using blockchain for secure management of transactions and data	Blockchain security in V2X
Tirunagari et al., 2022; Varga et al. (2025)	Accelerated technology adoption, infrastructure development, and reduced legal barriers	Development of laws and financial incentives for smart charging infrastructure	Policies and regulatory frameworks
Bhupathi and Chinta (2022); Rao et al. (2023)	Optimized decision-making, improved efficiency, and cost reduction	Combining AI, IoT, storage, and security technologies into comprehensive systems	Integrated smart systems

In terms of future opportunities, technologies such as advanced machine learning algorithms, broader IoT capabilities, and higher-efficiency energy storage can make a big difference in optimizing energy consumption and improving V2X interactions. Also, developing international standards and coherent support policies will provide a suitable platform for expanding these technologies.

Practical suggestions for advancing this area include the following:

1. Strengthening communication and energy infrastructure to support smart management systems.
2. Support and invest the development of intelligent algorithms and new storage technologies.
3. Develop legal frameworks and incentive policies to accelerate the adoption of smart charging technologies.
4. Increase cross-sector collaborations among industry, academia, and government institutions to develop comprehensive solutions.
5. Focus on cybersecurity and data protection in V2X interactions to preserve privacy and ensure system functionality.

By implementing these solutions, we can achieve a more sustainable and smarter future in energy and electric transportation, leading to sustainable economic and social development while preserving the environment.

References

Abir SMAA, Anwar A, Choi J, Kayes ASM. Iot-enabled smart energy grid: pplications and challenges. 21693536 IEEE Access 2021;9:50961–81. https://doi.org/10.1109/ACCESS.2021.3067331.

Acharya S, Dvorkin Y, Pandzic H, Karri R. Cybersecurity of smart electric vehicle charging: a power grid perspective. 21693536 IEEE Access 2020;8:214434–53. https://doi.org/10.1109/ACCESS.2020.3041074.

Ahmadzadeh S, Parr G, Zhao W. A review on communication aspects of demand response management for future 5G IoT- based smart grids. 21693536 IEEE Access 2021;9:77555–71. https://doi.org/10.1109/ACCESS.2021.3082430.

Ahmed W, Ansari H, Khan B, Ullah Z, Ali SM, Mehmood CAA, et al. Machine learning based energy management model for smart grid and renewable energy districts. 21693536 IEEE Access 2020;8:185059–78. https://doi.org/10.1109/ACCESS.2020.3029943.

Ali MH, Slaifstein D, Ibanez FM, Zugschwert C, Pugach M. IEEE computer society Russian federation power management strategies for vanadium redox flow battery and supercapacitors in hybrid energy storage systems. 2022/01. In: IEEE PES innovative smart grid technologies conference europe; 2022 1. https://doi.org/10.1109/ISGT-Europe54678.2022.9960352. 9781665480321, https://ieeexplore.ieee.org/xpl/conhome/1800214/all-proceedings 2022-.

Anisie A, Navarro JPJ, Antic T, Pasimeni F, Blanco H. Innovation landscape for smart electrification: Decarbonising end-use sectors with renewable power. 2023.

Aruna P, Vasan PV. Review on energy management system of electric vehicles. In: 2019 2nd international conference on power and embedded drive control (ICPEDC). IEEE; 2019. p. 371–4.

Barman P, Dutta L, Bordoloi S, Kalita A, Buragohain P, Bharali S, et al. Renewable energy integration with electric vehicle technology: a review of the existing smart charging approaches. Renew Sustain Energy Rev 2023;183:113518. https://doi.org/10.1016/j.rser.2023.113518. 13640321.

Bhupathi HP, Chinta S. Smart charging revolution: AI and ML strategies for efficient EV battery use. ESP J Eng Technol Adv 2022;2(2):154–67.

Bilal M, Oladigbolu JO, Mujeeb A, Al-Turki YA. Cost-effective optimization of on-grid electric vehicle charging systems with integrated renewable energy and energy storage: an economic and reliability analysis. 2352152X J Energy Storage 2024;100. https://doi.org/10.1016/j.est.2024.113170.

Bonfiglio A, Minetti M, Loggia R, Mascioli LF, Golino A, Moscatiello C, et al. Integrated vehicle-to-building and vehicle-to-home services for residential and worksite microgrids. 26246511 Smart Cities 2025;8(3). https://doi.org/10.3390/smartcities8030101.

Brown MA, Soni A. Expert perceptions of enhancing grid resilience with electric vehicles in the United States. Energy Res Social Sci 2019;57:101241. https://doi.org/10.1016/j.erss.2019.101241. 22146296.

Brown MA, Zhou S, Ahmadi M. Smart grid governance: an international review of evolving policy issues and innovations. 2041840X Wiley Interdiscip Rev Energy Environ 2018;7(5). https://doi.org/10.1002/wene.290.

Burnham A, Dufek EJ, Stephens T, Francfort J, Michelbacher C, Carlson RB, et al. Enabling fast charging – infrastructure and economic considerations. J Power Sources 2017;367:237–49. https://doi.org/10.1016/j.jpowsour.2017.06.079. 03787753.

Colak I, Irmak E, Tepe IF. A brief overview of future mobility and energy management strategies with vehicle-to-everything (V2X) concept on smart grid. 2024/01. In: 13th international conference on renewable energy research and applications, ICRERA 2024. Türkiye: Institute of Electrical and Electronics Engineers Inc; 2024 1. https://doi.org/10.1109/ICRERA62673.2024.10815234.

Dahiwale PV, Rather ZH, Mitra I. A comprehensive review of smart charging strategies for electric vehicles and way forward. 15580016 IEEE Trans Intell Transp Syst 2024;25(9):10462–82. https://doi.org/10.1109/TITS.2024.3365581.

Dasi S, et al. IoT-based intelligent energy management for EV charging stations. Power 2024;51(5).

Deb S, Goswami AK, Harsh P, Sahoo JP, Chetri RL, Roy R, et al. Charging coordination of plug-in electric vehicle for congestion management in distribution system integrated with renewable energy sources. 19399367 IEEE Trans Ind Appl 2020;56(5):5452–62. https://doi.org/10.1109/TIA.2020.3010897.

Domínguez-Barbero D, García-González J, Sanz-Bobi MA, Sánchez-Úbeda EF. Optimising a microgrid system by deep reinforcement learning techniques. 19961073 Energies 2020;13(11). https://doi.org/10.3390/en13112830.

Dong S and Guo S. "Advancements, challenges, and future trends of vehicle-to-grid technologies: a review," Energy Technol, p. 2500481.

Ferraz RSF, Ferraz RSF, Rueda Medina AC, Fardin JF. Multi-objective approach for optimized planning of electric vehicle charging stations and distributed energy resources. 14320487 Electr Eng 2023;105(6):4105–17. https://doi.org/10.1007/s00202-023-01942-z.

Ghani AA, Abdullah N, et al. Assessing technological-driven challenges and policies associated with electric vehicle (EV) adoption. 1879310X Transp Policy 2025;169:167–77. https://doi.org/10.1016/j.tranpol.2025.04.030.

Gong L, Cao W, Liu K, Yu Y, Zhao J. Demand responsive charging strategy of electric vehicles to mitigate the volatility of renewable energy sources. 18790682 Renew Energy 2020;156:665–76. https://doi.org/10.1016/j.renene.2020.04.061.

Hu J, Lin Y, Li J, Hou Z, Chu L, Zhao D, et al. Performance analysis of AI-based energy management in electric vehicles: a case study on classic reinforcement learning. Energy Convers Manag 2024;300:117964. https://doi.org/10.1016/j.enconman.2023.117964. 01968904.

Hussain SMS, Ustun TS, Nsonga P, Ali I. IEEE 1609 WAVE and IEC 61850 standard communication based integrated EV charging management in smart grids. 00189545 IEEE Trans Veh Technol 2018;67(8):7690–7. https://doi.org/10.1109/TVT.2018.2838018.

Irshad A, Alreshoodi M. SEMS-5G: a secure and efficient multi-server authentication scheme for 5G networks. 21693536 IEEE Access 2024;12:49062–77. https://doi.org/10.1109/ACCESS.2024.3381616.

Islam MM, et al. Improving reliability and stability of the power systems: a comprehensive review on the role of energy storage systems to enhance flexibility. IEEE Access 2024.

Jahan Jui J, Ahmad MA, Imran Molla MM, Rashid MIM. Optimal energy management strategies for hybrid electric vehicles: a recent survey of machine learning approaches. J Eng Res 2024;12(3):454–67. https://doi.org/10.1016/j.jer.2024.01.016. 23071877.

Karmaker AK, Hossain MA, Pota HR, Onen A, Jung J. Energy management system for hybrid renewable energy-based electric vehicle charging station. 21693536 IEEE Access 2023;11:27793–805. https://doi.org/10.1109/ACCESS.2023.3259232.

Khalid MR, Alam MS, Sarwar A, Jamil Asghar MS. A Comprehensive review on electric vehicles charging infrastructures and their impacts on power-quality of the utility grid. eTransportation 2019;1:100006. https://doi.org/10.1016/j.etran.2019.100006. 25901168.

Kong Q, Fowler M, Entchev E, Ribberink H, McCallum R. The role of charging infrastructure in electric vehicle implementation within smart grids. Energies 2018;11(12):3362. https://doi.org/10.3390/en11123362. 1996-1073.

Kumar GMS, Guo X, Zhou S, Luo H, Wu Q, Liu Y, et al. State-of-the-art review of smart energy management systems for supporting zero-emission electric vehicles with X2V and V2X interactions. 18790690 Renew Sustain Energy Rev 2025;208. https://doi.org/10.1016/j.rser.2024.115050.

Kumar A. A comprehensive review of an electric vehicle based on the existing technologies and challenges. 25784862 Energy Storage 2024;6(5). https://doi.org/10.1002/est2.70000.

Li S, Gu C, Zeng X, Zhao P, Pei X, Cheng S. Vehicle-to-grid management for multi-time scale grid power balancing. Energy 2021;234:121201. https://doi.org/10.1016/j.energy.2021.121201. 03605442.

Longo M, Foiadelli F, Yaïci W. Electric vehicles integrated with renewable energy sources for sustainable mobility. New Trends Electr Vehi Powertrains 2018;10:203–23.

Mansour SH, Azzam SM, Hasanien HM, Tostado-Véliz M, Alkuhayli A, Jurado F. Deep reinforcement learning-based plug-in electric vehicle charging/discharging scheduling in a home energy management system. 18736785 Energy 2025;316. https://doi.org/10.1016/j.energy.2025.134420.

Mao T, Zhang X, Zhou B. Intelligent energy management algorithms for EV-charging scheduling with consideration of multiple EV charging modes. 19961073 Energies 2019;12(2). https://doi.org/10.3390/en12020265.

Mojumder MdRH, Antara FA, Hasanuzzaman M, Alamri B, Alsharef M. Electric vehicle-to-grid (V2G) technologies: impact on the power grid and battery. Sustainability 2022;14(21):13856. https://doi.org/10.3390/su142113856. 2071-1050.

Momen H, Abessi A, Jadid S. Using EVs as distributed energy resources for critical load restoration in resilient power distribution systems. 17518687 IET Gener Transm Distrib 2020;14(18):3750–61. https://doi.org/10.1049/iet-gtd.2019.1561.

Moradewicz AJ. On/Off - board chargers for electric vehicles. 24499544 Prz Elektrotechn 2019;95(2):136–9. https://doi.org/10.15199/48.2019.02.30.

Park K, Moon I. Multi-agent deep reinforcement learning approach for EV charging scheduling in a smart grid. Appl Energy 2022;328:120111. https://doi.org/10.1016/j.apenergy.2022.120111. 03062619.

Rao PM, Jangirala S, Pedada S, Das AK, Park Y. Blockchain integration for IoT-enabled V2X communications: a comprehensive survey, security issues and challenges. 21693536 IEEE Access 2023;11:54476–94. https://doi.org/10.1109/ACCESS.2023.3281844.

Shamami MS, Alam MS, Ahmad F, Shariff SM, Alsaidan I, Rafat Y, et al. Artificial intelligence-based performance optimization of electric vehicle-to-home (V2H) energy management system. 26406438 SAE Int J Sustain Transp Energy Environ Policy 2020;1(2). https://doi.org/10.4271/13-01-02-0007.

Sharma G, Joshi AM, Mohanty SP. sTrade: blockchain based secure energy trading using vehicle-to-grid mutual authentication in smart transportation. 22131388 Sustain Energy Technol Assessments 2023;57. https://doi.org/10.1016/j.seta.2023.103296.

Subashini M, Vijayan S. Smart charging for zero emission vehicles – a comprehensive review. 17550084 Renew Energy Focus 2023;46:57–67. https://doi.org/10.1016/j.ref.2023.05.005.

Tan KM, Ramachandaramurthy VK, Yong JY. Integration of electric vehicles in smart grid: a review on vehicle to grid technologies and optimization techniques. 18790690 Renew Sustain Energy Rev 2016;53:720–32. https://doi.org/10.1016/j.rser.2015.09.012.

Tirunagari S, Gu M, Meegahapola L. Reaping the benefits of smart electric vehicle charging and vehicle-to-grid technologies: regulatory, policy and technical aspects. 21693536 IEEE Access 2022;10:114657–72. https://doi.org/10.1109/ACCESS.2022.3217525.

Varga P, Jászberényi ÁI, Pásztor D, Nagy B, Nasar M, Raisz D. How beyond-5G and 6G makes IIoT and the smart grid green—a survey. 14248220 Sensors 2025;25(13). https://doi.org/10.3390/s25134222.

Will C, Schuller A. Understanding user acceptance factors of electric vehicle smart charging. 0968090X Transport Res C Emerg Technol 2016;71:198–214. https://doi.org/10.1016/j.trc.2016.07.006.

Zhang T, Peng Q, Zeng S. Predicting EV charging demand in renewable-energy-powered grids using explainable machine learning. Sustainability 2025;17(9):4158. https://doi.org/10.3390/su17094158. 2071-1050.

CHAPTER TWELVE

Wind-to-vehicle (W2V): Unlocking the synergy between wind energy and EV flexibility

Cesar Diaz-Londono[1], Gibran David Agundis-Tinajero[2], Daogui Tang[3], Yupeng Yuan[3], Josep M. Guerrero[1]
[1]Center for Research on Microgrids (CROM), Huanjiang Laboratory, Shaoxing, Zhejiang, China
[2]Center for Global Cooperation, Danish Energy Agency, Copenhagen, Denmark
[3]School of Transportation and Logistics Engineering, Wuhan University of Technology, Wuhan, China

12.1 Introduction

Nowadays, the electrical systems around the world are undergoing a critical transition from fossil fuel-based to renewable energy-based generation due to environmental and sustainability concerns. This transition, also known as the green energy transition, has brought an accelerated introduction of distributed renewable energy resources to the electrical power grid, such as wind and solar energy, which not only changes the way conventional systems operate but has also brought new challenges to system modeling, analysis, control, and energy management due to the renewable energies stochastic nature, and their needed power electronic device interface (International Energy Agency, 2024).

Furthermore, in addition to the renewable energy resources introduction, another effort to cope with environmental issues has been the increasing adoption of electricity-based technologies, especially substituting those based on fossil fuels, which are widely used, this is the case for electrical vehicles (EVs). There are different types of EVs according to their engine technology: battery electric vehicles, plug-in-hybrid electric vehicles, hybrid electric vehicles, fuel cell electric vehicles, and extended-range electric vehicles (Togun et al., 2025).

In particular, 100% EVs, such as the battery electric vehicle, do not have an internal combustion engine and do not use any kind of liquid fuel, they have an autonomy that depends completely on the battery size (Larminie and Lowry, 2003). Current EV technologies include battery sizes that exceed the 100 kWh capacity, giving them an autonomy above 500 km

Renewable-to-Vehicle Smart Charging
ISBN: 978-0-443-36684-0
https://doi.org/10.1016/B978-0-443-36684-0.00005-9

per charge (Sanguesa et al., 2021). In addition to increasing battery capacity, advances in battery technology for EVs is making charging competitive with conventional petrol-based refueling time, which in turn has brought the enhancement of chargers infrastructure for fast charging, i.e., a power output higher than 22 kW and lower than 150 kW, and ultra-fast charging, i.e., a power output capable of delivering 150 kW or higher (International Energy Agency, 2025).

The combined penetration of renewable energy and EVs introduces significant uncertainty and variability to power systems. Wind energy, while clean and abundant, is intermittent and weather-dependent. Similarly, the growing deployment of fast and ultra-fast EV chargers results in high and sporadic demand peaks. When unmanaged, this dual stochasticity can cause voltage instability, line congestion, increased losses, and inefficient use of renewable resources. Hence, there is a critical need for coordinated energy management strategies that consider both wind generation and EV charging behaviors.

12.1.1 Literature review

The integration of EVs and renewable energy sources in distribution networks has attracted considerable research attention due to the associated operational and planning challenges. A significant portion of the literature has focused on the development of EV charging coordination strategies to mitigate their impact on grid performance.

Several studies have addressed the adverse effects of uncoordinated EV charging on voltage profiles, transformer aging, and peak load increases (Deilami et al., 2011; Manousakis et al.). To address these issues, various scheduling algorithms have been proposed, including centralized approaches (Diaz-Londono et al.), distributed solutions (Zhou et al., 2024), and price-based demand response schemes (Zanvettor et al., 2024). However, most of these works primarily focus on slow or standard charging, where the power levels are relatively low and the grid stress is moderate.

In contrast, the growing deployment of fast (22–150 kW) and ultra-fast (above 150 kW) EV charging stations introduces significant power surges in very short time intervals, posing greater risks to grid stability and requiring faster, more dynamic energy management solutions. Despite this trend, limited research has addressed the specific coordination challenges associated with high-power charging infrastructure.

On the renewable side, extensive studies have examined the variability and uncertainty of wind energy and its integration into power systems (Pappala et al., 2009; Ullah et al., 2024). While some works incorporate energy storage or flexible demand to absorb surplus wind energy, few explicitly leverage EVs as flexible loads aligned with wind generation availability.

Additionally, while the Vehicle-to-Grid (V2G) concept has been widely investigated (Kempton and Tomić, 2005; He et al.), it often assumes bidirectional chargers and user participation for discharging, which may not always be feasible at a large scale. Literature has proposed aligning EV charging with renewable availability (Wang and Cheng, 2020; Pan and Liu, 2022), but again, these studies often rely on small-scale test systems, do not incorporate the physical constraints of real distribution grids, or neglect the operational characteristics of fast-charging technologies.

Recent studies have further broadened the scope of coordinated energy management strategies combining EV charging with wind and other renewables. For example Su et al. (2025) proposes a multi-service dispatch framework integrating EVs to provide flexibility in network constraints, covering services beyond simple charging and discharging, and show marked reductions in grid stresses under uncertainty. Abdelsalam et al. (2025) develop an efficient charging coordination scheme specifically accounting for fast charging behaviour, forecasting EV usage, and aligning charging schedules with renewable availability. Wu et al. (2025) employ distributed deep reinforcement learning in multiple microgrids with distributed renewables to enhance resilience via optimal EV coordination, managing uncertainties in load, transport network, and renewables. Kumar et al. (2025) design a predictive control strategy incorporating EVs and an interline power flow controller for voltage/frequency regulation in a wind-integrated deregulated power system, obtaining significant improvement in dynamic performance. These works address some of the gaps in earlier literature, such as incorporating higher charging powers beyond slow residential levels, considering real network constraints, and accommodating large-scale stochastic renewables, bringing them closer to the assumptions and ambitions of our framework, which explicitly targets fast and ultra-fast charging technologies.

Although significant progress has been made in coordinating EV charging with renewable energy sources, a clear gap remains in the literature. To the best of our knowledge, the specific challenges associated with integrating fast and ultra-fast EV charging—characterized by high instantaneous power demand—with wind generation in distribution networks have not been thoroughly investigated. Most existing studies focus on slow or residential charging scenarios, overlooking the more demanding and impactful nature of high-power charging. Furthermore, there is a lack of comprehensive frameworks that integrate data-driven forecasting, optimization-based scheduling, and grid-level validation to ensure reliable and efficient operation under these conditions.

12.1.2 Chapter contributions

This work proposes an optimization-based framework to address these challenges by coordinating EV charging in accordance with wind power availability, with the aim of improving both grid stability and renewable energy integration. A central contribution of this work is the introduction of the *Wind-to-Vehicle (W2V)* concept, which complements existing Vehicle-to-Grid (V2G) strategies. While V2G focuses on discharging EVs to support the grid, W2V emphasizes the optimal charging of EVs using surplus wind energy, effectively treating EVs as flexible, controllable loads that absorb local renewable generation.

The proposed model incorporates the physical characteristics of the distribution network, wind turbine dynamics, and fast-charging EV profiles. It relies on forecasting techniques to anticipate base loads and EV connection patterns, which feed into an optimization model that schedules EV charging and discharging activities. The approach is evaluated through power flow simulations to assess the impact of different strategies on network performance.

The main contributions of this work are as follows:

- *A novel Wind-to-Vehicle (W2V) scheduling framework* that aligns EV charging profiles with wind energy availability, enhancing the self-consumption of renewable energy and reducing curtailment.
- *Integration of forecasting techniques* to predict base load and EV connection periods, enabling proactive scheduling under uncertainty.
- *Formulation of an optimization model* that generates coordinated charging and discharging profiles for EVs, considering both user requirements and system constraints.

- *Validation* via *power flow analysis*, which demonstrates that the proposed strategy reduces peak loads, avoids undervoltages, and maintains grid stability.
- *Comparative evaluation of three scenarios*: (i) uncontrolled EV charging without wind, (ii) uncontrolled EV charging with wind, and (iii) the proposed W2V strategy with wind, showing the technical advantages of coordinated operation.

12.1.3 Chapter outline

The remainder of this chapter is structured as follows. Section 12.2 introduces the problem definition and system architecture, including the distribution grid, wind generation, and EV charging models. Section 12.3 presents the proposed W2V energy management framework, covering forecasting, optimization, and power flow validation. Section 12.4 discusses the case studies and simulation results that demonstrate the effectiveness of the W2V strategy under different operating scenarios. Finally, Section 12.5 summarizes the main findings and highlights directions for future research.

12.2 Problem definition and system architecture

The increasing penetration of renewable energy sources and electric vehicles (EVs) in distribution grids introduces new operational challenges. In particular, wind energy, despite its environmental benefits, exhibits intermittent and unpredictable behavior due to its dependence on meteorological conditions. Simultaneously, the growing deployment of fast and ultra-fast EV chargers has intensified the stochasticity of power demand, as large charging loads may be connected or disconnected with little notice.

These dual sources of uncertainty, unpredictable wind generation and the sporadic high-power demand from EVs, complicate grid operation and planning. When not properly managed, they can lead to voltage instabilities, feeder congestion, increased power losses, and inefficient use of renewable resources. Therefore, advanced energy management strategies are required to ensure a balance between supply and demand, improve reliability, and maximize renewable energy utilization. Fig. 12.1 presents an overview of the proposed system architecture, highlighting the interaction between wind generation, the distribution network, and fast/ultra-fast EV chargers coordinated through local EMS units and the W2V strategy.

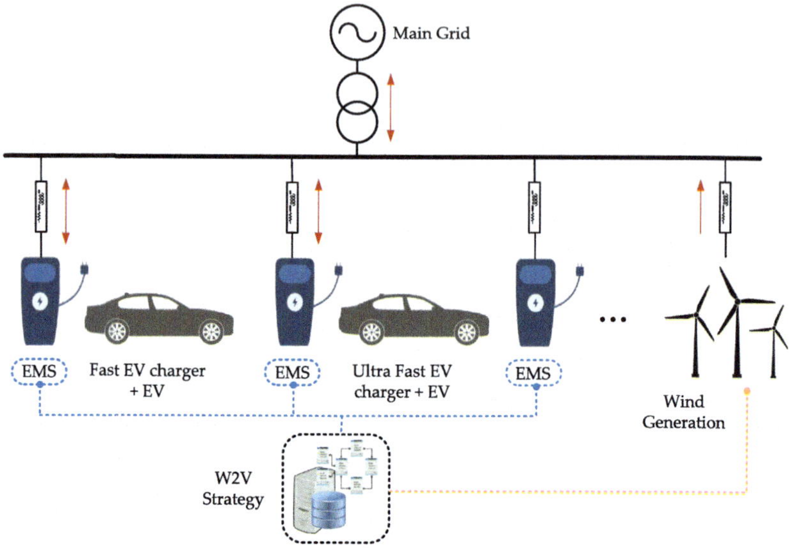

Figure 12.1 System architecture showing wind generation, distribution network, EV chargers, and the W2V strategy.

This work addresses these challenges by formulating an optimization-based coordination mechanism between wind generation and EV charging in a distribution network. The proposed framework seeks to:

- **Maximize the utilization of wind energy** by aligning EV charging schedules with periods of high wind availability, thereby minimizing renewable curtailment and enhancing sustainability.
- **Maintain network stability**, ensuring that voltage limits are respected under various operating scenarios.
- **Incorporate uncertainty** in both wind power generation and EV arrival/departure behavior through the use of machine learning-based forecasting techniques, which support a predictive scheduling phase.

The proposed model integrates the physical constraints of the distribution network, wind turbine characteristics, and the profiles of fast and ultra-fast EV charging. The scheduling framework is structured in two stages:

1. An optimization stage that aims to maximize the utilization of available wind energy for EV charging, while also improving the overall load profile of the distribution grid.
2. A power flow stage that ensures the feasibility of the proposed schedule by verifying grid constraints and balancing renewable energy use across the network.

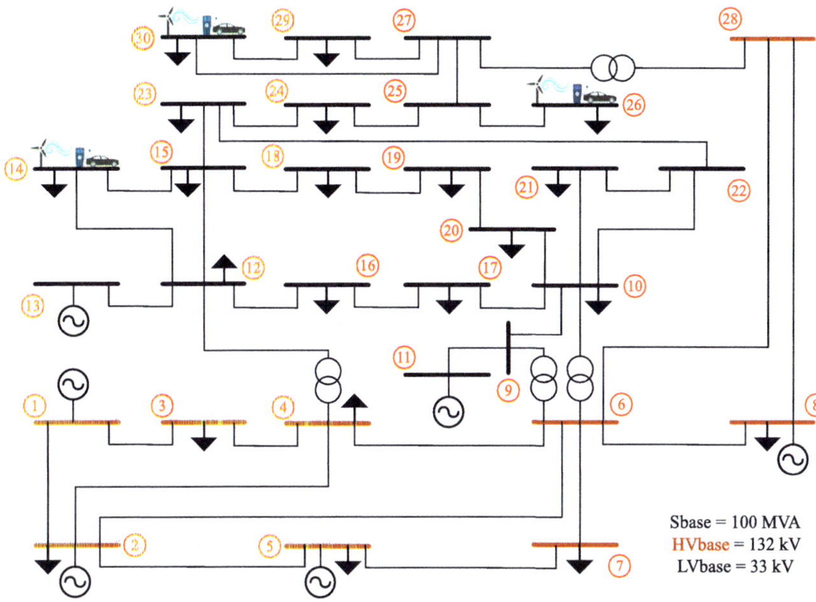

Figure 12.2 IEEE 30-bus single line diagram.

In the next subsection, we present the grid and wind power model description as well as the EV Chargers characteristics.

12.2.1 Description of the distribution grid and components

The IEEE 30-bus system single line diagram used for the case studies in this Chapter is shown in Fig. 12.2. The IEEE 30-bus system has two generators, and four synchronous condensers. This test system is part of the American Electric Grid as it was in 1961, and it has been widely used for planning studies and state estimation purposes (Peyghami et al., 2019). In this particular study, the IEEE 30-bus system is modified, such that nodes 14, 26, and 30 are assumed to be equipped with wind turbines and EV charging stations, as illustrated in Fig. 12.2. These nodes serve as representative locations for integrating renewable energy sources and flexible demand into the distribution network.

Table 12.1 summarizes the nominal bus data parameters selected for the case studies, which were taken from Saadat (1999). The line data parameters can be found in the same reference.

For the test cases, the loads in the high-voltage buses remain fixed; however, a load profile is used for the low-voltage buses active power load,

Table 12.1 IEEE 30-Bus system nominal bus data parameters.

Bus	Bus type	Voltage mag. (p.u.)	Angle	P load (MW)	Q load (Mvar)	P Gen (MW)	Q Gen (Mvar)	Qmin (Mvar)	Qmax (Mvar)	Injected Mvar
1	Slack	1.03	0	0	0	0	0	0	0	0
2	PV	1.01	0	21.7	12.7	40	0.0	−40.0	50.0	0
3	PQ	1.0	0	2.4	1.2	0	0	0	0	0
4	PQ	1.0	0	7.6	1.6	0	0	0	0	0
5	PV	1.015	0	94.2	19	0	0	−40	40	0
6	PQ	1.0	0	0	0	0	0	0	0	0
7	PQ	1.0	0	22.8	10.9	0	0	0	0	0
8	PV	1.015	0	30	30	0	0	−10	40	0
9	PQ	1.0	0	0	0	0	0	0	0	0
10	PQ	1.0	0	5.8	2.0	0	0	0	0	19
11	PV	1.02	0	0	0	0	0	−6	24	0
12	PQ	1.0	0	11.2	7.5	0	0	0	0	0
13	PV	1.02	0	0	0	0	0	−6	24	0
14	PQ	1.0	0	6.2	1.6	0	0	0	0	0
15	PQ	1.0	0	8.2	2.5	0	0	0	0	0
16	PQ	1.0	0	3.5	1.8	0	0	0	0	0
17	PQ	1.0	0	9	5.8	0	0	0	0	0
18	PQ	1.0	0	3.2	0.9	0	0	0	0	0
19	PQ	1.0	0	9.5	3.4	0	0	0	0	0
20	PQ	1.0	0	2.2	0.7	0	0	0	0	0
21	PQ	1.0	0	17.5	11.2	0	0	0	0	0
22	PQ	1.0	0	0	0	0	0	0	0	0
23	PQ	1.0	0	3.2	1.6	0	0	0	0	0

24	PQ	1.0	0	8.7	6.7	0	0	0	0	4.3
25	PQ	1.0	0	0	0	0	0	0	0	0
26	PQ	1.0	0	3.5	2.3	0	0	0	0	0
27	PQ	1.0	0	0	0	0	0	0	0	0
28	PQ	1.0	0	0	0	0	0	0	0	0
29	PQ	1.0	0	2.4	0.9	0	0	0	0	0
30	PQ	1.0	0	10.6	1.9	0	0	0	0	0

Table 12.2 Load profile data.

Hour	Load (p.u.)	Hour	Load (p.u.)
0	0.8711	12	0.8641
1	0.9362	13	0.8657
2	0.9742	14	0.8811
3	0.9715	15	0.8886
4	0.9348	16	0.8901
5	0.9050	17	0.9062
6	0.9400	18	0.9223
7	1.0000	19	0.9400
8	0.9555	20	0.9530
9	0.9492	21	0.9085
10	0.9202	22	0.8749
11	0.8838	23	0.8582

being the load value in Table 12.1 the base magnitude of the load profile. The low-voltage loads modeling is as follows,

$$\mathrm{LP}_n = L_n^{\mathrm{nominal}} * \mathrm{L}_{\mathrm{profile}}^{p.u.} \tag{12.1}$$

where L_n^{nominal} is the nominal active power load magnitude given in Table 12.1 for the bus n, $\mathrm{L}_{\mathrm{profile}}^{p.u.}$ is the load profile given in per unit, and LP_n is the 24-hour load profile in real values for bus n. The daily p.u. load profile is taken from the National Control Center - CNES. The load profile is given every hour, but an interpolation is made such that the samples are assumed to be taken every 5 minutes. Table 12.2 shows data from the load profile.

12.2.2 Modeling of wind power generation

Depending on the technical characteristics of the wind turbine generator (WTG), it is possible to estimate the WTG energy production by using its characteristic power curve. The power curve is a graphical representation of the WTG power output as a function of wind speed, and gives three key turbine operation points: (i) cut-in speed, that is, the minimum wind speed at which the turbine will inject useful power; (ii) rated wind speed, that is, the wind speed at which the turbine is generating at maximum power capacity; and (iii) cut-out speed, that is, the maximum wind speed at which the turbine is allowed to deliver power. This power curve is given by the manufacturers and is essential for planning, performance analysis, and maintenance strategies (Manwell and McGowan, 2010).

On the other hand, to emulate the variable nature of the wind, three approaches can be followed: 1) time-domain wind modeling; 2) historical wind speed data; or 3) statistical techniques using limited wind speed data. In the first approach, the wind behavior is defined by ramp, gust, and random functions; in the second approach, it is assumed that a large quantity of wind data has been collected by proper measurements and instrumentation; Finally, for the statistical approach, the frequency of occurrence of wind speeds is described by a probability distribution function (Manwell and McGowan, 2010).

In this work, the statistical approach is used due to its flexibility to emulate the wind behavior. For this purpose, the Weibull probability density function is used. This function has been widely applied in the literature for wind speed analysis, and is given by the following equation (Manwell and McGowan, 2010),

$$p(U) = \left(\frac{k_{sf}}{c}\right)\left(\frac{U}{c}\right)^{k_{sf}^{-1}} e^{-\left(\frac{U}{c}\right)^{k_{sf}}} \tag{12.2}$$

where k_{sf} is a shape factor, U is the wind speed, and c is a scale factor. Using both the WTG power curve and the wind distribution, the WTG output power can be calculated and used for the analysis calculations. In this way, Table 12.3 shows the power curve characteristics in p.u. of an 8 MW WTG selected to be used in the case studies.

Table 12.3 Power curve characteristics.

Wind (m/s)	Power (p.u.)
0	0
4 (cut-in wind speed)	0.0138
5	0.0750
6	0.1425
7	0.2375
8	0.3625
9	0.5194
10	0.7037
11	0.8938
12	0.9831
13	0.9962
14–25 (rated wind speed)	1.0000
>25 (cut-out wind speed)	0

12.2.3 Fast and ultra-fast EV charging behaviors

EV chargers and their associated vehicles are frequently represented in distribution-grid studies by a simplified, linear link between charging power and State of Charge (SoC). In that scheme the charger delivers a fixed power level from arrival until the requested departure SoC. A variant assumes a constant-power segment that tapers only after a predefined SoC threshold (Ziyat et al., 2023). Although both simplifications are adequate for slow and some fast chargers, they become inaccurate for fast and ultra-fast technologies. To obtain a more faithful description, we analyze six measured profiles drawn from two manufacturers: Tesla and BYD. The real-world data published in Fastned Support (2024) form the basis of our evaluation.

Vehicle and charger specifications: Fig. 12.3 compares three models from each brand. Fig. 12.3a summarizes, via bar charts, the peak power each vehicle can accept together with its battery capacity. The six models tolerate between 88 and 250 kW, while their usable capacities range from 50 to 86.4 kWh. Suitable charging equipment includes Tesla's 250 kW Supercharger V3 for the Tesla models and ABB's Terra 125 (125 kW) for the BYD vehicles (Diaz-Londono et al., 2023).

Power–SoC characterization: The available measurements do not start at 0% SoC nor reach 100%. They begin at higher initial SoC values and stop before full charge. To obtain complete curves from 1% to 100% SoC we linearly extrapolated the missing sections. The resulting profiles, plotted in Fig. 12.3b, highlight the diversity inherent in high-power charging: the energy-management system (EMS) of each EV reaches its maximum admissible power at different points—some immediately after plug-in, others near 40% SoC, and still others at intermediate levels. These curves correspond to optimal cell temperatures ($\approx 30°$ C); deviations either colder or hotter would lower the achievable power.

Implications for charging duration: Using the reconstructed power curves, we computed the time required to charge each vehicle from 1% to 100% SoC. Although real users typically plug in around 30% SoC (Diaz-Londono et al., 2023), the full-range calculation provides an upper-bound comparison. Fig. 12.3c shows that a complete charge spans 53–65 minutes, depending on model and brand. The corresponding SoC trajectories are given in Fig. 12.3d: SoC rises steeply at first, while power is high, and gradually flattens as the taper region is reached.

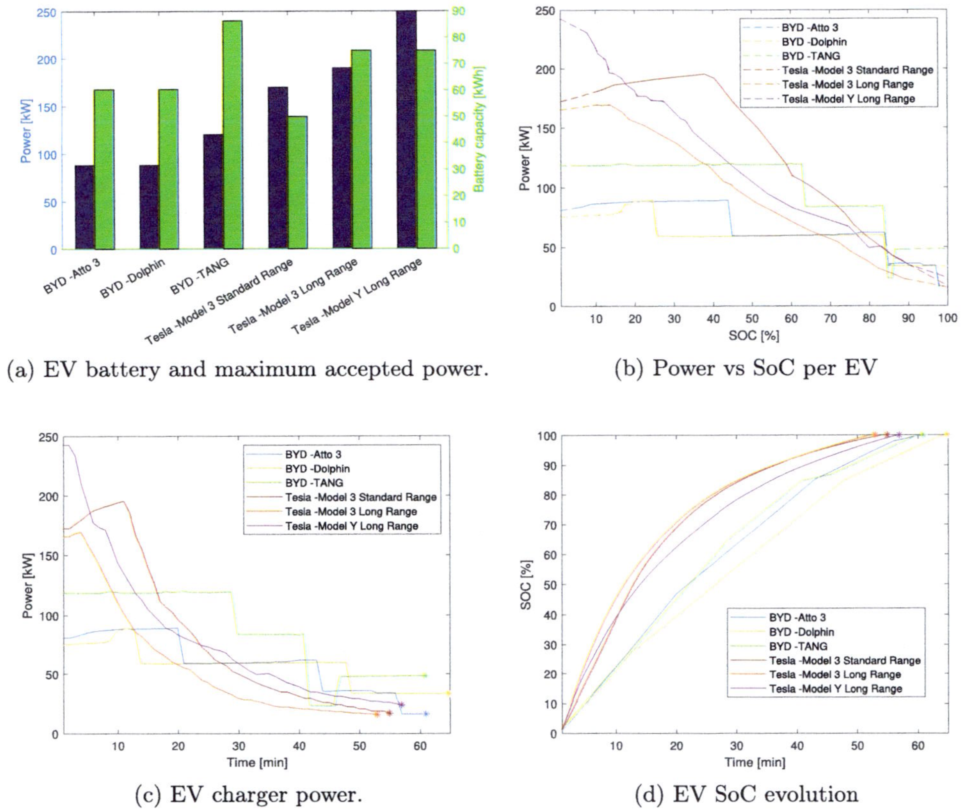

(a) EV battery and maximum accepted power.

(b) Power vs SoC per EV

(c) EV charger power.

(d) EV SoC evolution

Figure 12.3 EV characteristics for some BYD and Tesla models. (a) EV battery and maximum accepted power. (b) Power versus SoC per EV. (c) EV charger power. (d) EV SoC evolution.

12.3 Energy management framework of the W2V strategy

This section presents the proposed framework for the coordinated management of EV charging and wind energy integration in distribution grids, named the W2V strategy. The framework is designed to address the uncertainties and operational constraints introduced by high shares of wind energy and the increasing deployment of fast and ultra-fast EV chargers. The framework is composed of three main stages, as illustrated in Fig. 12.4.

- **Forecasting stage:** This initial step relies on machine learning models to predict key variables required for optimal scheduling. These include the connection times and durations of EVs and the overall demand on nodes of the distribution network.
- **Optimization stage:** Based on the forecasts, this stage solves a scheduling problem to determine the optimal charging power profiles for all connected EVs. The model explicitly considers the physical and operational characteristics of fast and ultra-fast chargers, ensuring feasible and efficient charging behavior aligned with grid and renewable availability.
- **Power flow validation:** In the final step, the scheduled charging actions are validated through a power flow analysis to verify that all

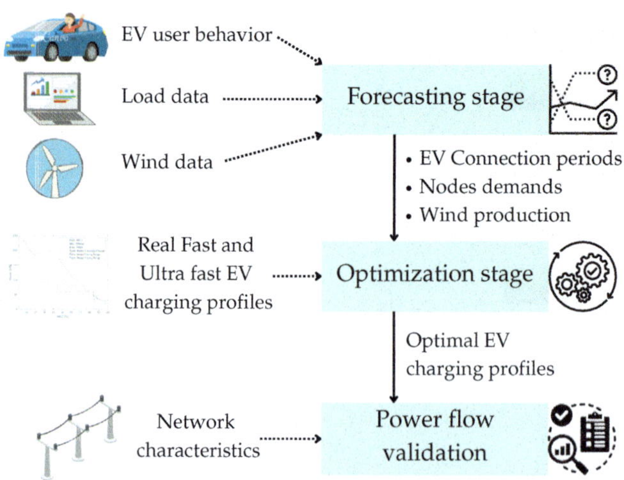

Figure 12.4 Methodological framework of the proposed W2V strategy, including forecasting, optimization, and power flow validation.

network constraints are satisfied, such as voltage limits, line capacities, and transformer ratings. This ensures that the resulting strategy is technically implementable in the real distribution grid.

The combination of predictive analytics, optimization, and network validation provides a robust tool for enhancing the reliability and efficiency of multi-energy systems with high wind penetration.

12.3.1 Machine learning forecasting

12.3.1.1 Occupancy prediction for EV charging stations

This subsection presents an approach for predicting the occupancy profile of individual EV chargers, as inspired by the methodology proposed in Koruturk et al. (2024). Due to the limited availability of datasets for fast and ultra-fast charging infrastructure, the analysis leverages historical data from slow charging events. Specifically, the model is trained using a publicly accessible dataset provided by the Electric Nation project (Electric Nation Customer Trial Final), which comprises nearly 50,000 charging sessions recorded in the United Kingdom between January 2017 and December 2018.

To ensure the reliability of the training dataset, multiple chargers within the dataset were analyzed to determine the best candidate. The charger with the highest data completeness, i.e., achieving a fill ratio of 90% was selected as the reference unit for generating the training and testing datasets.

The predictive model is designed to forecast the binary occupancy status of the charger, one time step at a time, based on time series data from the previous 14 days. The prediction task is thus formulated as a binary classification problem where the target is to predict the occupancy at the first time step of the test day using the historical occupancy sequence as input features.

The real occupancy dynamics of an EV charger can be modeled as:

$$y_{k-1} = f(\gamma_{k-1}), \tag{12.3}$$

where γ_{k-1} represents the latent user intent to utilize the charger at the previous time step, y_{k-1} is the observed occupancy state at that time, and $f(\circ)$ is an unknown function describing the mapping between them. The index k indicates discrete time steps.

The predictor's mathematical model is expressed as:

$$\hat{y}_k = \sigma(\overline{y}^{\mathsf{T}}), \tag{12.4}$$

where, $\widehat{y}_k$ denotes the predicted occupancy at time step k, $\overline{y}^\mathsf{T}$ is the row vector comprising past observed occupancy values used as input, and $\sigma(\circ)$ is the machine learning function that approximates the mapping from historical patterns to future occupancy.

Two machine learning models are implemented in this study: the XGBoost classifier and regressor. The classifier operates based on logistic regression and minimizes the binary cross-entropy loss:

$$J_{\text{class}} = -\frac{1}{N}\sum_{n=1}^{N}\left(y_n \log(p_n) + (1 - y_n)\log(1 - p_n)\right), \tag{12.5}$$

where N is the number of data points, y_n is the true occupancy label, and p_n is the predicted probability.

Alternatively, the regressor employs a linear regression approach and minimizes the mean squared error (MSE) loss:

$$J_{reg} = \frac{1}{N}\sum_{n=1}^{N}\left(y_n - \widehat{y}_n\right)^2, \tag{12.6}$$

where $\widehat{y}_n$ is the predicted continuous occupancy value for each instance n.

Fig. 12.5 illustrates an example of the model's performance. The binary occupancy profile predicted by the model (in blue) is compared against the actual observed data (in red). The figure highlights the model's capability to accurately capture the arrival a_j and departure d_j time of the EV_j, although slight timing mismatches are visible. This illustrates the inherent difficulty of

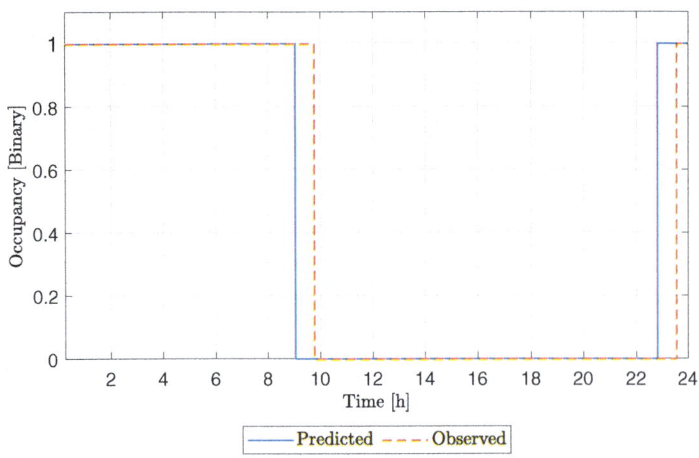

Figure 12.5 Example of the predicted EV occupancy profile.

capturing user behavior variability in EV charging applications but also confirms the potential of learning-based predictors to provide useful forecasts for smart energy management systems. These predicted arrival $\widetilde{a}_j$ and departure $\widetilde{d}_j$ times are used as key inputs in the scheduling process of EV charging, enabling the optimization of charging strategies while considering user availability constraints.

12.3.1.2 Distribution grid load demand forecasting

To accurately represent the non-flexible (baseline) load at the nodes of the distribution network, real-world consumption data is employed. Specifically, this study utilizes high-resolution load profiles obtained from the RTE dataset (RTE, 2025), which provides active power demand measurements.

Given the stochastic and time-dependent nature of residential and commercial electricity consumption, a data-driven forecasting approach is proposed to estimate the inflexible demand in future time steps. Anticipating future load levels is essential for both optimal energy scheduling and maintaining the operational reliability of the distribution grid, particularly under high renewable and EV penetration.

To this end, an artificial neural network (ANN) is implemented. The selected architecture is a feedforward ANN consisting of a single hidden layer equipped with a sigmoid activation function, trained to learn the nonlinear mapping between past demand values and future consumption. The model input comprises the most recent 144 time steps (equivalent to 24 hours of historical data with a 10-minute resolution), capturing diurnal load patterns and daily consumption variability. The target output is the predicted load demand $\widetilde{P}^l_{g,k}$ at each node g for the next k time slot, i.e., 10-minute interval.

The data set is partitioned into training and validation subsets using a split 70%–30%. The training set is used to optimize the ANN weights through backpropagation and stochastic gradient descent, while the validation set is employed to assess generalization performance and avoid overfitting. The predictor order (input sequence length) of 144 was found to provide the best performance based on empirical trials, balancing complexity and accuracy.

This short-term load forecasting strategy provides reliable demand estimates that are then used as inputs to the scheduling algorithm described in the next subsection.

12.3.2 Optimization formulation for the W2V strategy

The primary goal of the W2V scheduling strategy is to maximize the utilization of locally generated wind energy on the demand side, encompassing both the inflexible base load $\widetilde{P}^{l}_{g,k}$ (which is treated as a forecasted input, as previously described) and the flexible EV consumption (which is modeled as a decision variable in the optimization problem). Simultaneously, the schedule aims to allow EVs to supply energy back to the grid during periods of insufficient wind generation. Nevertheless, it remains essential that all EVs fulfill their charging requirements within their respective connection windows.

Consequently, the cost function can also be interpreted as the minimization of energy exchange with the main grid. On one hand, this encourages the use of local wind generation, thereby avoiding both curtailment and the injection of excess energy into the grid. The latter is considered suboptimal due to the absence of favorable regulatory frameworks and the typically lower compensation rates for energy exports.

The objective of the proposed cost function is to minimize the ℓ_2 norm of the mismatch between the available wind generation $P^{w}_{g,k}$ and the total power demand at each time step. Then, the total demand is composed of the inflexible load $\widetilde{P}^{l}_{g,k}$ and the aggregated EV charging and discharging power $\sum \phi_{g,k}$ at each node g and time slot k. Accordingly, the objective function is defined as:

$$\min \| P^{w}_{g,k} - \left(\widetilde{P}^{l}_{g,k} + \phi_{g,k}\right) \|_2 \tag{12.7}$$

The variable $\phi_{g,k}$ represents the aggregated power of all EV chargers connected to node g, which are collected in the set Φ_g. Accordingly, $\phi_{g,k}$ is defined as:

$$\phi_{g,k} = \sum_{i} P^{c}_{i,k} - P^{d}_{i,k} \forall i \in \Phi_g \tag{12.8}$$

To reduce the computational complexity of the optimization problem and improve tractability, particularly in the presence of binary and integer decision variables, the original quadratic cost function based on the ℓ_2 norm is reformulated using the ℓ_1 norm. While the ℓ_2 norm minimizes the square of the mismatch between wind generation and total demand, resulting in a smooth and differentiable objective function, it leads to a non-linear and potentially non-convex formulation when combined with other discrete constraints.

In contrast, the ℓ_1 norm, defined as the sum of the absolute values of the mismatch, allows the problem to be linearized. This is achieved by introducing auxiliary variables ($\Gamma_{g,k}$) that represent the absolute deviation at each node and each time step. The resulting linear formulation enables the use of efficient MILP solvers, which are more robust and scalable than their non-linear counterparts. Then, the cost function is reformulated as:

$$\min \sum_{k} \sum_{g} \Gamma_{g,k} \tag{12.9}$$

$$\text{s.t.} \Gamma_{g,k} \geq P^{w}_{g,k} - \left(P^{l}_{g,k} + \phi_{g,k}\right) \tag{12.9b}$$

$$\Gamma_{g,k} \geq - P^{w}_{g,k} + \left(P^{l}_{g,k} + \phi_{g,k}\right) \tag{12.9c}$$

Although the ℓ_1 norm does not penalize large deviations as strongly as the ℓ_2 norm, it offers a computationally efficient alternative that still captures the essential dynamics of the mismatch. To further ensure system reliability, a maximum allowable deviation ($\Gamma_{\max,g,k}$) is defined for each node equipped with EV chargers, limiting the extent to which the local demand can diverge from the available wind generation, forumalet as:

$$\Gamma_{g,k} \leq \Gamma_{\max,g,k} \tag{12.10}$$

The modeling of EV charger dynamics considers three distinct operating conditions, defined by the estimated arrival time $\widetilde{a}_j$, departure time $\widetilde{d}_j$, and a binary variable $\xi_{i,k}$ indicating whether an EV is connected to the i-th charger at time slot k. The first condition defines the state of charge (SoC) of the EV upon arrival. The second condition governs the evolution of the energy stored in the EV battery while the vehicle remains connected. In this phase, the SoC dynamics depend on the charging power $P^{c}_{i,k}$ and discharging power $P^{d}_{i,k}$, adjusted by their respective efficiencies η^{c} and η^{d}. The change in stored energy also accounts for the battery capacity E_j of EV$_j$ and the time step duration Δt. The third condition applies when no EV is connected or during the departure phase, at which point the SoC remains unchanged. Notice that the evolution of $x_{i,k}$, representing the state of charge of the charger (not the EV), follows the formulation in Diaz-Londono et al. (2026). These dynamics are formalized as follows:

$$x_{i,k+1} = \begin{cases} \mathrm{SoC}_{j,\widetilde{a}_j} & \text{if} \quad k = \widetilde{a}_j, \\ x_{i,k} + \dfrac{\Delta t}{E_j}\left(\eta^c P^c_{i,k} - \dfrac{P^d_{i,k}}{\eta^d}\right) & \text{if} \quad \widetilde{a}_j < k < \widetilde{d}_j, \\ 0 & \text{if} \quad \xi_{i,k} = 0 \vee k = \widetilde{d}_j. \end{cases} \tag{12.11}$$

The formulation enforces a constraint ensuring that, at the departure time $\widetilde{d}_j$, the SoC of each EV remains within a user-defined range. Specifically, the final SoC must lie between a minimum threshold $S\check{o}C_{j,d_j}$ and a maximum limit $\widehat{SoC}_{j,d_j}$, as specified by the EV owner:

$$S\check{o}C_{j,d_j} \leq x_{i,d_j} \leq \widehat{SoC}_{j,d_j} \tag{12.12}$$

The SoC at the charger is always non-negative and is constrained by the maximum capacity of the battery of the connected EV, as well as by a minimum threshold to prevent deep discharge, as follows:

$$x_{i,\min} \leq x_{i,k} \leq x_{i,\max} \tag{12.13}$$

The set of decision variables includes the charging ($P^c_{i,k}$) and discharging ($P^d_{i,k}$) power levels for each EV, as well as the binary variable $z_{i,k}$, which indicates whether the vehicle is actively charging or discharging during time slot k. To reflect the behavior of the EV Battery Management System (BMS), the model integrates the power–SoC relationship depicted in Fig. 12.3b. This relationship is enforced by dynamically constraining the maximum allowable power denoted as $P^c_{i,\max,SoC_{i,k}}$, based on the current SoC. These limits are applied symmetrically to both charging and discharging processes and are updated at each time step according to the battery's operating point. The resulting constraints are formulated as follows:

$$0 \leq P^c_{i,k} \leq P^c_{i,\max,SoC_{i,k}} z_{i,k} \tag{12.14}$$

$$0 \leq P^d_{i,k} \leq P^d_{i,\max,SoC_{i,k}} \left(1 - z_{i,k}\right) \tag{12.15}$$

Therefore, the full mathematical formulation of the EV scheduling problem is described below.

$$\min_{P^c_{i,k}, P^d_{i,k}, z_{i,k}, \Gamma_{g,k}} \sum_k \sum_g \Gamma_{g,k} \tag{12.16a}$$

$$\text{s.t. Eq.(12.8), Eq.(12.9b), Eq.(12.9c),} \tag{12.16b}$$

$$\text{Eq.(12.10), Eq.(12.11), Eq.(12.12),}$$

$$\text{Eq.(12.13), Eq.(12.14), Eq.(12.10)} \tag{12.16c}$$

$$\forall k = 1, 2, \ldots, H; i = 1, 2, \ldots, I;$$

$$j = 1, 2, \ldots, J; g = 1, 2, \ldots, G;$$

These optimized EV charging profiles resulting from the scheduling formulation are integrated into the distribution network model and used as input for the power flow analysis presented in the following section.

12.3.3 Power flow problem formulation

The power flow is one of the most common and frequently performed mathematical techniques used in practical power systems; it plays an important role in analysis, planning and designing the future expansion of power systems as well as for economic scheduling, monitoring, and control of existing systems, among others (Saadat, 2002).

In conventional power systems, the power flow studies are well-established with standard mathematical models and solving methodologies (Stevenson and Grainger, 2003). In those formulations, every bus falls into one type, for AC systems.

- Slack bus: One bus, known as slack or swing bus, is taken as reference where the magnitude and phase angle for the voltage are specified. This bus makes up the difference between the scheduled loads and the generated power that are caused by losses in network.
- Load buses: At these buses the active and reactive powers are specified. The magnitude and the phase angle of the bus voltage are unknown. These buses are called PQ buses.
- Regulated buses: These buses are the generator buses. At these buses, the real power and voltage magnitude are specified. The phase angles of the voltages and the reactive power are to be determined. The limits on the value of the reactive power are also specified. These buses are called PV buses.

In order to solve this problem, the power flow equations are required. Considering a typical bus of a power system network, using the system admittance matrix $\mathbf{Y}_{\text{bus}}$, the nodal equations can be written as follows,

$$\mathbf{I} = \mathbf{Y}_{\text{bus}}\mathbf{V} \tag{12.17}$$

where I is the vector of source currents injected into each bus and V is the vector of bus voltages, then for bus n, the n-th equation in Eq. (12.17) is,

$$I_n = \sum_{m=1}^{M} Y_{nm} V_m \tag{12.18}$$

where M is the number of buses in the system. The apparent power delivered to bus n is,

$$S_n = P_n + jQ_n = V_n I_n^* \tag{12.19}$$

then, substituting Eq. (12.18) in Eq. (12.19),

$$P_n + jQ_n = V_n \left[\sum_{m=1}^{M} Y_{nm} V_m \right]^* \tag{12.20}$$

using the following notation,

$$V_m = |V_m| e^{j\delta_m} \tag{12.21}$$

$$Y_{nm} = |Y_{nm}| e^{j\theta_{nm}} \tag{12.22}$$

Eq. (12.20) becomes,

$$P_n + jQ_n = |V_n| \sum_{m=1}^{M} |Y_{nm}||V_m| e^{j(\delta_n - \delta_m - \theta_{nm})} \tag{12.23}$$

furthermore, taking the real and imaginary parts of Eq. (12.23), it can be written as follows,

$$P_n = \sum_{m=1}^{M} |V_n||V_m||Y_{nm}| \cos(\delta_n - \delta_m - \theta_{nm}) \tag{12.24}$$

$$Q_n = \sum_{m=1}^{M} |V_n||V_m||Y_{nm}| \sin(\delta_n - \delta_m - \theta_{nm}) \tag{12.25}$$

these equations constitute a set of nonlinear algebraic equations in terms of the independent variables, voltage magnitude and phase angle, which can be used to solve the power flow problem.

The power flow problem solution can be computed with different numerical techniques, being one of the most used methods the conventional Newton–Raphson method. In this method, Eqs. (12.24) and (12.25) are expanded in Taylor's series about an initial estimate, and neglecting all higher-order terms results in the following set of linear equations (Saadat, 2002),

$$\begin{bmatrix} \Delta P_n \\ \Delta Q_n \end{bmatrix} = \begin{bmatrix} \mathbf{J}_1\mathbf{J}_2 \\ \mathbf{J}_3\mathbf{J}_4 \end{bmatrix} \begin{bmatrix} \Delta \delta_n \\ \Delta |\mathbf{V}_n| \end{bmatrix} \tag{12.26}$$

where the terms ΔP_n and ΔQ_n are the difference between the scheduled and calculated values. The Jacobian matrix gives the linearized relationship between small changes in voltage angle $\Delta\delta_n$ and voltage magnitude $\Delta|V_n|$ with small changes in real and reactive power ΔP_n and ΔQ_n (Saadat, 2002). Finally, the new estimates for bus voltages are,

$$\delta_n^{(k+1)} = \delta_n^{(k)} + \Delta\delta_n^{(k)} \tag{12.27}$$

$$|V_n^{(k+1)}| = |V_n^{(k)}| + \Delta|V_n^{(k)}| \tag{12.28}$$

Using the new voltages, P_n and Q_n are computed again and the process continues until ΔP_n and ΔQ_n are below the specified accuracy (Saadat, 2002; Stevenson and Grainger, 2003).

This section has presented the general power flow problem formulation, which serves as a foundation for evaluating the impact of EV charging schedules on a distribution network characterized including penetration of wind generation.

12.4 Case studies and simulations

This section presents the case studies designed to evaluate the proposed W2V concept. The assessment focuses on two main aspects: the performance of the optimization algorithm for EV scheduling, and its impact on the electrical grid, which is analyzed through power flow simulations. Accordingly, the section is structured into two subsections: one dedicated to analyzing the scheduling outcomes, and another focused on evaluating the resulting power flow in the distribution network.

The effectiveness of the W2V strategy is compared against an uncoordinated charging approach, referred to as the Minimum Time (MT) strategy. In the MT strategy, each EV charger attempts to charge the connected EV at the maximum allowable power until the battery reaches its full capacity $SoC_{i,\max}$, or earlier if the EV is disconnected before that point, presenting a typical uncontrolled charging behavior.

Three simulation scenarios are considered:

- **MT scenario:** Evaluates the impact of uncontrolled EV charging in a network without wind generation.
- **MT-W scenario:** Analyzes the effect of uncontrolled EV charging in the presence of wind generation.
- **W2V scenario:** Assesses the impact of the optimized EV charging (W2V strategy) in a network with wind generation.

The dataset used for simulating EV charging events includes diverse use cases, covering residential, workplace, and public charging patterns (Diaz-Londono et al., 2025). Furthermore, the simulation process is structured into two main steps, reflecting the separation between the optimization stage and the subsequent power flow evaluation:

In the *first step*, the scheduling optimization is executed based on forecasted information. Prior to solving the optimization problem, the base load at each node is predicted, as well as the arrival $\widetilde{a}_j$ and departure $\widetilde{d}_j$ times of EVs, which are inferred from historical occupancy data. For each forecasted charging event, an EV is randomly assigned from a pool of real EV profiles (as shown in Fig. 12.3), and its characteristics—such as battery capacity and efficiency—are used to parameterize the charging session. The initial SoC is defined as a random value proportional to the selected EV's capacity, while the desired final SoC is also sampled randomly to reflect user variability. With these inputs, the optimization algorithm computes the optimal charging and discharging power profiles $P_{i,k}^{c*}$ and $P_{i,k}^{d*}$ across time for each EV and charger. The resulting profiles are then aggregated by node, forming the input EV demand profiles that are used in the power flow analysis.

In the *second step*, the power flow simulation is performed using the electrical network model. The power flow is formulated using the network topology and generation and load parameters, and the optimized EV demand profiles obtained in the previous step are included as loads in the corresponding nodes. As illustrated in Fig. 12.2, EV charging stations are connected at nodes 14, 26, and 30. The aggregated EV power profiles at these nodes are added on top of the base profile loads, and the power flow equations are solved to calculate the nodes voltage magnitude and angle, and the active and reactive power injections throughout the grid. The results obtained clearly present the steady-state system behavior, allowing an impact assessment, in terms of grid performance and voltage stability, of the charging strategy under different operative scenarios.

12.4.1 Performance analysis of the EV scheduling strategy

This subsection analyzes the performance of the proposed W2V optimization scheduling strategy. The goal is to evaluate how effectively the W2V approach leverages wind generation while ensuring the fulfillment of EV charging requirements within the available connection windows. To this end, key performance indicators are considered, including the total energy supplied to EVs, the utilization of local wind generation, and the extent of vehicle-to-grid (V2G) operation. These metrics allow for a detailed comparison between smart W2V and uncoordinated MT charging behaviors, as represented by the MT and MT-W scenarios, and highlight the benefits of smart scheduling in the context of renewable energy integration.

The analysis begins from the perspective of a single EV charger. In Fig. 12.6, two examples of charging behavior are shown for the MT and W2V strategies. These examples illustrate the power profile defined by the charger and the corresponding SoC evolution tracked at the charger for the connected EV.

As observed, under the MT strategy, the charger operates at the maximum allowed power immediately after the EV is connected and stops once the target SoC at departure is reached. However, in many real-world cases, such as residential or workplace charging, EVs often remain connected for longer than the time strictly required to complete charging. The W2V strategy takes advantage of this extended connection window to schedule periods of charging and discharging in a flexible manner.

The figures also show the power limit available at each time step, which varies across vehicles and is not constant throughout the session. In

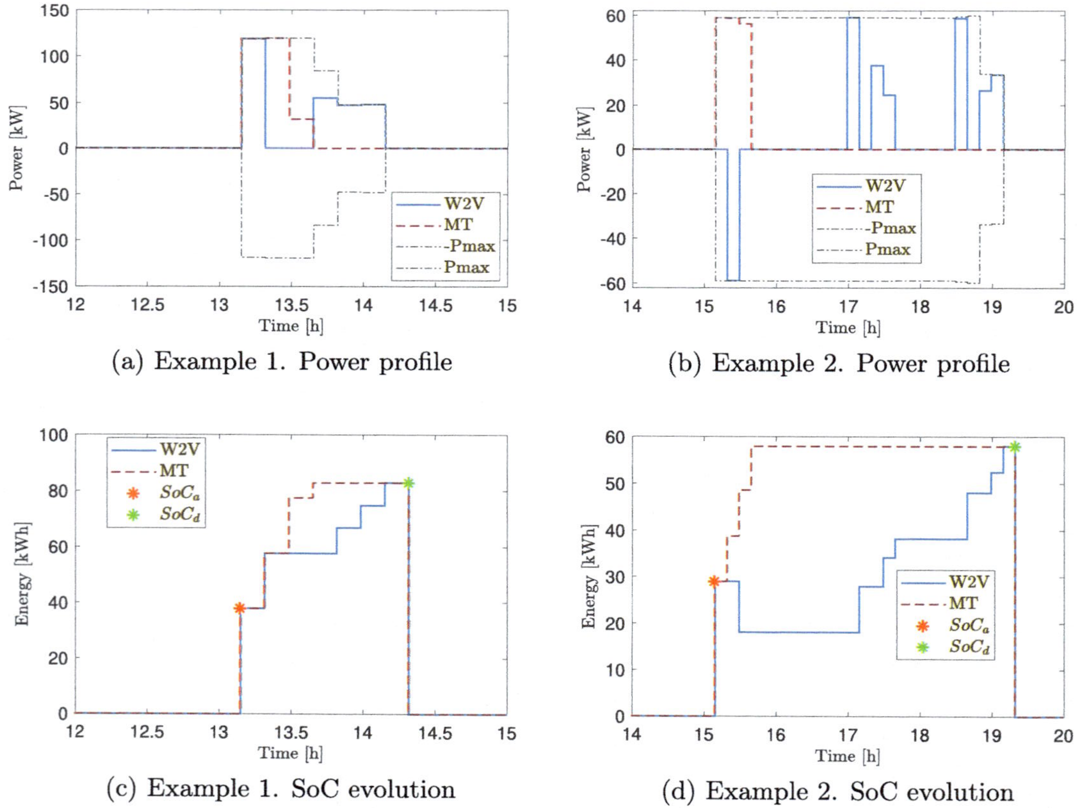

Figure 12.6 Behavior of the EV charger for a single vehicle connection under the MT and W2V strategies. Asterisks indicate the EV's SoC at arrival and departure. (a) Example 1. Power profile. (b) Example 2. Power profile. (c) Example 1. SoC evolution. (d) Example 2. SoC evolution.

Example 1, the W2V strategy only performs charging (i.e., positive power), but not always at the maximum rate. Instead, the charging power is adapted based on the aggregated demand and the conditions of the electrical node. In Example 2, the strategy first discharges the EV battery down to its minimum SoC limit at the beginning of the session, then recharges it in a staged manner. This discharge likely occurs during a period of wind shortage, in order to support local demand.

These examples demonstrate how the smart W2V scheduling strategy leverages flexibility at the individual charger level to optimize system-wide energy usage while respecting user constraints and grid conditions.

This analysis also includes the aggregated response of the W2V strategy. Fig. 12.7 illustrates the power dynamics at node 26. The figure displays several key profiles: the wind generation (*blue line*), which shows high variability with peaks reaching 9 MW; the inflexible load profile (*black line*), characterized by low variability; the EV charging demand scheduled using the W2V strategy (*yellow line*); the net power exchange with the grid under the MT strategy (*red line*); and the net power exchange with the grid under the W2V strategy (*green line*).

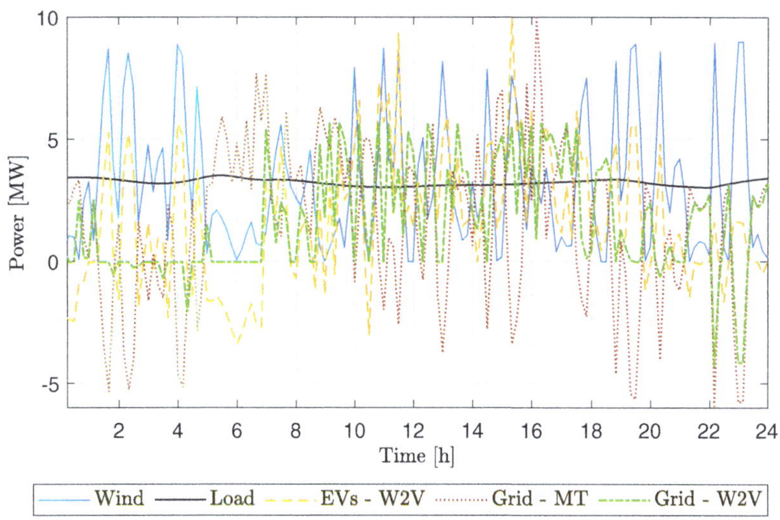

Figure 12.7 Power profiles at node 26 under different charging strategies, before power flow. The W2V strategy shows better alignment between local wind generation and EV demand, reducing grid exchange and peak load.

It can be observed that the MT strategy leads to several instances of negative power exchange, indicating excess energy that must either be curtailed or injected back into the grid due to the lack of coordination. Moreover, demand peaks under the MT strategy can exceed 10 MW, posing potential risks to the grid infrastructure. In contrast, the W2V strategy significantly reduces such peaks, maintaining demand levels around 5 MW, and minimizes the energy injected into the grid. This reflects a better alignment between wind generation and local consumption—primarily through smart EV scheduling.

Notably, during certain periods, such as between 5:00 and 7:00, the W2V strategy achieves nearly 100% self-consumption by discharging EVs to support the load when wind generation is low. Conversely, during periods of high wind production, EVs are scheduled to charge, effectively absorbing the excess renewable energy.

These results highlight the effectiveness of the W2V strategy in improving local renewable energy utilization, reducing grid stress, and balancing supply and demand through coordinated EV flexibility.

12.4.2 Impact assessment on the distribution grid

This subsection focuses on assessing the impact of the different EV charging strategies on the distribution grid operation. Using the aggregated charging profiles obtained from the previous scheduling step, power flow simulations are carried out to evaluate the resulting voltage magnitude and angle profiles and the power flows across the network. The analysis specifically examines the effect of smart (W2V) versus uncoordinated (MT and MT-W) charging in scenarios with and without local wind generation. The goal is to identify potential voltage drops that could provoke instabilities in the system and overall improvements in grid performance enabled by the proposed W2V strategy.

Fig. 12.8 shows the system voltage magnitude profiles with and without considering wind and the EVs charging strategies. Note in Fig. 12.8a that the consideration of the MT strategy, that is, the uncontrolled charging of EV, results in a critical voltage drop in nodes 26 and 30, having in the worst case a voltage magnitude below 0.94 p.u., which could induce the trip of the undervoltage relays, leading to line and/or load disconnections, and in the worst case the system instability. The voltage drop in nodes 26 and 30 is reduced when wind generation is considered, this can be seen in Fig. 12.8b. Note that in this scenario,

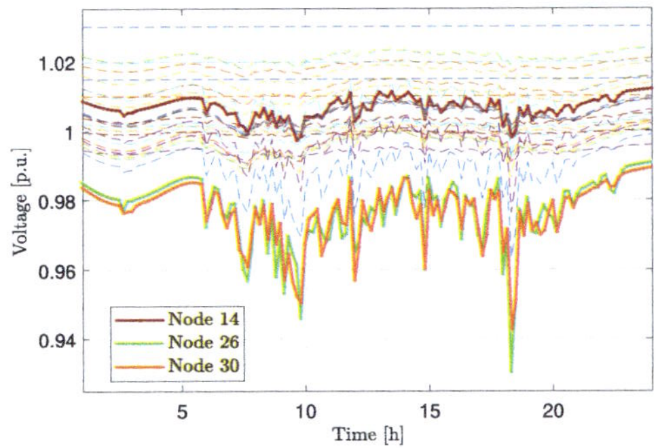

(a) MT strategy without wind

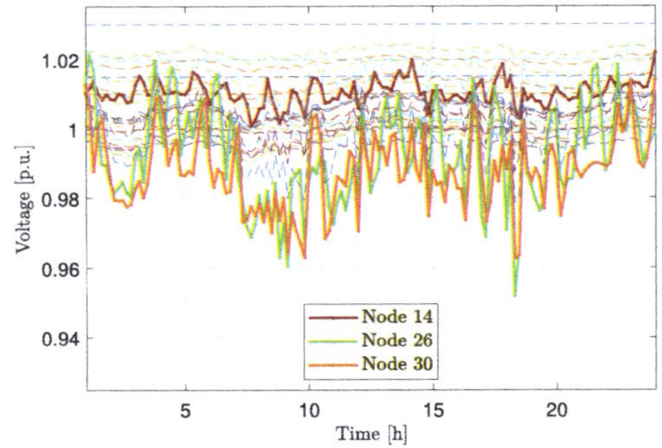

(b) MT-W: MT strategy with wind

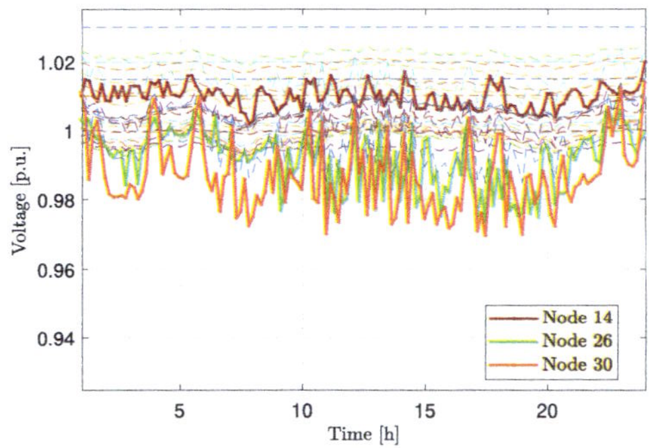

(c) W2V strategy with wind

Figure 12.8 Comparison of voltage magnitude profiles at all nodes under different EV charging strategies and wind generation conditions. (a) MT strategy without wind. (b) MT-W: MT strategy with wind. (c) W2V strategy with wind.

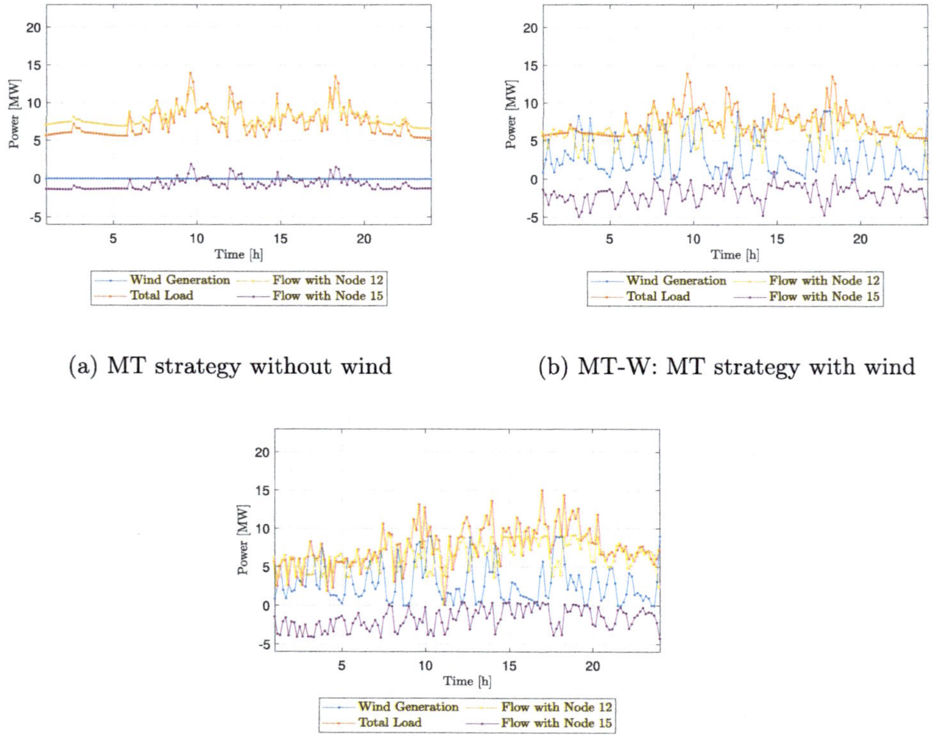

(a) MT strategy without wind

(b) MT-W: MT strategy with wind

(c) W2V strategy with wind

Figure 12.9 Active power profiles in node 14 with and without considering wind generation and the EVs charging strategy. (a) MT strategy without wind. (b) MT-W: MT strategy with wind. (c) W2V strategy with wind.

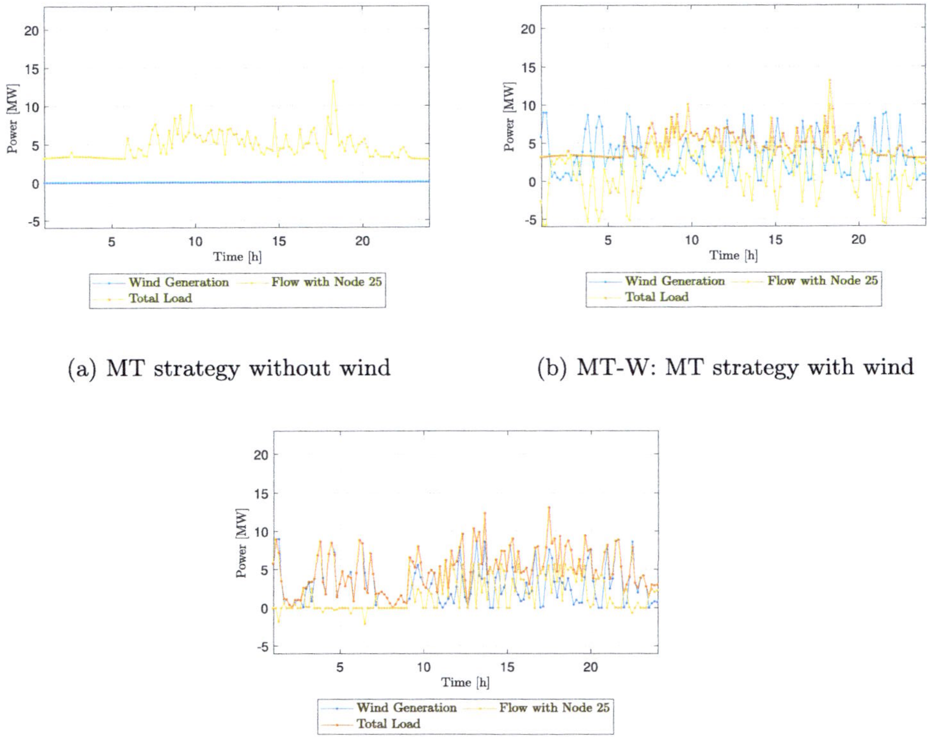

(a) MT strategy without wind

(b) MT-W: MT strategy with wind

(c) W2V strategy with wind

Figure 12.10 Active power profiles in node 26 with and without considering wind generation and the EVs charging strategy. (a) MT strategy without wind. (b) MT-W: MT strategy with wind. (c) W2V strategy with wind.

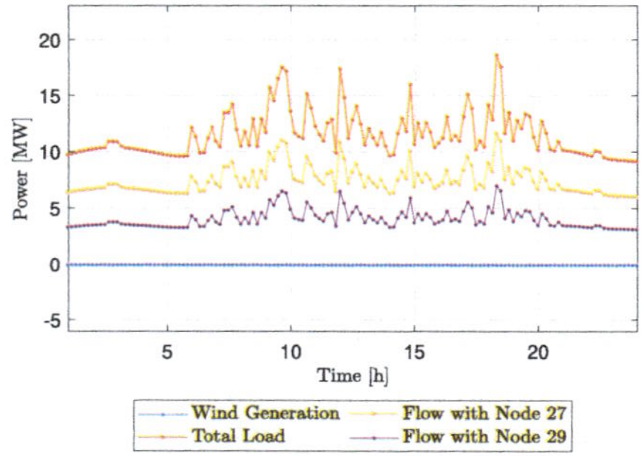

(a) MT strategy without wind

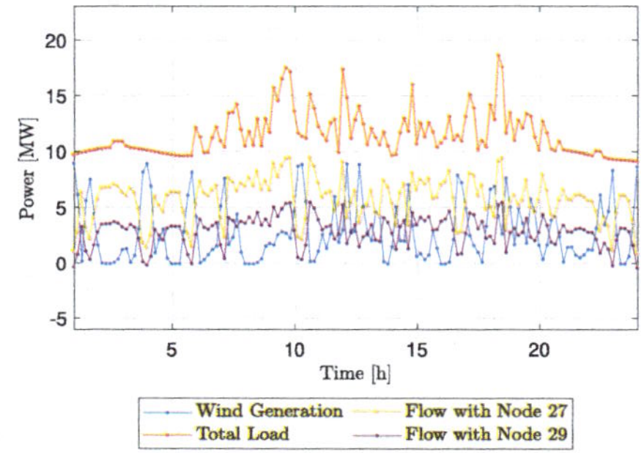

(b) MT-W: MT strategy with wind

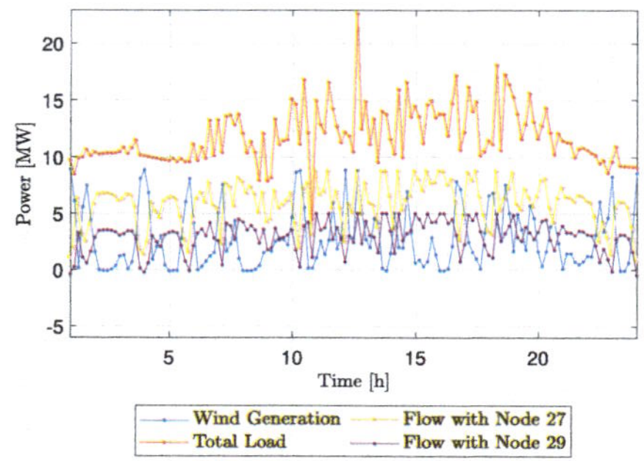

(c) W2V strategy with wind

Figure 12.11 Active power profiles in node 30 with and without considering wind generation and the EVs charging strategy. (a) MT strategy without wind. (b) MT-W: MT strategy with wind. (c) W2V strategy with wind.

the voltage drops below 0.96 p.u. but just in a small window of time. Although this change improves the system operation, the system voltage stability is still compromised. To cope with this issue, The W2V strategy, that is, the smart EV charging strategy, is included in the system operation as shown in Fig. 12.8c. Note that, by optimally shifting the EV loads, the system operation keeps the voltage magnitudes above 0.97 p.u., and avoids sudden voltage magnitude drops. This shows the effectiveness of the proposed W2V strategy and the need to consider modern strategies for charging in the real system operation.

Figures 12.9–12.11 show the active power profiles—both load and generation at nodes 14, 26, and 30, respectively, under different scenarios: without wind generation, with wind generation under the MT strategy, and with wind generation under the W2V strategy. Across all three nodes, the W2V strategy effectively shifts EV charging and discharging in time to better align with the availability of wind power. This alignment enhances the local utilization of renewable energy and contributes to the voltage improvements previously discussed. Moreover, the W2V strategy leads to smoother and more balanced power exchanges with neighboring nodes, supporting more stable and efficient grid operation.

Table 12.4 Comparison of parameters across different configurations on node 14.

Parameters	MT	MT-W	W2V
Minimum voltage [p.u.]	0.997	1.001	1.002
Variance [(p.u.)^2]	8.7×10^{-6}	1.2×10^{-5}	1.1×10^{-5}
Standard deviation [p.u.]	0.003	0.004	0.003
Energy inflow from node 12 [kWh]	192.15	148.73	153.81
Energy inflow from node 15 [kWh]	2.61	0.70	1.06
Energy outflow to node 12 [kW]	0.00	0.00	0.00
Energy outflow to node 15 [kWh]	19.49	44.72	41.42
Peak power entering from node 12 [kW]	11.99	10.01	9.22
Peak power entering from node 15 [kW]	1.95	1.47	0.53
Peak power exiting to node 15 [kW]	1.49	5.05	4.20

Finally, in Tables 12.4–12.6, key parameters are shown for the nodes 14, 26, and 30, respectively, where the voltage magnitudes, energy inflow and outflow, and peak powers, are compared for each case, with and without the consideration of wind generation and the smart EV strategy.

Table 12.5 Comparison of parameters across different configurations on node 26.

Parameters	MT	MT-W	W2V
Minimum voltage [p.u.]	0.930	0.952	0.974
Variance [(p.u.)^2]	9.0×10^{-5}	2.0×10^{-4}	7.7×10^{-5}
Standard deviation [p.u.]	0.009	0.014	0.009
Energy inflow from node 25 [kWh]	112.37	53.94	40.26
Energy outflow to node 25 [kW]	0.00	18.48	1.01
Peak power entering from node 25 [kW]	13.22	9.99	5.70
Peak power exiting to node 25 [kW]	0.00	5.91	2.06

Table 12.6 Comparison of parameters across different configurations on node 30.

Parameters	MT	MT-W	W2V
Minimum voltage [p.u.]	0.943	0.963	0.970
Variance [(p.u.)^2]	7.2×10^{-5}	1.3×10^{-4}	1.2×10^{-4}
Standard deviation [p.u.]	0.008	0.012	0.011
Energy inflow from node 27 [kWh]	176.61	132.52	136.62
Energy inflow from node 29 [kWh]	97.48	67.18	70.10
Energy outflow to node 27 [kW]	0.00	0.00	0.00
Energy outflow to node 29 [kWh]	0.00	0.36	0.42
Peak power entering from node 27 [kW]	11.62	9.42	8.76
Peak power entering from node 29 [kW]	7.03	5.50	5.07
Peak power exiting to node 29 [kW]	0.00	0.50	0.52

12.5 Conclusions

This chapter addressed a current operational challenge: maximizing renewable energy utilization while leveraging electric vehicle (EV) flexibility in distribution system operation. To this end, a Wind-to-Vehicle (W2V) strategy was introduced and analyzed in detail. The W2V approach enables flexible EV charging schedules that prioritize the absorption of surplus wind energy, thus reducing curtailment and improving overall grid performance.

To evaluate the effectiveness of the proposed strategy, the IEEE 30-bus system was adapted with realistic wind generation and base load profiles. The W2V framework incorporated a wide range of EV usage patterns as well as randomized connection events based on forecasted occupancy behavior. The analysis considered commercially available EV models from Tesla and BYD, which represent different battery capacities and charging capabilities commonly found in fast and ultra-fast charging infrastructure. These elements were combined into a schedule optimization model, which generated charging and discharging profiles for EVs that were later integrated into power flow simulations of the grid.

The simulation results demonstrated that, compared to uncoordinated or non-optimized charging strategies, the W2V approach significantly improved the utilization of available wind energy. Moreover, it enhanced voltage stability across the network, preventing undervoltage conditions that were observed in the baseline scenarios. Notably, the W2V strategy reduced power peaks, avoided critical voltage drops, especially at vulnerable nodes and achieved periods of full wind energy self-consumption by the EV fleet.

In summary, this work shows that the W2V strategy is a practical and effective solution for coordinating renewable energy and EV charging in modern power systems. It enables the grid to accommodate both the variability of wind generation and the high instantaneous power demand of fast and ultra-fast EV charging.

Future work should explore real-time implementations of this strategy, scalability to larger and meshed networks, integration with photovoltaic generation, and interoperability with other demand response mechanisms. The continuous evolution of EV battery technologies and high-power charging infrastructure will require increasingly adaptive and smart energy management solutions, W2V being a promising step in that direction. In addition, future work will focus on embedding a feedback loop between the scheduling and power flow stages, enabling the system to iteratively test multiple candidate charging solutions and select the one that both optimizes EV operation and guarantees convergence of the power flow equations. A further extension will include modeling battery degradation costs, which are currently neglected to maintain tractability, but whose nonlinear characteristics will be addressed in more advanced formulations to capture realistic impacts on EV scheduling.

References

Abdelsalam HA, Attia EM, Arzani A, Mahajan SM. Efficient charging coordination of electric vehicles: a consensus tracking control approach. Sustain Energy Grids Netw 2025;43:101771. https://doi.org/10.1016/j.segan.2025.101771.

Deilami S, Masoum AS, Moses PS, Masoum MAS. Real-time coordination of plug-in electric vehicle charging in smart grids to minimize power losses and improve voltage profile. IEEE Trans Smart Grid 2011;2(3):456–67. https://doi.org/10.1109/TSG.2011.2159816.

Diaz-Londono C, Vuelvas J, Gruosso G, Correa-Florez CA. Remuneration sensitivity analysis in prosumer and aggregator strategies by controlling electric vehicle chargers, Energies 15 (19). https://doi.org/10.3390/en15196913.

Diaz-Londono C, Motta M, Pareschi D, Gruosso G. Utilization analysis of rapid and ultra-rapid electric vehicle chargers in europe. In: 2023 IEEE vehicle power and propulsion conference (VPPC); 2023. p. 1–6. https://doi.org/10.1109/VPPC60535.2023.10403289.

Diaz-Londono C, Agundis-Tinajero GD, Maffezzoni P, Gruosso G, Guerrero JM. Evaluating the electrical network impact of ev charging strategies used by operators across residential, workplace, and public users. IEEE Open J Vehi Technol 2025;6:1721–35. https://doi.org/10.1109/OJVT.2025.3581803.

Diaz-Londono C, Orfanoudakis S, Vergara PP, Palensky P, Ruiz F, Gruosso G. Open source algorithms for maximizing v2g flexibility based on model predictive control. Elec Power Syst Res 2026;250:112082. https://doi.org/10.1016/j.epsr.2025.112082.

Electric Nation Customer Trial Final Report, https://electricnation.org.uk/.

Fastned support, ev. 2024. https://www.fastnedcharging.com/en.

He C, Peng J, Jiang W, Wang J, Du L, Zhang J. Vehicle-to-grid (v2g) charging and discharging strategies of an integrated supply–demand mechanism and user behavior: a recurrent proximal policy optimization approach, World Electr Vehic J 15 (11). https://doi.org/10.3390/wevj15110514.

International Energy Agency. World energy outlook 2024, licence: CC BY 4.0 (report); CC BY-NC-SA 4.0 (Annex A). 2024.

International Energy Agency. Global ev outlook 2025, licence: CC BY 4.0. 2025.

Kempton W, Tomić J. Vehicle-to-grid power implementation: from stabilizing the grid to supporting large-scale renewable energy. J Power Sources 2005;144(1):280–94. https://doi.org/10.1016/j.jpowsour.2004.12.022.

Koruturk YU, Diaz-Londono C, Ruiz F. Joint occupancy and load profile prediction for electric vehicle charging stations. In: 2024 IEEE 8th forum on research and technologies for society and industry innovation (RTSI); 2024. p. 283–8. https://doi.org/10.1109/RTSI61910.2024.10761602.

Kumar V, Kumar V, Dev A. Voltage and frequency regulation in wind penetrated deregulated power system using an electric vehicle and ipfc assisted model predictive controller. Sci Rep 2025;15:31109. https://doi.org/10.1038/s41598-025-16826-4.

Larminie J, Lowry J. Electric vehicle technology explained. Wiley; 2003.

Manousakis NM, Karagiannopoulos PS, Tsekouras GJ, Kanellos FD. Integration of renewable energy and electric vehicles in power systems: a review, Processes 11 (5). https://doi.org/10.3390/pr11051544.

Manwell ALRJF, McGowan JG. Wind energy explained: theory, design and application. 2nd ed. Wiley; 2010.

National Control Center – CNES, https://www.services-rte.com/en/view-data-published-by-rte/daily-load-curves.html.

Pan J, Liu T. Optimal scheduling for unit commitment with electric vehicles and uncertainty of renewable energy sources. Energy Rep 2022;8:13023–36. https://doi.org/10.1016/j.egyr.2022.09.087.

Pappala VS, Erlich I, Rohrig K, Dobschinski J. A stochastic model for the optimal operation of a wind-thermal power system. IEEE Trans Power Syst 2009;24(2):940–50. https://doi.org/10.1109/TPWRS.2009.2016504.

Peyghami S, Davari P, Fotuhi-Firuzabad M, Blaabjerg F. Standard test systems for modern power system analysis: an overview. IEEE Ind Electron Mag 2019;13(4):86–105. https://doi.org/10.1109/MIE.2019.2942376.

RTE - Réseau de Transport d'Électricité, Daily load curves. 2025. https://www.services-rte.com/en/home.html.

Saadat H. Power system analysis, no. v. 1 in McGraw-Hill series in electrical and computer engineering. WCB/McGraw-Hill; 1999. https://books.google.com.hk/books?id=NQH3vAEACAAJ.

Saadat H. Power system analysis, McGraw-Hill series in electrical and computer engineering. McGraw-Hill; 2002.

Sanguesa JA, Torres-Sanz V, Garrido P, Martinez FJ, Marquez-Barja JM. A review on electric vehicles: technologies and challenges. Smart Cities 2021;4(1):372–404. https://doi.org/10.3390/smartcities4010022.

Stevenson W, Grainger J. Power system analysis, McGraw-Hill series in electrical and computer engineering: power and energy. McGraw-Hill Education (India) Pvt Limited; 2003.

Su C, Wang Y, Strbac G. Coordinated electric vehicles dispatch for multi-service provisions: a comprehensive review of modelling and coordination approaches. Renew Sustain Energy Rev 2025;223:115977. https://doi.org/10.1016/j.rser.2025.115977.

Togun H, Basem A, Abdulrazzaq T, Biswas N, Abed AM, dhabab JM, et al. Development and comparative analysis between battery electric vehicles (bev) and fuel cell electric vehicles (fcev). Appl Energy 2025;388:125726. https://doi.org/10.1016/j.apenergy.2025.125726.

Ullah F, Zhang X, Khan M, Mastoi MS, Munir HM, Flah A, et al. A comprehensive review of wind power integration and energy storage technologies for modern grid frequency regulation. Heliyon 2024;10(9):e30466. https://doi.org/10.1016/j.heliyon.2024.e30466.

Wang W, Cheng Y. Optimal charging scheduling for electric vehicles considering the impact of renewable energy sources. In: 2020 5th Asia conference on power and electrical engineering (ACPEE); 2020. p. 1150–4. https://doi.org/10.1109/ACPEE48638.2020.9136545.

Wu Y, Cai T, Li X. Resilience driven ev coordination in multiple microgrids using distributed deep reinforcement learning. Sci Rep 2025;15:27038. https://doi.org/10.1038/s41598-025-12471-z.

Zanvettor GG, Fochesato M, Casini M, Lygeros J, Vicino A. A stochastic approach for ev charging stations in demand response programs. Appl Energy 2024;373:123862. https://doi.org/10.1016/j.apenergy.2024.123862.

Zhou H, Li W, Lin Z. Distributed optimal scheduling for evs charging and discharging: a penalty-based consensus approach. Int J Electr Power Energy Syst 2024;161:110194. https://doi.org/10.1016/j.ijepes.2024.110194.

Ziyat I, Gola A, Palmer PR, Makonin S, Popowich F. Ev charging profiles and waveforms dataset (ev-cpw) and associated power quality analysis. IEEE Access 2023;11:138445–56. https://doi.org/10.1109/ACCESS.2023.3340131.

CHAPTER THIRTEEN

A comprehensive review of algorithms for optimizing the location and sizing of charging stations for renewable-to-vehicle (R2V) charging

Hossein Jafari[1], **Sasan Azad**[2], **Alireza Sobbouhi**[2]
[1]Department of Electrical, Electronic, and Information Engineering, University of Bologna, Bologna, Italy
[2]Department of Electrical Engineering, Shahid Beheshti University, Tehran, Iran

13.1 Introduction

In contemporary society, a major challenge is the rapid increase in energy consumption from pollutant-emitting sources, driven by industrial expansion and modernization. The growing concerns surrounding global warming, air pollution, greenhouse gas emissions, and dwindling fossil fuel reserves highlight the urgent need to shift toward renewable energy sources (RESs). Making this transition is crucial for securing a sustainable future. A study from 2012 found that the transportation sector accounted for roughly 27% of global energy consumption and about 33.7% of greenhouse gas emissions (Fazelpour et al., 2014).

The environmental costs linked to traditional transportation systems cover various aspects—from public health expenses to reforestation efforts and even the restoration of cultural heritage sites damaged by acid rain. To address these challenges, considerable research and development efforts are focused on the construction of transportation systems that are both highly efficient and environmentally sustainable. Consequently, the electrification of transportation is increasingly recognized as a viable solution for reducing greenhouse gas emissions and improving air quality. EVs differ from internal combustion engines (ICEVs) in powertrain technology and efficiency. As a result, EVs provide a driving experience that is clearly different from ICEVs, highlighting the benefits of modern electric technology.

Renewable-to-Vehicle Smart Charging
ISBN: 978-0-443-36684-0
https://doi.org/10.1016/B978-0-443-36684-0.00015-1

Governments worldwide are introducing policies and initiatives to encourage EV adoption. Increasing public awareness of EV benefits and building a robust charging infrastructure are crucial for accelerating their adoption. These approaches are proving effective, as shown by the increasing acceptance of EVs within the community. According to the Global EV Outlook prepared by the Electric Vehicle Initiative (EVI) and the International Energy Agency (IEA) (Yong et al., 2015), the global stock of EVs surpassed 18,000 units by the end of 2012. Projections from the Edison Electric Institute (EEI) and the Institute for Electric Innovation (IEI) estimate that approximately 18.7 million EVs will be on the roads by 2030, necessitating around 9.6 million electric vehicle charging stations (EVCSs) to support this growing fleet (Bilal et al., 2020).

Wang et al. (2025) discusses integrating renewable energy into distribution networks, emphasizing energy efficiency and stability. It explores optimizing network configurations with renewable sources and EVs, using copula theory to manage uncertainties. An advanced multi-objective particle swarm optimization algorithm is introduced to determine the optimal capacities and locations of charging stations. A case study on the IEEE 33-node distribution network demonstrates how EVs can function as shared energy storage to mitigate uncertainties linked to renewable energy integration. A study in Choi and Bhakta (2025) explores the potential of utilizing a solar-tracking photovoltaic and hydrogen fuel cell system for electric vehicle charging stations (EVCSs) in South Korea's humid continental climate. The research examines the financial and technological factors related to charging stations, highlighting that efficiently obtaining energy from the grid can result in a substantial decrease in CO2 emissions over time.

The strategic positioning and sizing of Renewable Distributed Generators (RDGs) and EVCSs are of paramount importance, particularly in urban environments. The rise of EVs in cities transforms transportation and offers a significant chance to incorporate more renewable energy into our energy generation systems. The rise of EVs generates substantial additional electricity demand, which can be effectively met by RESs such as solar and wind power when integrated into the existing grid. This approach not only facilitates the decarbonization of urban transport but also addresses issues such as air and noise pollution, reduces dependence on fossil fuels, and promotes innovative solutions for urban mobility (IRENA, 2019).

Many cities are increasingly reliant on RDGs as a primary energy source. For instance, as highlighted in reference (Ulpiani et al., 2023), over 70% of qualified cities utilize solar thermal and solar photovoltaic systems for electricity generation. Furthermore, it has been shown that cities are actively conducting feasibility studies and initiating small-scale projects for local hydropower generation. Some cities have achieved nearly 100% renewable electricity usage, especially sourced from hydropower and other renewable options. Accordingly, the integration of RDGs and EVCSs offers a significant opportunity for urban areas to accelerate their transition toward sustainable and emissions-free environments. Cities can help achieve global goals for reducing carbon emissions and improving urban living conditions by integrating RESs with electrified transportation systems.

As a result, the adoption of EVs and RDG technologies has recently attracted considerable interest within the industry. This transition signifies a fundamental shift in the planning and operation of distribution systems. Given the increasing focus on this subject, numerous research studies have been conducted; however, much of the existing literature has primarily concentrated on identifying optimal locations for EVs while often overlooking the planning dimensions of both EVCSs and RDGs, which are critical to overall system performance and efficiency. This review begins with an initial bibliometric analysis to systematically clarify and investigate research trends. This is followed by a detailed content analysis and classification of prominent studies to present a clear picture of optimizing methods and facilitate future research. To the authors' knowledge, this study is the first to cover both bibliometric and content analyses of optimizing methods, offering comparative assessments and future recommendations. The objective is to provide a comprehensive overview of research activities over the past decade, ensuring that findings contribute to future research endeavors. This step-by-step guide for sizing and optimal placement, based on available literature, will help readers better understand and select the approaches best suited for optimizing in specific conditions.

13.2 Methodology

A comprehensive hybrid bibliometric and content analysis approach was considered for selecting, classifying, and reviewing the related literature. To do so, a primary bibliometric study was conducted. Bibliometric analysis provides a broad perspective on specific research areas, including significant participating countries, prominent institutions, authors, and the most influential journals and publications. In this regard, the initial data and related documents were provided from the Scopus database.

Scopus is considered a premier and reliable database since it covers most of the studies related to engineering topics and publishers.

The investigation in this review consists of the situation of global studies conducted in the past 14 years (2010–2024) in the field of optimizing the location and size of renewable-based charging stations. As a result, the following topic has been searched for deriving documents about optimizing renewable-based charging stations' size and location among the titles, abstracts, and keys: TITLE-ABS-KEY(Optimal AND (Location OR Sizing) AND (Renewable AND (Station OR Stations)) AND (Renewable AND to AND vehicle) OR (R2V)). Based on this, a total of 211 documents were retrieved and extracted for the bibliometric analysis. Fig. 13.1 illustrates the method for primary extraction and selection of related documents.

Then, the "Biblioshiny" function from the "Bibliometrix" in the R language was configured to perform bibliometric analysis. The bibliometric study in the research area was conducted involving the following aspects about the most important findings in optimizing techniques for EV charging stations.

(a) published articles and total citations;
(b) prominent authors, countries, and sources;
(c) More mentioned keywords and trend topics.

For each aspect, the number of publications, citations, co-citations, keywords, and the most important authors, sources, countries, and institutions were determined. Based on the extraction and selection method, in this study, a total of 211 documents were identified that were written by 720 authors. Furthermore, after excluding irrelevant results and those with informational redundancies, the bibliometric analysis reported that the documents are spread over 140 journal articles, 52 conference papers, three review articles, eight conference reviews, eight books/chapters, and 10 other surveys, as shown in Fig. 13.2. Table 13.1 presents more complete information about the primary selected documents in the research area.

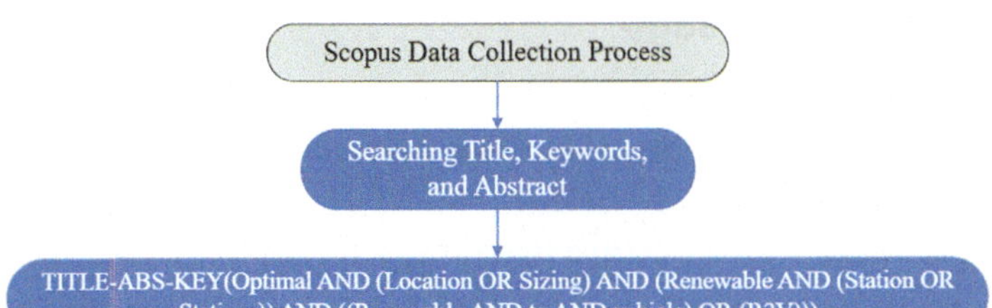

Figure 13.1 Primary document selection for the bibliometric evaluation.

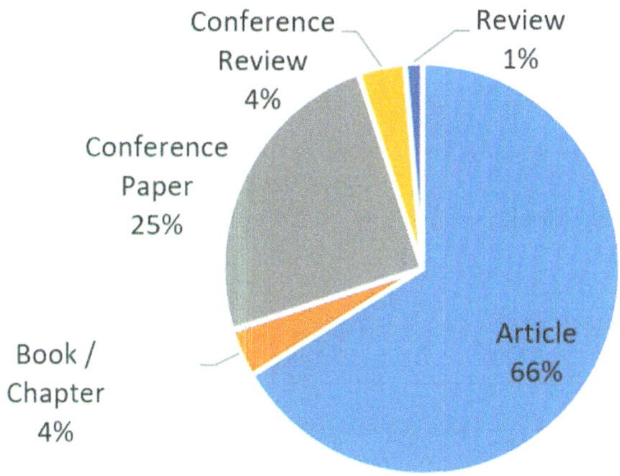

Figure 13.2 Most frequent types of documents about optimizing the location and size of renewable-based charging stations.

A review of all the documents followed the initial comprehensive bibliometric assessment of the chosen documents. Through this step, a content analysis method was applied to classify the most relevant and prominent studies. The results demonstrated that the documents could be classified into four primary categories: "Optimal Location," "Optimal Sizing," "Renewable Station," "Renewable to Vehicle," and "R2V." Each category was examined in detail, produced significant insights, emphasized areas where research is lacking, and offered suggestions for future investigations. The subsequent sections will primarily focus on presenting the bibliometric analysis results, followed by the content analysis findings, with a particular emphasis on the first two categories: "Optimal Location" and "Optimal Sizing."

13.3 Bibliometric analysis results

Bibliometric evaluation findings are presented in six subsections as follows.

13.3.1 Publication and citation trends

The number of publications related to optimizing the location and size of renewable-based charging stations has increased over the past approximately 14 years (2010–2024), according to the Scopus database. Fig. 13.3 illustrates the published articles (N) along with the average number of citations per publication (meanTCperArt) and year (MeanTCperYear), over the study period.

Table 13.1 Summary of bibliometric data for the selected documents in the current study.

Description	Results
Main information about data	
Timespan	2010:2024
Sources (Journals, Books, etc.)	131
Documents	211
Annual Growth Rate %	34.29
Document Average Age	3.47
Average citations per doc	22.27
Document contents	
Keywords Plus (ID)	1346
Author's Keywords (DE)	621
Authors	
Authors	720
Authors of single-authored docs	2
Authors Collaboration	
Single-authored docs	9
Co-Authors per Doc	3.95
International co-authorships %	33.18
Document Types	
Article	140
Book/Chapter	8
Conference Paper	52
Conference Review	8
Review	3

The overarching trajectory can be delineated into four distinct epochs:

(I) The Observational Phase (2010–17): This period marked the initial recognition of optimization in sizing and location challenges, within basic concepts (Tabandeh et al., 2022; Ferro et al., 2022; Amer et al., 2021).

(II) Rise of Global Awareness (2017-Present): In recent years, the proliferation of global development, particularly the extensive deployment

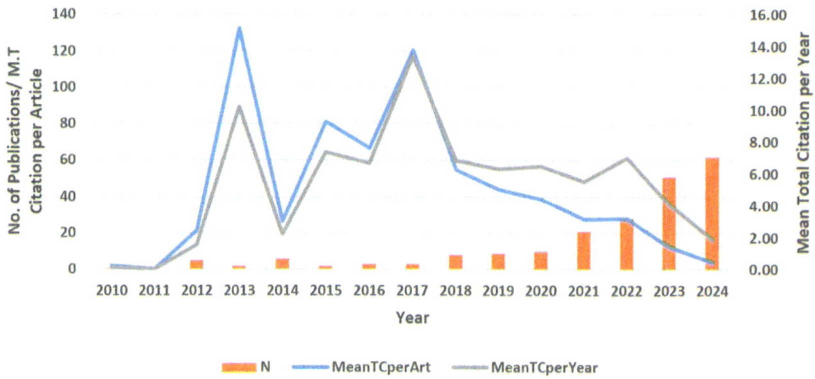

Figure 13.3 History of publications and citations in the research area.

of EVs in the USA—where Tesla is dominant—spurred a broader acknowledgment of the implementation of EVs. Consequently, after global attention to the challenge, efforts were directed toward gaining a comprehensive understanding of the phenomenon.

The primary focus of this study centers on documents published within the latter two stages, especially involving various methodologies for location and sizing optimization techniques. In 2013, there was a notable peak in the average number of citations per article, potentially attributable to the relatively low number of articles during the initial stage. Subsequently, a drop in the average number of citations per article occurred. One contributing factor to this drop is the increasing number of articles, which is contributing to a reduction in the average citations per article. The decline in citations per year can be attributed to a growing awareness of recent articles. This suggests that fewer articles are needed to achieve significant results in future research. However, even though a fluctuating trend was discernible in the number of citations per year or article, the number of published documents within the research field, spanning from 2017 to the present (Fig. 13.3), was significantly increased. This trend emphasizes the importance of the topic and has seen a considerable increase in recent years.

13.3.2 Source influence

Source analysis has been deployed to examine the influence of each publication source on the location optimization research field. In this context, Bradford's Law has been utilized. Bradford's Law divides sources based on article production in a field and distributes them into several zones based on cumulative publication frequency. Bradford's Law states that the number of journals in the second and third zones will be n and n^2 times the number of journals in the first zone, respectively. Therefore, if the number of journals in the core and intermediate zones is known, the total number of journals containing articles on a topic can be estimated. Bradford's Law introduces the most prominent journals as primary sources (Zone 1) in the literature on optimizing the location and size of renewable-based charging stations, as mentioned in Fig. 13.4.

According to the Scopus database, 211 documents have been collected from 131 sources. For the core sources, Energies appropriated the first rank with 13 articles, followed by IEEE Access (9), Applied Energy (7), Journal of Energy Storage (7), Energy (5), Sustainability (Switzerland) (5), and IEEE Transactions on Transportation Electrification (4).

13.3.3 The high-rate-contributed authors

Lotka's Law map was utilized to analyze the productivity of authors' publications in the current high-rate contributing authors. As shown in Fig. 13.5, the dotted lines show the authors' publication rate, and the solid lines illustrate the actual publication rate of the authors.

It can be observed that most of the authors contributed to only one article. It was observed that 638 researchers have published one article, which is proportional to 88.6% of all publications in determining the list of authors' published articles. Additionally, 68 researchers have published two articles, and 14 authors have published more than two articles, accounting for 9.4% and 2% of the total number of publications, respectively.

13.3.4 High-frequency keyword analysis

To grasp the main topics covered in the scholarly documents, the "Biblioshiny" tool was employed for retrieving and counting the most common keywords. Keywords that are mentioned more than 10 times were selected, and a word tree map, as illustrated in Fig. 13.6, was created to visualize them. "Electric Vehicle Charging," "Renewable Energy Resources," "Electric Vehicle," "Optimizing Algorithms," "Battery Storage," "Solar Energy," "Vehicle-to-Grid," and "Cost Benefit Analysis" were among the most commonly used research topic terms, comprising 19%, 10%, 8%, 6%, 4%, 4%, 3%, and 3% of the high-frequency keywords, respectively.

13.3.5 Mapping results of keyword co-occurrence

Keyword co-occurrence analysis was conducted using the VOSviewer network, which is depicted in Fig. 13.7.

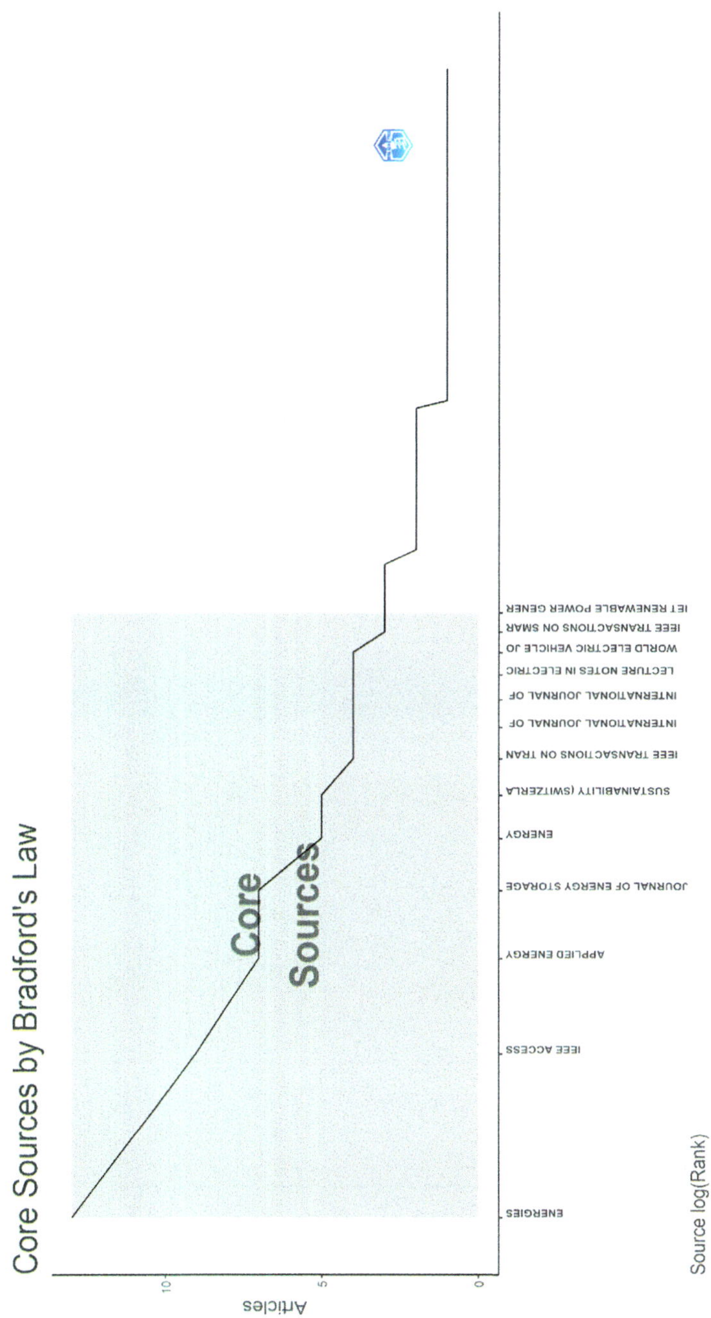

Figure 13.4 Distribution analysis of the core sources by Bradford's law.

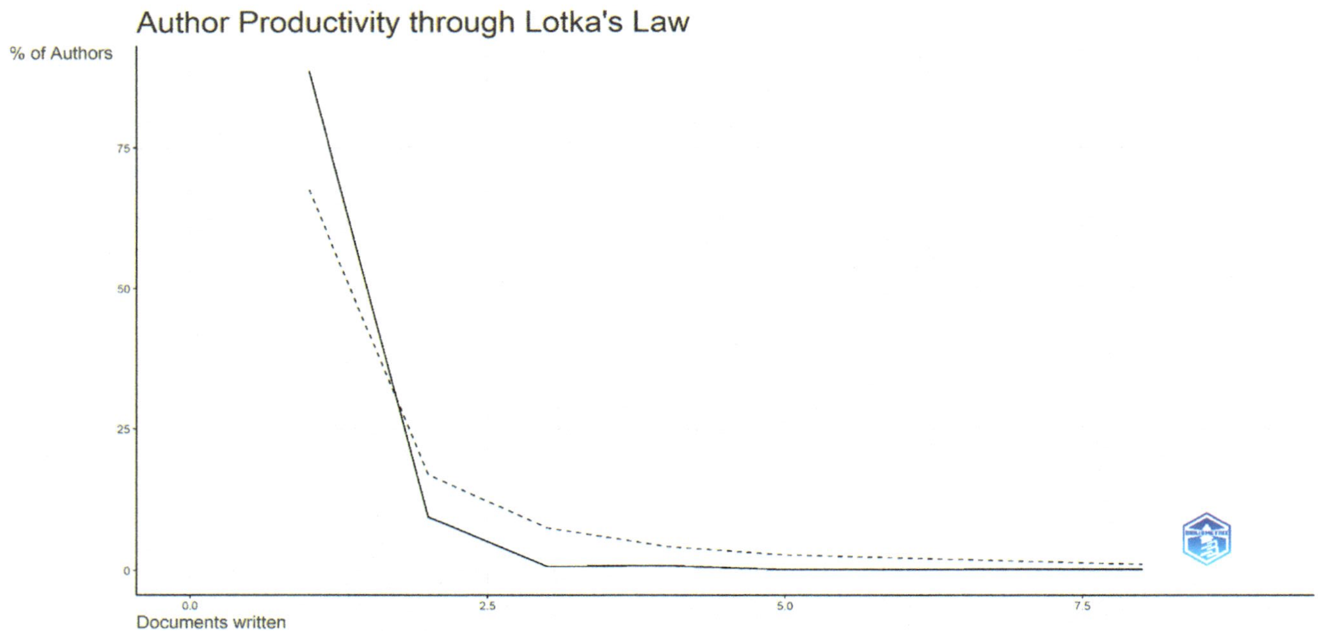

Figure 13.5 Frequency distribution of scientific productivity.

Figure 13.6 Word tree map of high-frequency keywords in the field.

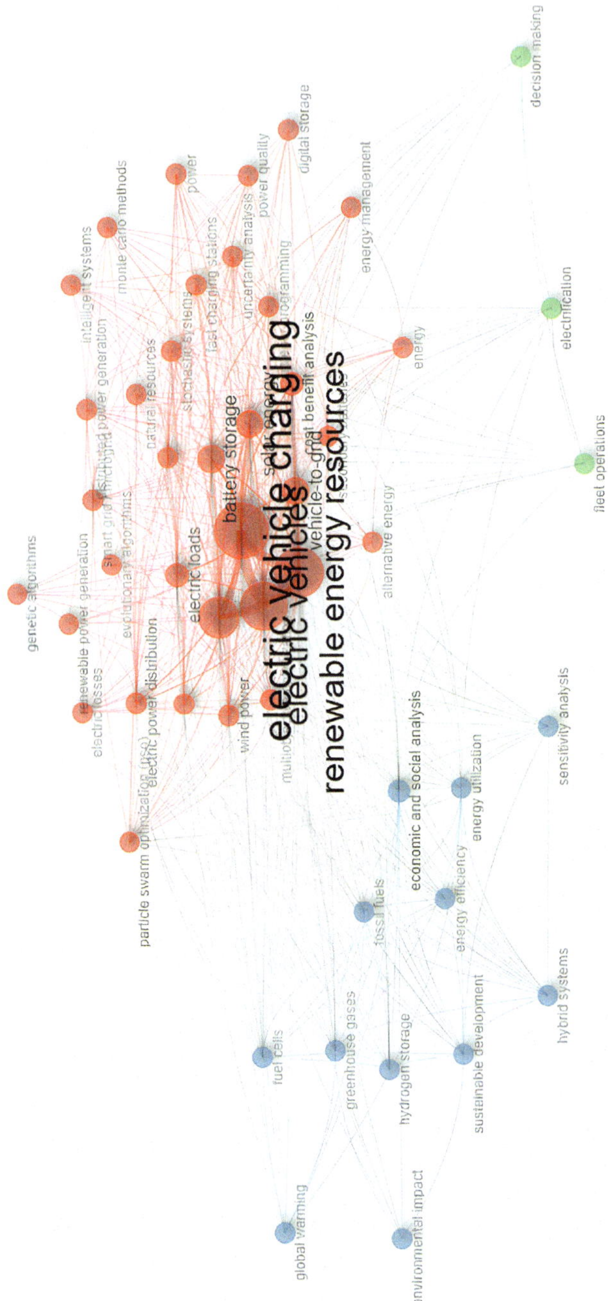

Figure 13.7 Bibliometric network maps of keywords.

As shown in Fig. 13.8, there are three major classified cluster keywords in the field of optimizing the location and sizing of EV charging stations. The blue cluster (on the right) has a lower intensity of keyword co-occurrence, including EVs and renewable energy resources. These and other important keywords in this cluster refer to the "electric vehicle charging" that can be applied to minimize dust impacts on the photovoltaic systems and improve cleaning efficiency.

As can be seen in the clustering network maps, the blue section correlates the most with the right side of the main (red) cluster, in which either environmental impacts (such as greenhouse gases and global warming) or economic and social analysis are located. This is because these preventive considerations can boost the efficiency of optimization by reducing greenhouse gases or fossil fuels. Therefore, these preventive considerations were set as one of the main classifications for further content analysis. The co-occurrence of factors such as sensitivity analysis, electric vehicle charging, and optimizing algorithms could also be observed. In this context, various "renewable energy resources" as well as "economic and social analysis" were the main focus of researchers for effective optimization of sizing. Fig. 13.8 depicts the chronological evolution of research topics, which is consistent with the clustering network map illustrated in Fig. 13.7. Prior to 2018, discussions on basic concepts of smart power grids predominantly revolved around a multi-objective optimization problem. Subsequently (2018–21), the focus shifted toward addressing EV charging stations, planning, energy storage systems, and electric automobiles. Recent years (2020-Present) have seen a surge in publications emphasizing sensitivity analysis and a range of optimizing algorithms to enhance the power quality of EV charging systems, as evidenced in Fig. 13.8. As a result, a classified content analysis review of these topics with a focus on "Optimizing techniques" was carried out, and the results are presented in the following sections. The bibliometric analysis revealed that meta-heuristic algorithms are currently the most prevalent approach for R2V station optimization, followed by exact methods. Motivated by these findings, Section 4 provides a detailed content analysis of each class of optimization algorithms, discussing their applications in R2V charging problems.

13.4 Optimization techniques

Optimizing the location and sizing of renewable-based charging stations for R2V smart charging requires various computational methods,

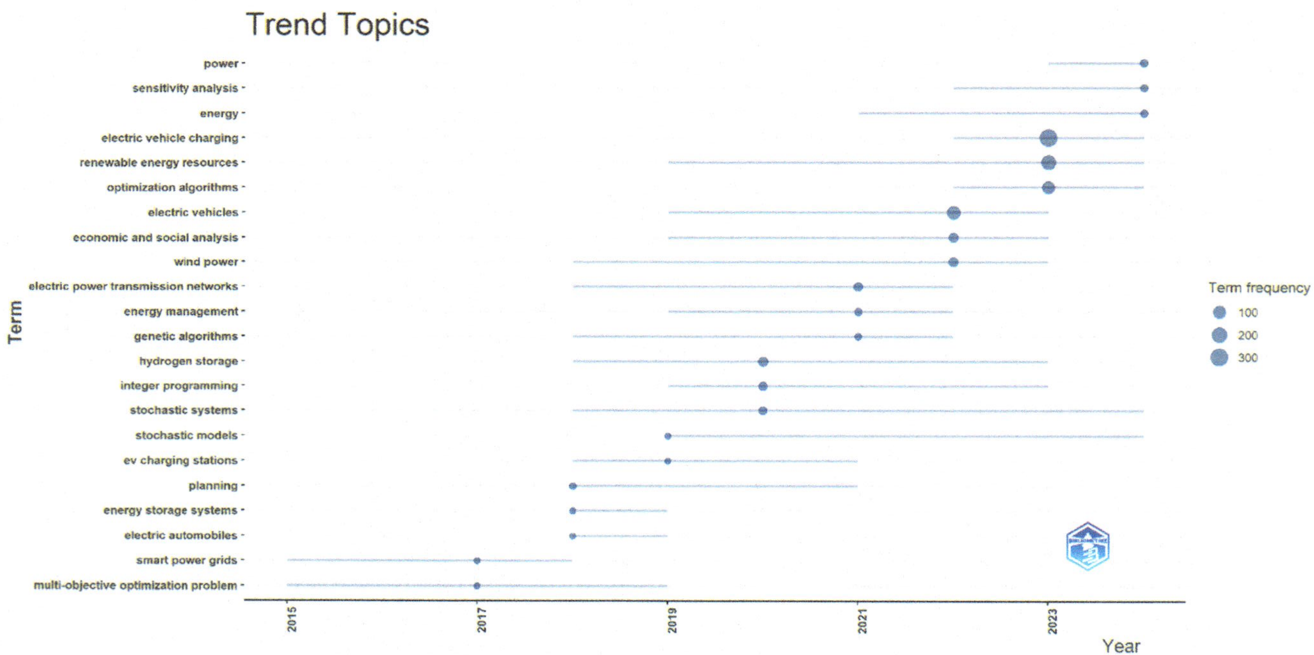

Figure 13.8 Top research trend keywords in the field of optimizing the location and sizing of the EV charging stations.

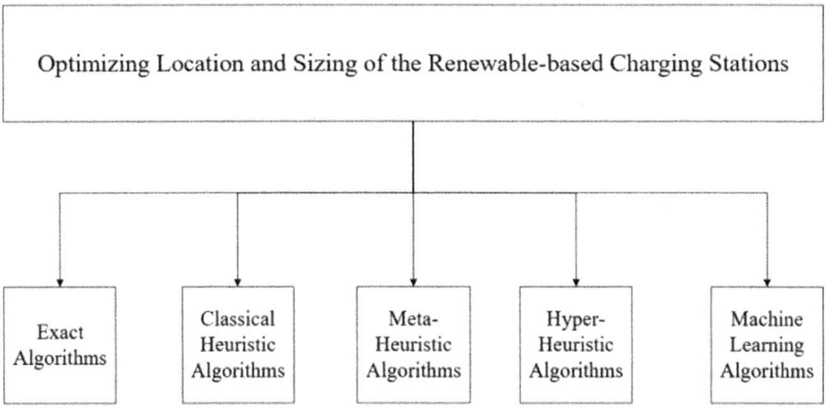

Figure 13.9 Classification of optimization methods for R2V charging station placement, organized by method type.

from deterministic to heuristic and machine learning techniques. This review examines these approaches, noting that complex real-world problems necessitate advanced algorithms. While exact methods can yield global optimal solutions, their NP-hard complexity makes them impractical for large-scale issues. Heuristic and metaheuristic algorithms are better suited for larger problems, offering near-optimal solutions efficiently. However, choosing the correct algorithm remains challenging due to sensitivity to parameters. Recently, hyper-heuristic algorithms have emerged, providing generality and efficiency by dynamically selecting or combining simpler heuristics. Machine learning also shows promise, enhancing algorithm design to produce superior solutions faster. Current algorithms are classified based on their approach, acknowledging that hyper-heuristic and machine learning methods are newer to R2V research and may require further classification in the future. Fig. 13.9 illustrates a taxonomy of these optimization methodologies for R2V charging station siting and sizing, categorized by the type of approach.

13.4.1 Exact algorithms

Exact algorithms provide a rigorous approach to finding optimal solutions for the R2V charging station placement and sizing problem. These include classical mathematical programming methods such as Linear Programming (LP) (Feng et al., 2024) and various mixed-integer formulations. Notably, Mixed-Integer Linear Programming (MILP) has been widely applied in this domain (Kumbhar et al., 2025; Dhakar et al., 2025; Oladigbolu et al., 2025; Bhatia and Das, 2025; Ghaithan et al., 2024; He et al., 2024; Er et al., 2024; Singh et al., 2024; Zhang et al., 2024; Saadati et al., 2023; Bao et al., 2023; Félix, 2020; Wu et al., 2022; Amer et al., 2021; Sa'adati et al., 2021), alongside Mixed-Integer Nonlinear Programming (MINLP) techniques (Shamani et al., 2025; Ghani and Farjah, 2025; Aduama et al., 2024; Ahmad and Kumar, 2024; Khazali et al., 2023; Ali et al., 2022; Tabandeh et al., 2022). Specialized exact methods like the ε-constraint multi-objective approach (Abdulnasser et al., 2025), Model Predictive Control (MPC) schemes (Cardona et al., 2023), bi-level optimization frameworks (Ferro et al., 2022), and location-allocation models (Li et al., 2025) have also been explored. Variants and extensions include linear/MPC-based optimization hybrids (Li et al., 2023), decomposition-based stochastic programming (Sabzian-Molaee et al., 2023), and even game-theoretic formulations such as a Stackelberg game model (Adil et al., 2021). Each exact method has unique characteristics. For example, branch-and-bound search can systematically explore the solution space to guarantee optimality, while dynamic programming breaks complex problems into simpler subproblems. Column-generation techniques can further reduce computational burden by introducing dual variables to refine the solution space iteratively.

This study jointly optimizes siting and sizing of integrated battery-swapping/charging stations (BSCS) by coupling a renewable microgrid (PV/wind) with a road network so placement, capacity, and power exchange co-evolve with traffic and grid constraints; applied to a real network, the mixed-integer formulation delivers investment/operating cost minima, practical design quantities (chargers, swappers, land), and shows renewables markedly cut distribution-system costs (PV $<$ wind $<$ PV $+$ wind) while network-aware placement better disperses charging demand for present and forecast flows (Kumbhar et al., 2025). Focusing on *mobile* charging stations (MCS), this paper frames a routing-and-scheduling problem that embeds distribution-network physics, time-varying tariffs, and operational limits of distributed energy resources (PV, WT, MT, BES) with bidirectional power exchange; the linearized MILP scales to IEEE 33/123-bus cases and demonstrates that honoring grid constraints materially shifts optimal tours and profitability, highlighting MCS as a flexible, sustainability-aligned complement to fixed infrastructure under diverse demand scenarios (Shamani et al., 2025). Targeting hydrogen R2V ecosystems, this work proposes a least-cost, fully electrified refueling-station framework that co-optimizes on-site electrolysis, LOHC storage (hydrogenation/dehydrogenation with thermal integration), and dispensing under coupled mass/heat/power constraints; scenario analyses show grid-connected LOHC systems can attain competitive delivered-H_2 costs and high capacity factors, with specific LOHC choices improving resilience to intermittency, underscoring near-term techno-economic viability of optimized designs (Dhakar et al., 2025).

These exact optimization algorithms offer several **strengths**. First, they serve to **verify model accuracy** by providing a benchmark solution for minor instances. They are guaranteed to find the *global optimum* for a well-formulated problem (given sufficient computation time), which is invaluable for model validation and as a gold standard against which heuristic methods can be compared. Additionally, exact solutions help illuminate the theoretical limits of system performance under idealized conditions. However, as the problem size grows, the **computational complexity** and solve times increase exponentially, often rendering pure exact methods intractable for realistic large-scale scenarios. In practice, R2V planning models with many variables (locations, capacities, time periods, etc.) become too slow or memory-intensive to solve to optimality. For this reason, exact algorithms are frequently augmented with heuristics or decomposition techniques to improve tractability, aiming to find near-optimal solutions in a reasonable time. This hybridization attempts to retain the solution quality of exact methods while extending to larger problem instances.

In the context of EV charging infrastructure planning, researchers have extensively used exact formulations (particularly MILP and MINLP solved with commercial solvers like CPLEX or Gurobi) to determine optimal charger placement and sizing. These studies, especially in the last 5 years, underscore some recurring **limitations** of a purely exact approach. One major issue is **poor scalability**. As the network size or the number of scenarios increases, even powerful MILP/MINLP solvers can become computationally infeasible, requiring many hours or drastic model simplifications to reach a result. This limits the applicability of exact methods to city- or regional-level planning and makes them unsuitable for real-time or iterative planning needs. A second challenge is the **heavy data requirement**—exact models demand extensive, high-fidelity input data (such as EV travel demand distributions, renewable generation profiles, and grid parameters). The optimal solutions obtained are highly sensitive to these input assumptions. Uncertainty in inputs is typically handled via scenario analysis or stochastic programming, but even then, the results can vary significantly with different scenario setups. If the assumed data (e.g., projected EV adoption or solar output) turns out to be inaccurate, an "optimal" solution may no longer perform well in reality. Table 13.2 summarizes recent studies that employed exact algorithms for R2V charging station optimization, highlighting their chosen solution methods and noted limitations. As shown, most works use MILP or MINLP solved by commercial optimization software, and they frequently report issues with scalability, data sensitivity, or restrictive assumptions.

Despite their rigor, exact approaches often rely on simplifying assumptions to remain tractable. Many models fix certain behaviors or parameters (e.g., assuming EV drivers follow deterministic charging patterns, using static electricity prices, or assuming perfect foresight of renewable output) to reduce complexity. Such assumptions mean that important real-world factors—for instance, stochastic driver behavior, temporal fluctuations in demand, or evolving policy incentives—are only coarsely represented or even ignored. This gap between model and reality can limit the realism of the results. Moreover, solutions obtained from exact optimization tend to be highly case-specific, optimized for the particular test network, data set, and scenarios used in the study. These solutions may not directly generalize to different cities or to future conditions without completely rerunning the optimization from scratch. In other words, many studies report one-off optimal solutions rather than broadly applicable design rules, so planners must be cautious in extrapolating these findings. Exact algorithm plans are also typically static—they produce a fixed infrastructure design that does not adapt easily to real-time changes or iterative updates in response to new data. Additionally, there is a lack of real-world field validation for most exact algorithm results: studies usually stop at simulation or offline analysis. Practical deployment challenges (such as data availability, computational resources on-site, regulatory and market constraints) are seldom discussed, so the true performance of these "optimal" plans in a live setting remains unproven.

In summary, exact algorithms offer a powerful foundation for computing optimal R2V charging station placement and sizing solutions, but they come with significant limitations in realism and scalability. Their high complexity, strict data demands, and idealized assumptions often constrain direct practical use. To leverage their strengths moving forward, researchers are increasingly looking at hybrid approaches—for example, combining exact solvers with heuristic or intelligent methods—to handle large-scale problems more efficiently. Another important direction is

Table 13.2 Recent studies have used exact optimization algorithms for R2V charging station placement and sizing.

Ref.	Year	Solution/Algorithm	Limitation
Kumbhar et al. (2025)	2025	MILP, solved with CPLEX	Complex model; results depend on data availability—necessitates iterative runs to converge.
Shamani et al. (2025)	2025	MINLP solved by CPLEX	Computationally intensive on large instances; relies on highly accurate input data.
Dhakar et al. (2025)	2025	MILP, scenario analysis	Does not scale well with scenario count; results are sensitive to scenario assumptions.
Abdulnasser et al. (2025)	2025	ε-constraint method + Pareto front generation	Very high modeling complexity; results depend on scenario setup, and the method is data-intensive.
Oladigbolu et al. (2025)	2025	CPLEX solver for MILP (exact approach)	Assumes fixed demand profile; focuses on technical factors while simplifying real-world user behavior and uncertainty.
Ghani and Farjah (2025)	2025	Two-stage: Analytical Hierarchy Process + MINLP (solver-based)	Uncertainty is handled by limited scenarios; it misses potential policy or behavior changes in reality.
Bhatia and Das (2025)	2025	MILP solved via CPLEX; scenario-based analysis	Model complexity relies on accurate uncertainty estimation, scalability for large MGs.
Ghaithan et al. (2024)	2024	MILP solved by CPLEX	Data requirements for RES profiles, solution scalability
Aduama et al. (2024)	2024	MINLP, Simulation, solved by Gurobi	High computational time, real-time data needs, scenario limitations.

(Continued)

Table 13.2 Recent studies have used exact optimization algorithms for R2V charging station placement and sizing.—cont'd

Ref.	Year	Solution/Algorithm	Limitation
He et al. (2024)	2024	MILP via Gurobi/CPLEX solvers	Assumes perfect data/scenarios, ignores certain market complexities.
Er et al. (2024)	2024	MILP via commercial solvers	Model is scenario/case-study dependent; may not generalize.
Ahmad and Kumar (2024)	2024	MINLP solved via optimization software	MINLP complexity, sensitive to input assumptions.
Singh, D'Arpino and Goveas (2024)	2024	MILP Optimization	Does not consider real-time uncertainty or dynamic user demand changes; the model relies on accurate cost and demand input.
Li et al. (2025)	2024	Location-Allocation Optimization	Static location planning (not dynamic); ignores real-time operational variability.
Feng et al. (2024)	2024	MCDA + Linear Programming	Requires expert judgment for criteria; data-intensive; not adaptive to fast-changing scenarios.
Zhang et al. (2024)	2024	Mixed-integer linear programming (MILP)	High computation for large-scale; results are sensitive to input data.
Raad et al. (2024)	2024	Hybrid p-robust, DRSO, with mixed-integer linear programming, Bender's decomposition	Complexity/scalability for large networks; needs accurate scenario data; real-time use is challenging.
Choi and Bhakta (2024)	2024	Mixed-integer programming; Info-Gap Decision Theory (IGDT) for risk/opportunity	Focus on a single system; robust to modeled uncertainties only; not for grid-tied settings.
Li et al. (2023)	2023	Linear/MPC-based optimization	Relies on forecast accuracy; implementation may be complex.
Sabzian-Molaee et al. (2023)	2023	Decomposition-based Stochastic Optimization	High computation, sensitive to scenario selection, and model scale.

Saadati et al. (2023)	2023	Bilevel MILP solved with optimization solvers	Assumptions about price elasticity, user response, and complexity.
Khazali et al. (2023)	2023	MINLP with advanced optimization (e.g., branch-and-bound)	Detailed battery degradation data required, computationally intensive.
Bao et al. (2023)	2023	MILP solved using commercial solvers	The model is for scheduling, not siting; practical implementation needs real-time data connectivity.
Félix (2020)	2023	MILP solved by commercial solver (exact optimization)	Grid constraints and costs may be region-specific; battery prices and lifespans can affect results.
Cardona et al. (2023)	2023	Model Predictive Control (MPC)	Model accuracy versus plant reality; battery under-sizing; focus on single demo site.
Saadati et al. (2022)	2022	Mixed-Integer Linear Programming, Scenario Analysis	A case study limited to a specific test system may not capture all uncertainties.
Cao et al. (2022)	2022	Stochastic Programming, Conic Optimization	High computational burden, resource problem complexity.
Ali et al. (2022)	2022	MINLP (solver-based: e.g., branch-and-bound)	Solving MINLP is resource-intensive and requires strong assumptions.
Mejia et al. (2022)	2022	MILP	Data requirement, increasing complexity for large systems.
Wu et al. (2022)	2022	MILP solved by commercial optimization software	Computationally demanding; reliant on the quality of emission and cost forecasts.
Tabandeh et al. (2022)	2022	MINLP solved by GAMS with branch-and-bound, Pareto optimization	Model/data heavy; regional grid topology; scenario dependence; complex to scale.
Ferro et al. (2022)	2022	Bi-level optimization with full AC power flow and user equilibrium constraints	Complex, data-intensive; assumes equilibrium models reflect real user behavior; single region.

(Continued)

Table 13.2 Recent studies have used exact optimization algorithms for R2V charging station placement and sizing.—cont'd

Ref.	Year	Solution/Algorithm	Limitation
Amer et al. (2021)	2021	Two-stage MILP solved with CPLEX	High computational complexity, scalability issues, and scenario reduction accuracy
Sa'adati, Jafari-Nokandi and Saebi (2021)	2021	MILP with iterative/bi-level solution	Convergence and computational time, bi-level model complexity
Adil et al. (2021)	2021	Stackelberg game, backward induction algorithm	Theoretical; does not implement a system; highlights challenges of SBG in a dynamic/multi-agent context

incorporating uncertainty and real-world variability into the models more robustly, so that solutions remain effective under a range of future conditions. By acknowledging the drawbacks of pure exact methods, the field is steering toward more resilient frameworks that still benefit from the precision of mathematical optimization. This means using exact algorithms judiciously (e.g., for smaller subproblems or as benchmarks) and focusing on techniques to overcome their weaknesses, such as automated decomposition, scenario reduction, and integration with data-driven or adaptive algorithms. Embracing these developments will help ensure that the theoretical optimums found by exact methods can be translated into practical, deployable solutions for sustainable EV charging infrastructure.

13.4.2 Classical heuristic algorithms

Classical heuristic algorithms provide more feasible, quicker solutions to complex combinatorial optimization problems compared to exact methods. In the context of R2V charging station planning, researchers have applied a variety of heuristic-based tools. Notable examples include simulation-based approaches like the HOMER energy modeling software (Nirbheram et al., 2025; Choi and Bhakta, 2025; Alhazmi, 2025; Khazali et al., 2024; Basem et al., 2024; Oladigbolu et al., 2023; Andrenacci et al., 2022; Strunz et al., 2021) and its updated version, HOMER Pro (Hossein et al., 2025; Gökçek et al., 2024; Karipoğlu et al., 2022; Al Wahedi and Bicer, 2022; Ekren et al., 2021), which use heuristic search and iterative simulation to size hybrid renewable systems. Other techniques range from fuzzy logic methods (e.g., a fuzzy Reliability–Availability–Operability, RAO-3, approach (Bhattacharjee et al., 2024)) and custom iterative search algorithms (Rizk-Allah et al., 2024), to simplified optimization procedures (e.g., a mixed-integer optimizer used in a heuristic fashion (Oladigbolu et al., 2023)), and multi-criteria decision analysis paired with heuristics like Analytic Hierarchy Process (AHP) (Andrenacci et al., 2022). These heuristic approaches do not guarantee a global optimum. Still, they can deliver good-enough solutions with significantly lower computation time, making them attractive for large-scale or fast-turnaround planning exercises. They are especially suitable for preliminary design, large energy system simulations, or scenario explorations where obtaining a strictly optimal answer is less critical than understanding general system behavior.

This paper investigates a PV–grid–hydrogen hybrid charging hub explicitly designed to serve both battery EVs and hydrogen fuel-cell vehicles under uncertain, time-varying arrivals; the authors aim to demonstrate that modeling multiple arrival distributions (random, uniform, normal) and co-optimizing component sizes with an improved search-space reduction (ISSR) method yields robust designs and stable costs. In the best random-arrival case, the optimal sizing is ~391 kW PV with a 54 kg H_2 tank (LCOE ≈ $0.3209/kWh), while sensitivity runs under uniform and normal arrivals remain resilient (LCOE ≈ $0.3164–0.3154/kWh); the study argues that such dual-fuel layouts hedge intermittency and grid constraints better than single-vector stations (Nirbheram et al., 2025). Targeting mixed energy demands at a tourist complex, this work seeks to show how an integrated hybrid renewable energy system (HRES)—coupling PV, wind, boiler/CHP, BESS, converters, and the main grid—can be planned to *simultaneously* meet electrical, thermal, and EV-charging loads across nine Iranian locations with diverse climates, using HOMER Grid for configuration and sensitivity analysis. The key findings are that optimal designs materially depend on local resource/climate profiles and grid presence; the framework quantifies trade-offs in NPC/LCOE, emissions, and capacity across locations, offering siting/sizing guidance for EVCS within multi-energy microgrids (Hossein et al., 2025). Focusing on highway corridors in Saudi Arabia, the author demonstrates a techno-economic design procedure—via HOMER Grid—for PV-powered fast-charging stations (200/300/500 kW) whose locations are chosen to address range anxiety and leverage local solar resources; the study's goal is to quantify feasible layouts, throughput, and costs under realistic driving/charging behavior. Results indicate that a 500 kW site yields ~3.21 GWh/year, the 300 kW option trades higher CAPEX for better efficiency, and a 200 kW site can charge ≈6100 vehicles/year; across selected roads the combined annual charging capacity exceeds 65,000 vehicles, with a reported total project cost of about USD 2.79 M—evidence that corridor-scale PV EVCS is technically and economically viable (Alhazmi, 2025).

Heuristic methods typically rely on expert experience, rules of thumb, and iterative improvement processes guided by feedback. For example, an analyst using HOMER or a similar tool will generate an initial design and then iteratively adjust system configurations (e.g., sizes of solar arrays, batteries, and number of chargers) to improve cost or performance metrics. An iterative search heuristic might run repeated simulations under different configurations, each time refining the decision variables based on previous outcomes. Compared to rigorous mathematical programming, heuristics are easier to implement and more adaptable to different problem variations or constraints. They require less formal problem definition and can incorporate custom rules or discontinuous constraints without much difficulty. Moreover, they often yield satisfactory results in far less time than exact methods, making them practical for early-stage planning or when an approximate solution is acceptable. Classic heuristics can be categorized into: (i) *construction methods* that build a feasible solution from scratch (for instance, HOMER's initial design based on load and resource inputs), (ii) *two-phase approaches* that combine simulation with some optimization or decision rules (HOMER's iterative optimization mode can be seen as this), and (iii) *improvement-focused iterations* that fine-tune a solution through parametric analysis or trial-and-error adjustments. While these approaches do not guarantee finding the absolute best configuration, they generally produce robust and *feasible designs* that meet most of the practical requirements. It is worth noting that some algorithms commonly termed "heuristic" in energy systems (like Genetic Algorithms or Particle Swarm Optimization) are actually meta-heuristics by strict definition. However, many recent works still refer to them broadly as heuristic methods. In the past 5 years, simple implementations of Genetic Algorithms (GA), Particle Swarm Optimization (PSO), and Simulated Annealing (SA) have indeed been used in R2V charging station studies (sometimes embedded within simulation tools) as straightforward optimization engines. These *classical heuristics* have achieved reasonable placement and sizing solutions, but often require substantial parameter tuning and may struggle as problem complexity grows. In large or rapidly evolving systems, even these heuristic optimizers can become computationally intensive, since they might need many iterations or populations to converge on a good solution. Their effectiveness is highly sensitive to algorithm parameters (e.g., population size in GA/PSO, cooling schedule in SA). Without careful tuning, a heuristic run may converge prematurely to a suboptimal point or wander inefficiently, yielding mediocre results. This parameter sensitivity limits the adaptability of classical heuristics—an algorithm configured for one scenario might perform poorly if the scenario changes, unless its parameters are readjusted.

Another concern is that while heuristics are valuable in simulated planning studies, their real-time performance or applicability is not always verified. Most classical heuristics work in a batch optimization mode rather than continuously adapting to streaming data. Further research is needed to speed up these algorithms and enable their use in near-real-time decision support (for instance, dynamically siting mobile charging units or adapting capacities as usage patterns change). Additionally, many heuristic-based studies rely on specific modeling platforms and detailed input datasets (for example, a particular region's driving patterns or solar irradiance profile). The solutions they produce are often limited to the scenarios or locations that have been tested. There is a clear need for more robust heuristic methods that can handle variability—for example, stochastic heuristics or extensive sensitivity analyses—to ensure solutions remain effective under different conditions.

Practical deployment of heuristic-derived solutions also faces challenges. Many advanced heuristic frameworks assume the presence of enabling infrastructure that may not exist in all regions. For example, some recent studies propose using blockchain-enabled smart contracts for EV charging and energy trading as part of the solution mechanism, which presupposes a sophisticated communication network and cybersecurity infrastructure in the grid. Implementing such technology at scale is non-trivial. The interconnected nature of smart charging networks raises cybersecurity risks (data integrity, privacy, and system control issues), and the robustness of classical heuristics against such threats is rarely evaluated. Many heuristic control schemes rely on a centralized controller managing multiple charging stations and vehicles, introducing single points of failure and reliability concerns (e.g., if communications fail or the central node is compromised). Furthermore, heuristic optimizations in literature often focus on a narrow set of objectives (for instance, minimizing cost, or maximizing reliability) or use weighted sums of objectives. This can omit important stakeholder preferences or broader impacts. In reality, decision-makers might want a portfolio of trade-off solutions (e.g., cost vs. environmental impact vs. user convenience), but classical heuristic runs typically return a single "optimal" point, thus limiting insight into trade-offs. Table 13.3 compiles a selection of recent papers that utilized classical heuristic algorithms for R2V station placement/sizing, with their approaches and limitations. These examples illustrate common issues like site-specific assumptions and a lack of generality in results.

Like any simplified approach, classical heuristics can oversimplify real-world complexities. For instance, many heuristic models of distribution networks neglect detailed electrical constraints such as voltage regulation or the effects of dynamic load interactions. These factors can significantly influence the feasibility of a charging station deployment, yet are often ignored to keep the problem manageable. To effectively move from paper studies to smart city implementations, a number of technical and regulatory hurdles must be addressed. The interplay with the power grid (voltage limits, protection, etc.), regulatory policies on energy sharing, and market

Table 13.3 Recent studies have used classical heuristic approaches for R2V charging station placement and sizing.

Ref.	Year	Solution/Algorithm	Limitation
Nirbheram et al. (2025)	2025	HOMER simulation, LCOE calculation	Real-world arrival data is hard to generalize; it ignores some practical H2 handling/infra issues.
Hossein et al. (2025)	2025	HOMER Pro, scenario simulation, cost optimization	Focused on the Iranian climate; assumes availability of all technologies everywhere; thermal integration challenges.
Choi and Bhakta (2025)	2025	HOMER simulation & scenario analysis	Local climate focus; tracker/tilt/maintenance assumptions; scaling may differ elsewhere.
Alhazmi (2025)	2025	HOMER Grid simulation evaluates 200/300/500 kW PV stations on major highways	Saudi climate/resource specific; focus on PV, no hybrid/backup modeled; only highways.
Bhattacharjee et al. (2024)	2024	Fuzzy Reliability, Availability, Operability (RAO-3) method	Fuzzification choices affect results; complexity for large networks.
Rizk-Allah et al. (2024)	2024	Iterative Search Algorithm	May not guarantee global optimality; time-consuming for large solution spaces.
Gökçek et al. (2024)	2024	HOMER Pro, scenario simulation	Only two regions/countries; LCOH sensitive to scale; operational constraints not deeply explored.
Khazali et al. (2024)	2024	HOMER simulation/optimization, scenario-based sensitivity	High initial cost; site-specific; current H2 cost is high; tech development needed for cost drop.

(Continued)

Table 13.3 Recent studies have used classical heuristic approaches for R2V charging station placement and sizing.—cont'd

Ref.	Year	Solution/Algorithm	Limitation
Basem et al. (2024)	2024	HOMER simulation; sensitivity analysis (inflation, PV/WT cost, solar/wind resource)	Site-specific; assumes static prices/demand; results not generalizable without local adjustment.
Oladigbolu et al. (2023)	2023	HOMER software simulation of PV, WT, ESS mixes; scenario/strategy analysis	Local meteorology key; optimal config depends on renewable potential; only one location.
Karipoğlu et al. (2022)	2022	HOMER Pro simulation & optimization	Site-specific data may limit generalization; limited to the Nigerian context; weather/data uncertainties.
Al Wahedi and Bicer (2022)	2022	HOMER Pro (derivative-free), scenario/sensitivity analysis	China-focused; reliability drops with larger load; omits grid/market constraints.
Andrenacci et al. (2022)	2022	HOMER optimization (simulation-based sizing), sensitivity analysis	Site/weather specific; capital cost high; scaling issues for other climates.
Strunz et al. (2021)	2021	Simulation & scenario analysis (HOMER or similar)	Assumes infrastructure sharing is feasible; implementation needs regulatory/social acceptance; HVDC cost.
Ekren et al. (2021)	2021	HOMER Pro, parametric/simulation optimization	Focused on one case study; scaling and load variations are not detailed.

mechanisms for pricing and incentives all need to be factored into the solutions.

Looking ahead, future research on R2V charging infrastructure is gravitating toward more integrated methods to overcome the limitations of stand-alone heuristics. One promising direction is the development of hybrid methods that combine classical heuristics with exact optimization or with machine learning tools. For example, a heuristic algorithm might rapidly generate a candidate solution, which is then fine-tuned by an exact solver, or conversely, an exact model might provide bounds that guide the heuristic search. Another trend is toward *uncertainty-aware optimization*, where heuristic algorithms incorporate stochastic elements or robust evaluation to ensure solutions perform well under variability (such as fluctuating renewable output or uncertain EV demand). In addition, researchers emphasize the importance of pilot projects and field demonstrations—deploying heuristic-optimized solutions in real networks to identify practical challenges (communications, control, user behavior) and iterating on the designs. By addressing these issues, the community aims to enhance the speed, adaptability, and reliability of heuristic optimization so that it can bridge the gap between theoretical planning and on-the-ground deployment.

13.4.3 Meta-heuristic algorithms

Meta-heuristic algorithms build upon classical heuristics by introducing stochastic search procedures and more sophisticated exploration–exploitation balances, which enable them to handle complex optimization problems more effectively within reasonable computation times. Meta-heuristics can broadly be classified into single-solution-based techniques and population-based techniques. Single-solution meta-heuristics improve a single candidate solution iteratively; for example, variants of Tabu Search were prominent in the early 2010s, and more recently, algorithms like the Jaya Optimization Method (Bhavana et al., 2025; Harsh and Das, 2022; Mohanty and Babu, 2020), the Ebola Search Optimization Algorithm (ESOA) (Zhu et al., 2024), or the Bald Eagle Search Algorithm (BESA) (Yuvaraj et al., 2024a) have been introduced. In contrast, population-based meta-heuristics maintain and evolve a pool of candidate solutions. This category includes well-known methods such as Particle Swarm Optimization (PSO) (Zheng and Zheng, 2025; Guindi and Kamel, 2024; Kumar et al., 2024; Gupta et al., 2024; Nareshkumar and Das, 2024) and Genetic Algorithms (GA) (Hisoglu et al., 2023; Jaganath et al., 2023; Leone et al.,

2022)—both widely used since the 2000s—as well as newer nature-inspired algorithms. Recent studies have explored a diverse array of novel population-based algorithms: Gray Wolf Optimizer (GWO, including a multi-objective version MOGWO) (Gönül et al., 2024a), Moth–Flame Optimizer (MFO) (Bilal et al., 2024), the Pufferfish Optimization Algorithm (POA) (Alanazi et al., 2024), the Giza Pyramids Construction Algorithm (GPCA) (Elomiya et al., 2024), various forms of Non-Dominated Sorting Genetic Algorithms (NSGA and NSGA-II) (Farjamipur et al., 2024; Tostado-Véliz et al., 2024; Gönül et al., 2024b; Li et al., 2023), the Honey Badger Algorithm (HBA) (Kaur et al., 2025), Spotted Hyena Optimization (SHOA) (Yuvaraj et al., 2024b), an Improved Multi-Objective Arithmetic Optimization Algorithm (MOIAOA) (Al-Sahlawi et al., 2024), Teaching–Learning-Based Optimization (TLBO) and its variants (Pagidipala and Sandeep, 2024; Krishnamurthy et al., 2023), a Chaotic Student Psychology-Based Optimization (CSPBO) (Balu and Mukherjee, 2023), a Self-adaptive Butterfly Optimization Algorithm (SABOA) (Pandraju and Janamala, 2022), the Levy Flight Honey Badger Algorithm (LFHBA) (Thumati et al., 2024), a Multi-Objective Differential Evolution Algorithm (MODEA) (Ali et al., 2023), the Gorilla Troop Optimizer (GTO) (Eid et al., 2022), the Political Optimization Algorithm (a differently-termed POA) (Dharavat et al., 2022), and the Archimedes Optimization Algorithm (AOA) (Zaki et al., 2021). This proliferation of meta-heuristic variants—many inspired by animal behavior, evolutionary processes, or physics—demonstrates the field's rapid innovation. Each algorithm introduces its own operators and parameters, but all share the goal of escaping local optima and efficiently searching a high-dimensional solution space.

This study proposes an Improved Human Evolutionary Optimization Algorithm (IHEOA) to co-optimize the *location* and *size* of hybrid fast-charging stations (HFCS) jointly deployed with PV, wind, capacitor banks, and D-STATCOMs on distribution feeders; the multi-objective goal is to cut losses, improve voltage stability, and lower GHG emissions under both G2V and V2G operations on IEEE 33-bus test cases. Methodologically, IHEOA adds an adaptive Gaussian random walk and a group-hopping strategy to strengthen exploration and avoid stagnation. Empirically, it outperforms baseline metaheuristics, achieving loss reductions of ~31% (G2V) and ~67% (V2G), AVD improvements of ~24% and ~32%, and GHG reductions of ~21% and ~96%, respectively—evidence that joint siting/sizing with reactive support meaningfully hardens the grid to fast-charging impacts

(Bhavana et al., 2025). A location-management framework for *distributed* EV charging stations integrates a Honey Badger Algorithm (HBA) for the charging operator's optimization with *smart contracts* that encode DR-program incentives and grid constraints from transmission and distribution operators. Architecturally, the system has prediction–optimization–control–blockchain layers and coordinates congestion/voltage-stability limits while responding to random EV arrivals and variable renewables; validation on a coupled transport–power testbed (incl. IEEE-33 network) shows the approach curtails renewable overproduction, stabilizes distribution operations, and improves operator profitability with modest management overheads (Kaur et al., 2025). Focusing on *layout planning* to boost renewable-energy consumption, this paper builds a two-stage model: (i) an optimization-configuration for a single PV–storage–charging station and (ii) a network layout model (load estimation, site selection, capacity sizing) solved by an *improved PSO* with randomized inertia, acceleration factors, and fast non-dominated sorting. Benchmarks show stronger convergence/accuracy than comparators; the worked example suggests per-station sizing around 500 kW PV and ~680 kW charger capacity with storage, and—critically—selects *seven* station sites as the cost-optimal network plan that both lowers construction cost and raises renewable uptake (Zheng and Zheng, 2025).

In recent years (c.2018–2024), there has been a focus on hybridizing algorithms and refining existing ones. Researchers often compare new meta-heuristics against established methods to quantify improvements in convergence speed or solution quality. For instance, enhanced versions of human evolutionary algorithms or improved variants of BESA have been benchmarked to highlight their strengths and weaknesses in terms of computational efficiency and robustness. The theoretical foundations of meta-heuristics are also being advanced, as seen in cross-domain applications and adaptations (even extending to emerging areas like blockchain-based smart contracts for energy). Given the plethora of algorithms, comprehensive comparative reviews are valuable to identify which techniques are most effective for particular problem characteristics. This active research area means that new meta-heuristic algorithms continue to appear, often with incremental improvements or specialized features, and it is crucial to match the algorithm to the problem context for best results.

For R2V charging station placement and sizing, meta-heuristics such as GA, Ant Colony Optimization (ACO), GWO, and PSO have become increasingly popular tools. They are commonly used to handle the multi-objective nature of the problem (e.g., minimizing cost while maximizing renewable usage and service level), where an exact solution might be impractical. Meta-heuristic approaches can search a vast solution space and provide a set of near-optimal solutions (for example, an approximation to the Pareto front in multi-objective cases). However, these methods face several challenges when applied to large-scale or real-time R2V optimization. First, many meta-heuristics are computationally demanding, especially in scenarios with numerous decision variables (many candidate sites, multiple time periods, etc.) or when a very high iteration count is needed for convergence. This can make them impractical for online applications or very large networks without significant simplification or parallelization. Their scalability to extremely complex grids (national-scale infrastructure, for example) remains uncertain, as most published studies test them on relatively modest case studies (e.g., a single city or a test distribution network).

Secondly, like simpler heuristics, meta-heuristics require careful parameter tuning and are not guaranteed to reach the true global optimum. They may converge prematurely to local optima if, say, the balance between exploration and exploitation is not well managed. The quality of results is thus highly sensitive to algorithm parameters (population size, mutation rate, cooling schedule, etc.) and even the initial population or starting solution. To mitigate this, multiple runs or hybrid algorithms (combining two meta-heuristics or adding problem-specific local search steps) are often employed, but that increases computational effort. Another limitation is that most meta-heuristic solutions in the literature are validated through simulation rather than deployment, meaning the algorithms optimize a static snapshot or a simulated environment. Real-world systems are dynamic—EV charging demand can vary unpredictably, and renewable output can fluctuate—whereas many meta-heuristic studies optimize a fixed scenario or a limited set of scenarios. As a result, the solutions might not maintain optimality or feasibility in a truly dynamic environment. Few studies have addressed maintaining *real-time consistency* (i.e., continuously updating the solution as conditions change), which is an important consideration for smart charging networks.

Moreover, meta-heuristic studies often assume that high-quality input data (from both the power grid and transportation system domains) is available and can be combined seamlessly. In practice, gathering and fusing such data (traffic flows, driver behavior patterns, renewable generation forecasts, grid constraints) is challenging, and uncertainties abound. Many implementations end up focusing on a subset of the problem (for instance, optimizing assuming an average daily load profile) due to limited data, which can limit the robustness of their outcomes. Solutions tend to be tailored to the specific data and scenario used; if applied elsewhere, their performance can degrade.

Table 13.4 provides an overview of recent meta-heuristic algorithm applications to the R2V charging station problem, including the type of algorithm used and the limitations reported. It reflects the breadth of techniques (from improved versions of classical algorithms to exotic new algorithms) and points out common issues like data dependence, scalability, and scenario-specific tuning.

A review of these studies reveals that many meta-heuristic applications simplify critical factors in the R2V optimization problem for tractability. For example, it is common to optimize assuming only one type of renewable energy (often solar PV) or to use a fixed EV charging demand profile, rather than capturing the full variability of multiple renewables and stochastic traffic patterns. Similarly, driver behavior and preferences are often modeled in a rudimentary way or not at all, and idealized conditions (such as perfect coordination or full information availability) are assumed. Such simplifications can undermine the validity of the results when applied to real systems, as they omit factors that could significantly affect station usage and economics (e.g., drivers' willingness to wait for charging, or fluctuations in renewable output). Furthermore, many "optimal" solutions proposed rely on advanced infrastructure and policy support that may not yet exist. For instance, some studies assume the presence of smart grids with real-time pricing, vehicle-to-grid support, or government incentives for renewable charging—favorable conditions that may not hold universally. They also often presuppose sophisticated EV and charger control capabilities (like every EV and charger being connected and dispatchable according to the algorithm's strategy), which in practice are not widely implemented in today's charging networks.

Another important limitation is the near-total lack of field trials or experimental validation of meta-heuristic optimized plans. The outcomes are typically specific to the test case (network topology, data, and assumptions) and site-specific, with no guarantee that the solution approach will work under different real-world conditions. Without pilot implementations, it is uncertain whether these optimized plans would maintain performance when faced with unmodeled issues like communication delays, driver non-compliance (e.g., not plugging in when expected), or equipment failures. Issues of network interaction (how multiple charging stations

Table 13.4 Recent studies have used meta-heuristic algorithms for R2V charging station placement and sizing.

Ref.	Year	Solution/Algorithm	Limitation
Bhavana et al. (2025)	2025	Improved Jaya Optimization Method	Only solar integration; user preference modeling approximated
Kaur et al. (2025)	2025	Honey Badger Algorithm (HBA) + smart contract system	Requires smart grid/smart contract infra; robustness to cyber risk
Zheng and Zheng (2025)	2025	Improved Particle Swarm Optimization (PSO)	Heavily model/data dependent; real-time adaptation not tested; only simulation-based validation
Gönül, Duman and Güler (2024a)	2024	Gray Wolf Optimizer (MOGWO), Multi-Objective Differential Evolution (MODE)	Scalability; parameter sensitivity; real system uncertainties
Bilal et al. (2024)	2024	Moth–Flame Optimizer (MFO)	Manual exploration for storage; cost assumptions
Alanazi et al. (2024)	2024	Pufferfish Optimization Algorithm (POA)	Weather/resource assumptions; only the specific region tested
Elomiya et al. (2024)	2024	Giza Pyramids Construction Algorithm (GPCA), compared with AHA, FPA, and WOA	Model input uncertainty; limited to specific climate/financial data
Yuvaraj et al. (2024b)	2024	SHOA, compared to BESA and PSO	Performance on larger, more realistic grids has not been tested; algorithm parameter tuning is needed
Al-Sahlawi et al. (2024)	2024	Improved Multi-Objective Arithmetic Optimization Algorithm (MOIAOA)	Results site-specific; high computational demand; only tested for one location

Guindi and Kamel (2024)	2024	Particle Swarm Optimization (PSO), Cuckoo Search Algorithm	Algorithm performance depends on initialization & tuning; scalability
Pagidipala and Sandeep (2024)	2024	TLBO, Non-dominated Sorting TLBO	Scalability, probabilistic modeling complexity
Zhu et al. (2024)	2024	Ebola Search Optimization Algorithm (ESOA)	Model relies on accurate renewable/traffic data; scenario-specific
Farjamipur et al. (2024)	2024	Improved Non-dominated Sorting Genetic Algorithm (NSGA)	Scalability, and it was only tested in simulation
Kumar et al. (2024)	2024	Particle Swarm Optimization (PSO); analytical & heuristic techniques	Dependent on weather/seasonal data, scenario assumptions
Gupta et al. (2024)	2024	PSO-based Optimal Power Flow (OPF) + Integrated Power Management (IPM)	Focused on 1 MG type; coordinated scheduling needs real data
Yuvaraj et al. (2024a)	2024	Bald Eagle Search Algorithm (BESA), Cuckoo Search Algorithm (CSA), comparative scenario analysis	Scalability, real-world complexity, focuses on the 33-bus test system, with limited ML integration
Tostado-Véliz et al. (2024)	2024	NSGA-II genetic algorithm	May miss the global optimum; the result depends on the GA parameters
Nareshkumar and Das (2024)	2024	Particle Swarm Optimization (PSO), Monte Carlo simulation	Requires extensive data, coupling may be difficult to implement in practice, focus on the 33-bus/25-node case
Gönül, Duman and Güler (2024b)	2024	NSGA-II (Non-dominated Sorting Genetic Algorithm II)	Site/data-specific; off-grid viability lower; detailed for one region/case only

(Continued)

Table 13.4 Recent studies have used meta-heuristic algorithms for R2V charging station placement and sizing.—cont'd

Ref.	Year	Solution/Algorithm	Limitation
Balu and Mukherjee (2023)	2023	Chaotic Student Psychology-Based Optimization (CSPBO), compared to SPBO/Harris Hawk	Model complexity, generalization to other systems.
Pandraju and Janamala, (2022)	2023	Self-adaptive Butterfly Optimization Algorithm (SABOA)	May require tuning for self-adaptation; scenario dependent
Thumati et al. (2024)	2023	Levy Flight Honey Badger Algorithm (LFHBA)	Dependent on the model data and scenario, real-time implementation has not been tested.
Krishnamurthy et al. (2023)	2023	TLBO	Model depends on test data; scalability.
Li et al. (2023)	2023	Non-dominated Sorting Genetic Algorithm II (NSGA-II)	Real-world implementation may face more uncertainty; the model depends on battery degradation accuracy, operational complexity.
Hisoglu et al. (2023)	2023	Genetic Algorithm (GA)	Heavily depends on expert weighting; only as good as input GIS data; dynamic factors not addressed.
Jaganath et al. (2023)	2023	Genetic Algorithm (GA), HOMER software	Focused on two locations; excludes large-scale grid effects; cybersecurity is only briefly mentioned.
Ali et al. (2023)	2023	Multi-Objective Differential Evolution Algorithm (MODEA)	Algorithm complexity, advanced control assumptions, no field trial data.

Eid et al. (2022)	2022	Gorilla Troop Optimizer (GTO)	A two-stage process may be computationally intensive, based on simulated profiles.
Dharavat et al. (2022)	2022	POA	Dependent on parameter selection and scenario realism, computation for large systems.
Harsh and Das (2022)	2022	JAYA algorithm + incentive-based DR aggregator scheme	Depends on aggregator reliability, tested only in simulation
Leone et al. (2022)	2021	Scalarization (to a single objective), solved with Genetic Algorithm (GA)	Scalarization can miss trade.-off info; it relies on test case, model input assumptions.
Mohanty and Babu (2020)	2021	JAYA Algorithm	Algorithm convergence; scenario dependence.
Zaki et al. (2021)	2021	Archimedes Optimization Algorithm (AOA), compared with PSO & CSA	Scenario-based, requires tuning; PV modeling simplifications.

might stress the local grid) and user acceptance (will EV drivers behave in the way the algorithm expects?) are frequently glossed over or treated in a cursory manner. This indicates an important gap between simulation results and practical deployment.

These limitations are driving new research directions. There is growing interest in developing faster, more efficient algorithms that can scale up or run in near-real time, potentially by leveraging parallel computing or by simplifying meta-heuristic operations without greatly sacrificing solution quality. Integrating machine learning techniques is another promising area—for example, using ML to dynamically adjust meta-heuristic parameters or to approximate the fitness function, thereby speeding up convergence. Additionally, explicitly incorporating uncertainties and human factors into the optimization is an active area: this includes robust or stochastic meta-heuristics that account for variability, and multi-agent simulations that model individual driver decisions in response to the charging system. The consensus in recent literature is that purely relying on meta-heuristics will not be enough; the future lies in hybrid approaches that combine the strengths of different techniques (exact, heuristic, and ML). Smarter algorithms—for instance, those that learn and adapt during the optimization process—are expected to produce solutions that are more robust and transferable to real-world smart charging networks. By continuing to refine these algorithms and addressing their current shortcomings, researchers aim to ensure that the next generation of optimization tools can reliably guide the deployment of R2V charging infrastructure under real-world complexities.

13.4.4 Hyper-heuristic algorithms

Despite the progress made with meta-heuristics, significant challenges remain in choosing the right algorithm and tuning its parameters for each new problem. Hyper-heuristic algorithms have been proposed as a higher-level approach to overcome these challenges by automating the selection and generation of heuristics. A hyper-heuristic can be thought of as a "heuristic to choose heuristics," operating in a two-layer framework. In the lower layer, a set of diverse low-level heuristics (LLH) is prepared—these could be simple constructive heuristics, local search moves, or even mini meta-heuristics tailored to different aspects of the problem. In the upper layer, a high-level strategy (HLH) monitors the search progress and decides which low-level heuristic to apply at each step, or how to combine them.

The high-level strategy may use rules, learning mechanisms, or meta-heuristic logic itself to manage the pool of LLHs. The result is an approach that can dynamically adapt the search behavior based on feedback, ideally producing high-quality solutions without the need for a human to manually fine-tune the process.

This paper advances EVEN (EV energy network) planning by (i) proposing a stochastic Markov (dis)charging–behavior model that embeds driver heterogeneity (range anxiety, price sensitivity), SoC, dynamic tariffs, and station geography to generate realistic spatiotemporal demand, and (ii) co-optimizing bidirectional charging-station (BCS) siting/sizing with a multi-objective deployment strategy that minimizes installation cost and drivers' electricity bills while guaranteeing QoS (waiting-time/accessibility) and voltage stability; a hybrid GA augmented with hill climbing and PSO improves search exploitation and evolution. Validations show a strong fit of the behavior model, effective parameter selection, and that enabling energy trading materially improves outcomes—with ultracapacitors emerging superior for trading—while policy simulations indicate feasible levers for driver profitability (Tang et al., 2025). Framed as distribution-system co-planning, this study targets simultaneous allocation of EVCS and renewable DGs (PV, wind) under uncertainty, seeking to reduce losses, improve voltage profiles, and lower operating costs. Uncertainty in RDG outputs is modeled via PDFs; candidate locations/sizes are screened with sensitivity factors and then optimized using a Fuzzy-Puzzle Optimization (FPO) algorithm. On an IEEE-33-bus microgrid across scenarios (single/multiple DGs, with/without EVs), coordinated placement and scheduling deliver notable loss/cost reductions and voltage enhancements, and comparative tests show FPO outperforming prior heuristic baselines on all major metrics (Thiruvenkadam et al., 2024). Focusing on PHEV-oriented networks with RES support, the authors cast CS planning as a multi-objective problem—building a cost-efficient charging network while respecting distribution-network operational attributes—and introduce a novel hybrid "Atom Search Woven Aquila Optimization" (AT-AQ) metaheuristic that blends Atom Search Optimization with Aquila Optimizer to mitigate local trapping and slow convergence. The framework explicitly incorporates station parameters (e.g., connectors, rated power) and accessibility/routing indices; experiments report that AT-AQ consistently minimizes the composite cost function better than comparators, underscoring its promise for practical CS siting/sizing with renewables (Subramaniam et al., 2023).

Current hyper-heuristic methods can be categorized by the strategy the high-level controller uses. Four broad categories are: random selection, greedy selection, meta-heuristic-based selection, and learning-based selection. In a *random-based* hyper-heuristic, the HLH picks among the available low-level heuristics uniformly at random or in some non-informed manner. This serves as a baseline and can occasionally find good solutions by chance, though it's generally less efficient. *Greedy-based* hyper-heuristics try whichever LLH gives the best immediate improvement: at each iteration, the algorithm may test several heuristics and choose the one that yields the largest gain in objective, which can be effective but may also lead to higher computation and risk of getting stuck in local optima. *Meta-heuristic-based* hyper-heuristics incorporate principles from existing meta-heuristics (like PSO, GA, etc.) at the high level—for example, using a genetic algorithm to evolve a sequence of LLH operations, or using PSO particles where each particle represents a particular heuristic to apply. These leverage proven meta-heuristic tactics to orchestrate the low-level moves. Lastly, *learning-based* hyper-heuristics employ machine learning, either offline or online, to inform which heuristic to use. They might analyze historical performance of heuristics on the problem (offline learning) or adapt in real-time as the search progresses (online learning), using reinforcement learning or other adaptive mechanisms. The aim of learning-based approaches is to improve over time, identifying which types of heuristics work best at different search stages or problem instances.

The hyper-heuristic concept was initially introduced by Cowling and later refined by Burke and others, emphasizing the idea of managing a set of heuristics rather than solving the problem directly. Recent research has proposed various hybrid hyper-heuristic models in the energy optimization domain. For example, one study developed an Atom Search Woven Aquila Optimization (AT-AQ) algorithm (Subramaniam et al., 2023), which hybridizes two nature-inspired methods (Atom Search and Aquila Optimizer) within a hyper-heuristic framework. Another introduced a Hybrid Fuzzy-Puzzle Optimization (FPO) algorithm (Thiruvenkadam et al., 2024), combining fuzzy logic with a puzzle-based heuristic approach. These complex hybrids illustrate the adaptability of hyper-heuristics—they can integrate domain knowledge (e.g., fuzzy rules based on expert insight) with iterative search to handle problem-specific challenges. In general, hyper-heuristics are valued for producing robust, generalizable solutions: by not being tied to one specific heuristic, they can, in principle, be applied to many types of optimization problems or changing scenarios with minimal retuning. This makes them attractive for complex, multi-factor problems like optimal EV charging station deployment, which involve technical, economic, and behavioral components.

However, applying hyper-heuristics to R2V charging station planning in recent years has revealed several limitations. Optimal placement and sizing of renewable-based EV charging stations is a multi-factor, multi-scale problem; hyper-heuristics have been tested in this area to handle its complexity. In practice, these approaches often require very comprehensive data from the real world (such as detailed driver behavior patterns, fine-grained market price signals, etc.) to perform well. The models guiding the LLHs or the decisions of the HLH can be quite elaborate. If high-quality, rich data is not available, researchers sometimes substitute assumptions or simplistic estimates, which can degrade the solution quality significantly. In essence, a hyper-heuristic is only as good as the low-level heuristics and feedback it has—with poor data or poorly designed LLHs, the high-level control cannot work miracles.

Additionally, hyper-heuristics tend to be highly parameterized and complex themselves. A given hyper-heuristic may have numerous parameters to set (for example, parameters controlling a learning mechanism, or thresholds for switching heuristics, etc.), as well as parameters within each LLH. Often, fuzzy hyper-heuristics require designing membership functions and rule bases. Success, therefore, depends on careful configuration, and setting all these parameters can be labor-intensive. If any key parameter is off, the entire approach might underperform or fail to find good solutions. This brittleness is a concern—the more complex the algorithm, the more ways it might need adjustment for a new problem instance.

The additional layer of search and control in hyper-heuristics also incurs a computational cost. By orchestrating multiple heuristics, hyper-heuristics can require significantly longer runtimes than using a single, well-chosen meta-heuristic. Some reported hyper-heuristic implementations have only been feasible on *small problem instances* (e.g., a limited number of candidate sites or scenarios) because scaling up led to prohibitively long execution times. Learning-based hyper-heuristics, in particular, might need a large number of iterations or training episodes to calibrate the high-level policy, which again increases computation. This makes it challenging to apply hyper-heuristics in real-time or very large-scale planning without further algorithmic efficiency improvements.

Another issue is that, while hyper-heuristics strive for generality, many proposed ones still show domain-specific limitations. A hyper-heuristic might perform excellently on the specific benchmark or test system for which it was developed (often outperforming single meta-heuristics there), but when tested on a different system or a scaled-up version of the problem, its performance can degrade. Novel algorithms may inadvertently be overfit to the features of the test cases used in development. Thus, an algorithm that works well for, say, one country's transportation and grid setup might need significant re-tuning or even structural changes to handle another country's system. Generalizing these approaches to new scenarios or larger networks often requires further development and testing, rather than the hyper-heuristic instantly working out of the box.

Researchers are aware of these challenges and are actively working to address them. Efforts include developing methods for automated parameter tuning (so the hyper-heuristic can self-adjust or be auto-configured for a new problem), incorporating real-time data feeds so the hyper-heuristic can adjust its strategy on the fly, and creating more lightweight or adaptive frameworks that trim unnecessary complexity. The ultimate goal is to harness the strengths of hyper-heuristics—their generality, flexibility, and high-level automation—while mitigating their current weaknesses, such as slowness and brittleness. For R2V charging station planning specifically, it will be crucial to prove that hyper-heuristic solutions are robust (i.e., they perform well even when reality deviates from assumptions), efficient enough to handle realistic problem sizes, and validated with empirical data before they can be trusted for real-world planning decisions. Each limitation of today's hyper-heuristics can be viewed as an opportunity for innovation. For example, if computational cost is a barrier, researchers might develop more parallelizable or problem-specific LLHs to speed up the search. If learning is slow, one might use transfer learning (leveraging knowledge from one region to another) to reduce training time. By continuing to refine hyper-heuristic techniques in these ways, future studies aim to produce algorithms that truly deliver on the promise of "plug-and-play" optimization for complex problems like renewable-based EV infrastructure planning.

Table 13.5 below lists a few recent hyper-heuristic algorithm applications for the R2V problem, along with their approaches and noted limitations.

13.4.5 Machine learning

The iterative search processes of meta-heuristic algorithms often generate a wealth of data (e.g., fitness evaluations, patterns of successful moves) that

Table 13.5 Recent studies have used hyper-heuristic algorithms for R2V charging station placement and sizing.

Ref.	Year	Solution/Algorithm	Limitation
Tang et al. (2025)	2025	Hybrid: Markov stochastic model + hybrid PSO/GA/Hill Climbing	Model complexity requires real behavioral/market data.
Thiruvenkadam et al. (2024)	2024	Hybrid Fuzzy-Puzzle Optimization Algorithm (FPO)	Computational complexity, dependent on proper fuzzy membership and parameter tuning.
Subramaniam et al. (2023)	2023	Atom Search Woven Aquila Optimization (AT-AQ): a hybrid of Atom Search & Aquila Optimizer	Novelty may hide untested pitfalls, a high computational burden.
Rao et al. (2023)	2023	Mexican Axolotl Optimization + Wild Horse Optimizer (MAOWHO, hybrid metaheuristic)	Model/algorithm complexity; generalizability.

traditional methods do not explicitly leverage for learning. Machine learning (ML) techniques have recently been applied to tap into this latent information and enhance optimization for R2V charging infrastructure. In particular, *deep learning* models and *reinforcement learning* (RL) algorithms are being explored to either improve the optimization process or to serve as decision-making agents directly. Radial Basis Function Networks (RBN) and other neural network architectures have been trained to approximate complex mappings in the system (Draz et al., 2024). A specialized Multi-Objective Neural Network Algorithm (MONNA) was proposed to handle multiple objectives in an integrated learning-optimization framework (Mehouachi et al., 2023). The idea is that ML algorithms can learn patterns from data (historical or simulated) to inform the search for optimal solutions, potentially speeding up convergence or avoiding poor regions of the search space.

Heo & Chang propose a city-scale siting–sizing framework for fast-charging stations that *explicitly* balances charging demand and renewable supply by training an Advantage Actor–Critic (A2C) deep-RL agent on geospatial layers (land use, existing stations, traffic counts, rooftop-PV potential) to learn placement and capacity decisions without preset weights; in a Chicago case, the model strategically concentrates stations in zones with both high traffic and high solar potential and—under a 1000-station scenario—achieves a best total-reward of ~74.3% while also estimating per-site capacities, outperforming small-area MCDM/path-planning approaches and highlighting remaining generalization limits (e.g., boundary effects, single-city training) (Heo and Chang, 2024). Draz et al. develop a techno-economic design and *online* energy-management scheme for isolated microgrid-based fast-charging stations powered 100% by RES (PV, wind) with high-power energy storage (flywheels, supercapacitors); after optimizing component sizes under resource/operational uncertainties, a radial-basis deep network (RBN) is trained/validated to dispatch in seconds, reproducing optimized outputs with very low errors and adapting to perturbed weather/load scenarios (e.g., shifting PV/WT splits and flywheel strings)—thereby cutting computation versus conventional optimizers while meeting reliability/cost KPIs (Draz et al., 2024). Myint et al. cast microgrid co-planning with EVCS as a multi-agent deep-RL problem on the IEEE-33-bus system—co-optimizing the *locations* and *sizes* of RES (PV, wind) and BESS to minimize losses and improve voltages under residential, commercial, and mixed loads—comparing MAA2C, MADDPG, and MASAC; across 150-episode runs, MASAC is most stable and consistently finds superior placements/sizes, demonstrating the promise of MADRL for distribution-level siting/sizing in EV-integrated grids (Myint et al., 2024).

Machine learning is typically categorized into supervised, unsupervised, and reinforcement learning. Of these, reinforcement learning—particularly deep reinforcement learning (DRL)—has gained traction for optimization problems because it involves an agent learning to make decisions through trial-and-error interactions with an environment. In the R2V context, one can imagine an RL agent learning policies for, say, how to allocate charging capacity or where to open a new station, by maximizing a reward (e.g., overall system efficiency or profit) over many simulated episodes. Techniques like Deep Q-Networks (DQN) (Wen et al., 2022) and Advanced Actor–Critic (A2C) methods have been applied to such problems. Model-free RL algorithms (which do not require an explicit model of the environment's dynamics), such as Q-learning and A2C, have shown promising results in complex decision environments, as they can learn effective strategies purely from feedback (rewards) by exploring different actions. Multi-agent DRL approaches have also been introduced, wherein multiple agents (for example, each representing a charging station or an EV) learn to make decisions in a shared environment. Such frameworks can capture the interactive, real-time nature of smart charging networks better than single-agent setups, potentially improving solution quality and speed in a distributed manner.

The integration of deep neural networks with RL has significantly advanced sequential decision-making capabilities. For instance, deep neural nets can approximate the value function or policy function in high-dimensional spaces, enabling RL to handle more complex state representations than was previously possible. Research in this domain has demonstrated the feasibility of using DRL for combinatorial optimization tasks; however, many applications so far are on relatively *static instances* of problems (fixed environment setups), and only a few have tackled fully *dynamic environments* where conditions change over time. Nonetheless, there is a clear trend toward using DRL and related architectures for dynamic path optimization, adaptive scheduling, and other sequential planning problems in the EV charging context. Early results indicate that these advanced ML-based approaches can achieve higher solution quality and faster convergence compared to certain traditional heuristics, especially as problem complexity grows.

Despite this promise, current ML-based approaches in R2V optimization share a number of common limitations. They are typically data-hungry, requiring extensive high-quality datasets for training. Acquiring such data (for example, detailed EV movement data, fine-resolution solar output data, etc.) can be difficult, and if the data are not representative, the learned models will be suboptimal. The models themselves (deep neural networks, multi-agent frameworks) are quite complex, raising issues of interpretability and trust. Planners may be wary of relying on a "black box" model whose decision logic is not transparent. Moreover, these models often need to be retrained or fine-tuned for each new context—an ML model trained on one city's data may not generalize to another city with different traffic patterns or grid characteristics. This lack of transferability means the approach might not save effort in new cases unless transfer learning or robust generalization techniques are used. Advanced DRL methods come with their own issues: training can be unstable (the learning process might diverge without careful tuning of learning rates, exploration rates, etc.), and the computational load for training is heavy. Even after training, the policies learned by RL are heuristic in nature—they do not guarantee an optimal solution, and their performance can degrade if the environment changes beyond what was in the training data.

Indeed, ML-based optimization methods remain largely heuristic in that they provide no guarantee of global optimality and typically require extensive tuning (of network architectures, hyperparameters, reward shaping in RL, etc.). The solutions they produce can be hard to explain (lacking interpretability), which can be a problem for gaining stakeholder acceptance or for diagnosing why a certain charging station was recommended by the model. Many of these trends underscore a reality: while ML offers powerful new tools to tackle the R2V station planning problem, practical application is currently limited by concerns about data availability, scalability, and transparency. Table 13.6 summarizes several recent applications of machine learning in R2V charging infrastructure optimization, including the approaches used and their limitations. These illustrate how ML is being woven into the research tapestry (from prediction tasks to direct control via RL) and the key challenges encountered.

Overall, the incorporation of ML into R2V charging station optimization is in its nascent stages, and studies to date reveal some overarching themes. Data and model complexity are prominent concerns: the success of ML models hinges on having rich datasets and managing very complex models, which not all planners or utilities may have the resources to obtain or run. Many ML-driven models for this problem are essentially prototypes, often demonstrated on simplified representations or subset problems. They highlight the *potential* for improved adaptivity and decision-making but also show that significant development is needed before such models can be routinely used in practice. There is also a trade-off between optimality and practicality—ML methods, like meta-heuristics, do not guarantee optimal solutions and can sometimes be outperformed by simpler optimization methods if not properly trained.

Moving forward, research should focus on bridging these gaps. One priority is addressing the data challenge: finding ways to train ML models with limited data (through transfer learning, synthetic data generation, or federated learning across regions) and ensuring data quality. Another is improving scalability—both of training algorithms (so that multi-agent DRL or large neural nets can be trained faster, possibly using cloud or distributed computing) and of the resulting policies (so they can handle growth in EV numbers or charging stations without retraining from scratch). Enhancing the transparency and interpretability of ML models will also be important, as decision-makers need to understand or trust the reasoning behind infrastructure plans generated by an AI. Hybrid models that combine ML with physical insights or optimization constraints are a

Table 13.6 Recent studies have used machine learning techniques for R2V charging station planning.

Ref.	Year	Solution/Algorithm	Limitation
Heo and Chang (2024)	2024	Advantage Actor-Critic Deep Reinforcement Learning (A2C DRL) + Geospatial Analysis	Data and computational demand, policy transferability, may require retraining for new cities.
Draz et al. (2024)	2024	Radial Basis Network (RBN) Deep Learning	Requires system-wide data, mostly off-grid, may not model grid interaction fully.
Myint et al, (2024)	2024	Multi-Agent Deep Reinforcement Learning: MASAC, MADDPG, MAA2C	RL instability, computational cost, applicability to larger/more complex grids.
Mehouachi et al. (2023)	2023	MONNA: Multi-Objective Neural Network Algorithm	Focus on Tunis city, generalizability to other locations, data dependency, and interpretability of NN.
Wen et al. (2022)	2022	Deep Q-Learning (Reinforcement Learning)	Requires large training data, model generalizability, and non-deterministic optimality.

promising route, as they can enforce known requirements (like grid constraints) while letting the ML handle uncertain or human-behavior aspects.

Future research avenues include using AI to predict future charging demand more intelligently and incorporate those predictions into planning models. For example, advanced algorithms could forecast where and when EV charging will be needed with high spatiotemporal resolution, enabling proactive and flexible planning of station deployment. Another important direction is the deeper integration of renewable energy and smart charging—using ML to optimize not just the placement of chargers but their operation in conjunction with renewable generation and energy storage, to minimize carbon footprint and grid impact. This could involve real-time control policies that adjust charging rates based on renewable availability, which ML (especially RL) is well-suited to handle. As these research threads progress, we anticipate that artificial intelligence (AI) and ML will play an increasingly significant role in improving the accuracy, adaptability, and efficiency of R2V charging infrastructure optimization models. The long-term vision is a smart planning system that continuously learns and updates EV infrastructure deployment plans in response to emerging data and trends, ultimately accelerating the transition to sustainable and well-optimized EV charging networks.

13.5 Conclusions and recommendations

Through a bibliometric analysis, we offered an extensive overview of optimal sizing for research papers on R2V charging stations covering the years 2010–2024. Our examination encompassed bibliometric data on yearly publication counts, subject areas, source journals, countries, institutions, and keywords. The findings indicated a significant increase in research publications related to optimal sizing for R2V charging stations since its introduction in 2010. At present, the annual publication rate is on the rise, reflecting that an increasing number of researchers are exploring this subject. India, China, the United States, the United Kingdom, and Iran lead in research output for this area. The top five journals by publication volume on optimal sizing for R2V charging stations are Energies, IEEE Access, Applied Energy, and the Journal of Energy Storage. By using keywords and co-citation network analysis, we identified the current research hotspots and trends in optimal sizing for R2V charging stations. The visualization map revealed that Meta-Heuristic algorithms currently represent the largest segment in optimal sizing for R2V charging station research, followed by Exact algorithms. Metaheuristic algorithms are the most prevalent for addressing optimal sizing for R2V charging stations, while hyper-heuristic and machine learning algorithms are gaining traction. Through a literature visualization map, we were able to ascertain the current landscape, hotspots, and emerging trends in optimal sizing for R2V charging station research. In addition, this review presented a comprehensive classification of R2V charging station sizing algorithms based on the latest research and taxonomy, providing detailed references to help practitioners understand the field.

Bibliometric visualization map analysis provides researchers with a systematic method to understand the evolution of research fields. It helps to identify potential knowledge within the research area and obtain the latest information using various sources of information (Chen, 2017). However, bibliometric techniques come with specific drawbacks. First, for experienced researchers, the outcomes of bibliometric techniques might be inflated due to contentious research material, adverse citations, etc (Taylor and Taylor, 2009). Second, conducting a co-occurrence analysis of keywords to monitor research themes and trends may be influenced by the search data. The establishment of search keyword parameters can significantly impact the ultimate search findings, and the range of data coverage might also be restricted by the search source (such as relying on a single data source). For further details on assessing the influence of search strategies on outcomes, readers may consult Chen (2017). A more thorough examination of the research subject through suitable search techniques, alongside contextual analysis and other facets of literature review, will yield more insightful analytical results (Chen, 2017).

Overall, in the past decade, research on optimal sizing for R2V charging stations has made significant progress and received increasing attention. With the deepening of research on sizing problems and the complexity of real location problems, more and more variants of algorithms have been introduced that are closer to practical situations. The development of optimization algorithms for the sizing of R2V charging station models has progressed quickly. However, despite economic growth and rising environmental protection demands, along with the integration of technologies like machine learning and cloud computing in logistics, research into optimal sizing and its solving algorithms for R2V charging stations remains inadequate. There are significant challenges in modeling and optimizing R2V stations that need urgent attention. Specifically, model construction must account for uncertainty factors in different R2V station variants, including demand, speed, and travel time, which are treated as uncertain parameters in the GVRP field. Nevertheless, factors associated with the sustainability of the supply chain network should not be overlooked (including social issues, customer readiness, and operational threats). Additionally, the objective function serves as a fundamental aspect of the mathematical models for R2V stations and holds significant consequences for forthcoming research on R2V stations. Currently, most research centers on single-objective optimization due to its simplicity, while multi-objective optimization, particularly for R2V station models with multiple constraints, has received less attention. Future studies should focus on incorporating multiple objectives and developing efficient algorithms for real-world scenarios, such as creating multi-objective sustainable R2V stations with mixed constraints. Unmanned aerial vehicles (UAVs) and unmanned ground vehicles, or drones, are increasingly being used to solve last-mile delivery issues.

Additionally, incorporating multiple features into realistic models remains a challenge in the research on optimal sizing for R2V charging stations. Most algorithms have high specificity and cannot be applied to other optimizing variants; hence, their universality is relatively low. Furthermore, most optimization algorithms target large, complex problems, but ensuring solution quality remains a significant challenge. Currently, heuristic and metaheuristic algorithms are the most popular methods.

Furthermore, emerging technologies like machine learning, reinforcement learning, distributed systems, the Internet of Vehicles (IoV), and new fuel technologies will be crucial for advancing R2V station sizing research. The effective application of machine learning, precise algorithms, and hyper-heuristic algorithms to create hybrid algorithms for enhancing the quality and robustness of solutions is a future research focus for optimizing R2V station challenges. As real-world scenarios become more complex and technology advances quickly, it's important to define different models of R2V stations clearly. We need to research these models and their solutions more deeply. Using hybrid algorithms and machine learning can help us find efficient solutions. Efficient planning of R2V charging infrastructure ensures that electric mobility growth is supported by clean energy, maximizing emission reductions. By optimally locating and sizing charging stations with renewables, one can minimize fossil fuel use and improve grid stability through distributed generation–outcomes very much in line with sustainable energy goals.

References

Abdulnasser G, Mohamed EEM, Shaaban MF, Ali A. A multi-objective strategic planning of smart energy hubs and hydrogen refueling stations toward net-zero emissions. Sustain Energy Grids Netw 2025;42. https://doi.org/10.1016/j.segan.2025.101690.

Adil M, Mahmud MAP, Kouzani AZ, Khoo SY. Energy trading among electric vehicles based on Stackelberg approaches: a review. Sustain Cities Soc 2021;75. https://doi.org/10.1016/j.scs.2021.103199.

Aduama P, Al-Sumaiti AS, Al-Hosani KH, El-Shamy AR. Optimizing quasi-dynamic wireless charging for urban electric buses: a two-scenario mathematical framework with grid and PV-battery systems. Heliyon 2024;10(15). https://doi.org/10.1016/j.heliyon.2024.e34857.

Ahmad A, Kumar O. Optimal sizing and allocation of solar based distributed generation and wireless charging station for transportation electrification. Sci Iran 2024;0(0). https://doi.org/10.24200/sci.2024.59140.6073.

Al Wahedi A, Bicer Y. Techno-economic optimization of novel stand-alone renewables-based electric vehicle charging stations in Qatar. Energy 2022;243:123008. https://doi.org/10.1016/j.energy.2021.123008.

Alanazi F, Bilal M, Armghan A, Hussan MR. A metaheuristic approach based feasibility assessment and design of solar, wind, and grid powered charging of electric vehicles. IEEE Access 2024;12:82599–621. https://doi.org/10.1109/ACCESS.2024.3411291.

Alhazmi Y. Techno-economic design analysis of electric vehicle charging stations powered by photovoltaic technology on the highways of Saudi Arabia. Energies 2025;18(2):315. https://doi.org/10.3390/en18020315.

Ali A, Shaaban MF, Awad ASA, Azzouz MA, Lehtonen M, Mahmoud K. Multi-objective allocation of EV charging stations and RESs in distribution systems considering advanced control schemes. IEEE Trans Vehicular Technology 2023;72(3):3146–60. https://doi.org/10.1109/TVT.2022.3218989.

Ali H, Francois B, Brotcorne L. Optimal siting and sizing of charging stations for autonomous electric vehicle fleet with local renewable energy generation. In: CIRED Porto Workshop 2022: E-mobility and power distribution systems, 2022; 2022, June. p. 1168–72.

Al-Sahlawi AAK, Ayob SM, Tan CW, Ridha HM, Hachim DM. Optimal design of grid-connected hybrid renewable energy system considering electric vehicle station using improved multi-objective optimization: techno-economic perspectives. Sustainability (Switzerland) 2024;16(6). https://doi.org/10.3390/su16062491.

Amer A, Azzouz MA, Azab A, Awad ASA. Stochastic planning for optimal allocation of fast charging stations and wind-based DGs. IEEE Syst J 2021;15(3):4589–99. https://doi.org/10.1109/JSYST.2020.3012939.

Andrenacci N, Di Monaco M, Tomasso G. Influence of battery aging on the operation of a charging infrastructure. Energies 2022;15(24):9588. https://doi.org/10.3390/en15249588.

Balu K, Mukherjee V. Optimal allocation of electric vehicle charging stations and renewable distributed generation with battery energy storage in radial distribution system considering time sequence characteristics of generation and load demand. J Energy Storage 2023;59:106533. https://doi.org/10.1016/j.est.2022.106533.

Bao Z, Li J, Bai X, Xie C, Chen Z, Xu M, et al. An optimal charging scheduling model and algorithm for electric buses. Appl Energy 2023;332:120512. https://doi.org/10.1016/j.apenergy.2022.120512.

Basem A, Elbarbary ZMS, Atamurotov F, Abdullayeva I, Reyimberganov A, Benti NE. Optimal sizing of PV/wind/diesel generator/battery hybrid system for supplying

electrical vehicle charging station under different load demands in Saudi Arabia. Int J Low-Carbon Technol 2024;19:2522–39. https://doi.org/10.1093/ijlct/ctae190.
Bhatia AA, Das D. Demand response strategy for microgrid energy management integrating electric vehicles, battery energy storage system, and distributed generators considering uncertainties. Sustain Energy Grids Netw 2025;41. https://doi.org/10.1016/j.segan.2024.101594.
Bhattacharjee B, Sadhu PK, Ganguly A, Naskar AK. Using fuzzy systems for optimal network reconfiguration of a distribution system with electric vehicle charging stations and renewable generation. Microsyst Technol 2024;30(10):1381–92. https://doi.org/10.1007/s00542-023-05543-5.
Bhavana K, Lalitha K, Revana G, Amaleswari R. An improved human evolutionary optimization algorithm for optimal integration of hybrid fast charging stations. Int J Intell Eng Syst 2025;18(4):813–25. https://doi.org/10.22266/ijies2025.0531.52.
Bilal M, Rizwan M, Rizwan M. Electric vehicles in a smart grid: a comprehensive survey on optimal location of charging station. IET Smart Grid 2020;3(3):267–79. https://doi.org/10.1049/iet-stg.2019.0220.
Bilal M, Bokoro PN, Sharma G, Pau G. A cost-effective energy management approach for on-grid charging of plug-in electric vehicles integrated with hybrid renewable energy sources. Energies 2024;17(16):4194. https://doi.org/10.3390/en17164194.
Cao X, Sun X, Xu Z, Zeng B, Guan X. Hydrogen-based networked microgrids planning through two-stage stochastic programming with mixed-integer conic recourse. IEEE Trans Autom Sci Eng 2022;19(4):3672–85. https://doi.org/10.1109/TASE.2021.3130179.
Cardona P, Costa-Castelló R, Roda V, Carroquino J, Valiño L, Serra M. Model predictive control of an on-site green hydrogen production and refuelling station. Int J Hydrogen Energy 2023;48(47):17995–8010. https://doi.org/10.1016/j.ijhydene.2023.01.191.
Chen C. Science mapping: a systematic review of the literature. J Data Info Sci 2017;2(2):1–40. https://doi.org/10.1515/jdis-2017-0006.
Choi Y, Bhakta S. Hybrid solar photovoltaic-wind turbine system for on-site hydrogen production: a techno-economic feasibility analysis of hydrogen refueling Station in South Korea's climatic conditions. Int J Hydrogen Energy 2024;93:736–52. https://doi.org/10.1016/j.ijhydene.2024.11.037.
Choi Y, Bhakta S. Optimal sizing of grid-tied hybrid solar tracking photovoltaic/hydrogen fuel cell energy systems for electric vehicle charging stations in South Korea: a techno-economic study. J Clean Prod 2025;486:144511. https://doi.org/10.1016/j.jclepro.2024.144511.
Dhakar M, Wang M, Rafiee A, Liu Y, Skryabin I, Khalilpour K. Sustainable mobility with renewable hydrogen: a framework for refueling station design and optimisation using liquid organic hydrogen carriers (LOHCs). Sci Rep 2025;15(1). https://doi.org/10.1038/s41598-025-91649-x.
Dharavat N, Sudabattula SK, Velamuri S, Mishra S, Sharma NK, Bajaj M, et al. Optimal allocation of renewable distributed generators and electric vehicles in a distribution system using the Political optimization algorithm. Energies 2022;15(18). https://doi.org/10.3390/en15186698.
Draz A, Othman AM, El-Fergany AA. Optimal techno-economic assessment of isolated microgrid integrated with fast charging stations using radial basis deep learning. Sci Rep 2024;14(1). https://doi.org/10.1038/s41598-024-70063-9.
Eid A, Mohammed O, El-Kishky H. Efficient operation of battery energy storage systems, electric-vehicle charging stations and renewable energy sources linked to distribution systems. J Energy Storage 2022;55:105644. https://doi.org/10.1016/j.est.2022.105644.

Ekren O, Hakan Canbaz C, Güvel ÇB. Sizing of a solar-wind hybrid electric vehicle charging station by using HOMER software. J Clean Prod 2021;279. https://doi.org/10.1016/j.jclepro.2020.123615.

Elomiya A, Křupka J, Simic V, Švadlenka L, Průša P, Jovčić S. An advanced spatial decision model for strategic placement of off-site hydrogen refueling stations in urban areas. eTransportation 2024;22:100375. https://doi.org/10.1016/j.etran.2024.100375.

Er G, Soykan G, Canakoglu E. Designing an optimal multi-energy system with fast charging and hydrogen refueling station under uncertainties. Sustain Energy Grids Netw 2024;39: 101403. https://doi.org/10.1016/j.segan.2024.101403.

Farjamipur M, Lotfi H, Nikkhah MH. Simultaneous optimal allocation of EVCSs and RESs using an improved genetic method. IET Circ Devices Syst 2024;2024(1). https://doi.org/10.1049/2024/4798197.

Fazelpour F, Vafaeipour M, Rahbari O, Rosen MA. Intelligent optimization to integrate a plug-in hybrid electric vehicle smart parking lot with renewable energy resources and enhance grid characteristics. Energy Convers Manag 2014;77:250–61. https://doi.org/10.1016/j.enconman.2013.09.006.

Félix PMC. Battery energy storage system optimal sizing in a battery electric vehicle fast charging infrastructure (Master's thesis, Instituto Politecnico do Porto (Portugal)). 2020.

Feng J, Guo P, Xu G, Xu G, Ning Y. An integrated decision framework for resilient sustainable waste electric vehicle battery recycling transfer station site selection. Appl Energy 2024;373:123897. https://doi.org/10.1016/j.apenergy.2024.123897.

Ferro G, Minciardi R, Parodi L, Robba M. Optimal planning of charging stations in coupled transportation and power networks based on user Equilibrium conditions. IEEE Trans Autom Sci Eng 2022;19(1):48–59. https://doi.org/10.1109/TASE.2021.3066001.

Ghaithan AM, Mohammed A, Al Hanbali A, Alshibani A. A mixed integer linear programming model for optimal sizing of hydrogen refueling station powered using a PV-grid system. Arabian J Sci Eng 2024:1–16.

Ghani R, Farjah E. A two-stage approach for determination of optimal size and site of distributed generations and charging stations in the distribution system, considering load growth and the electric vehicle penetration. Energy 2025;319:134953. https://doi.org/10.1016/j.energy.2025.134953.

Gökçek M, Paltrinieri N, Liu Y, Badia E, Dokuz AŞ, Erdoğmuş A, et al. Optimum sizing of hybrid renewable power systems for on-site hydrogen refuelling stations: case studies from Türkiye and Spain. Int J Hydrogen Energy 2024;59:715–29. https://doi.org/10.1016/j.ijhydene.2024.02.068.

Gönül Ö, Duman AC, Güler Ö. A comprehensive framework for electric vehicle charging station siting along highways using weighted sum method. Renew Sustain Energy Rev 2024a;199. https://doi.org/10.1016/j.rser.2024.114455.

Gönül Ö, Duman AC, Güler Ö. Multi-objective optimal sizing and techno-economic analysis of on- and off-grid hybrid renewable energy systems for EV charging stations. Sustain Cities Soc 2024:115. https://doi.org/10.1016/j.scs.2024.105846.

Guindi M, Kamel RM. Optimal location and sizing of renewable distributed generations and electric vehicle charging stations. Renew Energy 2024;235:121272. https://doi.org/10.1016/j.renene.2024.121272.

Gupta K, Achathuparambil Narayanankutty R, Sundaramoorthy K, Sankar A. Optimal location identification for aggregated charging of electric vehicles in solar photovoltaic powered microgrids with reduced distribution losses. Energy Sources, Part A Recovery, Util Environ Eff 2024;46(1):6289–304. https://doi.org/10.1080/15567036.2020.1745335.

Harsh P, Das D. Optimal coordination strategy of demand response and electric vehicle aggregators for the energy management of reconfigured grid-connected microgrid. Renew Sustain Energy Rev 2022;160:112251. https://doi.org/10.1016/j.rser.2022.112251.

He L, Ke N, Mao R, Qi W, Zhang H. From curtailed renewable energy to green hydrogen: infrastructure planning for hydrogen fuel-cell vehicles. Manuf Serv Oper Manag 2024; 26(5):1750–67. https://doi.org/10.1287/msom.2022.0381.

Heo J, Chang S. Optimal planning for electric vehicle fast charging stations placements in a city scale using an advantage actor-critic deep reinforcement learning and geospatial analysis. Sustain Cities Soc 2024;113:105567. https://doi.org/10.1016/j.scs.2024.105567.

Hisoglu S, Tuominen A, Huovila A. An approach for selecting optimal locations for electric vehicle solar charging stations. IET Smart Cities 2023;5(2):123–34. https://doi.org/10.1049/smc2.12058.

Hossein K, Behrooz V, Hosseinian SH, Lazaroiu GC, Pierluigi S. Prospective design and evaluation of a renewable energy hybrid system to supply electrical and thermal loads simultaneously with an electric vehicle charging station for different weather conditions in Iran. Smart Cities 2025;8(2):61.

IRENA. Innovation outlook: smart charging for electric vehicles. Abu Dhabi: International Renewable Energy Agency; 2019.

Jaganath MM, Ray S, Dev Choudhury NB. Eco-friendly microgrid carport charging station for electric vehicles (EVs). In: e-Prime - advances in electrical engineering, electronics and energy, 5; 2023. p. 100196. https://doi.org/10.1016/j.prime.2023.100196.

Karipoğlu F, Serdar Genç M, Akarsu B. GIS-based optimal site selection for the solar-powered hydrogen fuel charge stations. Fuel 2022;324:124626. https://doi.org/10.1016/j.fuel.2022.124626.

Kaur A, Singh P, Singh J. Renewable energy optimization for distributed EV charging stations using HBA and smart contracts. IEEE Internet Things J 2025;12(17):34700–7. https://doi.org/10.1109/jiot.2025.3554506.

Khazali A, Fraser E, Al-Wreikat Y, Sharkh S, Wills R, Cruden A. Optimal sizing of stand-alone energy systems incorporating battery capacity fading. In: IET Conference Proceedings; 2023. p. 87–92. https://doi.org/10.1049/icp.2023.3131.

Khazali A, Al-Wreikat Y, Fraser EJ, Naderi M, Smith MJ, Sharkh SM, et al. Sizing a renewable-based microgrid to supply an electric vehicle charging station: a design and modelling approach. World Electr Vehi J 2024;15(8). https://doi.org/10.3390/wevj15080363.

Krishnamurthy NK, Sabhahit JN, Jadoun VK, Gaonkar DN, Shrivastava A, Rao VS, et al. Optimal placement and sizing of electric vehicle charging infrastructure in a grid-tied DC microgrid using modified TLBO method. Energies 2023;16(4). https://doi.org/10.3390/en16041781.

Kumar A, Kumar S, Sinha UK, Bohre AK. Intricate DG and EV planning impact assessment with Seasonal variation in a three-phase distribution system. World Electr Vehi J 2024; 15(9). https://doi.org/10.3390/wevj15090425.

Kumbhar SS, Kalkhambkar VN, Bhakar R. Optimal location and sizing of coordinated battery swapping and charging station in power and road transportation networks. Energy Storage 2025;7(4). https://doi.org/10.1002/est2.70186.

Leone C, Longo M, Fernandez-Ramirez LM, Garcia-Trivino P. Multi-Objective optimization of PV and energy storage systems for ultra-fast charging stations. IEEE Access 2022;10:14208–24. https://doi.org/10.1109/ACCESS.2022.3147672.

Li Z, Sun Y, Yang H, Wang S, Shen Y, Wang X, et al. A multi-time scale coordinated control and scheduling strategy of EVs considering guidance impacts in multi-areas with uncertain RESs. Int J Electr Power Energy Syst 2023;154:109444. https://doi.org/10.1016/j.ijepes.2023.109444.

Li J, He S, Yang Q, Ma T, Wei Z. Optimal design of the EV charging station with retired battery systems against charging demand uncertainty. IEEE Trans Ind Inform 2023;19(3):3262–73. https://doi.org/10.1109/TII.2022.3175718.

Li R, Li H, Huang W, Tao H, Xu W, Tai N, et al. Accelerating green shipping with spatially optimized offshore charging stations. Nat Energy 2025:1–12.

Mehouachi I, Trojette M, Grayaa K. MONNA: multi-objective neural network algorithm for the optimal location of electric vehicle charging infrastructure in Tunis city. J Clean Prod 2023;431:139837. https://doi.org/10.1016/j.jclepro.2023.139837.

Mejia MA, Macedo LH, Munoz-Delgado G, Contreras J, Padilha-Feltrin A. Multistage planning model for active distribution systems and electric vehicle charging stations considering voltage-dependent load behavior. IEEE Trans Smart Grid 2022;13(2): 1383–97. https://doi.org/10.1109/TSG.2021.3125786.

Mohanty AK, Babu PS. Optimal placement of electric vehicle charging stations using JAYA algorithm. In: Recent advances in power systems: select Proceedings of EPREC 2020. Singapore: Springer Singapore; 2020. p. 259–66.

Myint KPY, Khortsriwong N, Deanseekeaw A, Boonraksa T, Boonraksa P, Marungsri B. Optimal design and sizing of microgrid with hybrid renewable energy sources using multi-agent deep reinforcement learning techniques. In: 2024 international conference on power, energy and Innovations (ICPEI). IEEE; 2024, October. p. 151–6.

Nareshkumar K, Das D. Economic planning of EV charging stations and renewable DGs in a coupled transportation-reconfigurable distribution network considering EV range anxiety. Electr Eng 2024:1–20.

Nirbheram JS, Mahesh A, Bhimaraju A. Feasibility study of a PV-grid-assisted charging station for electric and hydrogen fuel cell vehicles under uncertain arrivals. Energy 2025; 322:135449. https://doi.org/10.1016/j.energy.2025.135449.

Oladigbolu JO, Bilal M, Gupta S, Mujeeb A, Li L. A novel approach for optimal deployment of plug-in electric vehicles with integrated renewable energy sources. Electr Eng 2025; 107(4):3847–81. https://doi.org/10.1007/s00202-024-02720-1.

Oladigbolu JO, Mujeeb A, Imam AA, Rushdi AM. Design, technical and economic optimization of renewable energy-based electric vehicle charging stations in Africa: the case of Nigeria. Energies 2023;16(1). https://doi.org/10.3390/en16010397.

Pagidipala S, Sandeep V. Optimal planning of electric vehicle charging stations and distributed generators with network reconfiguration in smart distribution networks considering uncertainties. Measurement: Sensors 2024;36:101400. https://doi.org/10.1016/j.measen.2024.101400.

Pandraju TKS, Janamala V. Self-adaptive butterfly optimization for simultaneous optimal integration of electric vehicle fleets and renewable distribution generation. In: Congress on intelligent systems. Singapore: Springer Nature Singapore; 2022, September. p. 777–84.

Raad NG, Swain C, Ayantayo A, Yang B, Rajendran S. Hybrid regret-based p-robust and distributionally robust optimization models for electric vehicle charging station network design. Comput Ind Eng 2024;198:110709. https://doi.org/10.1016/j.cie.2024.110709.

Rao CSP, Pandian A, Reddy CR, Bajaj M, Jurado F, Kamel S. Optimal location of EV parking lot by MAOWHO technique in distribution system. In: 2023 5th global power, energy and communication conference (GPECOM). IEEE; 2023, June. p. 103–7.

Rizk-Allah RM, Hassan IA, Snasel V, Hassanien AE. An optimal standalone wind-photovoltaic power plant system for green hydrogen generation: case study for hydrogen refueling station. Results Eng 2024;22. https://doi.org/10.1016/j.rineng.2024.102234.

Saadati R, Jafari-Nokandi M, Saebi J, Hossein Hosseini S. Optimal location of fast charging stations alongside pricing design from distribution company viewpoint considering renewable energy resource: a modified capacitated flow-capturing location model. Sustain Energy Grids Netw 2023;34:100989. https://doi.org/10.1016/j.segan.2022.100989.

Sa'adati R, Jafari-Nokandi M, Saebi J. Allocation of RESs and PEV fast-charging station on coupled transportation and distribution networks. Sustain Cities Soc 2021;65:102527. https://doi.org/10.1016/j.scs.2020.102527.

Saadati R, Saebi J, Jafari-Nokandi M. Effect of uncertainties on siting and sizing of charging stations and renewable energy resources: a modified capacitated flow-refueling location model. Sustain Energy Grids Netw 2022;31:100759. https://doi.org/10.1016/j.segan.2022.100759.

Sabzian-Molaee Z, Rokrok E, Doostizadeh M. An optimal master-slave model for stochastic planning of AC-DC hybrid distribution systems. J Renew Energy Environ 2023;10(2): 27–38. https://doi.org/10.30501/jree.2022.323167.1309.

Shamani M, Akbari Foroud A, Rastgoo R. Optimal routing and scheduling of mobile charging station considering operational limitations of energy sources in distribution system. J Energy Storage 2025;117:116135. https://doi.org/10.1016/j.est.2025.116135.

Singh G, D'Arpino M, Goveas T. A techno-economic assessment of DC fast-charging stations with storage, renewable resources and low-power grid connection. Energies 2024; 17(16):4012. https://doi.org/10.3390/en17164012.

Strunz K, Kuschke M, Schilling S. Climate-friendly and socially inclusive AC-DC renewable energy system with overlay multi-terminal HVDC network (OVANET): solution with fully distributed optimization and infrastructure combination. Int J Electr Power Energy Syst 2021;132:106119. https://doi.org/10.1016/j.ijepes.2020.106119.

Subramaniam A, Singh LRSR, Zonghai C. Optimal planning and allocation of Plug-in Hybrid Electric Vehicles charging stations using a novel hybrid optimization technique. PLoS One 2023;18(7):e0284421. https://doi.org/10.1371/journal.pone.0284421.

Tabandeh A, Hossain MJ, Li L. Integrated multi-stage and multi-zone distribution network expansion planning with renewable energy sources and hydrogen refuelling stations for fuel cell vehicles. Appl Energy 2022:319. https://doi.org/10.1016/j.apenergy.2022.119242.

Tang Y, Liu W, Chau KT, Hou Y, Guo J. Stochastic behavior modeling and optimal bidirectional charging station deployment in EV energy network. IEEE Trans Intell Transp Syst 2025;26(5):6231–47. https://doi.org/10.1109/TITS.2025.3553513.

Taylor A, Taylor M. Operations management research: contemporary themes, trends and potential future directions. Int J Oper Prod Manag 2009;29(12):1316–40. https://doi.org/10.1108/01443570911006018.

Thiruvenkadam S, Raju J, Ra IH. Optimal microgrid structure with renewable distributed generators and electric vehicles utilizing hybrid fuzzy-puzzle optimization algorithm. In: International conference on advances in modern Age technologies for health and engineering science, AMATHE 2024; 2024. https://doi.org/10.1109/AMATHE61652.2024.10582254. http://ieeexplore.ieee.org/xpl/mostRecentIssue.jsp?punumber=10582048.

Thumati S, Vadivel S, Rao MVG. Optimal location and sizing of hybrid photovoltaic public charging stations in reconfigurable feeders using Levy Flight honey badger algorithm. Int J Intell Eng Syst 2024;17(1):278–87. https://doi.org/10.22266/ijies2024.0229.26.

Tostado-Véliz M, Horrillo-Quintero P, García-Triviño P, Fernández-Ramírez LM, Jurado F. Optimal sitting and sizing of hydrogen refilling stations in distribution networks under locational marginal prices. Appl Energy 2024:374. https://doi.org/10.1016/j.apenergy.2024.124075.

Ulpiani G, Vetters N, Shtjefni D, Kakoulaki G, Taylor N. Let's hear it from the cities: on the role of renewable energy in reaching climate neutrality in urban Europe. Renew Sustain Energy Rev 2023;183.

Wang G, Li H, Yang X, Lu H, Song X, Li Z, et al. Multi-objective site selection and capacity determination of distribution network considering new energy uncertainties and shared energy storage of electric vehicles. Electronics 2025;14(1):151. https://doi.org/10.3390/electronics14010151.

Wen Y, Fan P, Hu J, Song K, Wu F, Zhu X. An optimal scheduling strategy of a microgrid with V2G based on deep Q-learning. Sustainability 2022;14(16):10351. https://doi.org/10.3390/su141610351.

Wu T, Xiang W, Zhang X, Wang G, Qiu J, Xia S. Carbon-oriented expansion planning of integrated electricity-natural gas systems with EV fast-charging stations. IEEE Trans Transp Electrif 2022;8(2):2797–809. https://doi.org/10.1109/tte.2022.3151811.

Yong JY, Ramachandaramurthy VK, Tan KM, Mithulananthan N. A review on the state-of-the-art technologies of electric vehicle, its impacts and prospects. Renew Sustain Energy Rev 2015;49:365–85. https://doi.org/10.1016/j.rser.2015.04.130.

Yuvaraj T, Devabalaji KR, Kumar JA, Thanikanti SB, Nwulu NI. A comprehensive review and analysis of the allocation of electric vehicle charging stations in distribution networks. IEEE Access 2024a;12:5404–61. https://doi.org/10.1109/ACCESS.2023.3349274.

Yuvaraj T, Prabaharan N, De Britto CJ, Thirumalai M, Salem M, Alhuyi Nazari M. Dynamic optimization and placement of renewable generators and compensators to mitigate electric vehicle charging station impacts using the spotted Hyena optimization algorithm. Sustainability 2024b;16(19):8458. https://doi.org/10.3390/su16198458.

Zaki MA, Mahmoud T, Atia M, Osman ESAEA. Optimal sizing and sitting of electric vehicle charging station by using archimedes optimization algorithm technique. Int J Power Electron Drive Syst 2021;12(4):2557–69. https://doi.org/10.11591/ijpeds.v12.i4.pp2557-2569.

Zhang X, Gao X, Zheng Y, Xu Y, Chung E, Sayed AR. Low-carbon oriented collaborative planning of electricity-hydrogen-gas-transportation integrated system considering hydrogen vehicles. IEEE Trans Netw Sci Eng 2024;11(6):5515–25. https://doi.org/10.1109/TNSE.2023.3279994.

Zheng D, Zheng B. Optimization of electric vehicle charging facility layout considering the enhancement of renewable energy consumption capacity and improvement of PSO algorithm. Energy Inform 2025;8(1). https://doi.org/10.1186/s42162-025-00514-8.

Zhu G, Yan G, Garmroudi D. Optimizing solar-wind hybrid energy systems for sustainable charging stations and commercial applications: a two-stage framework with ebola-inspired optimization. Expert Syst Appl 2024;246:123180. https://doi.org/10.1016/j.eswa.2024.123180.

CHAPTER FOURTEEN

Self-sufficient smart hangar with mobile vertical metallic park for EVs: A 3D approach to urban charging and renewable energy integration

Ismail Baklouti[1], Muhammad Abdul Mujeebu[2], Shantharam Patil[3], Abdultawab M. Qahtan[4]
[1]National School of Engineers of Monastir, University of Monastir, Av. Ibn El Jazzar, Monastir, Tunisia
[2]Department of Building Engineering, College of Architecture and Planning, Imam Abdulrahman Bin Faisal University, Al-Dammam, Saudi Arabia
[3]Manipal School of Architecture and Planning, Manipal Academy of Higher Education, Manipal, Karnataka, India
[4]Architectural Engineering Department, College of Engineering, Najran University, Najran, Saudi Arabia

14.1 Introduction

The confluence of rapid urbanization and the escalating adoption of electric vehicles (EVs) presents significant challenges for modern cities. Achieving a sustainable transition requires not only accommodating more vehicles but also ensuring that their energy demand is met by clean, renewable sources. This imperative places the development of innovative EV charging infrastructure, powered by renewable energy, at the heart of smart city strategies (Lopez-Behar et al., 2019; Guerra and Daziano, 2020). However, traditional parking solutions often struggle to accommodate the increased vehicle density typical of densely populated urban environments, leading to traffic congestion, inefficient land utilization, and heightened stress for drivers seeking parking (Chomiak-Orsa et al., 2023). Simultaneously, the burgeoning EV fleet necessitates the development of charging options that are not only accessible and reliable, but also seamlessly integrated with renewable energy generation to decarbonize the transport sector (Fotouhi et al., 2019; Badia et al., 2020). Integrating the substantial electrical load of EV charging into existing power grids introduces further complexities, particularly concerning load management during peak hours,

Renewable-to-Vehicle Smart Charging
ISBN: 978-0-443-36684-0
https://doi.org/10.1016/B978-0-443-36684-0.00013-8

maintaining grid stability, and ensuring that the energy supplied is derived from sustainable sources (Zhang et al., 2020; Medved et al., 2023).

Recent technological advancements, especially in the domains of the Internet of Things (IoT), Digital Twins (DTs), and smart systems, offer promising pathways to address these multifaceted urban mobility challenges (Faliagka et al., 2024; Shirowzhan et al., 2020). Digital twins, functioning as dynamic virtual counterparts of physical assets or systems, provide powerful capabilities for real-time monitoring, predictive analytics, complex simulations, and operational optimization. These features enable the implementation of more intelligent and adaptive management strategies for urban infrastructure (Liu et al., 2023b; Sahu et al., 2025; Yu et al., 2024). Specifically within the context of urban EV infrastructure, DTs can be employed to optimize the allocation of limited parking spaces (Zou et al., 2023; Shrivastav et al., 2023), manage charging schedules with greater intelligence to align with grid conditions or user needs (Sahu et al., 2025; Canzaniello et al., 2024), effectively integrate fluctuating renewable energy sources like solar power (Brasil et al., 2024), Osório (2021), and ultimately enhance the overall efficiency of the system and the experience for the end-user (Conde et al., 2022; Mortazavi et al., 2023).

This chapter introduces the "Self-Sufficient Smart Hangar"—an innovative conceptual framework designed to proactively tackle the intertwined challenges of EV parking and charging within the spatial and infrastructural constraints typical of urban settings. By synergistically integrating a high-density Mobile Vertical Parking System (MVPS) within a dedicated metallic hangar structure, which also incorporates rooftop photovoltaic (PV) panels and on-site battery energy storage systems (BESS), this solution aims to establish a compact, sustainable, and highly intelligent hub dedicated to urban e-mobility. The operational coherence and optimization of the Smart Hangar are facilitated through extensive IoT connectivity and a comprehensive digital twin framework, allowing for sophisticated real-time control, predictive management, and seamless integration into the broader ecosystem of smart city services and infrastructure.

14.1.1 Evolution of urban EV charging

The infrastructure supporting urban EV charging has undergone a significant transformation, moving from rudimentary, isolated charging posts to highly sophisticated, interconnected networks. Initial deployments were primarily focused on providing basic access to charging, often located opportunistically in public parking areas or as standalone dedicated stations (Sun et al., 2020; Haorui et al., 2018). However, as the number of EVs on the road increased, inherent limitations such as charger availability conflicts, extended charging durations, and the cumulative impact on local power grids became increasingly apparent and problematic.

The introduction and refinement of smart charging technologies represented a critical turning point in this evolution. These technologies enabled advanced functionalities including dynamic load balancing across multiple chargers, implementation of time-of-use pricing to influence charging behavior, and remote monitoring and control of charging sessions (Xie et al., 2020). The integration of standardized communication protocols, such as the Open Charge Point Protocol (OCPP) (Guillemin et al., 2024), alongside ubiquitous IoT platforms, facilitated greater interoperability between different hardware manufacturers and centralized management systems (Gopal et al., 2023; Mohammadi and Rashidzadeh, 2021). Concurrently, the Renewable-to-Vehicle (R2V) concept gained traction, promoting the direct utilization of renewable energy, frequently generated locally via PV installations integrated into parking structures or adjacent buildings, to power the charging of EVs, thereby enhancing sustainability (Marinescu, 2021).

Further enhancements came with the integration of Battery Energy Storage Systems (BESS), which proved instrumental in managing the intermittency associated with renewable sources like solar power. BESS units allow for the buffering of generated energy, enabling services such as peak shaving (reducing peak demand from the grid) and potentially providing frequency regulation support [Dvořáček2019429, (Graber et al., 2020)]. Vehicle-to-Grid (V2G) technology emerged as another potentially transformative development, conceptualizing EVs not just as consumers of electricity but as mobile storage assets capable of injecting power back into the grid, thus offering grid stabilization services and creating potential economic benefits for EV owners (Caggiani et al., 2021; Ali et al., 2023).

Most recently, the application of Digital Twins (DTs) is poised to significantly enhance the management and operation of EV charging infrastructure (Eskandari et al., 2024). By creating a virtual replica of the entire charging ecosystem—encompassing EVs, chargers, renewable generation, energy storage, and grid interactions—DTs enable unprecedented levels of simulation, prediction, and real-time optimization. This facilitates improved efficiency, heightened reliability, and a superior user experience. The ongoing trajectory points towards increasingly integrated, intelligent, and potentially autonomous charging systems that function as integral components of the broader smart city infrastructure. Fig. 14.1 provides a conceptual overview of this evolutionary trajectory.

14.1.2 Challenges in dense urban environments

Establishing effective and widespread EV charging infrastructure within dense urban environments encounters a unique confluence of challenges. Primarily, the scarcity and high cost of land severely constrain the feasibility of deploying traditional, horizontally expansive parking lots equipped with numerous charging stations (Sikder et al., 2023). The inherent high population density within cities translates into a potentially massive and geographically concentrated demand for EV charging. This demand often peaks during specific times, such as evening hours when residents return from work, placing considerable strain on the local electricity distribution grid, which may not have been designed for such concentrated loads (Lazzeroni et al., 2021).

Furthermore, a significant proportion of urban dwellers reside in multi-unit dwellings (MUDs), such as apartment buildings or condominiums, which often lack private garages or easily accessible, dedicated parking spots suitable for the installation of personal home chargers (Cheng and Kontou, 2023). This demographic reality forces many urban EV drivers to rely predominantly on public or shared charging facilities. Consequently, issues of charger accessibility, ensuring consistent availability, managing queues during peak demand periods, and addressing the inconvenience of finding operational chargers become critical concerns for user adoption and satisfaction. The problem of "ICEing," (an acronym for Internal Combustion Engine-ing), where non-EVs improperly occupy charging spots, further compounds these availability challenges (Chen et al., 2024).

Grid limitations represent another substantial obstacle. Existing electrical infrastructure in older urban areas may possess insufficient capacity to accommodate the significant additional load imposed by widespread EV charging without risking overloads or voltage instability. Upgrading this infrastructure is often a complex and expensive undertaking (Yoon et al., 2024; Zhang et al., 2023). While integrating intermittent renewable energy sources like solar PV is environmentally desirable, it introduces further complexity into grid management. Ensuring a reliable power supply necessitates sophisticated control strategies and often requires the integration of

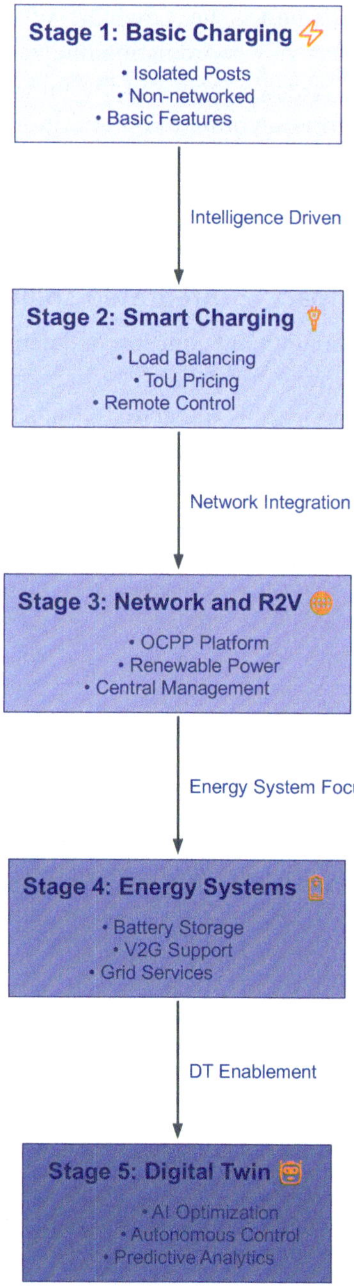

Figure 14.1 Evolution of urban EV charging infrastructure and technologies.

energy storage systems to buffer the variability of renewable generation (Bracco et al., 2015). Effectively coordinating fluctuating charging demand with variable renewable energy supply and the inherent constraints of the grid in real-time constitutes a complex optimization challenge that must be addressed for successful large-scale EV integration (Hussain et al., 2022). Fig. 14.2 provides a visual summary of these interconnected urban challenges.

14.1.3 Need for compact & integrated solutions

The inherent spatial constraints and concentrated demand patterns of dense urban areas strongly advocate for parking and charging solutions that are fundamentally compact and highly integrated (Boucetta et al. 2021). Traditional surface parking lots consume vast amounts of valuable urban land, an increasingly untenable approach in costly city centers. Automated vertical

Figure 14.2 Key challenges for EV charging in dense urban environments.

parking systems present a compelling alternative, leveraging vertical space to significantly multiply vehicle storage capacity within a minimal ground footprint, thereby alleviating parking pressure (Zhang and Ding, 2024).

Simultaneously, the necessity for convenient EV charging demands the direct embedding of charging capabilities within these parking solutions. Instead of requiring users to navigate separate facilities for parking and charging, integrated hubs that seamlessly combine automated parking with readily available charging infrastructure offer a more streamlined user experience and optimize the use of limited urban space (Ma and He, 2023; Tan and Kang, 2024). This deep integration facilitates highly efficient charging management; vehicles can potentially be charged automatically while stored, eliminating the need for drivers to manually relocate their vehicles between parking and charging spots.

This need for compact, integrated infrastructure marks a significant departure from conventional approaches. Standard solutions, such as retrofitting existing surface lots or multi-story garages with a limited number of fixed charging stations, fail to address the core issue of land scarcity and often create bottlenecks for EV drivers. While standalone automated parking systems and dedicated EV charging plazas have emerged as partial solutions, they typically address parking density and charging availability as separate problems. The "Self-Sufficient Smart Hangar" concept proposed in this chapter offers a holistic alternative by synergistically combining these functions. Unlike previous approaches, it integrates **(1) high-density automated vertical parking** to radically minimize land use, **(2) mobile robotic chargers** to flexibly service any vehicle, decoupling parking from charging, **(3) extensive on-site renewable energy generation** via rooftop PV, and **(4) a large-scale Battery Energy Storage System (BESS)** for energy autonomy and grid support. This deeply integrated, self-sufficient model represents a paradigm shift, creating a truly smart infrastructure node rather than just a place to park and charge.

Furthermore, the co-location and integration of local renewable energy generation, such as extensive rooftop solar PV arrays, and dedicated battery energy storage systems directly within these combined parking/charging hubs enable the creation of self-sufficient or, at minimum, grid-supportive micro-energy ecosystems. This approach significantly reduces dependence on potentially overloaded urban power grids, lowers operational energy costs through the self-consumption of locally generated renewable power, and inherently enhances the overall sustainability of the facility. The powerful synergy achieved through combining compact automated parking, fully integrated charging, and on-site renewable energy resources forms the cornerstone of efficient and sustainable urban mobility infrastructure, a concept embodied by the Smart Hangar approach detailed in this chapter. Fig. 14.3 offers a conceptual illustration of such integrated solutions.

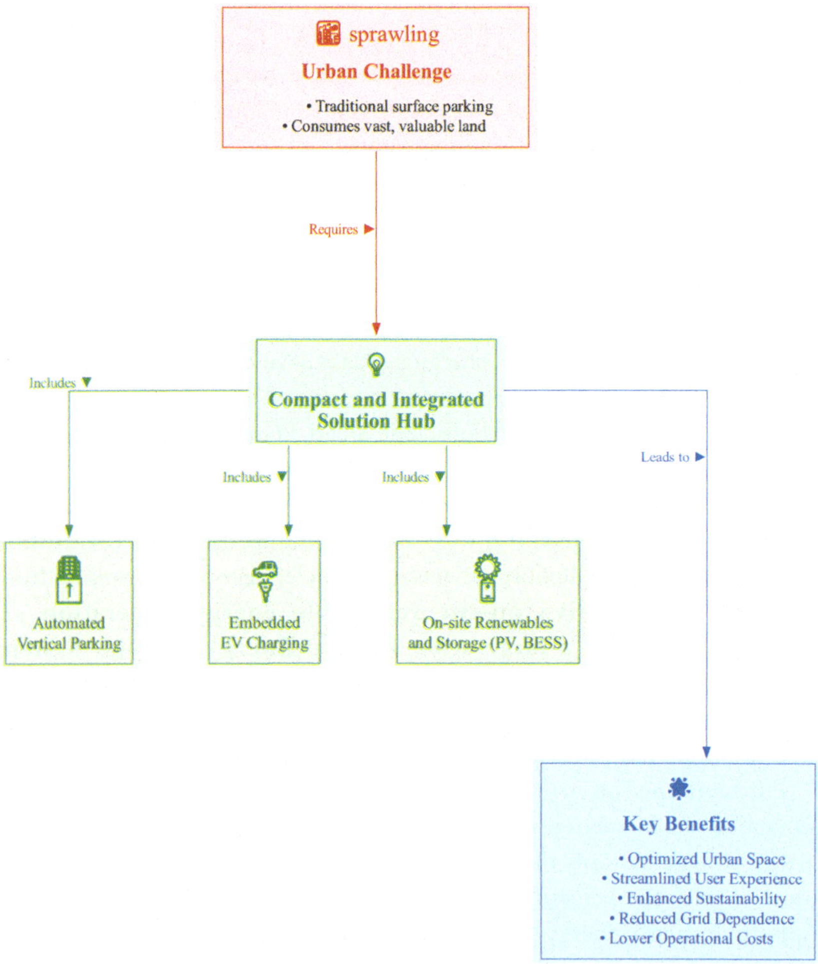

Figure 14.3 The need for compact and integrated urban EV parking and charging solutions.

14.1.4 Integrated operation of EVs and renewables for sustainability

The strategic integration of electric vehicles (EVs) with renewable energy sources (RES) represents a cornerstone strategy for achieving sustainable urban transportation systems and effectively mitigating the impacts of climate change (Nava and Greco, 2023). EVs, particularly when their charging is powered by electricity derived from clean sources such as solar or wind energy, provide a direct pathway to substantially curtail greenhouse gas (GHG) emissions originating from the transportation sector (Lefeng et al., 2025). This synergy between clean energy and clean transportation directly supports the ambitious carbon emission reduction targets established by numerous cities and nations globally (Qiu et al., 2019; Shang et al., 2024).

Beyond the crucial environmental benefit of emissions reduction, the integrated operation of EVs and RES, especially within localized energy ecosystems like microgrids or dedicated smart charging hubs, yields significant advantages for overall power system performance and stability (Mudiyanselage et al., 2024). The implementation of smart charging strategies enables the temporal shifting of EV charging demand. This allows charging activities to be aligned with periods of abundant renewable energy generation (e.g., peak solar hours) or times of low overall grid demand, thereby maximizing the utilization of intermittent renewable resources and concurrently reducing stress on the grid infrastructure (Liu et al., 2021). Furthermore, the potential deployment of Vehicle-to-Grid (V2G) technology enhances this symbiotic relationship. V2G transforms EVs into distributed energy storage assets capable of providing valuable ancillary services back to the grid, such as frequency regulation, peak load shaving, and voltage support, contributing significantly to grid stability and resilience.

The establishment of compact, integrated charging facilities powered predominantly by on-site renewable generation, as exemplified by the proposed Smart Hangar concept, perfectly embodies this integrated vision. Such facilities create localized ecosystems where clean energy generation is directly coupled with clean transportation needs, effectively minimizing energy losses associated with long-distance transmission and maximizing the potential for energy self-sufficiency. Within this complex interplay of energy flows and demands, Digital Twins emerge as critical enablers. They provide the necessary platform for sophisticated monitoring, forecasting, and optimization, ensuring that energy flows between PV generation, battery storage, EV charging loads, and potential grid interactions are managed optimally to achieve both environmental sustainability goals and efficient power system operation (You et al., 2021). Fig. 14.4 conceptually depicts this vital integrated operation.

This chapter is structured to provide a comprehensive overview of the Self-Sufficient Smart Hangar concept. The following sections will first provide a detailed overview of the hangar's architecture and key systems. Subsequently, the chapter delves into the technical specifics of the integrated charging and renewable energy infrastructure, including the mobile robotic charging, solar generation, and battery storage systems. The critical role of the IoT network and Digital Twin in enabling intelligent control and optimization is then explored. Finally, the chapter presents a cost-benefit analysis, examines case studies of the core technologies, and discusses future trends and development pathways for such advanced urban mobility infrastructure.

14.2 Smart hangar concept overview

The Self-Sufficient Smart Hangar initiative introduces a pioneering architectural and operational model, conceived to masterfully navigate the intricate challenges confronting electric mobility in contemporary urban landscapes. Paramount among these are the acute scarcity of parking real estate, the escalating demand for accessible and high-performance Electric Vehicle (EV) charging infrastructure, and the imperative for sophisticated integration with, or even augmentation of, existing power grids, particularly within the high-density fabric of modern cities.

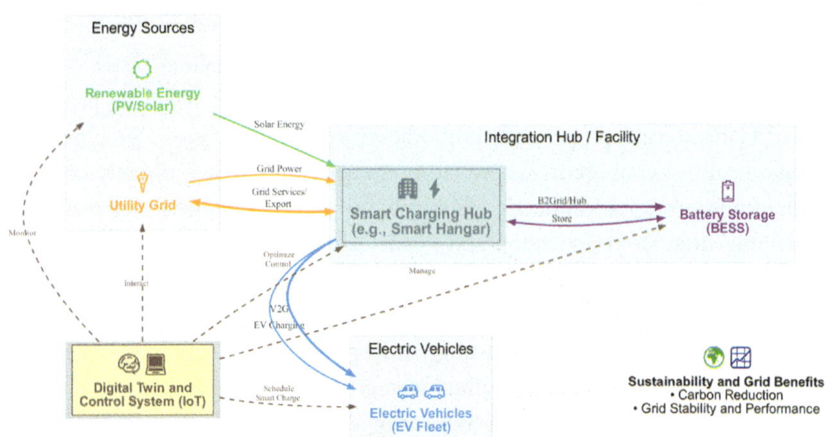

Figure 14.4 Integrated operation of electric vehicles and renewable energy sources for carbon reduction and grid performance improvement.

The conceptual bedrock of this facility is a resilient, multi-story metallic hangar. This structure is ingeniously designed to house not only a state-of-the-art automated Mobile Vertical Parking System (MVPS) but also to incorporate an advanced, deeply integrated EV charging ecosystem. This synergy facilitates a highly automated and energy-optimized operational flow. A critical enabler of its pronounced self-sufficiency is an expansive rooftop photovoltaic (PV) installation, engineered with a capacity to generate up to 1.872 MW (peak) of solar power. This substantial on-site generation is strategically buttressed by a high-capacity Battery Energy Storage System (BESS), offering 9.216 MW h of energy storage. The sophisticated choreography of these diverse energy assets and operational systems is orchestrated through a pervasive Internet of Things (IoT) network, with global oversight and continuous performance optimisation managed by a dynamic, data-driven Digital Twin (DT) framework.

The physical edifice is envisioned as a durable, potentially modular metallic framework, with planned dimensions of 120 m in length, 60 m in width, and a vertical elevation of 15 m. This architectural approach prioritizes streamlined construction methodologies while ensuring inherent adaptability to meet fluctuating urban demands through scalable design principles. The deliberate emphasis on verticality is a cornerstone of its strategy to profoundly optimize land utilisation, presenting a compelling and sustainable alternative to conventional, land-intensive horizontal parking paradigms. The hangar's rooftop is a multi-functional asset, hosting 2496 high-efficiency PV modules, each with a 750 W (peak) rating. This configuration is designed to maximize the utilization of the 7200 m^2 rooftop area, achieving a peak generation capacity of 1.872 MW. The BESS, comprising 320 modular blocks that collectively house 960 individual 48 V, 200 A h battery units, is typically sited at the structure's base. Its substantial 9.216 MWh energy storage capacity is strategically sized to capture and store approximately one full day of the hangar's solar generation (estimated at 9.36 MWh/day), enabling robust energy arbitrage and self-sufficiency. While a definitive capacity would result from a site-specific optimization analysis, this sizing provides a strong baseline for powering overnight charging and managing peak loads. This substantial energy reservoir is pivotal for capturing and storing surplus solar energy, while also serving to stabilize power delivery during periods of high concurrent EV charging demand, thereby enhancing operational reliability and enabling participation in valuable grid-support services.

Internally, the Smart Hangar's spatial and functional core is defined by its advanced MVPS. This system is structured across four principal automated lines, each designed to manage 320 vehicles, culminating in a comprehensive facility capacity of 1280 EV parking bays. The MVPS leverages sophisticated robotic technologies, including autonomous shuttles and high-speed lifts, to manage the three-dimensional transit of vehicles. EVs are inducted through one of eight clearly designated entry/exit portals (systematically labeled A1/A2 through D1/D2) and are subsequently navigated to their allocated parking slots by the automated system. This profound level of automation yields significant increases in parking density by obviating the need for internal driving aisles, concurrently bolstering vehicle security and elevating the user experience through expedited parking and retrieval processes. The EV charging needs are addressed by eight state-of-the-art Mobile Robotic EV chargers. These units, distinguished by their bi-directional (V2X) energy transfer capabilities, autonomously traverse dedicated double-pass corridors to service vehicles throughout the facility.

The operational acumen and intricate system-wide coordination of the Smart Hangar are fundamentally enabled by its extensive IoT network. This digital infrastructure acts as the facility's central nervous system, seamlessly interconnecting a vast array of sensors with a suite of responsive actuators. The Digital Twin, a high-fidelity, dynamic virtual representation of the physical hangar and its processes, is continuously updated by this rich stream of real-time data. This DT serves as the system's cognitive engine, facilitating superior real-time monitoring, enabling the simulation and rigorous evaluation of diverse operational scenarios (such as advanced smart charging protocols or adaptive parking assignment logic), supporting proactive predictive maintenance schedules, and meticulously orchestrating the complex, bidirectional energy exchanges between the PV array, BESS, aggregated EV loads, and the external power grid. This integrated, technology-forward philosophy is dedicated to establishing a paradigm of urban electric mobility infrastructure that is exceptionally efficient, decidedly user-focused, demonstrably sustainable, and actively grid-interactive.

The subsequent figures provide a visual narrative of this concept. Fig. 14.5 offers encompassing three-dimensional perspectives, elucidating the hangar's external architecture, the scale of its rooftop PV system, and key access points. These views are critical for understanding the facility's overall form and its integration with the immediate urban environment.

Delving deeper into the facility's internal configuration and core electrical infrastructure, Fig. 14.6 presents annotated cross-sectional and elevation

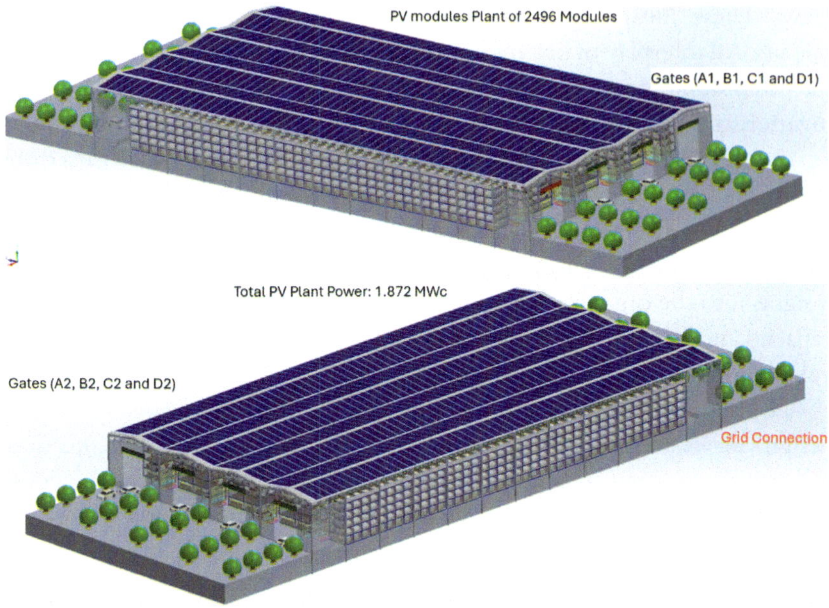

Figure 14.5 Isometric views of the smart hangar.

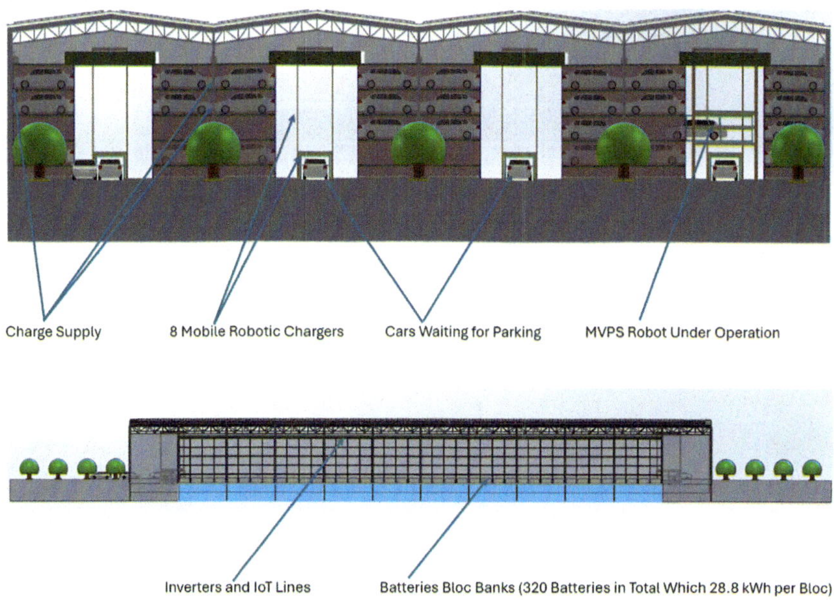

Figure 14.6 Internal and electrical layout details.

views. These illustrations are designed to clarify the spatial arrangement and operational interplay of key internal systems, including the MVPS in action, the positioning of mobile robotic chargers, and the conceptual layout of the inverter banks and the extensive battery energy storage system.

Finally, Fig. 14.7 provides a top-down plan view combined with a frontal operational cross-section. The plan view details the full extent of the rooftop PV array and its segmentation, alongside peripheral landscaping and designated vehicle access zones. The accompanying cross-section offers insights into the organization of the four main MVPS lines, illustrating their capacity and current operational status, thereby giving a clear picture of the vehicle processing capabilities of the hangar.

The detailed technical parameters underpinning the Smart Hangar's design are systematically presented in Table 14.1. These specifications collectively portray a facility meticulously engineered for high-density vehicle throughput, significant on-site energy generation, and robust energy storage capabilities, establishing it as a potentially transformative element of urban infrastructure. The substantial 1.872 MWp PV system, coupled with the 9.216 MW h BESS, forms the backbone of its energy self-sufficiency strategy, enabling considerable offsetting of grid energy

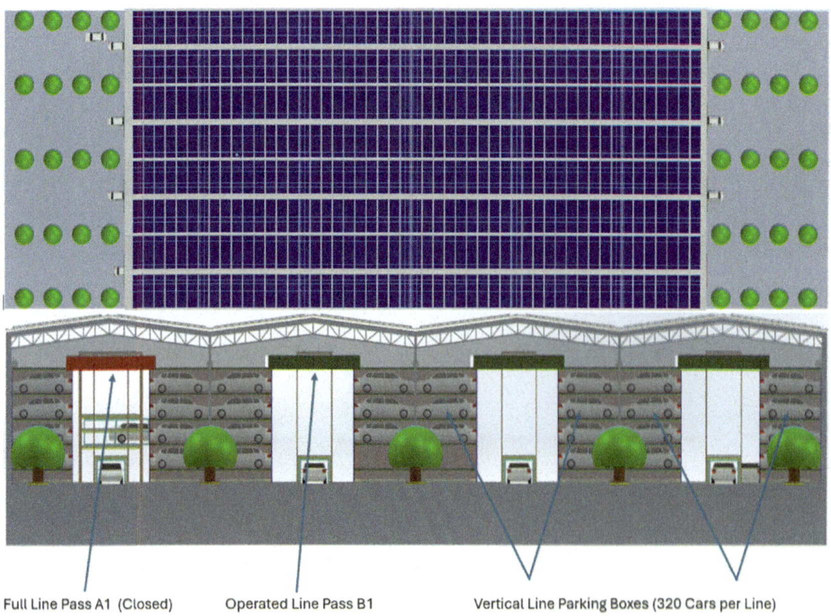

Figure 14.7 Overall plan and operational line configuration.

Table 14.1 Key technical specifications of the proposed smart hangar.

Parameter category	Specification	Value/Details
Hangar dimensions		
Length	Hangar footprint	120 m
Width	Hangar footprint	60 m
Height	Overall structure	15 m
Photovoltaic (PV) system		
Number of PV modules	Rooftop array	2496
Power per PV module	Individual module	750 W (peak)
Total PV installed power	System capacity	1.872 MW (peak)
Estimated daily output	PV generation	9360 kW h/day (at 5 peak sun hours/day)
Estimated annual output	PV generation	≈ 3.42 GW h/year
Inverter system		
Inverter unit capacity	Per inverter	100 kW (assumed peak capacity)
Number of inverters	System Configuration	19 units (Calculated as ⌈Total PV Power/Unit Capacity⌉ = ⌈1872/100⌉ = 19)
Inverter efficiency	Assumed	~ 96%
Battery energy storage system (BESS)		
Battery unit type	Individual unit	48 V, 200 A h lithium (9.6 kW h)
Batteries per block	Parallel Configuration	3 units
Number of battery blocks	Modular BESS	320 blocks (each 28.8 kW h)
Total number of battery units	System total	960 units
Total BESS energy capacity	System storage	9.216 MW h
Estimated BESS power	System output	Approx. 4.6 MW (assuming approx. 0.5C discharge rate)
System voltage (BESS)	Nominal	48 V
EV parking and charging system		
Total EV parking boxes	MVPS capacity	1280
Parking lines/Bays	Main MVPS lines	4 lines
Capacity per line	MVPS	320 vehicles
Mobile robotic chargers	EVSE units	8 (charger/discharger capability)
MVPS aisle design	Internal layout	Double pass per row (for robotic access)
EVs supported (operational metric)	Daily charging	1280 (assuming 10% of a 72.5 kW h battery charged via PV)

(Continued)

Table 14.1 Key technical specifications of the proposed smart hangar.—cont'd

Parameter category	Specification	Value/Details
IoT, control systems, and digital twin		
Sensors	Type and deployment	Extensive network covering occupancy, charging, energy systems, environmental conditions
Connectivity	Network infrastructure	Robust IoT network (e.g., LoRaWAN, 5G) for sensor/actuator communication
Control system	Centralized/Distributed	Advanced control algorithms for MVPS, charging, and energy management
Digital twin	Platform	Comprehensive virtual replica for monitoring, simulation, and optimization

consumption and facilitating participation in ancillary grid services. The parking capacity for 1280 EVs, managed by the highly automated MVPS and serviced by advanced V2X-enabled robotic chargers, underscores its role as a pivotal hub for urban e-mobility. The integration of a comprehensive IoT network and a Digital Twin framework further highlights the commitment to intelligent, adaptive, and optimized operations.

The profound environmental dividends anticipated from the Smart Hangar's sustained operation are detailed in Table 14.2. These benefits are not marginal but represent a substantial contribution toward achieving municipal and global decarbonization targets, concurrently enhancing urban air quality. The projected annual abatement of approximately 1640 t of CO_2 emissions is a direct outcome of the facility's core design philosophy, which champions the extensive use of solar-generated electricity for its internal energy requirements and for replenishing electric vehicles. This large-scale adoption of renewable energy directly displaces electricity that would otherwise be generated from carbon-intensive fossil fuels, thereby yielding a significant reduction in the overall carbon footprint attributable to both the hangar's operational lifecycle and the broader transportation ecosystem it actively supports. To contextualize this ecological impact, the CO_2 sequestration achieved is analogous to the annual carbon capture performance of a forest comprising approximately 27,000 mature trees. Moreover, when viewed through the lens of vehicular emissions, the transition to EVs charged with clean energy within the hangar is comparable to removing the annual emissions footprint of roughly 350 conventional internal combustion engine vehicles from the city's roadways. Such quantifiable metrics serve to powerfully underscore the transformative potential and critical importance of integrated renewable energy systems and

Table 14.2 Estimated CO_2 savings and environmental equivalents.

Aspect	Benefit/Value
CO_2 savings	~ 1640 t/year
Tree equivalent	~ 27,000 trees planted (equivalent sequestration)
Car equivalent	~ 350 conventional cars taken off the road annually (emissions avoided)

*Note: CO_2 savings are calculated based on the estimated annual PV generation of 3.42 GWh (from Table 14.1) displacing grid electricity. The calculation assumes a grid emission factor of 0.479 t CO_2/MWh, a representative average for the EU. Calculation: 3420 MWh/year * 0.479 tCO_2/MWh ≈ 1638 t/year. Tree and car equivalents are derived from standard conversion factors (e.g., EPA GHG Equivalencies Calculator). The actual savings will vary depending on the local grid's specific carbon intensity.

advanced electric mobility infrastructure in accelerating the indispensable shift toward creating more sustainable, resilient, and ultimately healthier urban environments for current and future inhabitants.

In addition to its commendable energy self-sufficiency and significant environmental contributions, the Smart Hangar concept yields compelling and potentially transformative advantages concerning the optimization of urban space and the enhancement of land-use efficiency. These aspects, critically evaluated in Table 14.3, are of paramount importance, particularly within the increasingly dense and land-constrained landscapes typical of modern metropolitan areas. A signal achievement of this innovative architectural design is its capacity to conserve approximately 1.7 hectares of valuable urban land. This figure represents the area that would typically be consumed by conventional, ground-level surface parking facilities designed to accommodate an equivalent fleet of 1280 vehicles. Such remarkable land conservation is realized through the intelligent exploitation of vertical dimensionality; the automated MVPS operates within a meticulously engineered 108 000 m^3 parking envelope, facilitating the efficient, high-density stacking of vehicles. Consequently, the Smart Hangar exhibits a land utilization efficiency estimated to be three to four times greater than that achievable with traditional parking infrastructures. This profound improvement in land efficiency transcends mere statistical abstraction, translating into concrete, actionable benefits for urban planners and developers. It effectively liberates substantial tracts of prime urban real estate, which can then be strategically repurposed for a multitude of vital community needs, including the development of much-needed public green spaces, the construction of affordable housing, the establishment of civic amenities, or the creation of commercial zones that invigorate the local economy. Ultimately, this sophisticated approach to parking infrastructure actively supports the evolution of more compact, pedestrian-friendly, and environmentally sustainable urban forms by mitigating the pervasive issue of urban sprawl,

Table 14.3 Estimated space utilization and urban efficiency benefits.

Aspect	Benefit/Value
Land Saved	~ 1.7 hectares saved (compared to traditional ground-level parking)
Volume used for parking	108 000 m^3 (efficient vertical EV parking volume)
Urban land efficiency	3× to 4× higher land efficiency than traditional parking

which is frequently exacerbated by the demand for extensive, land-intensive parking facilities.

The financial architecture underpinning the development and operationalization of a Self-Sufficient Smart Hangar, illustratively detailed with estimated cost ranges in Table 14.4, reflects a capital investment of considerable magnitude. This is an inherent consequence of the facility's advanced technological composition, its extensive automation, and its substantial physical scale. A granular analysis of the projected cost structure indicates that the most substantial capital allocations are demanded by the MVPS. This is attributable to the intricate amalgamation of sophisticated automation systems, advanced robotics, and the exacting structural engineering requisite for the safe, efficient, and high-throughput management of 1280 vehicles within a complex multi-story vertical array. The BESS also figures as a prominent cost center, a direct function of its extensive 9.216 MW h energy storage capacity, which is indispensable for enabling effective energy arbitrage, ensuring operational resilience, and facilitating grid support functionalities. Concurrently, the large-scale photovoltaic generation system and the primary hangar superstructure itself represent significant, though essential, upfront financial commitments. While these initial capital expenditures are undeniably high, their evaluation must be contextualized within a holistic, long-term value assessment and a comprehensive lifecycle cost analysis. The cost ranges presented are illustrative and synthesized from recent industry reports and market analyses (National Renewable Energy Laboratory, 2024; BloombergNEF, 2024; Market Data Forecast, 2025). The breadth of the ranges is intentional, reflecting significant variability based on geographical location, prevailing market dynamics, and specific technology choices—for instance, between standard and premium-grade MVPS automation, or the inclusion of advanced V2X-capable chargers versus unidirectional ones. The strategic emphasis on substantial on-site solar power generation is forecast to yield material reductions in operational energy procurement costs over the hangar's projected operational lifespan, thereby providing a valuable buffer against the inherent volatility of conventional energy markets. Furthermore, the facility's design inherently supports a diversified portfolio of potential revenue streams. These include, but are not limited to, standard and premium parking tariffs, value-added EV charging services, and potentially lucrative ancillary services supplied to the utility grid, made possible by the intelligent BESS and the V2X capabilities of the mobile robotic chargers. Such revenue diversity can significantly bolster the project's overall economic sustainability and expedite the return

Table 14.4 Illustrative estimated cost components for the smart hangar.

Component	Quantity/Scale	Estimated unit cost range (USD)	Estimated total cost range (USD)
Infrastructure & structure			
Hangar metallic structure & foundations	7200 m^2 (120 × 60 m)	\$200–\$400/m^2	\$1,440,000–\$2,880,000
Energy system			
PV modules	2496 units (1.872 MW_p)	\$0.30–\$0.50/W_p	\$561,600–\$936,000
PV inverters	19 units (total 1.9 MW)	\$10,000–\$18,000 per 100 kW unit	\$190,000–\$342,000
Battery energy storage system (BESS)	9.216 MWh (960 × 9.6 kWh)	\$250–\$450/kWh (installed)	\$2,304,000–\$4,147,200
Parking & charging automation			
Mobile vertical parking system (MVPS)	1280 parking spaces	\$18,000–\$35,000 per space	\$23,040,000–\$44,800,000
Mobile robotic EV chargers/dischargers	8 units	\$30,000–\$70,000 per unit	\$240,000–\$560,000
Control & management			
IoT network, sensors, control system & digital twin	System-wide (incl. extensive sensor suite)	Lump Sum	\$500,000–\$1,500,000
Miscellaneous			
Installation, commissioning, project management	Approx. 18%–28% of major equipment cost	N/A	\$5,089,608–\$15,418,228
Estimated total			**\$33,365,208–\$70,583,428**

Land acquisition costs are not included, Cost ranges are illustrative for a technologically advanced, large-scale project and subject to market conditions, specific technology choices, and geographical location, Unit costs are based on industry benchmarks and reports (e.g., NREL's Annual Technology Baseline for PV/BESS (National Renewable Energy Laboratory, 2024), BNEF's global energy storage outlook for BESS costs (BloombergNEF, 2024), and market analyses for automated parking systems (Market Data Forecast, 2025; Precedence Research, 2023)), "(peak)" after a unit indicates peak capacity.

on investment. Moreover, the profound environmental and urban efficiency benefits conferred by the Smart Hangar, although not always directly quantifiable in immediate monetary terms, embody substantial indirect economic advantages for the broader community, manifesting as improved public health outcomes, enhanced urban aesthetics, and increased overall livability. It remains crucial, however, to underscore that the presented cost figures are indicative and serve as high-level, illustrative estimates. The definitive project costs will invariably be shaped by detailed site-specific engineering assessments, prevailing market dynamics for construction materials and skilled labor, the selection of specific technology vendors and system integrators, and the ultimate terms negotiated through comprehensive procurement processes.

14.2.1 Mobile vertical parking system (MVPS)

The defining characteristic enabling the Smart Hangar's superior space efficiency and operational sophistication is its Mobile Vertical Parking System (MVPS). This system marks a radical departure from conventional multi-story garages, which are inherently limited by space-consuming ramps and manual driving aisles. Instead, the MVPS employs a highly coordinated assembly of automated robotic and mechanical systems to precisely induct, transport, and retrieve vehicles within the compact, multi-level hangar structure. The core engineering principle is multi-axis automation, integrating robust vertical lifts for Z-axis transit between parking tiers, agile horizontal shuttles or platforms for X-axis (along the bay) and Y-axis (into the slot) transfers within each tier, and integrated rotational mechanisms for optimal vehicle orientation.

The engineering of the specific MVPS conceptualized for the Smart Hangar, visually detailed in Fig. 14.8, illustrates a sophisticated robotic unit designed for complex vehicular management. The operational sequence begins when a user drives their vehicle onto the "EVs Parking System" platform at a designated ground-level entry bay. Once the vehicle is correctly positioned and secured (verified by integrated sensors that also capture vehicle dimensions for optimal slot assignment), the automated parking cycle commences.

The robotic unit, which forms the core of the MVPS shuttle, first prepares for movement. If the target parking slot requires the vehicle to be oriented differently than its entry position, the "Gear Motor (360 degrees Twist Axis)" is engaged. This powerful motor allows the entire platform

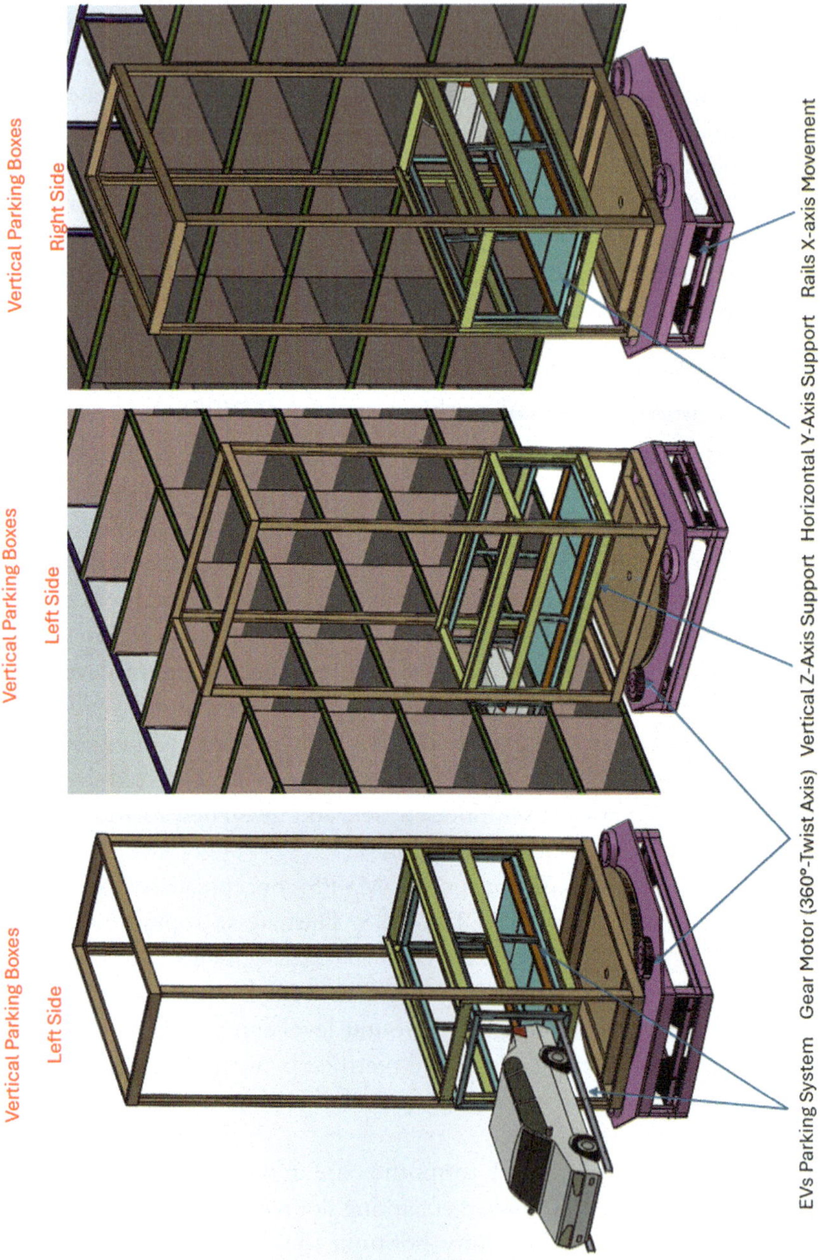

Figure 14.8 Detailed concept of the mobile vertical parking system (MVPS).

supporting the EV to rotate a full 360 degrees, ensuring the vehicle can be aligned perfectly for either left-side or right-side parking slots within the densely packed "Vertical Parking Boxes."

Following any necessary rotation, the unit utilizes its "Rails X-axis Movement" capability. These rails, running along the length of each parking aisle or bay, guide the entire shuttle horizontally to the column of parking slots where the assigned space is located. Upon reaching the target column, the "Vertical Z-Axis Support" system is activated. This system, likely employing a robust and reliable lift mechanism (such as synchronized chain drives, rack-and-pinion, or scissor lifts), elevates the platform—with the vehicle securely on it—to the designated vertical tier.

Once at the correct height and aligned with the target row, the final placement maneuver involves the "Horizontal Y-Axis Support." This mechanism allows the vehicle-laden platform to extend laterally, sliding precisely into the assigned parking slot, be it on the left or right side of the shuttle's central path. The reverse sequence is executed for vehicle retrieval. The platform extends from the slot, the unit descends or ascends to the exit level, travels along the X-axis rails to an exit bay, rotates if necessary to face the exit direction, and presents the vehicle to the user. This intricate choreography of rotational, linear (X, Y), and vertical (Z) movements, all orchestrated by a central control system informed by sensor data, allows for extremely high-density parking and efficient throughput.

This advanced level of multi-dimensional robotic automation drastically minimizes the per-vehicle spatial footprint compared to traditional parking methods, establishing the MVPS as an exemplary solution for severely land-constrained urban settings. The user experience is concomitantly streamlined, typically involving interactions via a mobile application or a dedicated kiosk at the transfer bay to initiate vehicle drop-off and retrieval requests. The flawless and dependable operation of such an intricate MVPS is critically contingent upon its seamless integration with the broader IoT network and the sophisticated Digital Twin framework. The Digital Twin, in this context, serves multiple vital functions: it provides continuous, real-time telemetry on the operational status, precise coordinates, and health of every vehicle and mechanical sub-component (including motors, actuators, lifts, shuttles, and sensors); it facilitates highly optimized and adaptive parking slot allocation strategies, considering dynamic variables such as vehicle dimensions, anticipated duration of stay, and specific EV charging requirements; it underpins a proactive, condition-based predictive maintenance regime by anlyzing diverse sensor data streams (e.g., monitoring motor current profiles, acoustic signatures, vibration patterns, and operational cycle counts) to preemptively identify potential component degradation or impending failures, thereby enabling targeted servicing interventions and minimizing unscheduled operational downtime. Furthermore, the Digital Twin environment enables exhaustive virtual commissioning and testing, where the complex control algorithms, safety interlock protocols, and intricate multi-axis mechanical movements can be rigorously validated and debugged within a simulated environment before any physical deployment, substantially de-risking the implementation phase and curtailing associated costs. This intelligent, fully automated, and exceptionally space-efficient parking system constitutes the robust mechanical and operational backbone of the Smart Hangar concept.

14.2.2 Renewable Energy Integration

A foundational principle underpinning the Self-Sufficient Smart Hangar's operational ethos is the profound and seamless integration of renewable energy systems. This is primarily actualized through an extensive photovoltaic (PV) installation, strategically deployed across the entirety of the hangar's considerable rooftop expanse (Bokhari, 2022). This deliberate design choice directly aligns with, and actively contributes to, broader municipal and global sustainable urban development imperatives by facilitating the generation of clean, emissions-free electricity precisely at the point of consumption. The locally harvested solar energy is then intelligently prioritized to satisfy the facility's diverse internal power requirements. These encompass the substantial energy demands of the sophisticated Mobile Vertical Parking System (MVPS) during its operational cycles, as well as the significant and often variable electrical load imposed by the concurrent charging activities of numerous Electric Vehicles (EVs) hosted within the hangar.

The architectural and engineering design of the rooftop PV system is meticulously optimized to maximize the annual energy yield. This optimization process carefully considers a range of critical factors, including the hangar's precise geographical coordinates (which dictate the intensity and duration of solar irradiance), its specific orientation relative to the sun's ecliptic path throughout the year, and the total unshaded surface area available on the rooftop (quantified in m^2). The direct current (DC) electricity generated by the solar panels offers versatile application pathways: it can be efficiently channeled via a Main DC Bus to directly power DC fast charging stations, minimizing conversion losses for compatible EVs. Alternatively, and more commonly for broader applicability, this DC power is converted into alternating current (AC) through high-efficiency inverters. This AC power then supplies, via a Main AC Bus and distribution panel, standard AC EV chargers (Level 2) and powers all ancillary building services, such as advanced LED lighting systems, climate control (HVAC), security systems, and the distributed network of control systems and sensors.

To adeptly manage the inherent intermittency and variability of solar power generation—a consequence of fluctuating weather conditions, cloud cover, and the natural diurnal (day-night) cycle—and to guarantee a consistently stable and reliable power supply crucial for uninterrupted hangar operations, a sophisticated Battery Energy Storage System (BESS) is deeply integrated within the facility's energy infrastructure. The BESS units are typically sited at the hangar's base or on specially reinforced, climate-controlled levels to optimize spatial utilization and facilitate effective thermal management, which is critical for battery longevity and performance. Within the hangar's intricate energy ecosystem, the BESS performs several indispensable functions. It engages in energy arbitrage, strategically accumulating surplus solar energy (either directly from the DC bus or via AC coupling) generated during periods of abundant sunlight and storing it for later dispatch during nighttime hours or when solar irradiance is diminished. It enables highly effective peak shaving by rapidly discharging stored energy during intervals of high aggregate EV charging demand, thereby substantially curtailing the peak power drawn from the external utility grid. This capability not only yields significant reductions in electricity procurement costs, particularly in jurisdictions with punitive demand charges, but can also potentially obviate the necessity for expensive and time-consuming upgrades to the grid connection infrastructure. Furthermore, the BESS significantly enhances the facility's energy independence and operational resilience by providing critical backup power during unforeseen grid outages. This ensures that essential functions, including the MVPS operations, emergency lighting, and basic EV charging services, can remain functional, maintaining service continuity. Optionally, contingent upon prevailing local energy market structures and conducive regulatory frameworks, the BESS could actively participate in advanced grid services. This might involve dynamically injecting stored energy into the grid or absorbing excess grid power to assist in maintaining grid frequency and voltage stability, thereby potentially unlocking additional revenue streams for the facility. The entire facility maintains a robust, bidirectional connection to the utility grid, managed through a dedicated Grid Connection Infrastructure. This infrastructure typically includes essential components such as Protection & Switchgear for safety and control, a Smart Meter & Grid Interface for monitoring and communication, and a Grid Transformer Substation to step voltage up or down as required for compatibility with the external network, enabling both import and export of AC power.

The synergistic and intelligent orchestration of the PV array and the BESS is paramount. This coordinated operation is fine-tuned to maximize the self-consumption of the on-site generated renewable energy, rigorously minimize the reliance on—and associated costs of—energy imported from the utility grid (especially during high-tariff peak periods), substantially bolster the facility's operational resilience against external power supply disruptions, and achieve a significant, measurable reduction in its overall carbon footprint. The Digital Twin plays an increasingly pivotal role in the sophisticated management of this complex energy equilibrium. It employs advanced predictive algorithms, leveraging machine learning models trained on historical data and real-time sensor inputs, to dynamically optimize the charging and discharging cycles of the BESS. These optimization decisions are continuously informed by multi-parameter forecasts, including anticipated solar generation levels, predicted EV charging demand patterns (influenced by time-of-day, day-of-week, and local events), and fluctuating real-time electricity pricing signals from the grid. Fig. 14.9 provides a conceptual illustration of the intricate energy flows and interdependencies within this highly integrated renewable energy system, showcasing the journey of energy from solar capture, through storage and conversion, to its ultimate utilization by various loads or interaction with the external grid via the specialized connection infrastructure.

14.2.3 Smart charging and IoT control

The advanced operational intelligence, dynamic responsiveness, and overall efficiency of the Self-Sufficient Smart Hangar are critically contingent upon the sophisticated and seamless integration of cutting-edge smart charging functionalities and a comprehensive, pervasively deployed Internet of Things (IoT)-based control architecture. This intricate web of interconnected devices and intelligent systems is typically orchestrated and continuously optimized through the facility's overarching Digital Twin framework. Smart charging, in this context, represents a significant evolution beyond rudimentary power delivery to EVs. It embodies a sophisticated suite of dynamic, adaptive strategies meticulously designed to

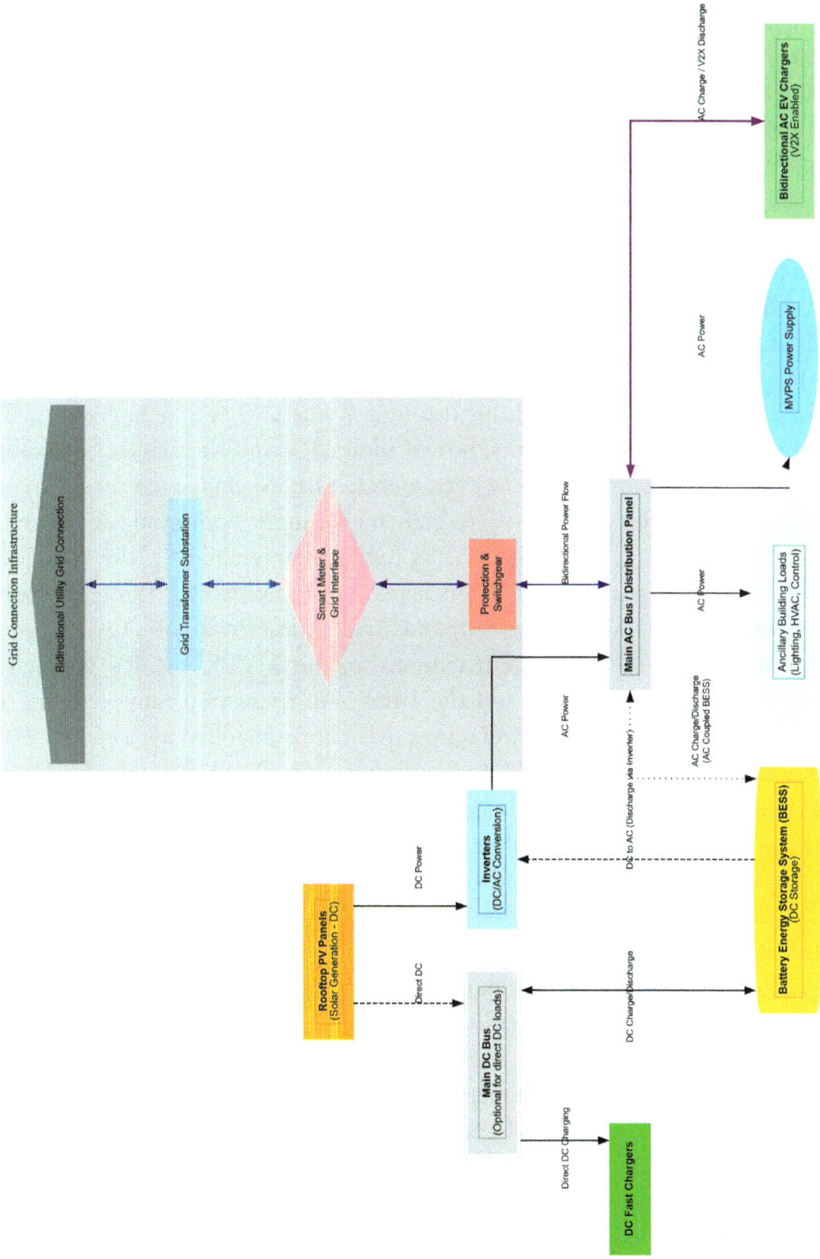

Figure 14.9 Conceptual diagram of renewable energy integration, flow, and grid connection infrastructure within the smart hangar.

optimize the entire charging process based on a rich tapestry of real-time data inputs and predictive analytical insights. These influencing factors include, but are not limited to, prevailing grid conditions (such as overall load levels, frequency stability, and localized transformer capacity), highly fluctuating energy pricing schemes (including time-of-use tariffs and real-time market prices), the instantaneous availability of on-site renewable energy from the PV system, specific user-defined preferences (e.g., required vehicle departure time, desired final state of charge, or cost sensitivity), and crucial considerations for preserving the long-term health and performance of EV batteries (Liu et al., 2023a).

The IoT network serves as the indispensable digital communication backbone, establishing robust and reliable connectivity between a highly diverse and distributed array of components. This includes numerous types of sensors deployed strategically throughout the facility (e.g., environmental sensors, occupancy detectors, power meters), various actuators controlling physical operations (e.g., MVPS motors, BESS charge controllers), the Electric Vehicles themselves (which increasingly communicate via standardized protocols like ISO 15118 for V2G interactions), the individual EV charging stations (often managed and monitored using industry-standard protocols such as the Open Charge Point Protocol—OCPP (Guillemin et al.)), the complex machinery of the MVPS, the PV generation system with its inverters, and the BESS. All these disparate elements are seamlessly interconnected and report to, or are controlled by, a central management platform, often hosted within or interfaced through the Digital Twin. Key constituent elements within this expansive IoT ecosystem include sensors for meticulously monitoring parking spot occupancy and availability; systems for obtaining critical EV battery data such as State of Charge (SoC) and, where possible, State of Health (SoH) (this information is frequently transmitted by the vehicle's onboard Battery Management System—BMS); devices for tracking individual charger status, real-time power flow, and energy consumption; instrumentation for measuring instantaneous PV generation and BESS energy levels and power capacity; smart meters for monitoring bidirectional power exchange with the utility grid; and sensors for capturing relevant ambient environmental data (like temperature and humidity, which can significantly affect battery charging efficiency and degradation rates). The EV chargers themselves, equipped with advanced communication interfaces (e.g., Ethernet, Wi-Fi, cellular), are essential, allowing them to receive dynamic commands from the central control system to precisely adjust charging power levels, implement variable charging rates, or adhere to specific, optimized charging schedules. The system is designed to incorporate a mix of AC (Level 2) chargers for standard overnight or longer-duration parking, and DC fast chargers for users requiring rapid top-ups. A sophisticated, often hierarchical (centralized with distributed intelligence) control system continuously processes the voluminous incoming data streams, executes complex multi-objective optimization algorithms, and dispatches precisely timed and calibrated control signals back to the EV chargers, the MVPS operational units, and the BESS power conditioning systems. Complementing this intricate backend infrastructure is an intuitive, user-centric interface, typically realized as a mobile application. This application empowers users to perform a range of actions, including reserving parking and charging sessions in advance, monitoring the real-time progress of their vehicle's charging, specifying personal charging preferences (such as desired departure SoC, maximum charging cost, or willingness to participate in demand response programs), and seamlessly handling secure payment transactions.

The Digital Twin, in this advanced operational context, functions as the supreme intelligence and coordination core of the entire Smart Hangar system. It continuously assimilates and processes the high-velocity, high-volume real-time data streams supplied via the IoT network. Within this dynamic, evolving virtual replica of the physical facility, highly sophisticated algorithms are perpetually executed to perform several mission-critical tasks. These include advanced demand forecasting, where historical operational data, temporal patterns (time-of-day, day-of-week, seasonality), local event calendars, weather forecasts, and other pertinent variables are ingested by machine learning models to accurately predict future parking occupancy rates and EV charging energy requirements. Smart charging scheduling algorithms then leverage these forecasts to optimize the charging plans for individual EVs or for the entire aggregated fleet of parked vehicles. These algorithms aim to concurrently achieve multiple objectives, such as minimizing overall operational energy costs by shifting load to off-peak or high-solar-availability periods, maximizing the direct utilization of on-site generated renewable energy, preventing overloads on the local grid connection or internal electrical infrastructure, and potentially enabling active participation in Vehicle-to-Grid (V2G) services, transforming parked EVs into a distributed energy resource (Honarmand et al., 2014). This often involves complex dynamic power allocation strategies among actively charging vehicles based on priority, battery status, and user settings. The Digital Twin also meticulously coordinates parking allocation in conjunction with the MVPS, intelligently assigning parking spots based on factors like specific charging power requirements (e.g., DC fast charge vs. Level 2 AC), anticipated duration of stay, and proximity to other system resources. Finally, and perhaps most critically, it performs holistic, end-to-end energy management. This involves dynamically balancing the intricate interplay of PV generation, BESS charging and discharging cycles, aggregate EV charging loads, building operational loads, and all interactions (import/export) with the external utility grid. This tightly integrated, data-driven, and intelligently controlled system elevates the Smart Hangar far beyond the status of a mere parking and charging facility, positioning it as a dynamic, responsive, highly optimized, and strategically valuable node within the increasingly complex and interconnected urban energy and transportation landscapes of the future. Fig. 14.10 provides a schematic overview of this smart charging architecture, highlighting the network of IoT devices and the central role of the control platform.

14.3 Integrated charging & renewable energy details

This section provides a more granular and detailed examination of the deeply symbiotic relationship established between the Electric Vehicle (EV) charging infrastructure and the comprehensive renewable energy systems—encompassing both photovoltaic (PV) generation and Battery Energy Storage System (BESS)—that are integral to the Self-Sufficient Smart Hangar. The primary objective herein is to meticulously delineate how these complex systems are not only designed but also dynamically operated in close concert. This coordinated approach aims to achieve maximal levels of environmental sustainability, superior operational efficiency, robust economic viability, and a harmonious, mutually beneficial integration with the broader external power grid, thereby positioning the Smart Hangar as a proactive and intelligent node within the urban energy landscape. The specific internal architecture, featuring eight distinct vertical parking box blocs, each potentially serviced by one of the eight dedicated mobile robotic chargers, informs the nuanced design and operational strategies discussed.

14.3.1 Charging infrastructure design & integration within an 8-bloc architecture

The architectural and operational paradigm of the EV charging infrastructure within the Smart Hangar is a cornerstone of its functionality, demanding a sophisticated reconciliation of diverse user charging

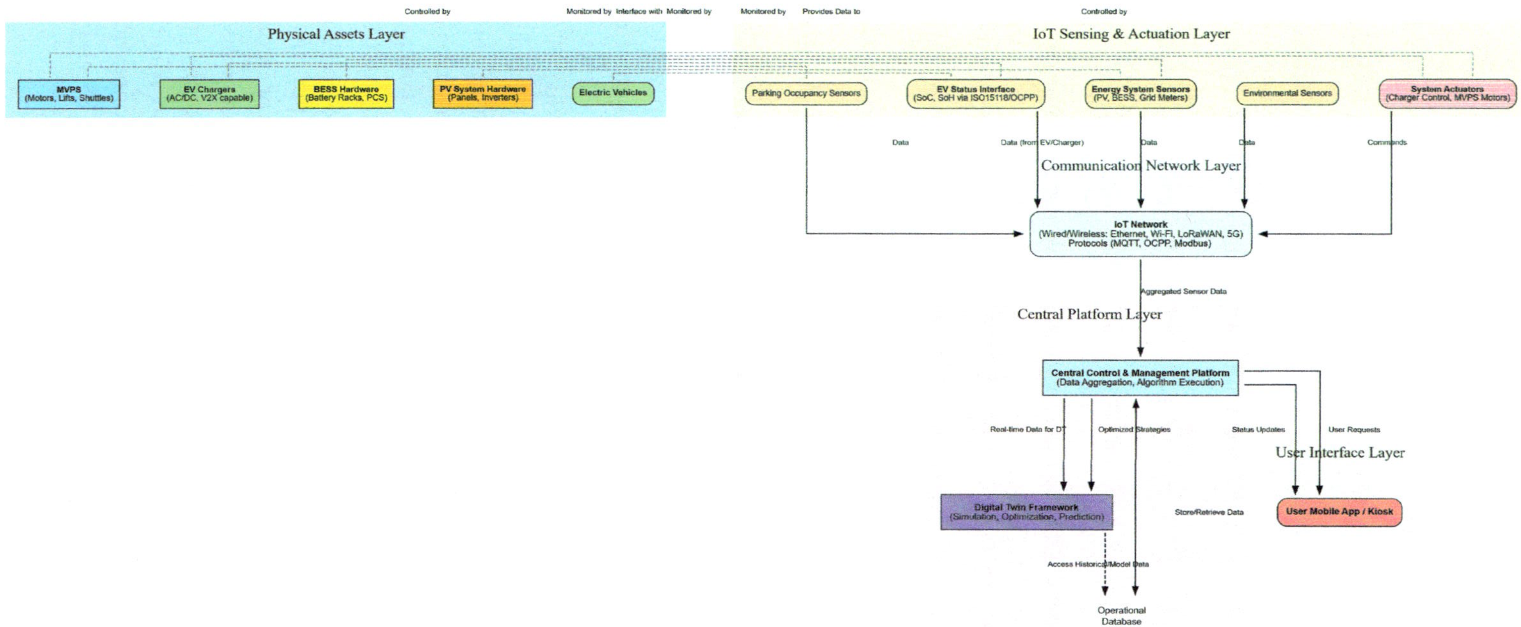

Figure 14.10 Conceptual smart charging architecture and IoT control system within the smart hangar.

expectations with the operational dynamics of the 8-bloc automated parking system and the fluctuating availability of integrated renewable energy resources. Several pivotal design considerations are paramount in shaping this distributed yet centrally orchestrated infrastructure. While the system relies on eight mobile robotic chargers for delivering power to EVs within the eight vertical parking box blocs, the underlying infrastructure must support this flexible model. This includes providing robust power and communication interfaces within each of the eight blocs, to which the mobile robots can dock or connect to draw power for charging EVs, and for their own operational energy if needed. The nature of charging (AC Level 2 vs. DC Fast Charging) will be determined by the capabilities of the mobile robots themselves, which could be designed as versatile units or a mix of specialized AC/DC robots. The decision on the robots' power delivery capacity (7 kW to 22 kW for AC, or 50 kW + for DC) and the number of robots with each capability would be based on detailed demand modeling, simulating user arrival patterns, dwell times, and desired states of charge across the eight distinct parking.

A critical operational question is the sufficiency of the eight mobile robotic chargers for the 1280-vehicle capacity. The system is designed to charge up to 8 EVs simultaneously. The adequacy of this number is predicated on a dynamic, sequential charging model managed by the Digital Twin. Given that most vehicles will be parked for extended durations (e.g., a full workday or overnight), the eight robots can service a large number of vehicles consecutively. The central control system optimizes the dispatch queue based on user-specified departure times, required state-of-charge, and real-time energy availability, ensuring all charging needs are met efficiently. This approach avoids the high cost and underutilization of installing a fixed charger at every spot. Furthermore, the system is inherently scalable; the number of mobile robotic chargers can be increased or decreased in the future based on evolving real-world demand patterns, providing long-term flexibility without major infrastructural changes.

The electromechanical integration of the mobile robotic chargers with the EVs stored within the eight vertical parking box blocs presents a sophisticated engineering endeavor. Each robot must be capable of autonomously navigating to a designated vehicle within any assigned bloc, precisely docking with the vehicle's charge port, and establishing a secure power and communication link. This entire process—including the robot's movement, connection, charging session management, and disconnection—must be fully automated and orchestrated by the central MVPS and charging management system, eliminating any need for manual intervention by users. Standardized communication protocols are crucial: the mobile robotic chargers themselves will communicate with the central system (likely via robust wireless links like Wi-Fi six or private 5G), which in turn manages their dispatch and monitors their status. The interface between the robot and the EV will adhere to standards like OCPP for backend communication regarding the charging session and ISO 15118 for direct, intelligent communication with the EV, enabling features like Plug & Charge and bidirectional power flow (V2X) if the robots are so equipped (Schurmann et al., 2017).

The internal electrical distribution network must be engineered to supply adequate power to interface points within each of the eight parking blocs, sufficient to support the simultaneous operation of multiple mobile robotic chargers at their rated capacities. While the robots distribute the charging load dynamically, the aggregate peak demand from all eight robots charging vehicles concurrently across the blocs must be managed. This necessitates a robust main distribution system, intelligent sub-panels feeding the blocs, and advanced smart charging algorithms to modulate the robots' power draw based on overall facility load, PV availability, BESS state, and grid conditions. This distribution network is intricately linked with the on-site PV generation and BESS for optimized energy sourcing. Scalability remains a key design tenet; the infrastructure should allow for the potential addition of more mobile robots or the upgrading of power delivery capacity to the parking blocs as EV charging technology and demand evolve. The user experience hinges on this automation: users interact via a mobile app to request charging, receive real-time updates on the robot's progress and charging status, and handle billing, all seamlessly integrated with the parking service (Bornioli et al., 2024).

The Digital Twin is instrumental in designing and optimizing this complex, distributed charging system. It allows for the simulation of robot movement patterns, charging task allocation algorithms across the eight blocs, power demand aggregation, and the overall efficiency of different robot-to-bloc assignment strategies. By modeling vehicle traffic and charging requests, the DT helps determine the optimal number and capabilities of the mobile robotic chargers needed to service the eight parking blocs effectively, ensuring high service levels without excessive capital investment in the robotic fleet or power infrastructure. Fig. 14.11 schematically illustrates this distributed charging concept, emphasizing power and

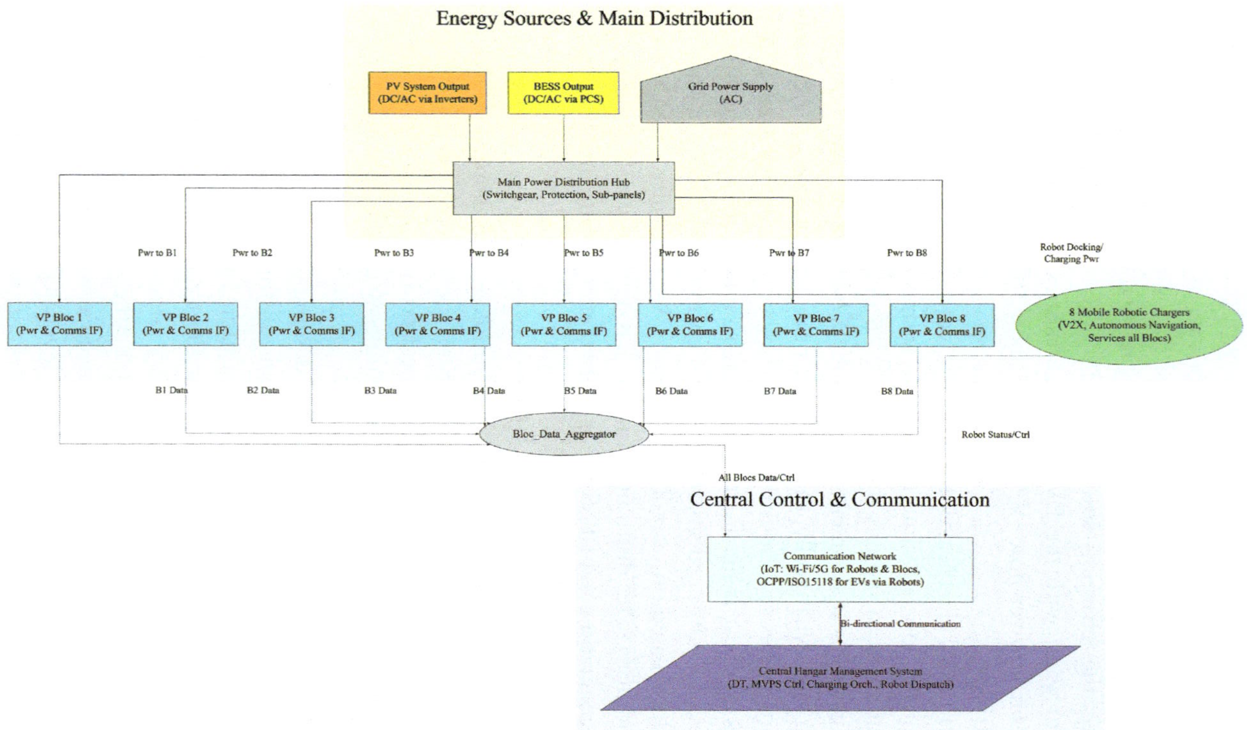

Figure 14.11 Conceptual schematic of the integrated charging infrastructure within the smart hangar, detailing power distribution to 8 vertical parking box blocs and interfaces supporting 8 mobile robotic chargers, alongside essential communication links.

data connections to the eight vertical parking box blocs and the role of the mobile robotic chargers.

14.3.2 Synergies of solar power, energy storage, & grid interaction

The strategic and co-optimized integration of a substantial rooftop solar PV generation system with a versatile, appropriately dimensioned BESS forms the energetic nucleus of the Smart Hangar's ambition for enhanced energy self-sufficiency, minimized operational carbon footprint, and beneficial grid interactivity. The PV system effectively repurposes the hangar's extensive rooftop area into a productive, distributed power generation asset, converting abundant incident solar irradiance directly into clean DC electricity. This locally generated, emissions-free energy is then managed with sophisticated flexibility: it can be used to directly power DC-compatible loads such as certain configurations of the mobile robotic chargers or contribute directly to charging the BESS. More typically, for broader distribution and compatibility with diverse loads, the raw DC power from the PV array is converted into AC by high-efficiency, grid-forming inverters. This conditioned AC power then supplies the power interface points within the eight parking blocs (for the mobile robotic chargers), supports Level 2 AC charging functionalities if any fixed points exist, and comprehensively meets the building's wide array of auxiliary power needs, including the MVPS, advanced LED lighting, efficient HVAC systems, and the distributed control and sensor networks.

The BESS operates as the indispensable energy reservoir and critical temporal smoothing element within this dynamic, integrated renewable energy ecosystem. Its primary role is to effectively decouple the often intermittent and unpredictable timing of solar energy generation from the equally variable and diverse patterns of energy consumption, which are predominantly driven by the aggregated EV charging demand across the eight blocs and the operational requirements of the hangar building itself. The meticulous sizing of the BESS, encompassing both its total energy storage capacity (expressed in kW h or MW h) and its maximum power rating (in kW or MW, which dictates the peak charge and discharge rates), is determined through a complex, multi-objective optimization calculus. This analytical process must holistically integrate precise forecasts of site-specific solar energy generation profiles (accounting for geographical location, seasonal variations, and diurnal patterns), granular predictions of aggregate EV charging load curves (factoring in diverse user behaviors, vehicle types, and charging patterns across the eight blocs), prevailing and anticipated utility grid electricity tariff structures (which frequently include substantial time-of-use price differentials and punitive demand charges linked to peak power consumption), and a rigorous techno-economic evaluation of the potential revenue streams and operational benefits derivable from active participation in sophisticated ancillary grid service programs (such as fast frequency response, capacity markets, or voltage regulation services) (Khalkhali and Hosseinian, 2019). A compelling and increasingly viable consideration in modern BESS design, particularly resonant with sustainability imperatives and circular economy principles, involves the strategic deployment of second-life EV batteries. While these repurposed batteries might exhibit a degree of capacity fade from their primary automotive use, they can offer a significantly more cost-effective and environmentally benign energy storage solution for less demanding stationary applications, such as those within the Smart Hangar (Piombo et al., 2025).

The intelligently orchestrated, synergistic operation of the PV generation system and the BESS unlocks a multifaceted array of substantial benefits. These advantages accrue not only to the hangar operator in terms of reduced energy costs and enhanced operational autonomy but also contribute positively to the stability, efficiency, and accelerated decarbonization of the wider surrounding utility grid. A paramount benefit is the considerable reduction of the facility's peak power demand imposed on the grid. By strategically discharging locally stored solar energy from the BESS during periods of high concurrent EV charging across the eight blocs or significant building operational load, the system dramatically curtails the maximum instantaneous power the hangar needs to import from the external utility. This sophisticated peak shaving capability directly mitigates the potential need for costly and often time-consuming upgrades to the local electricity distribution network infrastructure. Concurrently, it serves to reduce ongoing operational expenditures linked to high demand charges frequently levied by utility providers, thereby improving the facility's economic performance. Furthermore, the substantial on-site capacity for renewable generation coupled with dispatchable energy storage can materially enhance the local grid's ability to integrate higher penetrations of distributed renewable resources from other sources in the vicinity. This is achieved by alleviating localized grid stress, smoothing voltage fluctuations, and providing a degree of local energy balancing, effectively allowing the Smart Hangar to function as a valuable "grid-edge" resource. Advanced smart charging strategies, dynamically enabled and supported by the responsive PV and BESS assets, facilitate sophisticated load shifting functionalities. This empowers the system to strategically defer or advance significant portions of the EV charging demand (managed by the eight mobile robots) away from periods of system-wide grid peak load (when electricity is typically most expensive and carbon-intensive) toward times characterized by abundant on-site solar generation, lower overall system demand, or more favorable electricity tariffs. This contributes significantly to grid stability, optimizes energy procurement costs, and maximizes renewable energy utilization. With the integration of advanced, bidirectional power electronics (in the inverters/PCS and potentially in the mobile robotic chargers themselves) and sophisticated hierarchical control systems, the BESS—and prospectively, the aggregated fleet of connected EVs through Vehicle-to-Grid (V2G) paradigms—can actively provide a suite of valuable ancillary services. These may include rapid-response voltage support, dynamic frequency regulation, and provision of spinning or non-spinning reserve capacity to the utility grid, thereby materially enhancing its overall reliability, resilience, and its capacity to seamlessly integrate large volumes of intermittent renewable energy sources.

The Digital Twin platform is absolutely indispensable for the sophisticated orchestration of this intricate and highly dynamic energy ecosystem. It continuously employs advanced predictive control algorithms and machine learning models. These intelligent systems are perpetually refined and informed by high-resolution, real-time sensor data streams originating from all energy assets (PV, BESS, grid interface) and loads (EV chargers, MVPS, building systems), in conjunction with critical external data inputs such as precise meteorological forecasts (for solar prediction) and real-time electricity market pricing signals. These predictive and adaptive control strategies serve to continually optimize the complex operational decisions that govern the PV system's energy dispatch, the BESS's intricate charge and discharge cycles, and the aggregated EV charging schedules across the eight blocs. The overarching goal is to simultaneously satisfy multiple, often competing, operational objectives. These include minimizing total energy procurement costs, maximizing the direct self-consumption of on-site generated renewable energy, ensuring high levels of user satisfaction with regard to EV charging availability and speed, and fulfilling any contractual obligations related to the provision of ancillary services to the grid (Jiang et al., 2024). Fig. 14.12 provides a comprehensive visual representation of the complex, multi-directional energy flows and the intricate operational interactions that are continuously managed within this advanced, integrated

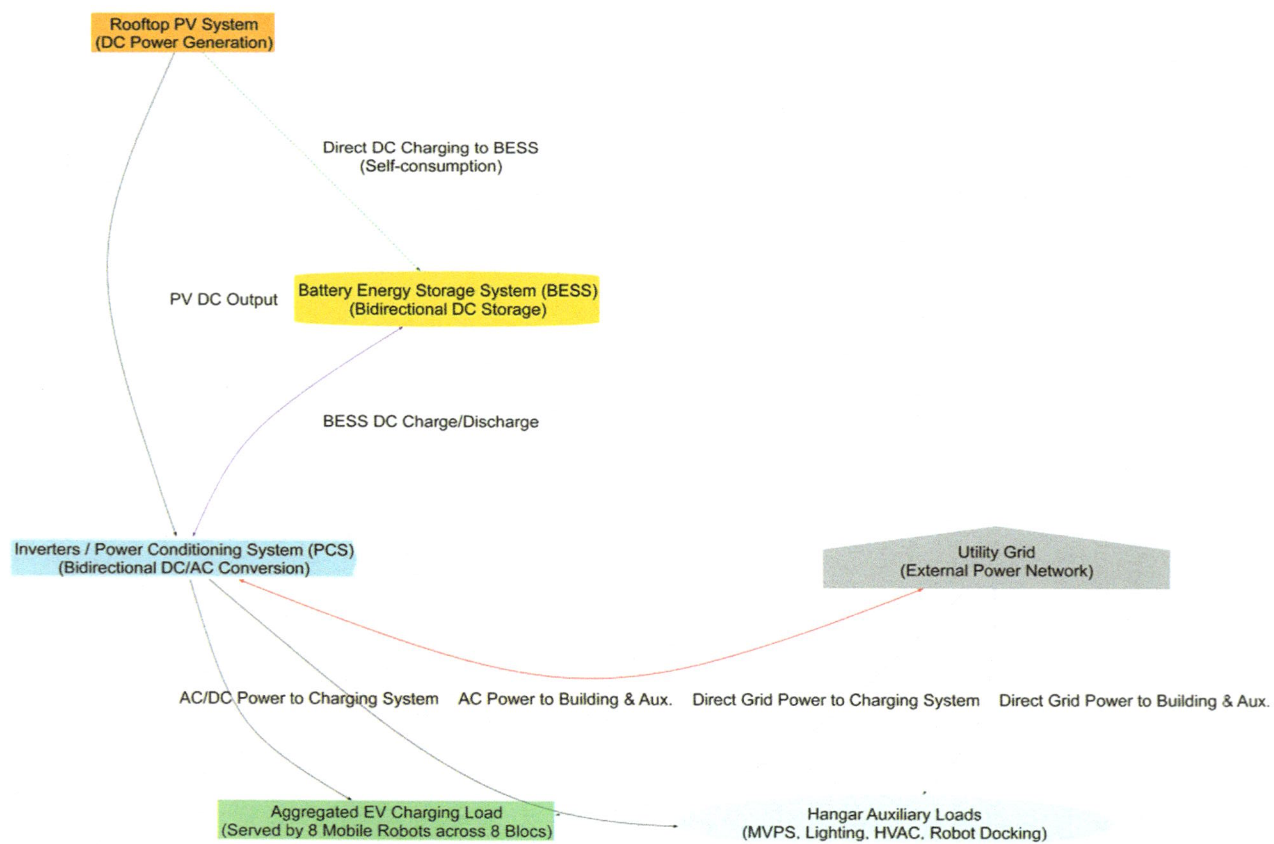

Figure 14.12 Conceptual energy flow diagram illustrating solar power generation, energy storage dynamics, aggregated EV charging load (from eight blocs serviced by eight mobile robots), and bidirectional grid interactions within the smart hangar.

solar power generation, high-capacity energy storage, and dynamically grid-interactive system, effectively highlighting its adaptive response capabilities to a wide spectrum of varying operational conditions and external stimuli.

14.4 IoT & control system architecture

The successful realization and sustained high-performance operation of the Self-Sufficient Smart Hangar's advanced capabilities are critically contingent upon a meticulously designed and robustly implemented Internet of Things (IoT) infrastructure. This infrastructure is inextricably coupled with a sophisticated, often hierarchical, multi-layered control system. This comprehensive architecture effectively serves as the central nervous system and cognitive engine of the entire hangar facility. It enables pervasive real-time data acquisition from a multitude of distributed sources, facilitates intelligent, data-driven decision-making through advanced analytics and optimization algorithms, ensures precise and timely actuation of physical systems, and supports seamless, secure integration with various external systems and stakeholders. This entire ecosystem is frequently augmented, managed, and visualized through a comprehensive Digital Twin (DT) platform, which acts as a dynamic virtual counterpart to the physical hangar.

14.4.1 System monitoring & control components: A layered perspective

The intricate architecture underpinning the Smart Hangar's monitoring and control capabilities can be effectively conceptualized as a series of interconnected, hierarchical layers, each tasked with performing specific, well-defined functions critical to the overall system operation.

At the foundational level resides the Physical Layer. This layer encompasses all tangible assets and electromechanical systems within the hangar. It includes the primary building structure, the complex and dynamic machinery of the Mobile Vertical Parking System (MVPS)—comprising numerous motors, precision drives, high-speed lifts, and autonomous shuttles—the distributed array of EV chargers (incorporating both AC Level 2 and DC Fast Charger typologies, serviced by the eight mobile robotic chargers across the eight parking blocs), the expansive rooftop PV installation, the high-capacity BESS units, and, transiently, the diverse fleet of Electric Vehicles (EVs) undergoing parking and charging processes. Crucially, this layer is also densely populated with an extensive network of embedded sensors and actuators. Diverse sensor modalities are strategically deployed to continuously monitor a wide spectrum of operational parameters. These include, but are not limited to, parking slot occupancy verification (employing technologies such as optical, ultrasonic, or magnetic sensors), the precise real-time position and operational status (e.g., speed, load, faults) of all critical MVPS components, critical thermal conditions within both the stationary BESS modules and the batteries of the EVs being charged, as well as the operating temperatures of charging equipment and power electronics. Electrical parameters, such as voltage, current, power factor, and energy consumption, are meticulously monitored at individual chargers, at the BESS interface, and at the main grid interconnection point. Ambient solar irradiance levels are continuously measured by pyranometers to inform PV generation forecasting models. Furthermore, sophisticated communication interfaces are leveraged to retrieve vital data directly from EVs, such as their State of Charge (SoC), State of Health (SoH), and specific user-defined charging preferences or requirements (Merkle et al., 2021). Complementing this sensory network, an array of actuators translates digital control signals into precise physical actions. These include the electric motors and variable frequency drives (VFDs) governing the MVPS movements, high-speed relays and solid-state contactors within the EV chargers (or robotic charging arms) that manage and modulate power flow to the vehicles, the sophisticated Power Conversion System (PCS) that controls the bi-directional charge and discharge cycles of the BESS, and potentially intelligent switchgear, smart meters, or remotely operable circuit breakers at the utility grid connection point, enabling dynamic interaction and control (Galkin et al., 2023).

Facilitating the vital information exchange between the distributed physical assets and the higher-level control and management systems is the Communication Layer, which constitutes the core of the IoT network. This layer characteristically employs a heterogeneous and resilient mix of communication technologies, carefully selected and deployed to meet the specific bandwidth, latency, reliability, and power consumption requirements of diverse connected devices and data streams. Robust wired connections, such as industrial-grade Ethernet (often employing protocols like Modbus TCP/IP or Profinet) or resilient CAN bus networks (Popa et al., 2022), are typically favored for stationary and mission-critical infrastructure components, including the primary MVPS control units, fixed charging infrastructure elements (if any), and BESS management systems.

Wireless technologies provide essential flexibility and scalability. High-bandwidth Wi-Fi (e.g., Wi-Fi 6/6E) is suitable for local data-intensive applications and communication with mobile elements like the robotic chargers. For broadly distributed, low-power sensors that transmit small data packets infrequently, Low-Power Wide-Area Networks (LPWANs) such as LoRaWAN or Narrowband IoT (NB-IoT) offer cost-effective and energy-efficient connectivity. Emerging private 5G network deployments within the hangar might also fulfill specialized roles, providing dedicated, high-bandwidth, ultra-low-latency communication channels, particularly pertinent for advanced vehicle-to-everything (V2X) communications or time-sensitive robotic control (Basnet and Ali, 2021; Rimal et al., 2022). The exchange of data across this network often relies on standardized, lightweight IoT messaging protocols like MQTT (Message Queuing Telemetry Transport) or CoAP (Constrained Application Protocol) for optimal efficiency and interoperability. Concurrently, domain-specific communication protocols such as OCPP (Open Charge Point Protocol) for comprehensive EV charger management and monitoring, and ISO 15,118 for advanced, secure communication between EVs and the charging infrastructure (enabling features like Plug & Charge and smart V2G negotiation), are indispensable for specialized functions. Throughout this complex communication infrastructure, ensuring robust end-to-end data security, including encryption, authentication, and access control, as well as safeguarding user privacy in compliance with relevant regulations, remains a paramount and ongoing design and operational concern (Naseri et al., 2023).

Optionally, but increasingly recommended for large-scale, latency-sensitive deployments, an Edge/Fog Computing Layer can be strategically interposed between the local communication network and the centralized cloud/control layer. This intermediate layer is responsible for performing localized data pre-processing, real-time analytics, and executing control logic that demands extremely low latency and high determinism. Examples include the fine-grained, closed-loop control of intricate MVPS movements, rapid safety interlock responses for EV chargers in anomalous conditions, or initial anomaly detection in sensor data streams. Edge computing devices (e.g., industrial PCs, dedicated microcontrollers) can significantly reduce the sheer volume of raw data transmitted to the cloud by performing aggregation, filtering, and preliminary analysis locally. This not only conserves network bandwidth but also enables the system to maintain essential or degraded-mode functionality even if connectivity to the central cloud platform is temporarily disrupted, thereby enhancing overall system resilience and fault tolerance.

The Cloud/Control Layer serves as the centralized intelligence and strategic command hub of the Smart Hangar. Typically hosted on scalable and resilient remote server infrastructure (either private cloud, public cloud, or a hybrid model), this layer houses the primary sophisticated control algorithms, extensive and high-performance databases for long-term historical data storage and retrieval, powerful data analytics engines for tasks such as demand forecasting, operational pattern recognition, and performance optimization, and, crucially, the comprehensive Digital Twin model of the entire facility. This layer receives aggregated, often pre-processed, data streams from the lower edge and communication layers. It is where computationally intensive tasks are executed, including complex multi-objective optimization routines for energy management and charging scheduling, large-scale simulations for "what-if" scenario analysis, and the training of machine learning models for predictive maintenance and demand forecasting. This layer is also responsible for issuing high-level strategic commands and setpoints back down to the edge computing devices or, in some cases, directly to the controllers of the physical assets. Various platforms designed for industrial IoT or smart city applications, such as FIWARE, or bespoke cloud-native solutions developed using microservices architectures, can provide the requisite robust and scalable infrastructure for this layer.

Finally, the Application/User Interface Layer provides the necessary interaction points for both end-users (EV drivers) and facility operators. This layer encompasses intuitive mobile applications that empower end-users to seamlessly manage their parking and charging experiences. This typically includes functionalities such as reserving parking slots and charging sessions in advance, monitoring the real-time status of their vehicle's charging progress, setting personal charging preferences (e.g., desired final SoC, preferred departure time, cost limits), receiving timely notifications, and securely handling payments for services rendered. For facility operators and maintenance personnel, this layer provides sophisticated dashboards, supervisory control and data acquisition (SCADA)-like interfaces, or dedicated control consoles. These tools enable comprehensive monitoring of overall system health and performance metrics, provide oversight of ongoing operations across the eight parking blocs and eight robotic chargers, facilitate the scheduling and tracking of routine and corrective maintenance activities, and allow for the configuration and fine-tuning of various system parameters and control strategies.

The Digital Twin predominantly operates and evolves within the Cloud/Control Layer, functioning as a dynamic, data-synchronized, high-fidelity virtual replica of the entire Smart Hangar ecosystem—its physical assets, processes, and operational environment. It provides operators with an integrated, real-time, and often multi-dimensional view of the system's holistic status and performance characteristics, thereby enabling more informed, proactive, and optimized decision-making and strategic management. Fig. 14.13 offers a high-level visual representation of this multi-layered IoT and control system architecture, illustrating the principal data flows and control signal pathways.

14.4.2 Optimization strategies & Smart Grid Integration for enhanced performance

The sophisticated, data-driven control system, significantly amplified by the analytical and predictive capabilities inherent in the Digital Twin, unlocks the full potential for implementing advanced optimization strategies within the Smart Hangar. These strategies are not only crucial for enhancing internal operational efficiency but also for enabling the facility's effective and mutually beneficial integration into the broader smart grid ecosystem. The primary optimization objectives are typically multi-faceted, revolving around the concurrent goals of minimizing the facility's total operational expenditures (OpEx), with a strong emphasis on reducing electricity procurement costs; maximizing the utilization of locally generated, clean renewable energy from the rooftop PV system; ensuring exceptionally high levels of user satisfaction by, for example, consistently guaranteeing that EVs reach their desired target state of charge by their requested departure times; while simultaneously minimizing any potential adverse impacts on the local utility grid (such as contributing to peak load or voltage instability) or, ideally, maximizing revenue streams and operational benefits derived from actively providing valuable grid support services.

Several key optimization strategies are dynamically implemented and continuously refined within the central control system, often leveraging complex algorithms and machine learning models hosted and executed within the Digital Twin environment. Smart Charging Scheduling is of paramount importance. This involves employing a range of sophisticated techniques, from established mathematical programming approaches (such as linear programming, mixed-integer linear programming, or dynamic

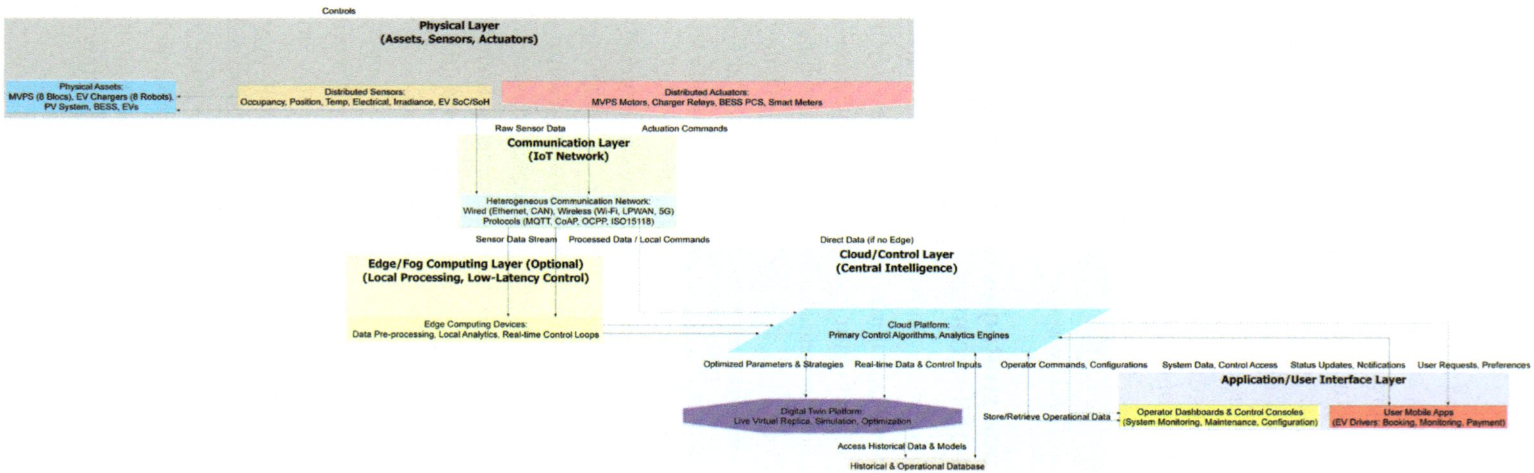

Figure 14.13 Multi-layered IoT and control system architecture for the smart hangar, showing data and control flow between physical assets, communication networks, edge/cloud platforms, and user interfaces.

programming for sequential decision making) to cutting-edge machine learning methodologies like deep reinforcement learning, which can learn optimal control policies through interaction with the system or its simulated counterpart. These advanced algorithms meticulously calculate and dynamically adjust the optimal charging power profile for each individual EV connected within any of the eight parking blocs throughout its entire parking duration. This intricate calculation process must harmoniously balance a multitude of interacting factors: the vehicle's precise arrival and user-specified (or predicted) departure times; its current State of Charge (SoC) upon arrival and the target SoC desired by the user; critical considerations for preserving long-term EV battery health and minimizing degradation (e.g., by avoiding excessively high charging rates when not necessary, or managing battery temperature) (Jafari and Byun, 2022); real-time and forecasted electricity prices, including complex time-of-use (TOU) tariffs or dynamic real-time pricing (RTP) signals from the utility; accurately predicted solar energy generation from the on-site rooftop PV system; and the stringent overall power capacity constraints of the Smart Hangar's internal electrical infrastructure and its primary grid connection point. The dispatch of the eight mobile robotic chargers to specific vehicles within the blocs is a direct output of this scheduling process.

Complementary to, and often integrated with, smart charging scheduling are strategies for Load Balancing and Peak Shaving. These strategies focus on intelligently distributing the aggregate EV charging load more evenly over extended periods or dynamically across multiple active robotic chargers and parking blocs. The core objective is to prevent the instantaneous total power demand of the facility from exceeding pre-defined operational limits or, more critically, from breaching the contracted capacity with the utility, which could trigger substantial financial penalties in the form of high demand charges. The BESS plays an absolutely pivotal role in these strategies, acting as a flexible energy buffer. It can rapidly absorb excess power demand during peak charging periods by discharging stored energy, and conversely, it can be strategically charged using surplus PV generation or low-cost grid electricity during off-peak periods (Prencuva et al., 2024). Even the Parking Allocation Optimization logic within the MVPS can be synergistically refined to support energy goals; the central control system might prioritize placing EVs with immediate or high-power charging requirements in blocs that are currently best serviced by available robotic chargers or that have optimal access to power infrastructure. Similarly, grouping vehicles with similar anticipated departure times within specific blocs or zones can streamline the MVPS retrieval process, thereby minimizing internal vehicle movements and associated energy consumption by the MVPS machinery.

Furthermore, dedicated BESS Operation Optimization algorithms continuously determine the most economically and technically advantageous charge and discharge strategy for the stationary battery system itself. This sophisticated strategy aims to achieve a dynamic balance across several objectives: maximizing the direct self-consumption of on-site generated PV energy (by storing surplus solar power that cannot be immediately used by EV charging or building loads); minimizing the net cost of energy imported from the utility grid (achieved by preferentially charging the BESS during periods of low electricity prices or high PV output, and discharging it to serve loads during high-price periods); and potentially generating additional revenue by actively participating in various grid ancillary service markets, such as providing fast frequency response (FFR) or capacity reserve, where permissible and economically viable. Looking toward more advanced grid integration scenarios, if Vehicle-to-Grid (V2G) Coordination is implemented (requiring V2G-capable EVs, robotic chargers, and bidirectional inverters/PCS), the Smart Hangar's control system would need to meticulously optimize the bidirectional power flow from the aggregated fleet of capable EVs. This involves complex decision-making regarding when, from which vehicles, and how much energy each EV should discharge back to support the facility's internal loads (Vehicle-to-Building, V2B) or export to the utility grid (V2G). Such decisions must carefully balance the potential revenue or grid stability benefits against critical factors like the impact on EV battery degradation due to increased cycling, ensuring that the user's primary mobility needs (i.e., sufficient charge for their next trip) are always unequivocally met, and adhering to user-defined V2G participation preferences (Movahedi et al., 2024).

Effective and dynamic integration with the broader smart grid necessitates robust, secure, and standardized communication channels with utility operators, distribution system operators (DSOs), or designated third-party energy aggregators (Kim et al., 2019). The Smart Hangar, by virtue of its significant dispatchable load (EV charging), on-site generation (PV), and flexible energy storage (BESS and potentially V2G-capable EVs), can function as an intelligent, highly flexible, and responsive grid-edge resource. It can dynamically adapt its net power consumption or injection profile in response to external grid signals. These signals might include direct demand response requests from the utility (e.g., asking the facility to temporarily

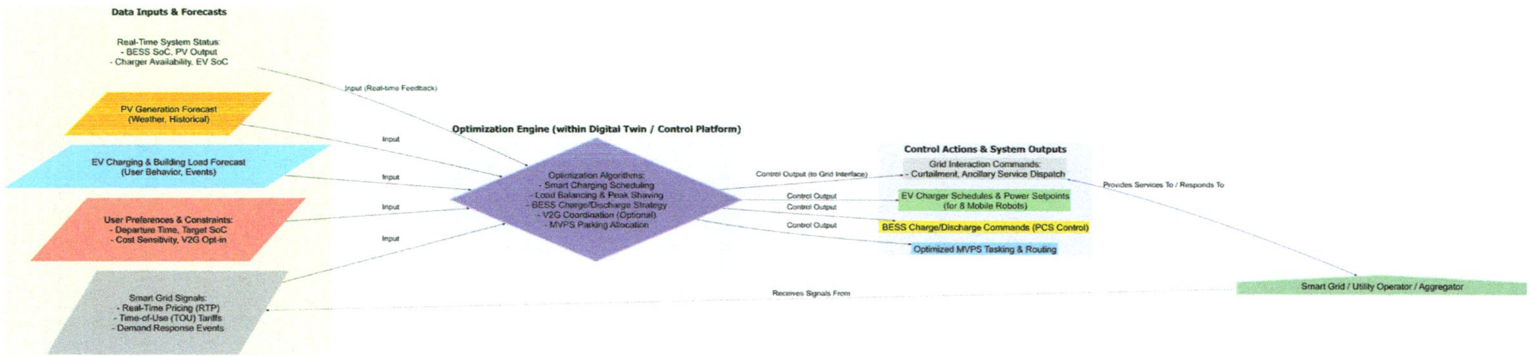

Figure 14.14 Conceptual diagram of optimization strategies and smart grid integration within the smart Hangar's control system.

curtail its aggregate load during periods of acute grid stress or generation shortfalls) or indirect incentives through dynamic pricing signals (e.g., TOU rates, critical peak pricing, real-time pricing) which economically incentivize shifting load to off-peak times or periods of high renewable energy availability on the grid. The Digital Twin greatly facilitates these complex grid interactions by enabling the system to perform rapid "what-if" simulations. It can model and evaluate the potential operational and economic impacts of different responses to received grid signals before committing to a specific course of action. This allows the system to determine the optimal strategy that concurrently satisfies internal operational goals (e.g., user charging needs, cost minimization) and external grid requirements or opportunities, thereby transforming the Smart Hangar from a passive consumer of electricity into an active, intelligent, and valuable participant that contributes positively to the overall stability, efficiency, resilience, and sustainability of the modern, evolving energy grid (Kohler et al., 2024; Daki et al, 2017). Figure 14.14 conceptually outlines these multifaceted optimization processes, the key data inputs, the control actions, and the dynamic interactions with the smart grid.

14.5 Economic feasibility and cost-benefit analysis

While the Self-Sufficient Smart Hangar paradigm presents a compelling suite of advanced technological solutions and substantial environmental benefits, its practical realization and widespread adoption are intrinsically linked to its demonstrable economic viability. A comprehensive and rigorous cost-benefit analysis is therefore indispensable to critically evaluate its feasibility, particularly when benchmarked against established conventional parking infrastructure and emerging EV charging solutions. This multifaceted analysis necessitates a meticulous examination of both the substantial initial capital investment required for development and deployment (CAPEX) and the ongoing, recurring operational expenditures (OPEX) anticipated over the facility's lifespan. These costs must then be judiciously weighed against the diverse spectrum of potential direct revenue streams, quantifiable operational savings, and significant, albeit sometimes less tangible, indirect economic and societal advantages.

14.5.1 Capital expenditure (CAPEX) breakdown

The upfront capital expenditure required to establish a Smart Hangar is significant, reflecting the sophisticated integration of advanced technologies and the construction of a substantial physical structure. The CAPEX encompasses several primary cost categories, each contributing materially to the total initial investment. A foundational cost factor is associated with Land Acquisition and Site Preparation. The price of acquiring suitably zoned urban or peri-urban land represents a highly variable and often dominant cost component, heavily influenced by geographical location, local real estate market dynamics, and site accessibility. Site preparation costs, including geotechnical surveys, demolition of existing structures (if any), and utility connections, also fall under this category. The construction of the multi-level Hangar Structure itself, typically a robust metallic framework optimized for vertical load-bearing and housing the MVPS, forms another major expenditure. The specific costs will depend on the chosen structural design, materials, height, and overall footprint (e.g., 120 m × 60 m × 15 m).

The Mobile Vertical Parking System (MVPS) represents a core technological investment and a significant portion of the CAPEX. This includes the procurement, fabrication, and installation of all mechanical components (such as high-capacity lifts, agile horizontal shuttles, precision drive mechanisms like chains or gears, robust electric motors, and potentially turntables), extensive sensor networks for vehicle detection and positioning, safety interlock systems, and the dedicated programmable logic controllers (PLCs) or industrial PCs required for the autonomous operation of the eight distinct vertical parking box blocs. The cost of the MVPS scales with the hangar's total vehicle capacity (1280 EVs), its vertical height, the number of independent operational lines/blocs (8 in this concept), and the selected level of technological sophistication and redundancy.

The integrated renewable energy and EV charging components also contribute substantially to the CAPEX. The Photovoltaic (PV) System cost encompasses the procurement of 2496 high-efficiency solar panels (1.872 MWp total), the 19 associated high-capacity inverters necessary for DC-to-AC power conversion, engineered rooftop mounting structures designed to withstand local environmental conditions, all necessary cabling and electrical balance-of-system components, and skilled installation labor. The Battery Energy Storage System (BESS), with its significant 9.216 MW h capacity, is frequently one of the most capital-intensive elements. This includes the cost of the 960 battery modules themselves (where the use of second-life EV batteries could potentially offer cost mitigation, albeit with considerations for remaining lifespan and performance), the sophisticated Power Conversion System (PCS) managing bidirectional charge/discharge cycles, essential thermal management systems (e.g., HVAC for the BESS enclosure) to ensure optimal battery performance and longevity, a comprehensive Battery Management System (BMS), and associated safety monitoring, fire suppression, and control hardware. The cost of the EV Charging Infrastructure, primarily supporting the eight mobile robotic chargers, includes the robots themselves (with their V2X capabilities), their docking/charging stations, and the power and communication interfaces within each of the eight parking blocs. The specific cost will depend on the robots' power delivery capabilities (AC/DC mix) and navigation sophistication. Finally, the IoT and Control System infrastructure incurs significant costs for the distributed network of sensors (occupancy, environmental, electrical, MVPS status), robust communication network hardware (gateways, routers, access points for wired and wireless technologies like Ethernet, Wi-Fi, 5G), potential edge computing devices for localized processing, scalable cloud computing resources (servers, databases, storage), the development or licensing of sophisticated software (including the comprehensive Digital Twin platform, central management system, and user applications), and extensive system integration, testing, and commissioning services. It is crucial to acknowledge that various government subsidies, targeted tax credits, or specific grants aimed at promoting renewable energy deployment, fostering the development of public EV charging infrastructure, or supporting innovative smart city projects can play a pivotal role in offsetting a substantial portion of these considerable initial investment costs, thereby enhancing the project's financial attractiveness. A conceptual breakdown is presented in Table 14.5.

14.5.2 Operational expenditure (OPEX) profile

Once commissioned and operational, the Smart Hangar will incur a range of ongoing, recurring expenses (OPEX) essential for its sustained functioning and service delivery. **Electricity Costs** represent a primary and highly visible OPEX component. These costs are for energy purchased from the utility grid, particularly during nighttime hours, periods of low solar PV generation (e.g., overcast days), or when the BESS is depleted and EV charging demand remains high. Sophisticated energy management strategies, dynamically orchestrated by the Digital Twin based on real-time data and predictive analytics, are absolutely crucial for minimizing these grid electricity procurement costs (Shojaei et al., 2022). Furthermore, utility demand charges, which are often levied based on the facility's highest peak

Table 14.5 Conceptual capital expenditure (CAPEX) categories for the smart hangar.

Category	Key components/considerations
Land acquisition	Prime urban/peri-urban location, zoning, accessibility
Site preparation	Surveys, demolition, utility hook-ups, groundwork
Hangar structure	Metallic framework (120 m x 60 m x 15 m), foundations, roofing, cladding
MVPS	Lifts, shuttles, motors, drives, sensors, PLCs for 8 blocs (1280 EVs)
PV system	2496 panels (1.872 MWp), 19 inverters, mounting, cabling, installation
BESS	9.216 MW h (960 units), PCS, BMS, thermal management, safety systems
EV charging system	8 mobile robotic chargers (V2X), docking stations, bloc interfaces
IoT & control system	Sensors, network hardware, edge/cloud computing, digital twin software, integration
Contingency	Typically 10%–20% for unforeseen expenses

power consumption within a billing period, can significantly inflate OPEX if not proactively managed through intelligent load balancing across the eight robotic chargers, dynamic peak shaving using the BESS, and strategic scheduling of deferrable loads.

Maintenance and Repair constitutes another significant recurring cost category. This encompasses the regular preventive and corrective upkeep of the MVPS's numerous mechanical and electrical parts (e.g., lubrication of chains/gears, inspection for wear-and-tear on rails and shuttles, motor health checks, sensor calibration), routine PV system maintenance (primarily involving periodic panel cleaning to maintain efficiency, inverter servicing), BESS health monitoring (including state-of-health assessments, thermal system checks, and potential eventual cell or module replacement toward the end of its operational life), servicing of the eight mobile robotic EV chargers (including their batteries, drive systems, and charging interfaces), and the ongoing maintenance of the extensive IT, communication, and control system hardware and software. The predictive maintenance capabilities enabled by the Digital Twin, which leverages sensor data and machine learning algorithms to anticipate potential equipment failures before they occur, can play a crucial role in optimizing maintenance schedules, minimizing unexpected downtime, and potentially reducing long-term maintenance expenditures compared to purely reactive or fixed-schedule preventative maintenance approaches.

Costs associated with Software, Data Management, and Connectivity must also be comprehensively factored into the OPEX budget. These include recurring fees for cloud platform subscriptions (e.g., for data storage, processing, hosting the Digital Twin), data transmission costs (especially if relying on public cellular networks like 5G for some IoT connectivity), software licensing fees for specialized control or analytics platforms, regular software updates and security patching, and the continuous implementation and monitoring of robust cybersecurity measures to protect the facility's operational technology (OT) and information technology (IT) systems from evolving threats (Naseri et al., 2023). While the Smart Hangar is designed for a high degree of automation, some level of skilled Human Labor will likely be indispensable. This may include on-site or remote operational oversight personnel, customer support staff (even if primarily digital), specialized technicians for non-routine maintenance and repair tasks, and administrative functions. Insurance, property taxes, and other miscellaneous administrative overheads also contribute to the overall OPEX. Table 14.6 outlines these key operational cost areas.

14.5.3 Revenue streams and economic benefits

The Smart Hangar is designed to generate revenue and offer compelling economic benefits through a diversified range of channels, leveraging its

Table 14.6 Conceptual operational expenditure (OPEX) categories for the smart hangar.

Category	Key components/considerations
Grid electricity	Energy imported from utility, demand charges, time-of-use tariffs
Maintenance & repair	MVPS (mechanical, electrical), PV (cleaning, inverters), BESS (health, cells), EVSE robots, IT systems
Software & data	Cloud subscriptions, data storage, software licenses, cybersecurity, connectivity fees
Labor	Operational oversight, specialized maintenance, customer support (remote/on-site)
Insurance	Property, liability, equipment
Property taxes	Based on assessed value and local rates
Miscellaneous	Consumables, administrative overhead, permits/licenses

advanced technology and strategic positioning within the urban mobility ecosystem. Direct revenue is primarily derived from Parking Fees charged to users for utilizing the automated parking service. These fees can be dynamically structured, potentially incorporating premium rates that reflect the enhanced convenience, superior security, and operational speed offered by the sophisticated MVPS, as well as demand-based pricing during peak periods. EV Charging Fees, representing the revenue generated from selling electricity to EV users, constitute another principal and growing income source. Sophisticated and dynamic pricing strategies can be implemented for charging services, varying the cost per kilowatt-hour (kW h) based on factors such as time-of-day, prevailing grid conditions, the level of charging power delivered (e.g., premium for DC fast charging), and the source of energy (e.g., discounted "green charging" using on-site PV). The potential for participation in Grid Services, actively facilitated by the intelligent BESS and, potentially, the aggregated V2G capabilities of the connected EV fleet (serviced by the 8 V2X-capable robotic chargers), offers an increasingly attractive additional revenue stream. This could involve earning direct payments or incentives from utility companies or grid operators for providing valuable services such as frequency regulation, operating reserves (spinning or non-spinning), capacity provision, or participation in formalized demand response programs that are coordinated with the utility or a third-party aggregator.

Beyond these direct transactional revenue streams, significant implicit economic benefits contribute to the overall financial attractiveness of the Smart Hangar. The exceptionally high parking density achieved by the vertical MVPS translates directly into a substantially Reduced Land Cost per Parking Space when compared to conventional sprawling surface lots or even traditional ramp-based multi-story garages. This is a particularly potent advantage in high-value, land-scarce urban real estate markets, where land often represents the single largest cost component for new developments. The pervasive high degree of automation inherent in the MVPS and robotic charging processes leads to markedly enhanced Operational Efficiency. This is primarily achieved by minimizing the direct labor costs typically associated with traditional parking facility management, such as attendants, valets, cashiers, and extensive security patrolling. Furthermore, the integration of cutting-edge technologies, the prominent use of on-site renewable energy generation, and the overall sustainable design principles can significantly lead to an Enhanced Asset Value and Brand Image. This makes the Smart Hangar an attractive and forward-looking investment, potentially commanding higher property valuations, attracting premium clientele, and enhancing its public perception and market positioning.

14.5.4 Comprehensive feasibility assessment & financial modeling

Determining the overarching economic feasibility and long-term financial sustainability of a specific Smart Hangar project necessitates a comprehensive, detailed, and highly localized techno-economic assessment. This rigorous evaluation involves the development and iterative refinement of sophisticated financial models. These models must accurately incorporate a wide array of site-specific data inputs, including precise local land acquisition and development costs, prevailing and projected electricity tariff structures (encompassing complex elements like time-of-use rates, demand charges, and any applicable renewable energy credits or feed-in tariffs), regional solar irradiance profiles for accurate PV generation forecasting, projected EV adoption rates and evolving charging behavior patterns within the intended service area, typical user parking durations and turnover rates, and, critically, the availability, scale, and eligibility criteria for relevant government incentives, subsidies, or green financing mechanisms. The Digital Twin, initially developed for operational management, can also serve as an exceptionally powerful analytical tool during the crucial planning and feasibility study phases. By enabling the simulation of a multitude of diverse operational scenarios (e.g., varying levels of charging demand across the eight parking blocs, fluctuations in PV generation due to weather, dynamic shifts in energy market prices) and facilitating sensitivity analyses by testing different financial assumptions (e.g., interest rates, component costs, revenue projections), the DT can provide robust, data-driven predictions of key financial performance metrics. These metrics include, but are not limited to, the project's Net Present Value (NPV), Internal Rate of Return (IRR), and the projected payback period (Royo et al., 2023a, 2023b). Ultimately, the strategic decision to invest in and proceed with a Smart Hangar deployment hinges on a conclusive determination that the projected long-term operational savings (primarily driven by substantially reduced grid electricity purchases due to high PV self-consumption and potentially optimized, predictive maintenance schedules) combined with the diversified portfolio of potential revenue streams, can robustly justify and provide an attractive return on the significant initial capital investment over the entirety of the project's planned operational lifespan.

14.6 Case studies and related implementations: validating core technologies

While a fully operational "Self-Sufficient Smart Hangar," integrating the entire described suite of advanced features into a single, cohesive entity, may currently represent a forward-looking, holistic concept, it is crucial to recognize that its foundational technological pillars have been extensively implemented, rigorously tested, and successfully validated across numerous real-world projects and dedicated research initiatives. A careful examination of these diverse yet related implementations provides compelling empirical evidence that robustly supports the inherent feasibility, tangible benefits, and transformative potential of the integrated Smart Hangar approach. These precedents collectively demonstrate the maturity and efficacy of the individual components, paving the way for their synergistic combination.

Automated and Vertical Parking Systems (AVPS/MVPS) have become indispensable solutions in dense urban conurbations. A notable example is the **Autostadt silo in Wolfsburg, Germany**, a fully automated "car tower" that demonstrates high-density vehicle storage and retrieval. In the United States, the **Serva Transport Systems' automated parking garage at The Railyards in Sacramento, California**, utilizes robotic valets to park cars more densely than conventional garages. These operational systems consistently prove the dramatic land-area savings and enhanced user convenience that form a cornerstone of the Smart Hangar concept. Contemporary research within this domain has increasingly focused on leveraging Digital Twin (DT) technology for improved operational management, real-time 3D visualization, proactive condition monitoring, and sophisticated performance optimization of these complex electromechanical systems.

The integration of **Photovoltaic (PV) Systems with Carports and Parking Structures** is now a commercially mature concept. For instance, **Rutgers University in New Jersey** hosts one of the largest solar carport installations in the U.S., generating 8 MW of power. Similarly, the **"Solar Roof" at the Audi plant in Ingolstadt, Germany**, showcases large-scale rooftop solar integration in an industrial context. These projects demonstrate the viability of generating substantial quantities of clean energy at the point of consumption, a core tenet of the Smart Hangar's self-sufficiency. A significant body of research concentrates on optimizing the sizing and configuration of PV arrays in conjunction with Battery Energy

Storage Systems (BESS) to maximize self-consumption and enhance facility resilience.

The strategic deployment of **Smart EV Charging Hubs** that integrate renewables, storage, and grid services is also gaining momentum. The **"Smart Solar Charging" project in Utrecht, Netherlands**, is a pioneering example. It combines thousands of solar panels with hundreds of bidirectional (V2G) chargers, allowing EVs to both charge on solar power and discharge to support the grid. This project validates the technical feasibility of the V2G and smart energy management functionalities envisioned for the Smart Hangar. Extensive ongoing research continues to refine advanced charging scheduling algorithms that incorporate multi-variable optimization criteria, including grid tariffs, renewable availability, and even long-term battery health.

The application of **Digital Twins for Comprehensive Infrastructure Management** is advancing rapidly. The **"Virtual Singapore"** platform is a prime example, creating a dynamic 3D digital model of the entire city-state to support planning, simulation, and operational management. On a more focused scale, the **Port of Rotterdam** uses a digital twin to monitor vessel movements, infrastructure health, and weather conditions in real-time to optimize port operations. These examples underscore the power of DTs to manage complex, interconnected systems, which is critical for orchestrating the Smart Hangar's diverse operations (O′Dwyer et al., 2020).

Finally, the overarching concept of Integrated Mobility Hubs is steadily gaining traction in progressive urban planning and smart city initiatives. These hubs aim to create fluid, seamless, and user-centric transitions between diverse transportation modes. They achieve this by strategically co-locating and operationally integrating public transit interchanges (e.g., train stations, bus terminals), shared mobility services (e.g., ride-sharing, car-sharing, e-scooters), bike-sharing schemes, and advanced EV charging facilities, often incorporating elements of automated parking and intelligent traffic management, typically all orchestrated within a cohesive, data-driven smart city framework. While the scope of these multi-modal hubs is often broader than the singular, focused functionality of the Smart Hangar concept, they share the fundamental objective of creating more efficient, convenient, user-friendly, and sustainable solutions to multifaceted urban mobility challenges.

In conclusion, although a direct, one-to-one, fully realized exemplar of the proposed Self-Sufficient Smart Hangar, encompassing every delineated feature, might still be on the developmental horizon, the extensively proven success, well-documented advantages, and technological maturity of its core constituent pillars—namely, automated vertical parking systems, integrated PV generation and battery energy storage, sophisticated smart EV charging algorithms, and comprehensive Digital Twin-based management systems—across a multitude of diverse and relevant applications provide exceptionally strong and compelling support for the overall concept's inherent feasibility. These precedents collectively affirm its significant potential to deliver transformative benefits in terms of land-use efficiency, energy sustainability, operational intelligence, and enhanced urban mobility.

14.7 Future trends and development pathways

The Self-Sufficient Smart Hangar concept, while embodying significant innovation, is not conceived as a static end-point but rather as a dynamic platform poised for continuous evolution. It exists and operates within a rapidly advancing technological landscape, and future developments are anticipated to further augment its diverse capabilities, deepen its sophisticated integration into complex urban systems, and significantly broaden its overall positive impact on cities. Key technological and strategic trends that are poised to shape the future trajectory of such integrated, intelligent parking and charging solutions include significant advancements in adaptable and resilient design paradigms, more profound and synergistic smart city integration, and the accelerated maturation and widespread adoption of bidirectional charging technologies, particularly Vehicle-to-Grid (V2G).

14.7.1 Adaptable designs, modularity, & deeper smart city integration

Future iterations of Smart Hangar designs are highly likely to embrace and champion core principles of modularity, scalability, and adaptability. This design philosophy will facilitate easier customization and tailoring of facilities to a wide spectrum of diverse urban site constraints (e.g., irregular plot shapes, height restrictions, existing underground infrastructure) and allow for more agile responses to fluctuating or evolving demand patterns for parking and EV charging. The increased utilization of prefabricated, standardized structural and system components, coupled with inherently scalable system architectures, could significantly streamline construction processes, reduce on-site assembly times, and enable more economically viable phased deployment strategies. This would allow facilities to grow organically in capacity or functionality as local EV adoption rates increase or as new service demands emerge. Furthermore, ongoing innovations in high-performance, sustainable building materials (such as advanced composites, recycled steel, or low-carbon concrete) and the adoption of cutting-edge, energy-efficient structural design principles (e.g., optimized for natural light, passive ventilation, or green facades) could further diminish the environmental footprint associated with both the construction and long-term operation of these vital urban infrastructures.

A more profound and transformative trend is the anticipated deeper, more intricate integration with broader Smart City platforms and interconnected urban ecosystems. The prevailing paradigm of isolated, siloed operation of individual urban infrastructure components is progressively yielding to a vision of highly interconnected, data-sharing systems that coordinate their actions and resources to achieve holistic, city-wide management objectives. Within this evolving context, the Smart Hangar's Digital Twin is envisioned not merely as an internal management tool but as an active, intelligent node within this larger, distributed network. It will be capable of secure, bidirectional communication and data exchange with diverse city-wide systems, including those managing real-time traffic flow, municipal urban planning databases (e.g., for zoning, land use), utility grid operator platforms (for energy forecasting and demand response), and potentially other mobility service providers (such as public transit authorities or ride-sharing platforms). Achieving this level of sophisticated interoperability necessitates unwavering adherence to open, standardized data models (e.g., NGSI-LD for context information management) and the widespread adoption of secure, well-documented Application Programming Interfaces (APIs), which collectively facilitate seamless, reliable, and secure information exchange across heterogeneous systems. For instance, real-time traffic congestion data sourced from city-wide traffic management systems could dynamically influence parking availability information displayed to users or adjust optimal routing guidance to the Smart Hangar. Conversely, aggregated and anonymized data pertaining to EV charging demand patterns, on-site renewable energy generation, and BESS operational status from the hangar could provide invaluable inputs for city-level energy planning, grid reinforcement strategies, and emergency preparedness (Sabet and Farooq, 2023). The Smart Hangar thus evolves from a standalone, self-contained facility into an active, contributing, and dynamically responsive component of a distributed, intelligent network of energy and mobility resources. This evolution enhances overall urban resilience against disruptions, improves operational efficiency across multiple sectors, and fosters a more adaptive and responsive urban environment (Golinska-Dawson and Sethanan, 2023; Othman et al., 2023). This integrated vision propels cities toward truly data-driven, interconnected urban ecosystems, as conceptually illustrated in Fig. 14.15.

14.7.2 Maturation of V2X technologies and advanced bidirectional charging

Vehicle-to-Everything (V2X) technology, encompassing Vehicle-to-Grid (V2G), Vehicle-to-Building (V2B), and Vehicle-to-Home (V2H), which collectively enable Electric Vehicles to not only consume electrical energy but also to intelligently supply stored power back to the grid, a building, or other loads, represents a significant and transformative future potential for facilities like the Smart Hangar, particularly those equipped with a fleet of 8 V2X-capable mobile robotic chargers. While V2X implementation currently faces certain practical hurdles—including concerns about potential impacts on EV battery degradation from increased cycling, the need for universally adopted communication and interoperability standards, the establishment of clear and supportive regulatory frameworks, and viable economic models—the underlying technology is rapidly maturing, and its strategic importance is increasingly recognized. Key future trends in this domain include the finalization, widespread industry adoption, and regulatory endorsement of comprehensive V2X communication and interoperability standards. Particular emphasis is placed on further extensions and refinements to protocols like ISO 15118 (specifically parts related to bidirectional power transfer and smart charging negotiation), which are absolutely essential for enabling secure, reliable, and interoperable bidirectional energy flow between diverse EV models, chargers (including mobile robotic units), and grid management systems.

The development and deployment of increasingly sophisticated and adaptive control algorithms will be critical for effectively managing and optimizing complex V2X operations at scale within the Smart Hangar environment. These algorithms, likely leveraging advanced artificial intelligence (AI), machine learning (ML) techniques (such as reinforcement learning for adaptive control), and distributed optimization methods, will predominantly reside within or be orchestrated by the Digital Twin framework. Their core function will be to intelligently and dynamically optimize bidirectional energy transfer schedules for the aggregated fleet of V2X-capable

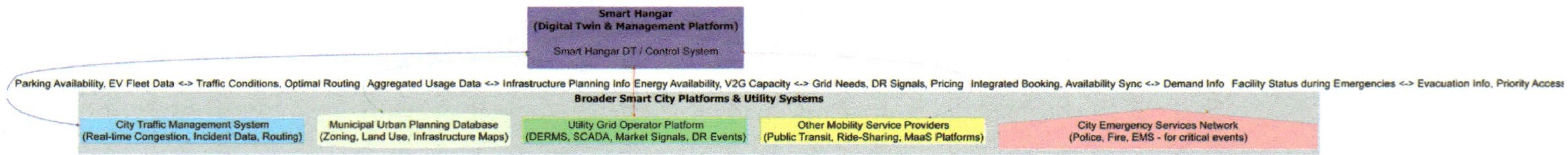

Figure 14.15 Conceptual diagram of future integration pathways for the smart hangar within broader smart city ecosystems, showcasing bidirectional data exchange and coordination with diverse urban management platforms.

EVs. This intricate optimization process must holistically consider a complex and often conflicting set of variables: real-time and forecasted grid ancillary service requirements (e.g., demand for frequency regulation, spinning reserves, or peak load reduction), dynamic electricity market pricing signals (for both energy purchase and sale), the potential nuanced impact of V2X operations on the long-term health and longevity of individual EV batteries, and, paramountly, ensuring that the primary mobility needs and user-defined preferences (e.g., minimum required SoC at departure, willingness to participate in V2G programs) of each EV owner are always unequivocally met and respected.

Furthermore, the strategic integration of distributed V2X capabilities (from the EV fleet) with the hangar's own centralized BESS offers substantially enhanced operational flexibility, energy management sophistication, and overall system resilience. The stationary BESS can effectively buffer and smooth V2X operations, aggregate the collective energy response of multiple participating EVs to present a more substantial and reliable resource to the grid, or provide essential grid services independently when the EV fleet is unavailable, insufficient in number, or its owners are unwilling to participate. The Digital Twin can play a crucial role by simulating various coordination strategies and optimizing the combined, synergistic response of both the distributed EV fleet and the centralized stationary BESS to maximize overall benefits while adhering to all operational constraints. The broader economic viability and widespread adoption of V2G and other V2X services also critically depend on the emergence and regulatory approval of innovative and equitable business models. These models must fairly compensate EV owners for making their battery capacity available and for any associated battery degradation, as well as appropriately reward facility operators like the Smart Hangar for aggregating these distributed resources and providing valuable services to the grid. Exploring, validating, and scaling these novel business models through well-designed pilot projects, supportive regulatory reforms, and open market participation mechanisms will be essential for unlocking the full potential of V2X. Finally, the overarching concept of mobile and flexible energy storage could extend beyond fixed V2G installations connected via specific chargers. Future scenarios might involve the strategic dispatch of the Smart Hangar's V2X-capable mobile robotic charging units to provide localized grid support or emergency power in nearby areas, or even facilitate coordinated Vehicle-to-Vehicle (V2V) energy transfer sessions within the facility itself. This could offer unparalleled charging flexibility and energy sharing

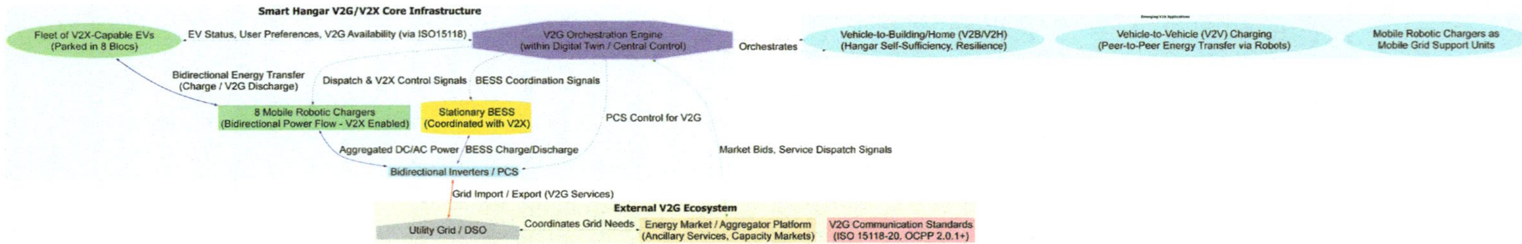

Figure 14.16 Conceptual illustration of future advancements in V2G/V2X technologies within the smart hangar, highlighting bidirectional power flows, energy market interactions, and coordinated control of EV fleet and BESS for grid services.

capabilities, all managed and optimized via sophisticated, AI-driven Digital Twin platforms (Kabir et al., 2021). As EV battery technology continues its trajectory of improvement (higher energy density, longer cycle life, lower cost), as the costs of bidirectional power electronics decrease, and as regulatory landscapes adapt to embrace distributed energy resources, V2X integration is poised to transition from a niche capability to a standard, integral feature. This will further transform advanced EV parking and charging hubs like the Smart Hangar into highly dynamic, interactive, and strategically invaluable assets within the continuously evolving, increasingly intelligent, and decarbonized energy grid, as conceptualized in Fig. 14.16.

14.8 Conclusion

The intensifying pressures of urbanization, coupled with the global imperative for sustainable transportation, necessitate transformative solutions for electric vehicle (EV) infrastructure. Traditional approaches to parking and charging often falter in dense urban settings, constrained by severe space limitations, increasing grid strain, and inherent logistical complexities. In response, this work has presented the "Self-Sufficient Smart Hangar," a comprehensive conceptual framework meticulously designed to address these multifaceted challenges through the synergistic integration of multiple advanced technologies. By combining an automated Mobile Vertical Parking System (MVPS)—potentially structured around distinct operational blocs such as the 8-bloc concept discussed—with substantial on-site renewable energy generation via rooftop PV, integrated high-capacity battery storage (BESS), pervasive IoT connectivity enabling features like mobile robotic charging, and an overarching Digital Twin (DT) management system, the Smart Hangar aims to provide a robust and adaptable blueprint for the development of highly efficient, environmentally sustainable, and exceptionally intelligent urban EV hubs.

The proposed Smart Hangar concept offers a compelling and multifaceted suite of advantages, primarily derived from its deeply integrated and intelligently orchestrated design. Its most immediate and visually striking benefit is exceptional space efficiency, decisively achieved through the MVPS, which dramatically increases parking capacity per unit of valuable land area when compared to conventional surface lots or ramp-based garages—a critical factor in high-density, high-value urban locations. Environmental sustainability is significantly advanced by the integral combination of rooftop PV generation and substantial BESS capacity. This synergy transforms the facility from a mere energy consumer into a decentralized power plant, enabling considerable on-site generation and optimized utilization of clean, renewable energy. By powering both facility operations and extensive EV charging activities with solar energy, the Smart Hangar minimizes its reliance on often fossil-fuel-based grid power, playing a direct and measurable role in decarbonizing the urban transport and energy sectors.

Beyond its internal operational efficiencies, the Smart Hangar is conceived to function as a proactive and beneficial component of the wider energy grid. Through sophisticated smart charging strategies dynamically managed across its charging fleet (e.g., the eight mobile robots), effective peak load shaving and load shifting facilitated by the BESS, and the significant future potential for Vehicle-to-Grid (V2G) services, the facility can transition from being merely a passive electrical load to an active, flexible, and intelligent asset. Such capabilities allow it to contribute positively to local grid stability, aid in the smoother integration of intermittent renewable energy sources on a larger regional scale, and potentially offer valuable ancillary services. Operational efficiency is inherently achieved through the pervasive automation of parking and charging processes, which significantly reduces ongoing labor requirements. Concurrently, the advanced DT framework enables highly optimized energy management strategies and facilitates predictive maintenance routines based on real-time condition monitoring, potentially lowering long-term operational costs and maximizing system uptime and reliability.

For end-users, the system promises a demonstrably enhanced and seamless user experience. The fully automated parking process effectively eliminates the stress and time associated with searching for available spaces, while integrated smart charging ensures vehicles are adequately and optimally charged according to user needs and facility capabilities, potentially alleviating range anxiety and providing a frictionless interaction from arrival to departure. Fundamentally, the robust performance and adaptability of the system lie in its sophisticated data-driven management capabilities. The pervasive IoT network and the comprehensive DT platform provide unprecedented levels of real-time system-wide monitoring, enable sophisticated, anticipatory control, and facilitate continuous operational optimization, allowing the Smart Hangar to operate adaptively and intelligently based on dynamic internal conditions, external environmental factors, and predictive analytical insights. Collectively, these diverse advantages position the Smart Hangar as a holistic, integrated, and forward-thinking solution uniquely suited to address the complex and evolving demands of modern urban e-mobility infrastructure, not merely as a parking solution, but as a critical enabler of a sustainable, renewable-powered urban future.

The potential impact of successfully implementing Self-Sufficient Smart Hangars extends significantly beyond the immediate operational benefits for users and facility operators, resonating at a broader urban and societal scale. On an urban planning level, the drastic reduction in land required for vehicle parking, directly enabled by the high-density MVPS, can liberate substantial tracts of valuable city space. This recovered land can then be strategically repurposed for other essential urban needs, such as the development of affordable housing, the creation of public green spaces and recreational areas, or the establishment of community facilities and local amenities, thereby contributing directly to the creation of more livable, vibrant, and equitable urban environments. The core strategic focus on maximizing on-site renewable energy generation and implementing highly efficient energy management strategies directly supports and accelerates the achievement of municipal, national, and international climate action goals. By minimizing reliance on grid electricity and actively promoting the use of clean energy for the transportation sector, these advanced facilities contribute tangibly to reducing overall greenhouse gas (GHG) emissions.

Furthermore, the inherent ability of Smart Hangars to alleviate strain on the local electricity distribution grid, to manage peak loads with high efficacy, and to potentially offer a suite of ancillary grid services enhances the resilience, stability, and overall efficiency of the urban energy infrastructure. This capability becomes increasingly critical as cities worldwide grapple with the dual challenges of rapidly electrifying key sectors like transportation and heating while simultaneously striving to integrate ever-higher penetrations of variable and often distributed renewable energy sources into their power systems. The deep and pervasive integration of digital technologies, particularly IoT, advanced data analytics, and comprehensive Digital Twins, aligns perfectly with the broader, overarching vision of developing truly Smart Cities. In such digitally enabled urban environments, interconnected infrastructure components and data-driven insights are systematically leveraged to optimize the delivery of urban services, improve the management of critical resources, enhance civic engagement, and ultimately elevate the overall quality of life for all residents.

In essence, the Smart Hangar concept represents far more than just an advanced, automated parking garage; it embodies a tangible and significant stride toward a more sustainable, intelligent, and resilient urban future. It envisions a new generation of transportation infrastructure that is inherently compact, demonstrably resource-efficient, primarily powered by clean, on-site renewable energy, and seamlessly integrated into both the digital and physical fabric of the modern, data-rich city. While the detailed economic feasibility of any specific Smart Hangar project will necessitate careful and nuanced consideration of local conditions, prevailing costs, and available incentives, the underlying technological components and system integration principles are rapidly advancing and maturing, bringing this innovative and compelling vision progressively closer to widespread practical realization. Continued, concerted efforts in targeted research, focused development, industry-wide standardization, and strategically implemented pilot project deployments are all essential to further refine the intricate design parameters, optimize the sophisticated control strategies, and fully demonstrate the long-term operational viability and profound transformative benefits of this integrated, intelligent approach to shaping the future of urban EV infrastructure.

References

Ali M, Kaddoum G, Li W-T, Yuen C, Tariq M, Poor HV. A smart digital twin enabled security framework for vehicle-to-grid cyber-physical systems. IEEE Trans Inf Forensics Secur 2023;18:5258–71. https://doi.org/10.1109/TIFS.2023.3305916.

Badia B, Berry RA, Wei E. Investment in ev charging spots for parking. IEEE Trans Netw Sci Eng 2020;7(2):650–61. https://doi.org/10.1109/TNSE.2019.2922594.

Basnet M, Ali MH. Exploring cybersecurity issues in 5g enabled electric vehicle charging station with deep learning. IET Gener Transm Distrib 2021;15(24):3435–49. https://doi.org/10.1049/gtd2.12275.

BloombergNEF. H1 2024 global energy storage outlook, Tech. rep., Bloomberg Finance L. P. April 2024.

Bokhari A. Smart parking management for electrical vehicles: solar parking lots, 10; 2022. p. 27–32. https://doi.org/10.5194/isprs-annals-X-4-W3-2022-27-2022.

Bornioli A, Vermeulen S, Mingardo G. Can app notifications reduce charging hogging? an experiment with electric vehicle drivers. Case Stud Transp Policy 2024;15. https://doi.org/10.1016/j.cstp.2023.101143.

Boucetta M, Hossain NUI, Jaradat R, Keating C, Tazzit S, Nagahi M. The architecture design of electrical vehicle infrastructure using viable system model approach. Systems 2021;9(1). https://doi.org/10.3390/systems9010019.

Bracco S, Delfino F, Pampararo F, Robba M, Rossi M. A dynamic optimization-based architecture for polygeneration microgrids with tri-generation, renewables, storage systems and electrical vehicles. Energy Convers Manag 2015;96:511–20. https://doi.org/10.1016/j.enconman.2015.03.013.

Brasil JAT, Giacomoni MH, Papagiannakis A, Johnson D, Mohagheghpour V, Tupas C, et al. Assessment of lid performance through integration of permeable pavements with digital twins. 2024. p. 254–67. https://doi.org/10.1061/9780784485477.022.

Caggiani L, Prencipe LP, Ottomanelli M. A static relocation strategy for electric car-sharing systems in a vehicle-to-grid framework. Transp Lett 2021;13(3):219–28. https://doi.org/10.1080/19427867.2020.1861501.

Canzaniello M, Amitrano S, Prezioso E, Giampaolo F, Cuomo S, Piccialli F. Leveraging digital twins and generative ai for effective urban mobility management. 2024. p. 146–53. https://doi.org/10.1109/CyberSciTech64112.2024.00032.

Chen W, Wang X, Wu M. Intelligent parking service system design based on digital twin for old residential areas. Electronics (Switzerland) 2024;13(23). https://doi.org/10.3390/electronics13234597.

Cheng X, Kontou E. Estimating the electric vehicle charging demand of multi-unit dwelling residents in the united states. Environ Res Infrastruct Sustain 2023;3(2). https://doi.org/10.1088/2634-4505/acde06.

Chomiak-Orsa I, Hauke K, Perechuda K, Pondel M. The use of digital twin in the sustainable development of the city on the example of managing parking resources, 225; 2023. p. 2183–93. https://doi.org/10.1016/j.procs.2023.10.209.

Conde J, Munoz-Arcentales A, Alonso A, Huecas G, Salvachua J. Collaboration of digital twins through linked open data: architecture with fiware as enabling technology. IT Professional 2022;24(6):41–6. https://doi.org/10.1109/MITP.2022.3224826.

Daki H, El Hannani A, Aqqal A, Haidine A, Dahbi A. Big data management in smart grid: concepts, requirements and implementation. J Big Data 2017;4(1). https://doi.org/10.1186/s40537-017-0070-y.

Eskandari A, Malekizadeh M, Peyravi M, Nedaei A, Kolahi M, Erdogan N, et al. Electric mobility and beyond: advancing charging infrastructure optimization with digital twins. 2024. https://doi.org/10.1016/B978-0-443-14070-9.00012-3.

Faliagka E, Christopoulou E, Ringas D, Politi T, Kostis N, Leonardos D, Tranoris C, Antonopoulos CP, Denazis S, Voros N. Trends in digital twin framework architectures for smart cities: A case study in smart mobility. Sensors 2024;24(5). https://doi.org/10.3390/s24051665.

Fotouhi Z, Hashemi MR, Narimani H, Bayram IS. A general model for ev drivers' charging behavior. IEEE Trans Veh Technol 2019;68(8):7368–82. https://doi.org/10.1109/TVT.2019.2923260.

Galkin N, Yang C-W, Vyatkin V. Automatic generation of charging point's digital twin for virtual commissioning of their automation systems. IEEE Open J Ind Electron Soc 2023; 4:14–26. https://doi.org/10.1109/OJIES.2022.3230214.

Golinska-Dawson P, Sethanan K. Sustainable urban freight for energy-efficient smart cities—systematic literature review. Energies 2023;16(6). https://doi.org/10.3390/en16062617.

Gopal S, Gupta P, Sharma M, Kaushal D, Joshi S, Sharma B. Iot enabled e-vehicles for developing smart transportation system. 2023. p. 1063–8. https://doi.org/10.1109/ICAICCIT60255.2023.10466202.

Graber G, Calderaro V, Galdi V, Piccolo A. Battery second-life for dedicated and shared energy storage systems supporting ev charging stations. Electronics (Switzerland) 2020;9(6). https://doi.org/10.3390/electronics9060939.

Guerra E, Daziano RA. Electric vehicles and residential parking in an urban environment: Results from a stated preference experiment. Transport Res Transport Environ 2020;79. https://doi.org/10.1016/j.trd.2020.102222.

Guillemin S, Choulet R, Guyot G, Hing S. Electrical vehicle smart charging using the open charge point interface (ocpi) protocol. Energies 2024;17(12). https://doi.org/10.3390/en17122873.

Haorui J, Manhao L, Bowen Z, Xin Z. Improved set covering location model for charging facility deployments435; 2018. https://doi.org/10.1088/1757-899X/435/1/012019.

Honarmand M, Zakariazadeh A, Jadid S. Optimal scheduling of electric vehicles in an intelligent parking lot considering vehicle-to-grid concept and battery condition. Energy 2014;65:572–9. https://doi.org/10.1016/j.energy.2013.11.045. https://www.scopus.com/inward/record.uri?eid=2-s2.0-84893007852&doi=10.1016%2fj.energy.2013.11.045&partnerID=40&md5=4249124388bac0e291ac3b6376a04f6d.

Hussain A, Bui V-H, Kim H-M. Deep reinforcement learning-based operation of fast charging stations coupled with energy storage system. Elec Power Syst Res 2022;210. https://doi.org/10.1016/j.epsr.2022.108087.

Jafari S, Byun Y-C. Prediction of the battery state using the digital twin framework based on the battery management system. IEEE Access 2022;10:124685–96. https://doi.org/10.1109/ACCESS.2022.3225093.

Jiang J, Li Y, Li Y, Li C, Yu L, Li L. Smart transportation systems using learning method for urban mobility and management in modern cities. Sustain Cities Soc 2024;108. https://doi.org/10.1016/j.scs.2024.105428.

Kabir ME, Sorkhoh I, Moussa B, Assi C. Joint routing and scheduling of mobile charging infrastructure for v2v energy transfer. IEEE Trans Intell Veh 2021;6(4):736–46. https://doi.org/10.1109/TIV.2021.3063221.

Khalkhali H, Hosseinian SH. Multi-stage stochastic framework for simultaneous energy management of slow and fast charge electric vehicles in a restructured smart parking lot. Int J Electr Power Energy Syst 2019;116. https://doi.org/10.1016/j.ijepes.2019.105540.

Kim Y-M, Jung D, Chang Y, Choi D-H. Intelligent micro energy grid in 5g era: Platforms, business cases, testbeds, and next generation applications. Electronics (Switzerland) 2019;8(4). https://doi.org/10.3390/electronics8040468.

Kohler C, Kersten R, Schopf M. Cloud-based digital twin for distribution grids: what is already available today. IEEE Power Energy Mag 2024;22(1):72–80. https://doi.org/10.1109/MPE.2023.3336255.

Lazzeroni P, Caroleo B, Arnone M, Botta C. A simplified approach to estimate ev charging demand in urban area: An italian case study. Energies 2021;14(20). https://doi.org/10.3390/en14206697.

Lefeng S, Chunxiu L, Jingrong D, Cipcigan L. External benefits calculation of sharing electric vehicles in case of chongqing china. Util Policy 2025;64. https://doi.org/10.1016/j.jup.2020.101021.

Liu S, Xia X, Cao Y, Ni Q, Zhang X, Xu L. Reservation-based ev charging recommendation concerning charging urgency policy. Sustain Cities Soc 2021;74. https://doi.org/10.1016/j.scs.2021.103150.

Liu S, Ni Q, Cao Y, Cui J, Tian D, Zhuang Y. A reservation-based vehicle-to-vehicle charging service under constraint of parking duration. IEEE Syst J 2023a;17(1):176–87. https://doi.org/10.1109/JSYST.2022.3145155.

Liu X, Fu Z, Qiu S, Zhang T, Li S, Yang Z, Liu X, Jiang Y. Charging private electric vehicles solely by photovoltaics: A battery-free direct-current microgrid with distributed charging strategy. Appl Energy 2023b;341. https://doi.org/10.1016/j.apenergy.2023.121058.

Lopez-Behar D, Tran M, Mayaud JR, Froese T, Herrera OE, Merida W. Putting electric vehicles on the map: a policy agenda for residential charging infrastructure in Canada. Energy Res Social Sci 2019;50:29–37. https://doi.org/10.1016/j.erss.2018.11.009.

Ma J, He Q. Studying the placement of ev charging stations in parking facilities in cities. 2023. https://doi.org/10.1109/IECON51785.2023.10312727.

Marinescu C. Design consideration regarding a residential renewable-based microgrid with ev charging station capabilities. Energies 2021;14(16). https://doi.org/10.3390/en14165085.

Market Data Forecast. Global automated parking system market size, share, trends, and growth forecast report (2025 to 2033), Tech. rep., Market Data Forecast. April 2025.

Medved D, Bena L, Oliinyk M, Dzmura J, Mazur D, Martinko D. Assessing the effects of smart parking infrastructure on the electrical power system. Energies 2023;16(14). https://doi.org/10.3390/en16145343.

Merkle L, Pöthig M, Schmid F. Estimate e-golf battery state using diagnostic data and a digital twin. Batteries 2021;7(1):1–22. https://doi.org/10.3390/batteries7010015.

Mohammadi F, Rashidzadeh R. An overview of iot-enabled monitoring and control systems for electric vehicles. IEEE Instrum Meas Mag 2021;24(3):91–7. https://doi.org/10.1109/MIM.2021.9436092.

Mortazavi F, Shkedova O, Feuerhake U, Brenner C, Sester M. Voxel-based point cloud localization for smart spaces management, 48; 2023. p. 325–32. https://doi.org/10.5194/isprs-archives-XLVIII-1-W1-2023-325-2023.

Movahedi H, Pannala S, Siegel J, Harris SJ, Howey D, Stefanopoulou A. Extra throughput versus days lost in v2g services: Influence of dominant degradation mechanism. J Energy Storage 2024;104. https://doi.org/10.1016/j.est.2024.114242.

Mudiyanselage MW, Aghdam FH, Mahdi Kazemi-Razi S, Chaudhari K, Marzband M, Ikpehai A, et al. A multiagent framework for electric vehicles charging power forecast and smart planning of urban parking lots. IEEE Trans Transp Electrif 2024;10(2): 2844–57. https://doi.org/10.1109/TTE.2023.3289196.

Naseri F, Kazemi Z, Larsen PG, Arefi MM, Schaltz E. Cyber-physical cloud battery management systems: Review of security aspects. Batteries 2023;9(7). https://doi.org/10.3390/batteries9070382.

National Renewable Energy Laboratory. 2024 transportation annual technology baseline update, Tech. rep. U.S. Department of Energy, National Renewable Energy Laboratory; November 2024.

Nava A, Greco L. The impact of last-mile logistics: a case study on the optimisation of commercial fleets through the European Union, 56; 2023. p. 2371–6. https://doi.org/10.1016/j.ifacol.2023.10.1209.

O'Dwyer E, Pan I, Charlesworth R, Butler S, Shah N. Integration of an energy management tool and digital twin for coordination and control of multi-vector smart energy systems. Sustain Cities Soc 2020;62. https://doi.org/10.1016/j.scs.2020.102412.

Othman A, Kaddoum G, Evangelista JVC, Au M, Agba BL. Digital twinning in smart grid networks: interplay, resource allocation and use cases. IEEE Commun Mag 2023;61 (11):120–6. https://doi.org/10.1109/MCOM.001.2200823.

Piombo G, Faraji Niri M, Marco J. A novel application-aware retired lithium-ion batteries regrouping approach to enable improved second life. J Clean Prod 2025;489. https://doi.org/10.1016/j.jclepro.2024.144526.

Popa L, Berdich A, Groza B. Cartwin—development of a digital twin for a real-world in-vehicle can network. Appl Sci 2022;13(1). https://doi.org/10.3390/app13010445.

Precedence Research. Automated parking system market - global industry analysis, size, share, growth, trends, regional outlook, and forecast 2023-2032, Tech. rep., Precedence Research. 2023.

Prencuva B, Cicek A, Erdinc O, Erdinc FG. A cost minimization oriented decision making model for corporation of railway systems and electric bus charging stations. 2024. p. 45–50. https://doi.org/10.1109/EPECS62845.2024.10805529.

Qiu Y, Zhou P, Sun H. Assessing the effectiveness of city-level electric vehicle policies in China. Energy Policy 2019;130:22–31. https://doi.org/10.1016/j.enpol.2019.03.052.

Rimal BP, Kong C, Poudel B, Wang Y, Shahi P. Smart electric vehicle charging in the era of internet of vehicles, emerging trends, and open issues. Energies 2022;15(5). https://doi.org/10.3390/en15051908.

Royo B, Politaki D, Gonzalez JN, Batalla A. An emerging and innovation transport solution: towards transforming parking lot to urban consolidation centre: Madrid living lab. In:

Lecture notes in intelligent transportation and infrastructure part F1378; 2023. p. 1179–91. https://doi.org/10.1007/978-3-031-23721-8_95.
Royo B, Politaki D, Gonzalez JN, Batalla A. Digital twin opportunities and benefits in last-mile logistics for madrid value case, 72; 2023. p. 1693–9. https://doi.org/10.1016/j.trpro.2023.11.642.
Sabet S, Farooq B. Energy-smart transportation systems: role of connectivity. In: Automation, big data, and machine learning, 1-4; 2023. https://doi.org/10.1007/978-3-030-97940-9_126.
Sahu D, Sinha P, Prakash S, Yang T, Rathore RS, Wang L. A multi objective optimization framework for smart parking using digital twin pareto front mdp and pso for smart cities. Sci Rep 2025;15(1). https://doi.org/10.1038/s41598-025-91565-0.
Schurmann D, Timpner J, Wolf L. Cooperative charging in residential areas. IEEE Trans Intell Transport Syst 2017;18(4):834–46. https://doi.org/10.1109/TITS.2016.2590960.
Shang W-L, Zhang J, Wang K, Yang H, Ochieng W. Can financial subsidy increase electric vehicle (ev) penetration—evidence from a quasi-natural experiment. Renew Sustain Energy Rev 2024;190. https://doi.org/10.1016/j.rser.2023.114021.
Shirowzhan S, Tan W, Sepasgozar SM. Digital twin and cybergis for improving connectivity and measuring the impact of infrastructure construction planning in smart cities. ISPRS Int J GeoInf 2020;9(4). https://doi.org/10.3390/ijgi9040240.
Shojaei S, Beiza J, Abedinzadeh T, Alipour H. Optimal energy and reserve management of a smart microgrid incorporating parking lot of electric vehicles/renewable sources/responsive-loads considering uncertain parameters. J Energy Storage 2022;55. https://doi.org/10.1016/j.est.2022.105540.
Shrivastav CS, Masola A, Capodieci N, Cavicchioli R. Parking lots management and visualization in the smart city - digital twin context. 2023. p. 185–92. https://doi.org/10.1109/MetaCom57706.2023.00045.
Sikder SK, Nagarajan M, Mustafee N. Augmenting ev charging infrastructure towards transformative sustainable cities: An equity-based approach. Technol Forecast Soc Change 2023;196. https://doi.org/10.1016/j.techfore.2023.122829.
Sun Z, Gao W, Li B, Wang L. Locating charging stations for electric vehicles. Transp Policy 2020;98:48–54. https://doi.org/10.1016/j.tranpol.2018.07.009.
Tan BQ, Kang K. A data-driven framework for smart parking and charging. 2024. p. 587–95. https://doi.org/10.1061/9780784485484.057.
Xie R, Wei W, Wu Q, Ding T, Mei S. Optimal service pricing and charging scheduling of an electric vehicle sharing system. IEEE Trans Veh Technol 2020;69(1):78–89. https://doi.org/10.1109/TVT.2019.2950402.
Yoon D-H, Seo H, Lee J, Kim Y. Online electric vehicle charging strategy in residential areas with limited power supply. IEEE Trans Smart Grid 2024;15(3):3141–51. https://doi.org/10.1109/TSG.2023.3339788.
You M, Wang Q, Sun H, Castro I, Jiang J. Digital twins based day-ahead integrated energy system scheduling under load and renewable energy uncertainties. Appl Energy 2021; 305. https://doi.org/10.1016/j.apenergy.2021.117899.
Yu J-U, Cho K-S, Park S-W, Son S-Y. Digital twin system framework and implementation for grid-integrated electric vehicles. Energies 2024;17(24). https://doi.org/10.3390/en17246249.
Zhang K, Ding Z. Virtual simulation design and debugging of lift-and-transverse stereo garage based on the digital twin. Appl Sci 2024;14(9). https://doi.org/10.3390/app14093896.
Zhang C, Zhao X, Shahidehpour M, Li W, Wen L, Yang Z. Reliability assessment of coordinated urban transportation and power distribution systems considering the impact of

charging lots. IEEE Access 2020;8:30536–47. https://doi.org/10.1109/ACCESS.2020.2973035.

Zhang Y, Yang J, Li Q, Deng W, Xie X. Special transformer sharing mode: utilizing special transformers of buildings to supply power for charging stations. IET Gener Transm Distrib 2023;17(5):1021–34. https://doi.org/10.1049/gtd2.12464.

Zou Y, Ye F, Li A, Munir M, Hjelseth E, Sujan S. A digital twin prototype for smart parking management. 2023. p. 250–7. https://doi.org/10.1201/9781003354222-32.

CHAPTER FIFTEEN

The effect of renewable-to-vehicle (R2V) smart charging on greenhouse gas emission reduction

Basir Darsareh[1], Sasan Azad[1], Hamdi Abdi[2]
[1]Department of Electrical Engineering, Shahid Beheshti University, Tehran, Iran
[2]Department of Electrical Engineering, Faculty of Engineering, Razi University, Kermanshah, Iran

15.1 Introduction

The climate crisis is getting worse quickly, mostly because we're releasing more and more greenhouse gases. This really highlights how crucial it is to drastically cut down on carbon emissions across everything we do globally. Two big parts of the problem are transportation and making electricity, as both still rely heavily on fossil fuels. For instance, in the United States, transportation is the biggest source of emissions, and it's the second-biggest worldwide (Kühnbach et al., 2020, 2021). Down in the Dominican Republic, just the electricity sector and road transport together create over 60% of the country's greenhouse gases, showing us that we need comprehensive and interconnected solutions to tackle this (Quevedo et al., 2024).

The switch to electric vehicles is becoming a key way to cut down on emissions from cars and trucks. Experts predict that by 2035, around half of all new cars sold globally will be electric, showing a big move away from traditional gas-powered engines (Will et al., 2024). But here's the catch: how "green" electric cars really are depends entirely on where the electricity to charge them comes from. In places where power plants still burn a lot of fossil fuels, electric cars might just be shifting emissions around instead of actually reducing them (Barman et al., 2023). Recent life-cycle analysis demonstrates that grid carbon intensity, driving intensity (usage patterns), and vehicle efficiency interact nonlinearly to determine whether an EV yields net CO_2 reductions, so site-specific evaluation is essential (Liu et al., 2025). This really underscores the need to ramp up renewable energy

Renewable-to-Vehicle Smart Charging
ISBN: 978-0-443-36684-0
https://doi.org/10.1016/B978-0-443-36684-0.00011-4

sources like solar and wind power, which are getting cheaper and make a strong case for pairing clean electricity with electric transportation (Vollmuth et al., 2024).

In this particular scenario, Renewable-to-Vehicle (R2V) smart charging stands out as a significant advancement. Unlike traditional smart charging methods that usually focus on cutting costs or managing grid demand, R2V is all about lining up charging times with when renewable energy is most plentiful (Wang et al., 2022a,b). This coordination boosts the environmental advantages of electric vehicles by making sure their power comes from low-carbon sources. R2V systems usually bring together renewable energy, electric vehicles equipped with advanced communication tech, and smart control software. They often use setups like shared DC-links, which allow for various charging options, including Grid-to-Vehicle, Vehicle-to-Grid, and R2V itself (Wang et al., 2022b). Sometimes, energy from nearby renewable sources, like rooftop solar panels, can be sent straight to vehicle batteries without needing extra conversion steps. This cuts down on energy losses and keeps charging going even if the main power grid goes down. By carefully planning when to charge, R2V cuts down on using electricity from fossil fuels and gets around the drawbacks of simpler smart charging, which might just move charging times around without necessarily lowering emissions (Yilmaz and Krein, 2013).

Even with all the potential upsides, putting R2V strategies into action isn't easy. There are quite a few hurdles to overcome. Technically speaking, we're dealing with issues like making sure different charging standards can work together, ensuring power gets converted efficiently, and worrying about how these practices might affect battery lifespan over time. On the money side, things get complicated too. The initial investment can be huge, and figuring out solid, workable business models isn't straightforward, which makes deployment even trickier. Then there's the regulatory picture – in many places, the rules haven't quite caught up to properly value the environmental benefits, like cutting carbon emissions, and the flexibility services that R2V vehicles can offer. And we can't forget the human element. This adds another layer of complexity, bringing in worries from drivers about things like 'range anxiety' (fear of running out of charge) and privacy concerns, not to mention questions about making sure charging infrastructure is available fairly to everyone.

The idea behind R2V definitely has strong backing from scientific research. For instance, Kühnbach and his team looked at controlled charging in Germany and found that strategies relying only on price signals can sometimes backfire, leading to things like more pollution or sudden surges in electricity demand (Kühnbach et al., 2021). Meanwhile, Quevedo and colleagues showed how important electric mobility is for shaping power sector emissions in the Dominican Republic (Quevedo et al., 2024). Will and others also explored how weaving renewable energy into charging plans can help achieve nearly zero-carbon results (Will et al., 2024). By looking at all these studies together, it becomes clear that the actual environmental impact of electrifying transportation really depends on a mix of factors: how people use their vehicles, how the electrical grid is set up, and, importantly, how 'clean' the electricity powering those vehicles actually is.

Past research has already delved into the intricate connections between electric vehicles, renewable energy, and cutting down emissions. For instance, studies have shown that charging EVs in a controlled way, matching when they charge with when clean energy is available, can lower both costs and carbon footprints (Kühnbach et al., 2020, 2021). On the flip side, other research has pointed out the dangers of just letting EVs charge whenever or only when prices are low, as this might accidently lead to more emissions (Quevedo et al., 2024). And when it comes to exploring ways EVs can send power back to the grid (Vehicle-to-Grid) or work both ways, the findings confirm that EVs can be really helpful for the grid if the right rules and incentives are in place (Will et al., 2024; Barman et al., 2023).

All these earlier discoveries set the stage for what we'll cover in this chapter, and you can find a more thorough look at the related studies in Section 15.2. This chapter is designed to take you from the basics right through to real-world uses. After the intro, Section 15.2 looks at the most important studies on smart charging, weaving renewables into the energy mix, and the impact on greenhouse gases. Then, Section 15.3 dives into the practical and money-saving strategies for getting renewable energy to power EVs. Section 15.4 brings in real-world examples, showing how this works in different places like isolated island grids and well-established markets on continents. Section 15.5 pulls together the main takeaways, and Section 15.6 wraps things up with a discussion and ideas for future research. Overall, these sections work together to create a clear story that connects the theory, the evidence we have, and what it all means for policy. A visual summary of the chapter's motivation, methodology, and contributions is provided in Fig. 15.1.

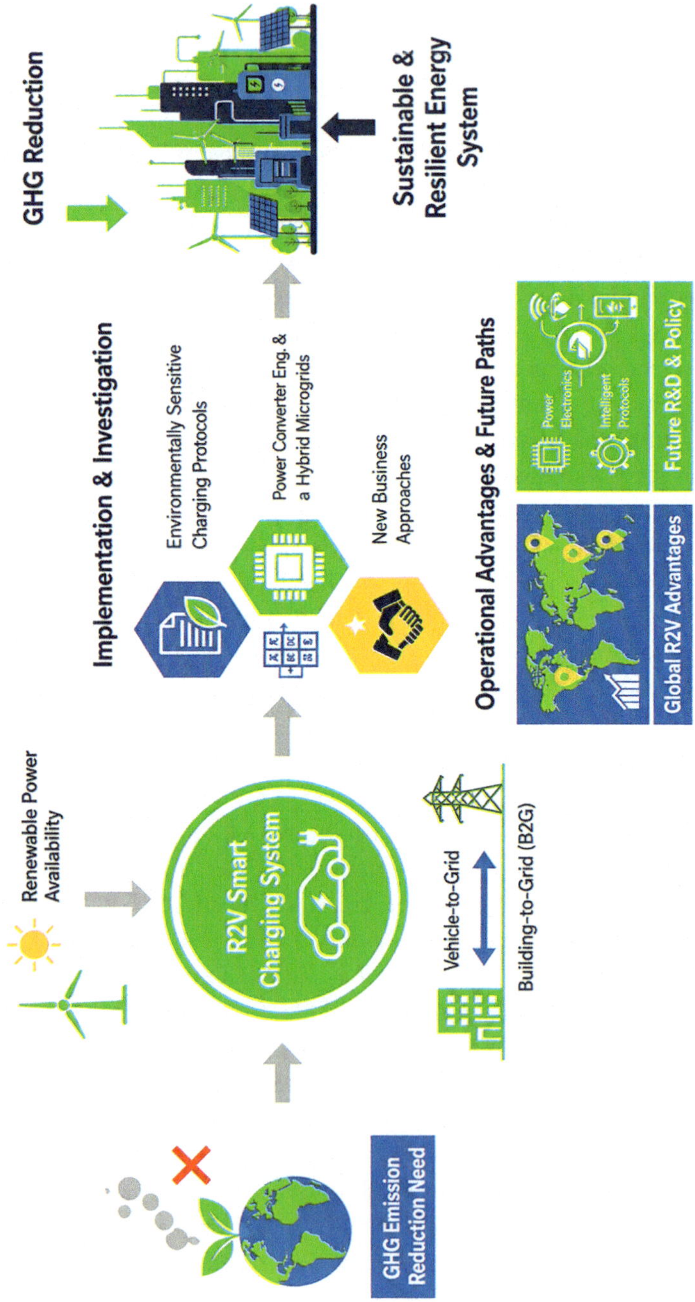

Figure 15.1 Book chapter's graphical abstract.

15.2 Literature review on smart charging and GHG emission reduction

The environmental performance of smart charging strategies is shaped by how charging interacts with temporal variations in electricity generation and the carbon intensity of the grid. Early studies demonstrated that unmanaged or price-driven charging can lead to unintended emission increases by shifting demand into fossil-heavy hours, highlighting the need for approaches that explicitly prioritize renewable utilization (Kühnbach et al., 2020, 2021).

Building on this, recent contributions have integrated system-level modeling with life-cycle analysis to quantify emission outcomes under Renewable-to-Vehicle paradigms. Wohlschlager et al. combined an hourly energy-system model with a prospective life-cycle assessment to evaluate Vehicle-to-Grid (V2G) strategies between 2025 and 2045 (Wohlschlager et al., 2024). Their findings showed that in medium-term scenarios (around 2035), V2G can reduce marginal operational emissions by up to 30% compared to uncontrolled charging, while also substituting as much as 60% of the stationary storage capacity otherwise required. The study employed an emission allocation formula in which hourly charging and discharging power are multiplied by instantaneous grid carbon intensity, demonstrating the importance of temporally resolved accounting. These results confirm that the greatest environmental benefits of R2V accrue when carbon intensity signals are directly embedded in control algorithms.

Distribution-level perspectives further clarify the operational challenges of scaling R2V. Mastoi et al. provided a comprehensive taxonomy of charging and discharging strategies and quantified their effects on low-voltage feeders (Mastoi et al., 2023). They observed that under uncontrolled charging at 40%–60% penetration, voltage unbalance can exceed 5% and total harmonic distortion can surpass IEEE-recommended limits. Their work also highlighted the tradeoff between grid support and battery health, applying degradation cost models such as $C_{\text{deg}} = \frac{\Delta Q}{Q_{\text{nom}}} . C_{\text{battery}}$, where capacity fade translates directly into economic penalties. Aggregator-based scheduling reduced overload risk by 15%–20%, illustrating the operational value of coordinated control.

Looking at things from both a market and policy viewpoint, Micari and Napoli studied European pilot projects and found that electric vehicle (EV) fleets could supply 3–5 GW of flexibility to help keep grids stable by 2030

(Micari and Napoli, 2024). They stressed that making this happen will depend on two main things: new tech breakthroughs—like wide-bandgap semiconductor converters that slash energy losses from about 4% down to nearly 1%—and clear rules that actively reward cutting carbon emissions. Their review even suggested ways to calculate revenue based on how many services these EV fleets provide together, really hammering home the connection between what the tech can do and what the market should pay for it.

When it comes to categorizing charging strategies, El-Bayeh and colleagues pitted unidirectional, bidirectional, and multi-vector approaches against each other using a well-organized decision matrix (El-Bayeh et al., 2021). Their findings revealed that unmanaged charging, when adopted by 30% of users, can cause peak demand to jump by nearly 20%. In contrast, coordinated charging helps cut that peak by about 15%. Even though Vehicle-to-Grid (V2G) strategies are quite complex, they offer the most potential for providing additional grid services, which confirms why they should be a priority when planning for a cleaner energy future.

Moreover, as shown by Alatawneh and Torok in 2025, the environmental perks of battery EVs really depend on the local power grid. For instance, driving in places like Sweden, which uses clean energy, only creates about 15 g of CO_2 per kilometer during use. But in areas like Poland, where the grid relies heavily on fossil fuels, that number jumps to over 150 g per kilometer. Their analysis applied the formulation $E_{\mathrm{BEV}} = \frac{E_{\mathrm{grid}}}{\eta_{\mathrm{charging}} \cdot \eta_{\mathrm{drivetrain}}}$, underscoring how grid emission factors dominate net outcomes. This evidence supports the view that R2V must be evaluated through geographically specific case studies, and that carbon-aware scheduling is indispensable to ensuring positive climate impacts (Alatawneh and Torok, 2025).

Recent comprehensive reviews further consolidate the methodological and technical foundations relevant to Renewable-to-Vehicle systems. A broad survey summarizes a taxonomy of charging control architectures and optimization objectives (centralized, decentralized, hierarchical), catalogs commonly used solution methods (MPC, MILP/MIQP, metaheuristics, game-theoretic approaches), and emphasizes battery degradation as a principal economic barrier that must be internalized in revenue-stacking models (Aghajan-Eshkevari et al., 2022). Complementing this, a power-quality–focused review synthesizes work on active/reactive control, converter topologies, and distribution-level impacts (voltage unbalance, harmonics, transformer loading), supporting the need for multi-scale control and power-electronics design choices when integrating large fleets of electric vehicles (Amiri et al., 2024).

When you look at all the research together, it's clear that not every smart charging plan actually cuts down on emissions. Only the ones that respond to carbon levels and work alongside renewable energy sources bring consistent improvements. It can be economically smart for vehicles to take part in several services, but you have to factor in wear and tear costs and create business plans that match both financial and environmental goals. Ultimately, how well these strategies work really depends on the specific situation, like how "dirty" the power grid is, what policies are in place, and how people use the system. Although this section highlights some important findings, Table 15.1 gives a broader overview of the research field, covering more studies from various places and using different methods.

15.3 Operational and economic strategies for emission reduction

For Renewable-to-Vehicle smart charging to really take off on a big scale, it really relies on two key, interconnected factors: how well it actually works in practice, and whether it makes good financial sense. Operational effectiveness means ensuring electric vehicles get charged up preferentially whenever there's renewable energy available. It also means making sure these vehicles can provide valuable services that help keep the power grid stable. Economic viability, on the other hand, is all about creating enough incentive for everyone involved to actually invest in and adopt these Renewable-to-Vehicle technologies. Making the operational side work well requires smart control strategies that pay close attention to real-time details like how clean the grid's energy is at any given moment, forecasts for renewable energy production, and any limits the electricity network might have. When it comes to economics, you need business models that can generate income from multiple sources and handle costs fairly, like the wear and tear on EV batteries. In the sections that follow, we'll bring together the main operational strategies, the necessary hardware and control systems, and the economic frameworks. All of these elements together determine whether large-scale Renewable-to-Vehicle programs can actually make a noticeable difference in cutting down greenhouse gas emissions. An overview of the major challenges limiting effective R2V integration for GHG emission reduction is provided in Fig. 15.2.

Table 15.1 Summary of key literature on smart charging, V2G, R2V, and GHG emissions.

References ID/Year	Core Focus/Contribution to R2V and GHG reduction	Methodology highlights	Key findings on GHG reduction, RES integration, and/or economic aspects	Identified challenges or limitations
Kühnbach et al. (2020)	Impact of EV charging on German household electricity prices and GHG emissions.	Modeling of EV diffusion, system load (eLOAD), electricity market (MiPU), controlled versus uncontrolled charging scenarios.	Controlled charging reduces household electricity prices (improved grid utilization outweighs generation cost increase); reduces CO_2 emissions via load shifting and generation mix changes.	Focus on private EVs; specific to German market conditions.
Kühnbach et al. (2021)	"Avalanche effects" of price-optimized EV charging; DR mitigation strategies.	Macro-energy system modeling; simulation of EV groups with dynamic DR signals for 2030 and 2050.	Price-optimized DR can cause avalanche effects; disaggregated EV groups with dynamic signals avoid this, improve RE integration, reduce peak residual load.	Raises issues of equitable cost savings among EV groups; model idealizations.
Barman et al. (2023)	Comprehensive review of RE integration with EV technology via smart charging.	Literature review covering RE sources, enabling tech (storage, power electronics, smart grid), smart charging strategies, challenges.	GHG reduction from EVs depends on electricity source; highlights importance of RE integration; discusses technical/systemic challenges.	As a review, relies on existing primary research; breadth may limit depth on specific sub-topics.

Vollmuth et al. (2024)	Profitability and GHG emissions of V2G multi-use combinations (smart/bidirectional charging).	Optimization modeling for flexible EV marketing; analysis of use cases (peak shaving, PV self-consumption, spot market, FCR) for Germany (today & 2030).	Smart charging profitable today; V2G has high future profit potential (esp. Multi-use like PV self-consumption + spot market) but needs standards/providers. GHG impacts linked to RES facilitation.	Achievable EV user numbers for use cases may be limited by market/user constraints, not just tech availability. Specific GHG quantification per use case detailed in main paper.
(Wang et al., 2022a,b)	Impact of electrifying communities (EEMs, DERs including EVs) on energy, emissions, grid.	URBANopt modeling of a mixed-use all-electric community in Denver.	Overnight EV charging with high grid carbon intensity can increase emissions in multifamily buildings; batteries reduce peak demand.	EEM/PV impact on peak demand was limited in this specific case study.
Wang et al. (2022b)	Feasibility of real-time carbon emission responsive EV charging control in buildings.	Simulation of rule-based EV charging controllers responsive to grid carbon intensity; metrics: carbon reduction, SOC at departure, peak demand.	Last-minute SOC compensation can increase emissions; significant carbon reductions (up to 12.7%) possible by removing compensation and managing SOC.	Rule-based control is preliminary; predictive control is a future step.

(Continued)

Table 15.1 Summary of key literature on smart charging, V2G, R2V, and GHG emissions.—cont'd

References ID/Year	Core Focus/Contribution to R2V and GHG reduction	Methodology highlights	Key findings on GHG reduction, RES integration, and/or economic aspects	Identified challenges or limitations
Yilmaz and Krein (2013)	Review of V2G impacts on distribution systems and utility interfaces.	Literature review of V2G concepts, services, benefits, components, technologies, challenges.	V2G improves grid efficiency, stability, reliability; enables ancillary services, RES integration. Highlights economic benefits and emission reductions.	Battery degradation, communication needs, infrastructure costs, social/political obstacles are key challenges. Dated but foundational.
Alatawneh and Torok (2025)	BEV use-stage emissions in EU.	Comparative analysis, grid-intensity formula.	15 gCO_2/km (Sweden) versus >150 gCO_2/km (Poland); grid mix decisive.	Regional variability; user behavior not fully modeled.
Wohlschlager et al. (2024)	V2G effects on emissions (system + life-cycle).	Energy system + pLCA, hourly carbon intensity formula.	Up to 30% emission reduction by 2035; BEVs can replace 60% of stationary storage.	Benefits shrink in highly renewable grids; allocation complexity.
Mastoi et al. (2023)	Charging/dispatch strategies, grid & battery impacts.	Review + grid simulations; degradation cost model.	Uncontrolled charging → >5% voltage unbalance; aggregators cut overloads 15%–20%.	Battery wear, high infra/ICT needs.
Micari and Napoli (2024)	V2G for flexibility & policy.	Review + EU pilot data; flexibility revenue model.	EVs can provide 3–5 GW flexibility; wide-bandgap converters boost efficiency (~99%).	Missing remuneration, regulatory gaps, cybersecurity.

El-Bayeh et al. (2021)	Taxonomy of charging/discharging strategies.	Systematic review + decision matrix.	At 30% EV share → unmanaged +20% peaks, coordinated −15%; V2G = high potential.	Complexity, interoperability, high costs.
Aghajan-Eshkevari et al. (2022)	Taxonomy of charging/control, optimization methods, battery degradation	Literature review; taxonomy of control (centralized/decentralized/hierarchical); survey of optimization methods (MPC, MILP/MIQP, metaheuristics)	Battery degradation is a key economic barrier; hierarchical/aggregator approaches enable emissions- and cost-optimized charging	Review paper -use primary sources for numeric models and case data
Amiri et al. (2024)	Power-quality and converter/control strategies for EV fleets (active/reactive support)	PQ-focused review; converter topology & control examples; prototype evidence	EV fleets can provide voltage/frequency support; PQ issues (voltage unbalance, harmonics) emerge at medium–high penetration	PQ focus (limited direct GHG/economic quantification); use to strengthen technical sections only

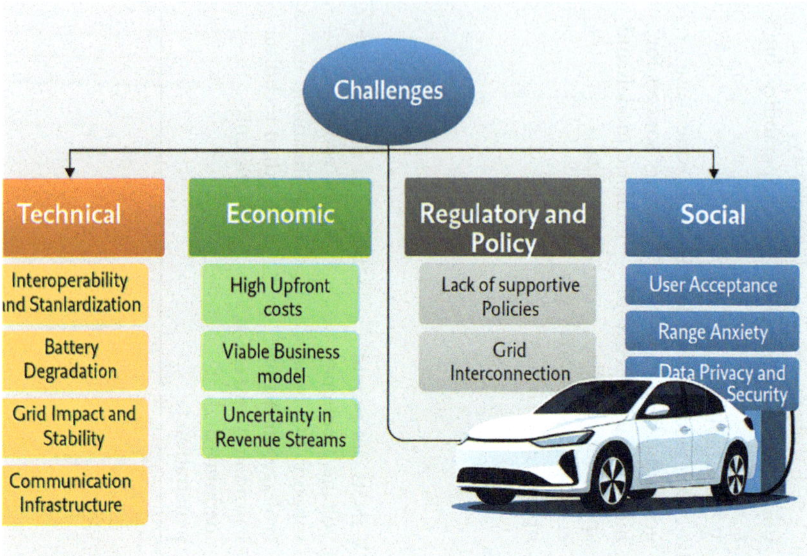

Figure 15.2 Challenges in R2V integration for GHG emission reduction.

Carbon-responsive charging aligns charging power with periods of low grid carbon intensity, converting charging control from a static scheduling problem into a dynamic, emissions-driven optimization. Rule-based implementations have demonstrated substantial impact: Wang et al. developed a controller that classifies each interval as clean, intermediate, or unclean according to real-time emissions signals and curtails or suspends charging during high-carbon intervals; removing last-minute "top-off" actions in that scheme reduced charging-related CO_2 emissions by up to 12.7% without materially affecting user mobility (Kühnbach et al., 2020). Systemic analyses extend this logic: Wohlschlager et al. use an hourly coupling of energy-system modeling and prospective life-cycle assessment to allocate charging and discharging to the instantaneous grid intensity I_t, employing the emission accounting relation $E_{V2G} = \sum_t \left(P_{\text{charge},t} I_t - P_{\text{discharge},t} I_t\right)$ to demonstrate that vehicle-to-grid operations can reduce marginal operational emissions substantially in medium-term decarbonization scenarios (Wohlschlager et al., 2024). These results show that carbon-aware policies and signals, whether provided by system operators, aggregators, or third-party platforms, are necessary to realize the full environmental potential of Renewable-to-Vehicle programs.

Price-optimized charging and demand-response mechanisms remain important because many vehicle owners are cost-sensitive; however, price signals must be designed to avoid synchronized responses that create new peaks and increase emissions. Recent review evidence emphasizes that charging behavior is multi-dimensional (time, location, SoC preferences, socio-economic traits) and that tariff design must reflect these behavioral heterogeneities to avoid rebound or avalanche effects (Shariatzadeh et al., 2025). Kühnbach et al. described the "avalanche effect," in which fleets responding to an identical low-price window produce steep, concentrated demand surges; their group-disaggregation demand-response mechanism offsets price windows for subgroups to flatten aggregate demand and preserve consumer savings while avoiding fresh peaks (Kühnbach et al., 2021). This insight has practical implications for Renewable-to-Vehicle systems that may wish to overlay price and carbon signals: multi-signal architectures that co-optimize emissions, cost, and network constraints (via offset windows, randomized start times, or controlled ramp rates) are required to balance competing objectives.

Bidirectional charging adds a qualitatively different set of operational capabilities because it enables vehicles to discharge power back to homes, buildings, or the grid. The foundational review by Yilmaz and Krein cataloged the suite of ancillary services that bidirectional vehicles can provide, peak shaving, frequency regulation, operating reserves, and voltage/reactive power support, and highlighted the technical and degradation tradeoffs that accompany intensive cycling (Yilmaz and Krein, 2013). More recent modeling and pilot studies corroborate these service values: controlled discharging timed with renewable surpluses can substitute for expensive, carbon-intensive peaker generation and reduce curtailment, while participation in frequency regulation markets yields high-value, short-duration revenue streams. From a system perspective, Wohlschlager et al. further show that V2G fleets can substitute a substantial fraction of required stationary storage capacity under certain scenarios, enhancing the overall system's ability to integrate variable renewable energy (Wohlschlager et al., 2024). Also, Liu et al. show that compact, low-energy EVs retain emission advantages across many grid scenarios, whereas high-energy vehicles may lose that advantage under high driving intensity and carbon-intensive grids, underscoring the need for fleet- and user-aware R2V policies and control strategies (Liu et al., 2025).

The hardware architecture that underpins Renewable-to-Vehicle functionality is evolving to reduce conversions and increase efficiency. Multi-port power converters and three-port integrated topologies (TPIT), which couple distributed photovoltaic DC outputs directly to a common

DC bus shared with vehicle batteries while retaining AC connectivity to the grid, minimize conversion losses and permit DC-to-DC charging during outages (Barman et al., 2023). Wide-bandgap semiconductors such as silicon carbide and gallium nitride materially improve conversion efficiency and thermal performance; Micari and Napoli note that these devices can reduce converter losses from a few percent to near-single-percent levels, improving round-trip efficiency and increasing the carbon-saving potential of direct renewable charging (Micari and Napoli, 2024). Hardware choices, therefore, feed directly into the emissions accounting performed by carbon-responsive controllers.

Practically, modern chargers can do more than simply move energy; they can actively help the grid by injecting or absorbing both active and reactive power to support voltage and frequency. Converter choices — for example, multiport and three-port topologies or high-efficiency bidirectional inverters — together with simple local rules like droop control or reactive-power setpoints, allow fleets of vehicles to provide meaningful voltage regulation and phase balancing at the feeder level. That said, if charging is left uncoordinated, high EV penetration can raise voltage unbalance and harmonic distortion beyond acceptable limits, so phase-aware scheduling, proper filtering, and adaptive protection are essential design elements (Amiri et al., 2024). A practical control pattern that appears across the literature is a hierarchical arrangement: fast local controllers handle immediate grid support (droop/frequency response) while a supervisory model-predictive or optimization layer coordinates energy, emissions, and battery health across the aggregated fleet (Aghajan-Eshkevari et al., 2022). Because converter hardware, local droop rules, and supervisory algorithms interact closely, these engineering choices directly shape both power, quality outcomes and the net emission benefits of R2V systems, so they should be treated as first-order design decisions in any deployment.

Predictive and machine-learning driven control strategies enhance traditional rule-based methods by weaving in forecasts for renewable energy output, electricity price movements, and how users are likely to behave. They do this using a technique called receding-horizon optimization. Techniques like Model Predictive Control and learning-based schedulers allow these systems to juggle multiple goals at once, balancing factors like cost, emissions, and battery wear. Real-world tests have shown these advanced methods can lead to combined financial and environmental benefits of roughly 10%–15% compared to simpler, static approaches, especially in situations with fluctuating prices and inconsistent renewable energy supply (Vollmuth et al., 2024; Micari and Napoli, 2024). However, these predictive controllers depend heavily on accurate short-term forecasts and strong penalties to account for battery aging. Mastoi et al. specifically point out that the cost of battery degradation must be factored into the calculations to prevent simply shifting those costs onto the vehicle owners. A commonly cited economic expression for degradation cost is $C_{\mathrm{deg}} = \left(\frac{\Delta Q}{Q_{\mathrm{nom}}}\right).C_{\mathrm{battery}}$, which links capacity fade to replacement cost and should be integrated into aggregator or settlement mechanisms (Mastoi et al., 2023).

Linking electric vehicles with stationary battery storage makes operations more flexible and cuts down on wasted renewable energy. Using layered storage setups lets you store extra renewable power in stationary batteries and then use it later for faster EV charging or to help the grid, which helps balance out the differences between when energy is made and when it's needed throughout the day and year. Wohlschlager shows that this hybridization can reduce the need for dedicated stationary capacity and enhance the net emission reduction per vehicle when coordinated optimally; Mastoi reports that coordinated dispatch with BESS can lower feeder overload and voltage stress relative to uncoordinated EV charging.

Economic viability rests on the ability to stack revenues from multiple services while accounting for costs and risks. Vollmuth et al. demonstrate that single-purpose, unidirectional smart charging yields modest returns but that revenue stacking—aggregating arbitrage, frequency regulation, reserve provision, and self-consumption services—can generate significantly higher annual revenues per vehicle under favorable market (Vollmuth et al., 2024). A simple revenue aggregation expression used in the literature is $R = \sum_s P_s.\pi_s$ where P_s is the capacity delivered for service $sand\pi_s$ its remuneration price; Micari and Napoli caution that the existence of market access, low bid thresholds, and clear settlement rules is a prerequisite to realize these revenues. Importantly, the revenue potential must be weighed against battery degradation costs and the capital premium of bidirectional charging hardware; current estimates place bidirectional chargers at a 30%–50% price premium over unidirectional units, and battery cycle costs remain non-negligible.

Stakeholder cost-benefit analyses really highlight both the distributional hurdles and the overall benefits of these systems. Looking at it from the owner's side, taking part in controlled charging and V2G (Vehicle-to-

Grid) activities can actually lower their net energy costs and open up new ways to make money, especially if warranty programs help cover the wear and tear on the battery from frequent charging cycles (Vollmuth et al., 2024). From the utility's perspective, smartly managing how renewable energy interacts with vehicles can help delay the need for costly network upgrades and cut down on energy procurement expenses. Research suggests that getting groups of EVs to work together can significantly postpone a large portion of distribution investments, particularly in areas with lots of EVs, sometimes by tens of percent on specific parts of the grid, and it can also reduce the amount of renewable energy that has to be wasted, making the most of the renewable energy capacity that's already installed (Waldron et al., 2022). On a broader societal level, the benefits include a collective reduction in greenhouse gases and improvements in local air quality, which in turn can lead to better public health. However, putting a dollar value on these kinds of societal benefits requires solid systems for measuring, reporting, and verifying the impact. However, the success of incentive schemes and V2G participation depends on users' willingness and QoS concerns (charging convenience, predictable SoC at departure). Shariatzadeh et al. synthesize pilot evidence showing that financial incentives alone are often insufficient; privacy, perceived convenience, and transparent QoS guarantees substantially influence actual uptake and participation rates (Shariatzadeh et al., 2025).

While there's a lot of potential, a bunch of overlapping challenges are making it hard to really scale things up. Technically speaking, the different communication standards for charging (like ISO-15,118, OCPP, CHAdeMO, CCS, and the new ones popping up) don't all play nicely together. This makes it tricky to group chargers or offer new services and actually raises costs (Commision, 2021). Cybersecurity and data privacy are huge worries too, because detailed info about how people charge their cars can reveal a lot about their habits. You really need solid security measures like safe key handling, firmware updates that you know are legit, and systems to spot anything fishy (Micari and Napoli, 2024). Our power grids also need smarter protection and real-time checks to stop transformers from getting overloaded and to manage electricity flowing back from EVs. Mastoi points out things like uneven voltage and harmonics as real technical headaches, especially as more EVs join the mix. From an economic standpoint, fuzzy and inconsistent market rules, the big upfront money needed, and small companies struggling to get a foothold are all putting a damper on investment. Vollmuth suggests things like lower bid limits, settlement systems that work better for aggregators, and payments based on actual flexibility provided to make investing less risky. And let's not forget the social side of things—worries about running out of charge, concerns over privacy, and not everyone having equal access to charging stations—all these things make it harder for people to adopt this technology. This really points to needing solutions that focus on the user, like promises to keep EVs charged to a minimum level, and specific programs aimed at apartment buildings and low-income neighborhoods.

We can get operational performance and economic benefits working together through policy and market changes. For instance, requiring new charging stations and vehicles to be ready for smart charging and vehicle-to-grid (V2G) capabilities, as some recent proposals suggest, can prevent us from getting stuck with outdated technology. Also, using dynamic pricing and special rates for peak times that specifically account for flexibility and carbon emissions will guide electric vehicles to participate in supporting renewables (Aghajan-Eshkevari et al., 2022). Financial tools can also help—things like subsidies for two-way chargers, payments based on proven greenhouse gas reductions, and payments for pooling flexible resources—can make it easier to get started and help pilot projects grow into full-scale operations. Simplifying the process to connect new systems, making technical rules consistent, and investing in modern communication and metering infrastructure (like 5G, LPWAN, advanced meters, and smart transformers) are essential to build a solid foundation for secure and scalable deployment (Mastoi et al., 2023; Micari and Napoli, 2024).

In synthesis, operational strategies for Renewable-to-Vehicle systems must combine carbon-responsive scheduling, price-aware demand response, bidirectional dispatch, and predictive control, all supported by efficient hardware and layered storage, to maximize greenhouse-gas reductions while preserving network stability. Economic frameworks that enable revenue stacking, internalize battery degradation costs, and provide clear remuneration for flexibility are indispensable for adoption. This section contributes a unified technical and economic synthesis: it maps the quantitative impacts and tradeoffs (e.g., 12.7% emission reductions from carbon-responsive rules, substantial medium-term V2G emission gains under hourly pLCA, 15%–20% feeder overload reductions from aggregation) and identifies the precise areas—standards harmonization, degradation accounting, market access reform, and cybersecurity—where targeted research and policy action will unlock the largest marginal benefits. The next section uses practical applications to illustrate how these operational and market strategies play out in differing institutional and grid contexts.

Table 15.2 Comparative analysis of R2V operational and economic strategies.

Strategy	Description & key technologies involved	Primary GHG reduction mechanism	Economic viability (potential revenue Streams/Cost savings, key cost drivers)	Key operational/ Technical challenges	Key economic/ Policy challenges
Real-time carbon-responsive charging (unidirectional)	EV charging adjusted based on real-time grid carbon intensity signals. Requires smart chargers, communication link to grid operator/ aggregator providing carbon data, control algorithms.	Minimizes charging during high-carbon grid periods; maximizes charging when RES share is high.	Consumer: Potential bill savings if carbon intensity correlates with low prices. Grid: Reduced system emissions. Costs: Smart charger, data services.	Accurate real-time carbon data, predictive algorithms, managing SOC at departure versus emissions.	Lack of widespread carbon-signal tariffs; user acceptance of flexible SOC.
Price-optimized charging with advanced demand response (unidirectional)	EV charging shifted to low-price periods, using DR tactics (e.g., EV group disaggregation, dynamic signals) to avoid "avalanche effects." requires smart chargers, price signals, DR management platform.	Indirectly reduces GHG if low prices align with high RES/low-carbon generation. Avoids creating new peaks that might be met by fossil fuels.	Consumer: Significant electricity bill savings. Grid: Load shifting, peak reduction, improved RE integration. Costs: Smart charger, DR platform.	Complexity of DR coordination, ensuring price signals reflect true grid stress/RES availability, avoiding rebound peaks.	Design of effective DR programs, equitable benefit distribution among users.
Bidirectional R2V/V2G for grid services & RES optimization	EVs charge from RES (grid or local) and discharge to grid/ home to provide services (peak shaving,	Maximizes RES use for charging; displaces fossil-fuel peaker plants with stored EV energy;	Consumer: Revenue from grid services, maximized PV self-consumption, energy arbitrage.	Battery degradation management, interoperability standards, complex control and	High upfront cost of bidirectional chargers, lack of standardized V2G tariffs/

	frequency regulation, arbitrage) or maximize RES self-consumption. Requires V2G-capable EVs, bidirectional chargers, advanced control/aggregation platforms.	enables higher overall RES penetration by providing grid flexibility.	Grid: Reduced peak capacity needs, ancillary service provision, deferred investments. Costs: Bidirectional charger, battery degradation, aggregation platform.	aggregation, cybersecurity, grid interconnection.	market access, regulatory uncertainty, defining battery ownership/liability.
Direct R2V from local RES (e.g., PV-to-ev DC charging)	EVs charge directly from on-site RES (e.g., solar PV) often via DC–DC converters, minimizing grid interaction for charging. Requires local RES, specialized chargers/inverters (e.g., TPIT).	Maximizes direct use of 100% local RES for charging, minimizing transmission losses and grid carbon intensity impact.	Consumer: Very low/zero marginal cost for charging energy (post-RES investment), energy independence. Grid: Reduced load if charging is off-grid. Costs: RES installation, specialized R2V charger.	Matching EV charging demand with local RES availability, potential need for local storage if RES is intermittent, charger technology maturity.	Higher initial investment for RES and specialized chargers; limited scalability if reliant solely on individual local RES.

Table 15.2 summarizes and contrasts the principal R2V strategies in terms of emissions mitigation potential, economic benefits, and implementation challenges.

15.4 Practical applications and empirical insights

After covering the theory, strategies, and economic angles, it's really important to look at how Renewable-to-Vehicle (R2V) smart charging actually works in the real world. Practical Applications from various settings—like small island grids and established continental markets—show us both the potential to cut greenhouse gas emissions and the technical or economic hurdles that need solving. These practical examples let us test ideas in different situations and draw lessons that can help with wider adoption.

Island power systems are often known for their high electricity prices, reliance on imported fuel, and the unpredictable nature of renewable energy sources. These unique challenges make them perfect places to test and develop 'R2V' (resilience and reliability through virtualization) ideas. In the Dominican Republic, for example, Quevedo and his team used a specific planning tool called OSeMOSYS to see how getting lots of electric vehicles on the road could help the country meet its goals for cutting carbon emissions (Quevedo et al., 2024). What they found was that if you want more electric cars, you need to phase out coal power plants sooner and, at the same time, invest heavily in solar and wind energy. In their vision of a low-carbon future, making transportation electric led to more than a 15% drop in overall power system emissions compared to just doing things the way they always have. This really highlights that adopting these R2V approaches can bring about big changes across the entire energy landscape, especially if the investments are planned thoughtfully.

Similar findings are coming from Cyprus too. Researchers like Taliotis and his team showed that using smart charging, which aligns with solar power production during the middle of the day, really helps integrate more renewable energy (Taliotis et al., 2020). In situations where lots of electric vehicles are on the road, planning when they charge helped add another 380 MW of solar power capacity by 2035, cut overall system costs by about 3%, and reduced combined power and transport emissions by 7.5%. Further reviews confirm that pairing smart charging schedules with some modest energy storage consistently performs better than just letting cars charge whenever they want, especially in isolated island settings (Manousakis et al., 2023). On San Pietro Island in Sardinia, Vargiu and colleagues looked into the cost differences between deep decarbonization paths. They found that while a plan connecting to the main grid for sustainable growth reached 72% renewable energy at a cost of €48.8 per MWh, a completely self-sufficient plan relying only on local storage ended up costing over €500 per MWh (Vargiu et al., 2022). These results really highlight the economic challenges of aiming for total energy independence compared to the more cost-effective approach of integrating with the grid. Adding to these studies, Flessa and Fragkiadakis introduced something called the IntE3 ISL framework. This framework connects energy system planning with economic modeling, helping to account for things like tourist demand, reliance on imported fuels, and storage needs (Flessa et al., 2023). Their findings showed that Vehicle-to-Grid (V2G) technology, used alongside stationary storage, could cut down imported fuel use by 40% and lower consumer electricity bills by 10%, demonstrating the real socioeconomic benefits of implementing V2G on islands.

While islands provide a controlled setting to test algorithms and configurations, insights from more established interconnected systems highlight the challenges of scaling up. In Germany, for instance, controlled charging has demonstrated the ability to cut household electricity costs by 2%–4% and slash sector emissions by as much as 7%. However, follow-up research flagged the danger of "avalanche effects," where consumers reacting in unison to price signals can generate new demand spikes. To tackle this, dividing electric vehicles into smaller groups and sending staggered signals can boost renewable energy use while keeping the grid stable. Expanding on these insights, an analysis of the economic upside of Vehicle-to-Grid (V2G) applications shows that while one-way smart charging is already making a small profit, offering multiple bidirectional services could bring in over €1600 per vehicle each year by 2030 (Marzouk, 2025). Yet, the absence of common technical standards and the limited number of vehicles equipped for V2G continue to be major hurdles for broader adoption.

Putting it all together, these practical applications highlight some key takeaways. Basically, if electric vehicle charging is left unmanaged, it can really mess with integrating renewable energy and can drive up peak demand. On the flip side, smart scheduling can really boost both environmental and economic results. Where you roll out these solutions really matters: strategies that work for island grids, like soaking up extra solar power midday, are quite different from what's needed in bigger continental markets, like integrating vehicles into multiple markets (V2G). To figure all this out, energy system modeling—whether using tools like OSeMOSYS, IntE3 ISL, or others—is absolutely crucial for testing scenarios and guiding policy decisions. Plus, getting utilities, regulators, and consumers on the same page is vital to really reap the benefits, and any successful pilot projects need clear plans to scale them up. In the end, these real-world experiences confirm that turning road transport into a flexible energy resource (R2V) is not only technically doable and good for the environment, but its success really hinges on tailoring it to the local context and having coordinated policy support. A structured overview of selected case studies on R2V smart charging implementation, covering GHG mitigation potential, RES integration effects, and economic performance, is provided in Table 15.3.

15.5 Conclusion

To cut down on emissions from transportation and power plants, we need to connect electric vehicles (EVs) more closely with renewable energy sources. One way to do this is through Renewable-to-Vehicle (R2V) smart charging, which syncs EV charging times with when renewable energy is available and how the power grid is doing. There's proof, like the example from Cyprus, that strategies that consider carbon emissions and allow EVs to send power back to the grid (Vehicle-to-Grid or V2G) can reduce greenhouse gases, make the grid more flexible, and help use more renewable energy (Kühnbach et al., 2020, 2021; Quevedo et al., 2024).

However, these good outcomes rely on smart planning. If charging happens at the wrong times or is only guided by prices, it can actually increase emissions and put stress on the grid (Kühnbach et al., 2020; Will et al., 2024). So, effective R2V systems need to find a balance between cutting emissions, keeping the grid stable, and meeting what drivers need. In the big picture, R2V is a crucial piece of the puzzle for a future with less carbon pollution. It turns EVs into valuable resources that can help both the power and transport industries by bringing together the right technology, rules, and how people behave.

15.6 Future works

The success of smart charging for vehicles powered by renewable energy isn't just about tech improvements—it also hinges on progress in the market, policy, and social areas. On the tech front, newer chargers using

Table 15.3 Overview of R2V smart charging practical applications and implementations.

Case study (region) reference ID/Year	Primary R2V/Smart charging application & objectives	Key Technologies/ Methodologies used	Reported/ Projected GHG reduction impact	Impact on RES integration & grid	Economic outcomes (Costs/ Benefits)	Key lessons learned & practical challenges
Dominican Republic (Quevedo et al., 2024)	EV transition for power sector decarbonization, meeting NDC targets.	OSeMOSYS modeling, scenarios for EV penetration (public/ private fleets).	27% national GHG reduction goal by 2030; implies accelerated coal plant conversion.	Facilitates RES integration as part of power sector decarbonization.	Significant short-term investments for long-run decarbonization benefits.	Importance of strategic planning for EV transition in developing island nations; linking transport to power sector goals.
Cyprus (Taliotis et al., 2020)	EV deployment for RES integration and emission reduction in an isolated grid.	OSeMOSYS modeling (REF, ET, SC scenarios), smart charging optimization.	Overall energy system emissions reduced by 7.5% (SC scenario vs. REF by 2035).	Higher solar PV deployment (380 MW more in SC); RES generation 7% higher (SC vs. ET); limited RES curtailment.	Electricity costs slightly lower in SC scenario (112 €/MWh vs. 116 €/MWh in REF by 2035); annual savings up to 34 million €.	Smart charging crucial for maximizing RES and cost benefits in isolated grids; need for charging infrastructure and supportive policies (e.g., ToU tariffs).
San Pietro island, Sardinia (Vargiu et al., 2022)	Energy cost assessment of future scenarios with high RES and potential EV integration.	Enhanced OSeMOSYS model, LCOE calculation for BAU, CPP, SG, SSR scenarios.	SG scenario: CO2 emissions 2.5 kton/y (vs. 10.0 kton/y in BAU by 2050). SSR: Carbon neutral.	SG scenario: 71.9% RES penetration. SSR: 100% RES.	SG LCOE (48.8 €/MWh) lower than BAU (49.1 €/MWh) by 2050. SSR LCOE very high (531.5 €/MWh) due to storage.	RES reduces LCOE long-term; full energy independence is costly with current storage tech; grid interconnection vital for cost-effectiveness.

(*Continued*)

Table 15.3 Overview of R2V smart charging practical applications and implementations.—cont'd

Case study (region) reference ID/Year	Primary R2V/Smart charging application & objectives	Key Technologies/ Methodologies used	Reported/ Projected GHG reduction impact	Impact on RES integration & grid	Economic outcomes (Costs/ Benefits)	Key lessons learned & practical challenges
Germany (Kühnbach et al., 2020)	Impact of private EVs on household electricity prices and grid.	Modeling (ALADIN, eLOAD, MiPU), controlled versus uncontrolled charging.	Controlled charging reduces CO2 emissions from power sector.	Controlled charging improves grid utilization, can avoid distribution grid expansion investments.	Reduces household electricity prices by 2030 with controlled charging.	Controlled charging is key to economic and environmental benefits in mature markets.
Germany (Kühnbach et al., 2021)	"Avalanche effects" from price-optimized EV charging; DR mitigation.	Macro-energy system modeling, dynamic DR signals for EV groups.	Effective DR facilitates RE integration.	Prevents new load peaks from synchronized EV charging; flattens residual load.	DR reduces EV owner electricity costs, but equitable distribution is a challenge.	Sophisticated DR needed to manage large-scale smart charging with price signals.
Germany (Vollmuth et al., 2024)	Profitability and GHG impact of V2G multi-use cases.	Optimization modeling for EV use cases (peak shaving, PV self-consumption, spot market, FCR).	GHG impact linked to RES facilitation and displacement of carbon-intensive peak power.	V2G enhances grid flexibility and RES integration.	Smart charging profitable now; V2G offers high future profit (e.g., PV self-consumption + spot market V2G: ~1660 €/ev/a today, ~2080 by 2030).	V2G rollout needs standards and provider scaling; multi-use is key for V2G economics. User uptake for specific use cases may be limited.

materials like silicon carbide or gallium nitride could make things more efficient. Meanwhile, predictive control and AI can help optimize things like costs, emissions, and battery health in real-time. Things like making different systems work together (interoperability) and keeping data secure (cybersecurity) are crucial too; without open standards, smooth communication, and safe platforms, R2V won't get beyond the testing phase. Research should also look into how virtual models (digital twins), public standards (open benchmarks), and flexible designs can speed up deployment and cut costs. All these innovations need to be backed by helpful rules that value services like grid flexibility, create fair pricing, and focus on outcomes rather than strict commands.

It's just as crucial to focus on the human and systemic aspects of future research. Shifting from pilot projects to widespread adoption needs strategies that fit the specific grid conditions and mobility habits of different regions, whether they're bustling cities, remote villages, or isolated island power systems. Paying close attention to engaging users, protecting data privacy, and making sure everyone has fair access will decide if R2V becomes a common solution or just a specialized one. Research should also look into how R2V can work together with other areas, like smart buildings, small-scale power grids, public transportation fleets, and maybe even hydrogen energy, to create comprehensive plans for reducing carbon emissions. In the end, the next phase of R2V research needs to blend new ideas in technology and software with policies and business plans that include everyone, making sure the advantages of flexible, renewable-powered charging are available to all and support a sustainable society.

References

Aghajan-Eshkevari S, Azad S, Nazari-Heris M, Taghi Ameli M, Asadi S. Charging and discharging of electric vehicles in power systems: an updated and detailed review of methods, control structures, objectives, and optimization methodologies. Sustainability 2022;14(4):2137. https://doi.org/10.3390/su14042137.

Alatawneh A, Torok A. Use-stage emissions of conventional, battery electric, and autonomous vehicles in Europe. Case Stud Transp Policy 2025;20:101398. https://doi.org/10.1016/j.cstp.2025.101398.

Amiri MM, et al. A review on utilization of electric vehicles for mitigating the power quality issues in power systems. Power quality—new insights. IntechOpen; 2024. https://doi.org/10.5772/intechopen.1003592.

Barman P, Dutta L, Bordoloi S, Kalita A, Buragohain P, Bharali S, et al. Renewable energy integration with electric vehicle technology: a review of the existing smart charging approaches. Renew Sustain Energy Rev 2023;183:113518. https://doi.org/10.1016/j.rser.2023.113518.

Commision, C.P.U., SB 350 standard review programs-annual transportation electrification evaluation CPUC. 2021.

El-Bayeh CZ, Alzaareer K, Aldaoudeyeh A-MI, Brahmi B, Mohamed Z. Charging and discharging strategies of electric vehicles: a survey. World Electr Vehic J 2021;12(1):11. https://doi.org/10.3390/wevj12010011.

Flessa A, Fragkiadakis D, Zisarou E, Fragkos P. Developing an integrated energy–economy model framework for islands. Energies 2023;16(3):1275. https://doi.org/10.3390/en16031275.

Kühnbach M, Stute J, Klingler AL. Impacts of avalanche effects of price-optimized electric vehicle charging - does demand response make it worse? Energy Strategy Rev 2021;34. https://doi.org/10.1016/j.esr.2020.100608.

Kühnbach M, Stute J, Gnann T, Wietschel M, Marwitz S, Klobasa M. Impact of electric vehicles: will German households pay less for electricity? Energy Strategy Rev 2020; 32:100568. https://doi.org/10.1016/j.esr.2020.100568.

Liu K, Liu F, Guo C. Carbon reduction potential of private electric vehicles: synergistic effects of grid carbon intensity, driving intensity, and vehicle efficiency. Processes 2025;13 (6):1740. https://doi.org/10.3390/pr13061740.

Manousakis NM, Karagiannopoulos PS, Tsekouras GJ, Kanellos FD. Integration of renewable energy and electric vehicles in power systems: a review. Processes 2023;11(5):1544. https://doi.org/10.3390/pr11051544.

Marzouk OA. Summary of the 2023 report of TCEP (tracking clean energy progress) by the International Energy Agency (IEA), and proposed process for computing a single aggregate rating. In: E3S web of conferences. EDP Sciences; 2025.

Mastoi MS, Zhuang S, Munir HM, Haris M, Hassan M, Alqarni M, et al. A study of charging-dispatch strategies and vehicle-to-grid technologies for electric vehicles in distribution networks. Energy Rep 2023;9:1777–806. https://doi.org/10.1016/j.egyr.2022.12.139.

Micari S, Napoli G. Electric vehicles for a flexible energy system: challenges and opportunities. Energies 2024;17(22):5614. https://doi.org/10.3390/en17225614.

Quevedo J, Herrera Moya I, Mariano-Hernandez D, Sbriz-Zeitun G, Cannone C, Howells M, et al. Decarbonising the power sector of the Dominican Republic: an approach from electric mobility transition. Energy Strategy Rev 2024;52:101357. https://doi.org/10.1016/j.esr.2024.101357.

Shariatzadeh M, Lopes MAR, Antunes CH. Electric vehicle users' charging behavior: a review of influential factors, methods and modeling approaches. Appl Energy 2025;396: 126167. https://doi.org/10.1016/j.apenergy.2025.126167.

Taliotis C, Fylaktos N, Partasides G, Gardumi F, Sridharan V, Karmellos M, et al. The effect of electric vehicle deployment on renewable electricity generation in an isolated grid system: the case study of Cyprus. Front Energy Res 2020;8. https://doi.org/10.3389/fenrg.2020.00205.

Vargiu A, Novo R, Moscoloni C, Giglio E, Giorgi G, Mattiazzo G. An energy cost assessment of future energy scenarios: a case study on san Pietro island. Energies 2022;15(13): 4535. https://doi.org/10.3390/en15134535.

Vollmuth, P., et al., Publication 3 Pub3: prospects of electric vehicle V2G multi-use: profitability and GHG emissions for use case combinations of smart and bidirectional charging today and 2030. Prospects of use cases and multi-use of smart electric vehicle charging and discharging from the user's perspective, 2024. 371: p. 123679.

Waldron J, Rodrigues L, Gillott M, Naylor S, Shipman R. The role of electric vehicle charging technologies in the Decarbonisation of the energy grid. Energies 2022;15(7): 2447. https://doi.org/10.3390/en15072447.

Wang J, El Kontar R, Jin X, King J. Electrifying High-Efficiency Future Communities: Impact on energy, emissions, and grid. Adv Appl Energy 2022a;6:100095. https://doi.org/10.1016/j.adapen.2022.100095.

Wang J, et al. Feasibility study of real-time carbon emission responsive electric vehicle charging control in buildings. Golden, CO (United States): National Renewable Energy Lab.(NREL); 2022b.

Will C, Zimmermann F, Ensslen A, Fraunholz C, Jochem P, Keles D. Can electric vehicle charging be carbon neutral? Uniting smart charging and renewables. Appl Energy 2024; 371:123549. https://doi.org/10.1016/j.apenergy.2024.123549.

Wohlschlager D, Kigle S, Schindler V, Neitz-Regett A, Fröhling M. Environmental effects of vehicle-to-grid charging in future energy systems – a prospective life cycle assessment. Appl Energy 2024;370:123618. https://doi.org/10.1016/j.apenergy.2024.123618.

Yilmaz M, Krein PT. Review of the impact of vehicle-to-grid technologies on distribution systems and utility interfaces. IEEE Trans Power Electron 2013;28(12):5673–89. https://doi.org/10.1109/TPEL.2012.2227500.

Index

Note: 'Page numbers followed by *f* indicate figures and *t* indicate tables.'

D

E

Q

R

W

X

Printed and bound by CPI Group (UK) Ltd, Croydon, CR0 4YY
14/09/2026
02216410-0007